THE
TUTORIAL ALGEBRA

(ADVANCED COURSE)

BASED ON THE *ALGEBRA* OF RADHAKRISHNAN

BY

WILLIAM BRIGGS, LL.D., M.A., B.Sc., F.R.A.S.

AND

G. H. BRYAN, Sc.D., F.R.S.

HONORARY FELLOW OF ST. PETER'S COLLEGE, CAMBRIDGE, AND SMITH'S PRIZEMAN
LATE PROFESSOR OF MATHEMATICS IN THE UNIVERSITY COLLEGE OF NORTH WALES

Authors of "The Right Line and Circle (Coordinate Geometry)"
"Matriculation Mechanics and Hydrostatics," etc.

PREFACE.

THE present volume is intended to form an advanced course in Algebra suitable for Examinations similar in scope to the Intermediate and B.A. Examinations for London University and the Higher Local Examinations. The treatment is very thorough, and special pains have been taken to deal with the difficulties of the average student. The more elementary portions of the subject such as are required for the London Matriculation and the Oxford and Cambridge School Certificate Examinations are dealt with in *School Algebra*, by A. G. Cracknell, M.A., B.Sc.

The simple properties of inequalities are treated at a very early stage, the applications of these properties to the comparison of ratios, arithmetic and geometric means and the like being left till subsequent chapters. A chapter has been devoted to the important properties of Zero and Infinity, which are often but scantily understood by beginners. The Theory of Quadratic Equations and Maxima and Minima are dealt with at some length in the endeavour to familiarise the student with the notion of a function. A new method of treating annuities without employing the formula for a geometrical progression has

PREFACE.

been introduced. The subdivision of Permutations and Combinations into two separate chapters will, it is hoped, be helpful in preventing confusion. One of the more important features of this work is the chapter on Graphical Representation, which, as well as much of the later portion of the book, is the work of J. H. Grace, F.R.S., late Fellow of St. Peter's College, Cambridge.

In preparing this book we have been largely indebted to the Algebra of Radhakrishnan, a book of unique merits in which the author has been very successful in expounding what is intelligible to the average student from the work of the greatest masters of the subject.

The chapter on Limiting Values and the Theory of Infinite Series (in its present form) and the Appendix on Simultaneous Equations are the work of Mr. C. E. Wright, M.Sc.

TABLE OF CONTENTS.

CHAP.		PAGE
I.	The Theory of Indices	1
II.	Surds	12
III.	Inequalities	34
IV.	Ratio	40
V.	Proportion	58
VI.	Variation	73
VII.	Equations reducible to Quadratics	87
VIII.	Simultaneous Quadratic Equations with Two Unknown Quantities	104
IX.	Simultaneous Quadratics with Three Unknowns	113
X.	Problems leading to Quadratics	120
XI.	Zero and Infinity	126
XII.	The Theory of Quadratic Equations	133
XIII.	Maxima and Minima	158
XIV.	Arithmetical and Geometrical Progression	171
XV.	Harmonical Progression	189
XVI.	Some other Simple Series	202
XVII.	Logarithms	215
XVIII.	Interest and Discount	237

CONTENTS.

CHAP.		PAGE
XIX.	Annuities	257
XX.	Permutations	277
XXI.	Combinations	304
XXII.	Mathematical Induction	329
XXIII.	The Binomial Theorem for a Positive Integral Index	334
XXIV.	Imaginary and Complex Quantities	355
XXV.	Rational Functions	376
XXVI.	Graphic Representation of Functions	392
XXVII.	Limiting Values and Elementary Theory of Infinite Series	422
XXVIII.	The Binomial Theorem for Fractional and Negative Indices	458
XXIX.	Exponential and Logarithmic Series	498
XXX.	Summation of Series	532
XXXI.	Determinants	552
XXXII.	Continued Fractions	572
XXXIII.	Elimination	588
XXXIV.	Indeterminate Equations	597
XXXV.	Partial Fractions	603
	APPENDIX. On the Solution of Equations	614
	ANSWERS	619

CHAPTER I.

THE THEORY OF INDICES.

1. The Index-Laws for positive integral exponents.—According to the definition of a *power* in elementary algebra, a^m is defined as the product of m factors each equal to a, and m the index of the power is necessarily a positive integer, quantities like a^0, $a^{\frac{1}{2}}$, $a^{-\frac{1}{2}}$, a^{-3} being meaningless. We shall now show how the principles of symbolical algebra enable us to assign meanings to quantities with fractional, zero, or negative indices.

We shall first state and prove the 'Index-Laws' for positive integral exponents.

2. To prove that
$$a^m \times a^n = a^{m+n}$$
when m and n are positive integers.

By definition we have,
$$a^m = a \times a \times a \times \ldots\ldots \text{to } m \text{ factors};$$
and
$$a^n = a \times a \times \ldots\ldots \text{to } n \text{ factors}.$$
$$\therefore a^m \times a^n = (a \times a \times a \times \ldots\ldots \text{to } m \text{ factors})$$
$$\times (a \times a \times \ldots\ldots \text{to } n \text{ factors})$$
$$= a \times a \times a \times \ldots\ldots \text{to } (m+n) \text{ factors}$$
$$= a^{m+n}.$$

TUT. ALG. II.

THE THEORY OF INDICES.

We can now extend the law to three or more factors, and show that
$$a^m \times a^n \times a^p \times \ldots\ldots = a^{m+n+p+\cdots}$$
when m, n, p......are positive integers.

For $a^m \times a^n \times a^p = (a^m \times a^n) \times a^p = a^{m+n} \times a^p = a^{m+n+p}$; and so on.

3. To prove that
$$\frac{a^m}{a^n} = a^{m-n} \text{ when } m > n, \text{ and } = \frac{1}{a^{n-m}} \text{ when } m < n,$$
m and n being positive integers.

$$\frac{a^m}{a^n} = \frac{a \times a \times a \times \ldots\ldots \text{to } m \text{ factors}}{a \times a \times \ldots \text{to } n \text{ factors}}.$$

CASE I. When $m > n$, this expression is equal to
$$(a \times a \times a \times \ldots\ldots \text{to } m - n \text{ factors})$$
$$\times \frac{a \times a \times \ldots \text{to } n \text{ factors}}{a \times a \times \ldots \text{to } n \text{ factors}}$$
$$= a \times a \times a \times \ldots \text{ to } m - n \text{ factors} = a^{m-n}.$$

[*Otherwise thus.*—Since $m - n$ is positive, therefore by § 2,
$$a^{m-n} \times a^n = a^{m-n+n} = a^m, \therefore a^m \div a^n = a^{m-n}.]$$

CASE II. When $m < n$ the expression is equal to
$$= \frac{a \times a \times \ldots \text{to } m \text{ factors}}{a \times a \times \ldots \text{to } m \text{ factors}} \times \frac{1}{a \times a \times \ldots \text{to } n - m \text{ factors}}$$
$$= \frac{1}{a \times a \times a \ldots \text{to } n - m \text{ factors}} = \frac{1}{a^{n-m}}.$$

[*Otherwise thus.*—Since $n - m$ is positive, therefore
$$a^{n-m} \times a^m = a^{n-m+m} = a^n, \therefore a^m \div a^n = 1/a^{n-m}.]$$

THE THEORY OF INDICES.

4. To prove that $(a^m)^n = a^{mn} = (a^n)^m$, where m and n are positive integers.

$(a^m)^n = a^m \times a^m \times a^m \times \ldots$ to n factors

$= a^{m+m+m+\ldots \text{ repeated } n \text{ times}}$ [§ 2]

$= a^{mn}$.

Similarly $(a^n)^m = a^{mn}$, $\therefore (a^m)^n = (a^n)^m$.

Thus the nth power of the mth power of a

$=$ the mth power of the nth power of a,

both being equal to the mnth power of a.

It should be noticed that this law is merely a special case of § 2, in which $m, n, p\ldots$ are all equal to m, and there are n factors.

5. To prove that $(ab)^m = a^m b^m$, where m is a positive integer.

$(ab)^m = (ab) \times (ab) \times (ab) \times \ldots$ to m factors

$= (a \times a \times a \times \ldots$ to m factors$)$

$\times (b \times b \times b \times \ldots$ to m factors$)$

$= a^m \times b^m$.

Thus the mth power of a product is equal to the product of the mth powers of its factors.

6. To prove that

$\left(\dfrac{a}{b}\right)^m = \dfrac{a^m}{b^m}$ **where m is a positive integer.**

The proof is similar to that of § 5.

7. The Principle of the Permanence of Equivalent Forms.

The equation $a^m \times a^n = a^{m+n}$ has been proved from the fundamental laws of algebra to be true for positive integral values of m and n. We shall now *assume that this Index-Law is true for all values of the indices,* whether positive or negative, fractional, integral or zero, and deduce from this law interpretations of a^n for fractional, zero, and negative values of n.

The principle that we are guided by in extending our definition of *index* is an important one, known as the Principle of the **Permanence of Equivalent Forms.**

This principle consists in the assumption that a law of algebra which admits of proof subject to certain limitations is true generally provided that the removal of the limitations is not *incompatible* with the truth of the law.

Ex. 1. Find the meaning of $a^{\frac{1}{2}}$.

Since the formula $a^{m+n} = a^m \times a^n$ has been accepted as true for all values of m and n, $a^{\frac{1}{2}} \times a^{\frac{1}{2}} = a^{\frac{1}{2}+\frac{1}{2}} = a^1 = a$. Thus $a^{\frac{1}{2}}$ is one of the square roots of a.

Ex. 2. Find a meaning for a^{-1}.

From the assumption $a^m \times a^n = a^{m+n}$ we have $a^2 \times a^{-1} = a^{2+(-1)} = a^{2-1} = a$, whence $a^{-1} = a/a^2 = 1/a$.

8. Interpretation of fractional indices.

To prove that $a^{\frac{m}{n}}$ represents the nth root of a^m, where m and n are any positive integers.

Since the Index-Law has been assumed as true for all values of the indices, we have

$$\left(a^{\frac{m}{n}}\right)^n = a^{\frac{m}{n}} \times a^{\frac{m}{n}} \times a^{\frac{m}{n}} \times \ldots \text{to } n \text{ factors}$$
$$= a^{\frac{m}{n}+\frac{m}{n}+\frac{m}{n}+\ldots \text{ to } n \text{ terms}} = a^{\frac{m}{n} \times n} = a^m.$$

Therefore $a^{\frac{m}{n}}$ must be the nth root of a^m; that is,

$$a^{\frac{m}{n}} = \sqrt[n]{a^m}.$$

Hence *in a fractional index, the numerator indicates the power and the denominator indicates the root.*

THE THEORY OF INDICES. 5

9. Interpretation of the index zero.
To prove that a^0 represents unity.

Since $a^m \times a^n = a^{m+n}$ has been assumed as true for all values of m and n, we substitute 0 for m, and obtain

$$a^0 \times a^n = a^{0+n} = a^n, \quad \therefore a^0 = \frac{a^n}{a^n} = 1.$$

Therefore *any finite quantity* with zero index equals unity*.

10. Interpretation of negative indices.
To prove that a^{-n} represents the reciprocal of a^n.

In the equation $a^m \times a^n = a^{m+n}$ which we have assumed to be true for all values of m and n, let $-n$ be substituted for m.

We get $\quad a^{-n} \times a^n = a^{-n+n} = a^0 = 1$,

$$\therefore a^{-n} = \frac{1}{a^n}.$$

That is, a^{-n} means the reciprocal of a^n.

Since $a^{-n} = 1/a^n$, and $a^n = 1/a^{-n}$, it is always possible in fractional expressions to transfer any power of a quantity from the numerator to the denominator and *vice versâ*, by changing the sign of its index.

Caution. It should be carefully noted that we have not actually *proved* from the fundamental laws of algebra that $a^0 = 1$, or $a^{1/n} = \sqrt[n]{a}$, but that we have deduced the meanings in the previous three articles from the Index-Law for positive integers, by applying the Principle of the Permanence of Equivalent Forms. This is a peculiarity of symbolical algebra, which assigns to symbols that are unintelligible in arithmetical algebra, meanings which permit of their obedience to the fundamental laws of algebra.

11. In the three preceding sections the Index-Law $a^m \times a^n = a^{m+n}$ was assumed to be true for all values of m and n. We shall now *conversely* show that the index law can be established universally from the definitions given above, and that these definitions are not inconsistent with the other laws of indices established in §§ 2-6.

---

* 0^0 is indeterminate, and is not necessarily equal to unity.

THE THEORY OF INDICES.

12. To prove that $a^m \times a^n = a^{m+n}$ for all values of m and n.

CASE I. If m and n be positive integers, this is proved in § 2.

CASE II. Let m and n be positive fractions, and be denoted by $\dfrac{p}{q}$ and $\dfrac{r}{s}$ respectively, where p, q, r and s are positive integers.

Then $a^m \times a^n = a^{\frac{p}{q}} \times a^{\frac{r}{s}} = k$, suppose.

$$k^{qs} = \left(a^{\frac{p}{q}} \times a^{\frac{r}{s}}\right)^{qs} = \left(a^{\frac{p}{q}}\right)^{qs} \times \left(a^{\frac{r}{s}}\right)^{qs} \quad [\S\ 5]$$

$$= \left\{\left(a^{\frac{p}{q}}\right)^q\right\}^s \times \left\{\left(a^{\frac{r}{s}}\right)^s\right\}^q \quad [\S\ 4^*]$$

$$= \{(\sqrt[q]{a^p})^q\}^s \times \{(\sqrt[s]{a^r})^s\}^q$$

$$= (a^p)^s \times (a^r)^q$$

$$= a^{ps} \times a^{qr} = a^{ps+qr} \quad [\S\ 2].$$

Thus by definition of fractional exponents

$$k = a^{\frac{ps+qr}{qs}} = a^{\frac{ps}{qs}+\frac{qr}{qs}} = a^{\frac{p}{q}+\frac{r}{s}} = a^{m+n},$$

which establishes the law for positive fractional values of m and n.

COR. Hence $a^m \div a^n = a^{m-n}$, where m and n are *positive fractions*, for $a^{m-n} \times a^n = a^{m-n+n} = a^m$.

CASE III. Let *one* of the indices, say m, be *negative*, and the other *positive*. Let $m = -p$ where p is a positive number, integral or fractional.

Then, by the foregoing corollary

$$a^m \times a^n = a^{-p} \times a^n = \frac{1}{a^p} \times a^n = \frac{a^n}{a^p}$$

$$= a^{n-p} = a^{m+n}.$$

---

* Since q and s are positive integral exponents, and
$$a^{mn} = (a^m)^n = (a^n)^m$$
when m and n are positive integers.

THE THEORY OF INDICES.   7

Case IV. Let both m and n be negative and be denoted by $-p$ and $-q$, where p and q are positive.

$$a^m \times a^n = a^{-p} \times a^{-q} = \frac{1}{a^p} \times \frac{1}{a^q} = \frac{1}{a^{p+q}} = a^{-p-q} = a^{m+n}.$$

Thus $a^m \times a^n = a^{m+n}$ *universally*.

Cor. 1. We can similarly prove that
$$a^{m+n+p} = a^m \times a^n \times a^p$$
for all values of m, n and p.

Cor. 2. Since $a^m = a^n \times a^{m-n}$ is true for all values of m and n, we have $a^m \div a^n = a^{m-n}$ for all values of m and n.

13. To prove that $(a^m)^n = a^{mn}$ for *all* values of m and n.

Case I. Let n be a positive integer.

Then, whatever m may be,
$$(a^m)^n = a^m \times a^m \times a^m \ldots \text{to } n \text{ factors}$$
$$= a^{m+m+m+\ldots \text{ to } n \text{ terms}} = a^{mn} \qquad [\S\ 12].$$

Case II. Let n be a positive fraction $= \dfrac{p}{q}$, p and q being any positive integers; then, whatever m may be,

$$(a^m)^n = (a^m)^{\frac{p}{q}} = k \text{ suppose.}$$

Therefore by Case I., since q is a positive integer
$$k^q = \left\{(a^m)^{\frac{p}{q}}\right\}^q = (a^m)^{\frac{p \times q}{q}}$$
$$= (a^m)^p = a^{mp}.$$

Hence taking the qth root of both the sides, we have by definition of a fractional index
$$k = a^{\frac{mp}{q}} = a^{mn}.$$

8   THE THEORY OF INDICES.

CASE III. Let n be negative and $= -p$, where p is a positive number, fractional or integral.

Whatever m may be, we have by Cases I and II
$$(a^m)^n = (a^m)^{-p} = \frac{1}{(a^m)^p} = \frac{1}{a^{mp}} = a^{-mp} = a^{mn}.$$

Hence $(a^m)^n = a^{mn} = (a^n)^m$ *universally*.

14. To prove that $(ab)^m = a^m b^m$ for *all* values of m.

CASE I. If m be a positive integer, this is proved in § 5.

CASE II. Let m be a positive fraction $= \frac{p}{q}$, where p and q are positive integers.
$$(ab)^m = (ab)^{\frac{p}{q}} = k \text{ suppose}.$$

Therefore, by § 13 and Case I, since q is a positive integer
$$k^q = \left\{(ab)^{\frac{p}{q}}\right\}^q = (ab)^p$$
$$= a^p b^p$$
$$= \left(a^{\frac{p}{q}} b^{\frac{p}{q}}\right)^q.$$

Hence, taking the qth root, we have $k = a^{\frac{p}{q}} b^{\frac{p}{q}}$.

Hence $(ab)^m = a^m b^m$, where m is a positive fraction.

CASE III. Let m be negative and equal to $-p$ where p is a positive number, integral or fractional. Then by definition and Cases I and II
$$(ab)^m = (ab)^{-p} = \frac{1}{(ab)^p} = \frac{1}{a^p b^p} = \frac{1}{a^p} \cdot \frac{1}{b^p}$$
$$= a^{-p} b^{-p} = a^m b^m.$$

Hence $(ab)^m = a^m b^m$ *universally*.

COR. 1. Thus $(abc...)^m = a^m b^m c^m ...$ for *all* values of m.

COR. 2. Since $(ab)^m = a^m b^m$ is true for *all* values of a, b and m,
$$\therefore (a^x b^y c^z ...)^m = a^{xm} b^{ym} c^{zm} ....$$

THE THEORY OF INDICES.

15. To prove that $\left(\dfrac{a}{b}\right)^m = \dfrac{a^m}{b^m}$ **for *all* values of** m.

By § 14, $\left(\dfrac{a}{b} \times b\right)^m = \left(\dfrac{a}{b}\right)^m \times b^m$,

that is $\qquad a^m = \left(\dfrac{a}{b}\right)^m \times b^m, \quad \therefore \left(\dfrac{a}{b}\right)^m = \dfrac{a^m}{b^m}.$

16. It is now obvious that roots of quantities may be represented either by indices, or by radicals.

Thus the square root of a may be represented by either $a^{\frac{1}{2}}$ or $\sqrt{a}$, and the nth root of a by $a^{1/n}$ or by $\sqrt[n]{a}$.

Hence the properties established above may also be proved without using fractional indices.

Ex. 1. Express the square root of the 5th power of a in both the notations.—The expressions are $\sqrt{a^5}$ and $a^{\frac{5}{2}}$.

Ex. 2. Prove that $x^{b-c} \times x^{c-a} \times x^{a-b} = 1$, when x is not equal to zero.

The expression $= x^{b-c+c-a+a-b} = x^0 = 1.$

Ex. 3. Simplify $\dfrac{\left\{(x^p)^{\frac{1}{q}} \times (x^m)^{\frac{1}{n}}\right\}^{qn}}{\{\sqrt[p]{y^q} \times \sqrt[n]{y^n}\}^{pm}}.$

The expression $= \dfrac{\left(x^{\frac{p}{q}} \times x^{\frac{m}{n}}\right)^{qn}}{\left(y^{\frac{q}{p}} \times y^{\frac{n}{m}}\right)^{pm}} = \dfrac{x^{pn} \times x^{mq}}{y^{mq} \times y^{pn}} = \dfrac{x^{mq+np}}{y^{mq+np}} = \left(\dfrac{x}{y}\right)^{mq+np}.$

Ex. 4. Simplify $\dfrac{1}{1+x^{a-b}} + \dfrac{1}{1+x^{b-a}}.$

The expression $= \dfrac{x^b}{x^b+x^a} + \dfrac{x^a}{x^a+x^b} = \dfrac{x^b+x^a}{x^b+x^a} = 1.$

Note 1. $a^{\frac{1}{2}}$ is taken to represent both the positive and the negative square roots, while the radical sign $\sqrt{}$ in $\sqrt{a}$, $\sqrt[3]{a}$... is generally assumed to denote the arithmetical root. Thus $\sqrt{3}$ will represent only the positive root unless the symbol $\pm$ is prefixed to it. Hence the two equations $a + \sqrt{b} = c$ and $a \pm \sqrt{b} = c$ are not identical.

Note 2. The identity $(a^{\frac{1}{2}})^2 = (a^2)^{\frac{1}{2}}$ is not true except when the arithmetical roots of a and a^2 are taken; otherwise, we should, for example, have $+4$ equal to ± 4, when $a = 4$.

EXERCISES I.

1. Express both with indices and as radicals,

(i) the 5th root of the product of the square root of x and the cube root of y.

(ii) the mth root of the pth power of the quantity obtained by dividing the square root of m by the cube root of n.

2. Express $a^{-1}bc + ab^{-3}c^2 + a^{-1}b^{-1}c^{-1}$, with positive indices.

3. Express $\dfrac{1}{a} + \dfrac{2}{a^2} + \dfrac{3}{a^3}$ with negative indices, so as to remove all powers from the denominators.

Simplify the following expressions :

4. $[a^{\frac{1}{3}}\{a^{-\frac{1}{2}}b^{-\frac{1}{3}}(a^2b^2)^{\frac{2}{3}}\}^{-\frac{1}{2}}]^6$.

5. $(a^{-1})^{-2} \times (a^{-\frac{1}{2}})^2$.

6. $[\{(2^{-1})^{-1}\}^{-1}]^{-1}$.

7. $\sqrt[13]{\{x^3y^2\sqrt[3]{[xyz^4\sqrt{(x^{-1}y^{-2}z^{-3})}]}\}^{12}}$.

8. $\dfrac{(2^{p+1})^q}{(2^{q+1})^p} \times \dfrac{2^{2p}}{2^{2q}} \times 2^q$.

9. $\dfrac{1+ax^{-1}}{a^{-1}x^{-1}} \times \dfrac{a^{-1}-x^{-1}}{a^{-1}x - ax^{-1}} \div \dfrac{ax^{-1}}{x-a}$.

10. $\sqrt[3]{(4a^{-1}b^2c^{\frac{1}{2}})} \times \sqrt[4]{(12a^3b^{-\frac{2}{3}}c^2)} \div \sqrt[12]{(108a^{-3}b^2c^{-4})}$.

11. $\left\{\left(p+\dfrac{1}{q}\right)^p \left(p-\dfrac{1}{q}\right)^q\right\} \div \left\{\left(q+\dfrac{1}{p}\right)^p \left(q-\dfrac{1}{p}\right)^q\right\}$.

12. Calculate the value of the expression $(4^{\frac{1}{2}} \times 4^{\frac{2}{3}} \times 4^{\frac{3}{5}}) \div (4^{\frac{4}{15}})$.

13. Multiply

(i) $x^n + x^{\frac{1}{2}n} + 1$ by $x^{-n} - x^{-\frac{1}{2}n} + 1$.

(ii) $a^{\frac{3}{4}} + a^{\frac{1}{2}}b^{\frac{1}{2}} + a^{\frac{1}{4}}b + b^{\frac{3}{2}}$ by $a^{\frac{1}{4}} - b^{\frac{1}{2}}$.

14. Divide

(i) $(a^{\frac{7}{3}} - a^2b^{-\frac{2}{3}} - a^{\frac{1}{3}}b + b^{\frac{1}{3}})$ by $(a^{\frac{1}{3}} - b^{-\frac{2}{3}})$.

(ii) $3x^{\frac{4}{3}} + 4x + 6x^{\frac{2}{3}} - 4x^{\frac{1}{3}} + 3$ by $3x^{\frac{2}{3}} - 2x^{\frac{1}{3}} + 1$.

(iii) $x^{3m} - x^{2m}y - x^{2m}z + x^my - x^mz - y^2 + z^2$ by $x^m - y - z$.

THE THEORY OF INDICES.

15. Extract the square root of
$$4a - 12a^{\frac{1}{2}}b^{-\frac{1}{2}}c^{-\frac{1}{3}} + 9b^{-1}c^{-\frac{2}{3}}.$$

16. Simplify $\sqrt[3]{\left\{\dfrac{x}{y} - \dfrac{y}{x} + 3\left(\dfrac{y^{\frac{1}{3}}}{x^{\frac{1}{3}}} - \dfrac{x^{\frac{1}{3}}}{y^{\frac{1}{3}}}\right)\right\}^{-2}}.$

17. Show that
$$\frac{(x^2+1)^{\frac{1}{2}} + (x^2-1)^{\frac{1}{2}}}{(x^2+1)^{\frac{1}{2}} - (x^2-1)^{\frac{1}{2}}} + \frac{(x^2+1)^{\frac{1}{2}} - (x^2-1)^{\frac{1}{2}}}{(x^2+1)^{\frac{1}{2}} + (x^2-1)^{\frac{1}{2}}} = 2x^2.$$

18. Find the value of $a^{\frac{11}{16}}[a\{a(a^{\frac{1}{2}})^{\frac{1}{2}}\}^{\frac{1}{2}}]^{\frac{1}{2}}$ if $a=16$.

19. Simplify
$$\frac{1}{1+x^{a-b}+x^{a-c}} + \frac{1}{1+x^{b-c}+x^{b-a}} + \frac{1}{1+x^{c-a}+x^{c-b}}.$$

20. If $a^{m^n} = (a^m)^n$, find m in terms of n.

21. Show that $\dfrac{a^{x-y}+xy^{-1}+yx^{-1}+a^{y-x}}{xy^{-1}a^{x-y}+2+x^{-1}ya^{y-x}} = \dfrac{ya^{x-y}+x}{xa^{x-y}+y}.$

22. If $a = x + \sqrt{(x^2+1)}$, show that $x \times \frac{1}{2}(a-a^{-1})$.

23. Prove that $(a^m)^n$ and $(a^n)^m$ have at least one value in common, whatever m and n may be. [See Note 2, § 16.]

24. Prove that if $a^n = 1$, then either $a=1$, or $n=0$.

25. Resolve $75^{\frac{1}{3}}$ and $150^{\frac{1}{4}}$ into prime factors with fractional indices.

CHAPTER II.

SURDS.

17. Definition.—Surds are roots of arithmetical numbers which cannot be exactly determined, and which are not therefore expressible by an integer or a finite fraction.

Thus $\sqrt{2}$, $\sqrt[3]{4}$, $\sqrt{7\frac{1}{2}}$ are surds.

$\sqrt{2\frac{1}{4}}$ and $\sqrt[3]{64}$ are arithmetical numbers under surd forms, since their values can be exactly determined.

$\sqrt{(\sqrt{3}+7)}$ is not a surd but a surd expression. The surd expression $\sqrt[3]{\sqrt{2}}$ can be reduced to a surd, since it equals $\sqrt[6]{2}$ which is a surd according to the definition.

Though surds are inexpressible either as integers or as fractions, the value of a surd can be obtained to any degree of accuracy, and we shall therefore be justified in treating surds as arithmetical quantities and assuming them to be subject to the same algebraic laws which apply to integers and fractions.

18. Irrational quantities.—Quantities involving surds are sometimes known as **irrational** quantities, 'irrational' literally meaning 'having no ratio.'

Thus $5+\sqrt{6}-\sqrt{7}$, $\sqrt{10}\div\sqrt{3}$, $x^2+\sqrt{5}\sqrt[3]{x}$ are irrational quantities.

An algebraical expression like $\sqrt{(a^2+6a)}$ is often called *irrational*, though a may have such a value (say 2) as makes $\sqrt{(a^2+6a)}$ an arithmetical number. Hence to determine whether an algebraically irrational quantity like $\sqrt{m}$ is arithmetically irrational, we should have to know the value of m.

19. Classification of surds.

DEFINITIONS. The **degree** or **order** of a surd is denoted by its root-symbol, or surd-index*.

Thus $\sqrt{a}$, $\sqrt[5]{b}$ and $\sqrt[n]{c}$ are of the second, fifth and nth orders respectively.

Surds of the second order are often called **quadratic surds**, those of the third order **cubic surds**, and those of the fourth, **biquadratic surds**.

Surds are said to be of the **same order** or **of different orders** according as their root-symbols or surd-indices are the same or different.

Thus $\sqrt{m}$, $m^{\frac{3}{2}}$, $5\sqrt{(pq)}$, $a\sqrt{(rs)}$ and $k^{\frac{1}{2}}$ are of the same order; but $\sqrt[3]{m}$ and $\sqrt[5]{m}$ are of different degrees or orders.

Equiradical surds are those that arise from the extraction of the same root.

Thus $a^{\frac{2}{5}}$, $a^{\frac{3}{5}}$, $a^{\frac{4}{5}}$, $a^{\frac{8}{5}}$ are all equiradical with $a^{\frac{1}{5}}$. But $a^{\frac{2}{5}}$ and $b^{\frac{1}{5}}$ are not equiradical, nor are $a^{\frac{1}{2}}$ and $a^{\frac{2}{3}}$ unless they be written as $a^{\frac{3}{6}}$ and $a^{\frac{4}{6}}$.

Surds which consist of a rational and a surd factor are called **mixed** surds, while surds which have no rational factor are called **entire surds**.

Thus $5\sqrt{2}$ is a mixed surd; $\sqrt{3}$ is an entire surd.

Monomial surds are surds consisting of a single term only.

Thus $\sqrt{3}$, $4\sqrt{6}$, $\sqrt{5} \times \sqrt{7}$.

A binomial surd is the sum or difference of two monomial surds, or of a rational number and a monomial surd.

Thus $\sqrt{3}+\sqrt{5}$; $7+\sqrt{6}$.

---

* In this chapter we only deal with the arithmetical values of surds expressed with indices, such as $a^{\frac{3}{4}}$, $a^{-\frac{1}{2}}$.

Trinomial surds are similarly defined.
Thus $1+\sqrt{2}+\sqrt{3}$; $\sqrt{2}+\sqrt{5}-7$.

A surd is considered to be in its **simplest form** when the root-symbol is the smallest possible and the quantity under the root is integral, and as small as possible.

A surd expressed with an index is considered to be in its simplest form only when the index is rendered positive and less than unity.

Thus $\sqrt{18}=\sqrt{(9\times 2)}=3\sqrt{2}$ (the simplest form).
$\sqrt[4]{25}=\sqrt{5}$ (the simplest form).
$\left(\dfrac{1}{2}\right)^{-2\frac{1}{2}} = \dfrac{1}{\left(\frac{1}{2}\right)^{2\frac{1}{2}}} = \dfrac{2^{2\frac{1}{2}}}{1^{2\frac{1}{2}}} = 2^{2\frac{1}{2}} = 2^2 \cdot 2^{\frac{1}{2}} = 4 \cdot 2^{\frac{1}{2}}$ (the simplest form).

Two or more surds are said to be **like** or **similar** when, on being reduced to their simplest forms, they have the same surd-factor, the other factors being rational.

Surds which are not like or similar are said to be **unlike** or **dissimilar**.

Thus $5\sqrt{2}$ and $3\sqrt{2}$ are similar, $\sqrt[3]{54}$ and $\sqrt[3]{16}$ are also similar, since they are equal to $3\sqrt[3]{2}$ and $2\sqrt[3]{2}$.
But $\sqrt{2}$ and $\sqrt{6}$ are dissimilar; $\sqrt{2}$ and $\sqrt[3]{2}$ are also dissimilar.

Reduction and transformation of surds.

20. To reduce a rational quantity to the form of a surd of any given order.

Ex. Express 4 in the form of a surd of the 4th order, and a^2b in the form of a surd of the nth order.

$$4=(4^4)^{\frac{1}{4}}=\sqrt[4]{256}$$

$$a^2b=\{(a^2b)^n\}^{\frac{1}{n}}=\sqrt[n]{a^{2n}b^n}.$$

21. To transform a mixed surd into an entire surd.

Ex. Express $3ab^2c\sqrt[5]{/r^3}$ as a complete surd.

The given expression $=\sqrt[5]{(3ab^2c)^5}\times \sqrt[5]{/r^3}=\sqrt[5]{(243a^5b^{10}c^5r^3)}$.

SURDS.

22. To reduce surds to their simplest forms.

Ex. Reduce $\sqrt[3]{16a^5(b^3c^4 - c^7)}$ to its simplest form.

The expression $= \sqrt[3]{8a^3c^3 \times 2a^2(b^3c - c^4)} = 2ac\sqrt[3]{2a^2c(b^3 - c^3)}$.

Ex. 2. Reduce $5\sqrt{\tfrac{1}{3}}$ to its simplest form.
$$5\sqrt{\tfrac{1}{3}} = 5\sqrt{(\tfrac{1}{9} \times 3)} = \tfrac{5}{3}\sqrt{3}.$$

[NOTE. Hence, if the quantity under the root is a fraction, its numerator and denominator must be multiplied by such a number as will allow us to remove the latter from under the root.]

Ex. 3. Transform $\sqrt{(a^2b)}$ into another surd with b for coefficient.
$$\sqrt{a^2b} = b\sqrt{a^2b/b} = b\sqrt{a^2b/b^2} = b\sqrt{a^2/b}.$$

23. To transform a surd of any order into a surd of a different order.

Ex. Express $\sqrt[5]{a}$ as a surd of the 17th order.
$$\sqrt[5]{a} = a^{\tfrac{1}{5}} = \left(a^{\tfrac{17}{5}}\right)^{\tfrac{1}{17}} = \sqrt[17]{a^{\tfrac{17}{5}}}.$$

24. To transform surds of different orders into surds of the same order.

RULE. *Express these radicals as surds with indices and reduce the indices to equivalent fractions with a common denominator* [or better, the *least common denominator*].

Ex. Transform $\sqrt[3]{a}$, $\sqrt[5]{b^3}$ and $\sqrt[6]{c}$ into surds of the same order.

Here $\sqrt[3]{a} = a^{\tfrac{1}{3}};\ \sqrt[5]{b^3} = b^{\tfrac{3}{5}};\ \sqrt[6]{c} = c^{\tfrac{1}{6}}.$

Now the L.C.M. of 3, 5, 6 $= 30$.

$\therefore\ \sqrt[3]{a} = a^{\tfrac{10}{30}};\ \sqrt[5]{b^3} = b^{\tfrac{18}{30}};\ \sqrt[6]{c} = c^{\tfrac{5}{30}}.$

Thus the required surds are $\sqrt[30]{a^{10}}$, $\sqrt[30]{b^{18}}$ and $\sqrt[30]{c^5}$.

25. To compare surds of different orders.

RULE. *Transform the surds into surds of the same order, and then arrange them according to the magnitude of the numbers under the roots.*

Ex. Compare $\sqrt{5}$, $\sqrt[3]{24}$ and $\sqrt[4]{11}$.

Since the L.C.M. of 2, 3 and 4 is 12, we have to express the given surds as surds of the 12th order.

SURDS.

$$\sqrt{5} = 5^{\frac{1}{2}} = (5^{12})^{\frac{1}{12}} = (5^6)^{\frac{1}{12}} = {}^{12}\!\sqrt{15,625}.$$

Similarly $\quad \sqrt[4]{24} = (24^3)^{\frac{1}{12}} = {}^{12}\!\sqrt{13,824};$

and $\quad \sqrt[3]{11} = (11^4)^{\frac{1}{12}} = {}^{12}\!\sqrt{14,641}.$

Thus $\quad \sqrt{5} > \sqrt[3]{11} > \sqrt[4]{24}.$

26. To find the sum of difference of like simple surds.

RULE. *Reduce the surds to their simplest forms and add or subtract.*

Ex. Simplify $\quad \sqrt{243} + \sqrt{75} - \sqrt{12} - \sqrt{\tfrac{1}{3}}.$

The expression $= \sqrt{81 \times 3} + \sqrt{25 \times 3} - \sqrt{4 \times 3} - \sqrt{\tfrac{1}{3} \times 3}$

$\qquad = 9\sqrt{3} + 5\sqrt{3} - 2\sqrt{3} - \tfrac{1}{3}\sqrt{3} = 11\tfrac{2}{3}\sqrt{3}.$

NOTE. *The sum or difference of unlike surds cannot be simplified.*

27. To find the product of simple surds.

Ex. Find the product of $3\sqrt{2}$, $4\sqrt[3]{3}$ and $5\sqrt[4]{5}$.

The L.C.M. of 2, 3 and 4 is 12.

$\therefore 3\sqrt{2} = 3{}^{12}\!\sqrt{2^6} = 3{}^{12}\!\sqrt{64};$

$\qquad 4\sqrt[3]{3} = 4{}^{12}\!\sqrt{3^4} = 4{}^{12}\!\sqrt{81};$

and $\qquad 5\sqrt[4]{5} = 5{}^{12}\!\sqrt{5^3} = 5{}^{12}\!\sqrt{125}.$

$\therefore 3\sqrt{2} \times 4\sqrt[3]{3} \times 5\sqrt[4]{5} = 3 \times 4 \times 5\,{}^{12}\!\sqrt{64 \times 81 \times 125} = 60\,{}^{12}\!\sqrt{648,000}.$

28. To find the product or quotient of two or more compound surds.

RULE. *Proceed as in multiplying or dividing compound rational algebraical expressions.*

Ex. Find the square of $3\sqrt{2} + 5\sqrt[3]{9}.$

$(3\sqrt{2} + 5\sqrt[3]{9})^2 = (3\sqrt{2})^2 + (5\sqrt[3]{9})^2 + 2 \cdot (3\sqrt{2}) \cdot (5\sqrt[3]{9})$

$\qquad = 18 + 25\sqrt[3]{81} + 30\sqrt{2}\sqrt[3]{9},$

$= 18 + 75\sqrt[3]{3} + 30\sqrt[6]{2^3 \times 9^2} = 18 + 75\sqrt[3]{3} + 30\sqrt[6]{648}.$

[The product of $(a + \sqrt{b})(c + \sqrt{d})$ cannot contain any surd except $\sqrt{b}$, $\sqrt{d}$, and $\sqrt{bd}$; and the product $(a + \sqrt{b})(a + \sqrt{c} + \sqrt{d})$ cannot contain any surd except $\sqrt{b}$, $\sqrt{c}$, $\sqrt{d}$, $\sqrt{(bc)}$ and $\sqrt{(bd)}$.]

EXERCISES II.

1. Express a^4b^4 as a cube root.
2. Reduce ab^2c^3 to the form of an mth root.

Transform into entire surds :

3. $5\sqrt{7}$.   4. $2\sqrt[3]{3}$.   5. $a^2b\sqrt[5]{c}$.   6. $3\sqrt[5]{2}$.

7. $\frac{3}{4}\sqrt{\frac{5}{17}}$.   8. $\frac{a}{b}\sqrt[m]{\frac{c}{d}}$.   9. $\frac{a^n}{b^r}\sqrt[s]{\left(\frac{b^m}{a^{2n}}\right)}$.

10. $\frac{x+y}{x-y}\sqrt{\frac{x-y}{x^3+y^3}}$.

Reduce to their simplest forms :

11. $\sqrt{125}$.   12. $\sqrt[3]{768}$.   13. $a\sqrt{(b^7c)}$.   14. $\sqrt[3]{\frac{1}{5}}$.
15. $3\sqrt{\frac{2}{3}}$.   16. $5\sqrt[3]{5\frac{2}{5}}$.   17. $(72)^{\frac{3}{5}}$.   18. $(1\frac{1}{8})^{-\frac{1}{2}}$.
19. $(30\frac{3}{8})^{-\frac{2}{5}}$.   20. $5\sqrt[3]{(-1536)}$.   21. $\sqrt{(mx^2-2mx+m}$.

Express as surds of the 5th order with positive indices :

22. $ab^{\frac{2}{3}}$.   23. $a^2c \div a^{-7}$.   24. $a^{-\frac{1}{3}}b^{-\frac{2}{3}} \div c^{-4}$.

25. Express $ab^{\frac{2}{3}}$ and $a^{-2}b^2c^{\frac{1}{3}}$ as surds of the 7th order with negative indices.

Express as surds of the mth order with positive indices :

26. $\sqrt{(ab^3)}$.   27. x^{-n}.   28. $\frac{x^{-2}}{y^{\frac{1}{3}}z}$.   29. $\sqrt[3]{(a^2c)}$.

30. Change the index of $\sqrt[3]{(a+b)^2}$ to $\frac{5}{6}$.
31. Express $\sqrt{5}$ with 6 as coefficient.
32. Express $\sqrt{\{a^2bc(a-b)\}}$ with abc as coefficient.
33. Express $\sqrt[5]{(64a^2b)}$ with $2a$ as coefficient.
34. Show that $\sqrt{12}$, $5\sqrt{\frac{1}{27}}$, $\sqrt{48}$ and $7\sqrt{147}$ are similar surds.
35. Show that $\sqrt{125}$, $\sqrt{\frac{1}{20}}$, and $\sqrt[4]{400}$ are like surds.

SURDS.

36. Show that $3\sqrt[3]{4}$, $\frac{1}{3}\sqrt[3]{\frac{1}{54}}$, $(250)^{-\frac{1}{3}}$ and $(500)^{\frac{1}{3}}$ are similar.

37. Show that $(8)^{-\frac{1}{5}}$, $(4)^{\frac{1}{5}}$ and $(16{,}384)^{\frac{1}{10}}$ are like surds.

Express as surds of the same lowest order:

38. $\sqrt[3]{a}$ and $\sqrt{b}$. **39.** $\sqrt{2}$, $\sqrt[3]{3}$ and $\sqrt{4}$.

40. $\sqrt[6]{5}$, $\sqrt{11}$ and $\sqrt[3]{7}$. **41.** $5\frac{1}{2}$, $\sqrt[3]{7}$, and $\sqrt[4]{3}$.

42. $\sqrt[3]{9}$, $\sqrt[6]{5}$ and $\sqrt[8]{7}$.

Compare:

43. $\sqrt{5}$ and $\sqrt[3]{11}$. **44.** $\sqrt{14}$, $3\sqrt[3]{2}$, and $2\sqrt[3]{6}$.

Arrange in order of magnitude:

45. $\frac{1}{2}\sqrt{3}$, $\frac{1}{3}\sqrt{5}$ and $\frac{1}{4}\sqrt{7}$. **46.** $\sqrt{5}$, $2\sqrt[3]{\frac{3}{2}}$ and $3(4\frac{1}{2})^{-\frac{1}{6}}$.

Find the values of:

47. $\sqrt{24} + \sqrt{54} - \sqrt{96}$. **48.** $7\sqrt[3]{54} + 3\sqrt[3]{16} - 5\sqrt[3]{128}$.

49. $\sqrt[3]{128} + \sqrt[3]{250} - \sqrt[3]{\frac{1}{4}}$.

50. $5\sqrt{243} + 4\sqrt{\frac{1}{27}} - 5\sqrt{\frac{1}{48}} + \sqrt{\frac{1}{3}}$.

51. $8\sqrt{\frac{3}{4}} - \frac{1}{2}\sqrt{12} + 4\sqrt{27} - 2\sqrt{\frac{3}{16}}$.

52. $2\sqrt{54} + 3\sqrt{294} - \sqrt{\frac{2}{3}} - \sqrt{\frac{27}{50}}$.

53. $2\sqrt{(a^5x^3)} - 5\sqrt[3]{(a^6x^2)}$.

Find the values of the following products:

54. $7\sqrt{5} \times 8\sqrt{10} \times 3\sqrt{2}$.

55. $\sqrt[3]{7} \times \sqrt{2} \times \sqrt[6]{5}$. **56.** $2\sqrt{24} \times 3\sqrt[4]{18} \times 4\sqrt[6]{24}$.

57. $\sqrt[3]{3} \times \sqrt[4]{5}$. **58.** $(2+\sqrt{3})(\sqrt{3}-7)$.

59. $(\sqrt[3]{3}+1)\sqrt[3]{9} - \sqrt[3]{3}+1)$.

60. $(\sqrt{x}+\sqrt{y})(\sqrt[4]{x}+\sqrt[4]{y})(\sqrt[4]{x}-\sqrt[4]{y})$.

61. $(x^2+1)(x^2+\sqrt{2}x+1)(x^2-1)(x^2-\sqrt{2}x+1)$.

62. $\frac{1}{8}\{2\sqrt{(5+\sqrt{5})} - \sqrt{10+\sqrt{2}}\}$
$\times \frac{1}{8}\{2\sqrt{(5+\sqrt{5})} + \sqrt{10-\sqrt{2}}\}$.

63. Show that there are $n-2$ distinct roots that are equiradical with $\sqrt[n]{a}$.

29. Rationalisation of surds. DEFINITION.—
When the product of two surds or surd expressions is rational, each is called a **rationalising factor** of the other.

Ex. Suppose we require to find the value of $\dfrac{\sqrt{5}+\sqrt{3}}{\sqrt{2}}$ to 3 places of decimals. If the expression is kept in its present form, we shall have to extract the square roots of the three numbers 5, 3, 2 each to at least 4 places of decimals, and obtain the quotient of the sum of the first two of these by the last by *long division*. We can save ourselves one-half of this trouble, if we multiply the numerator and the denominator of the given expression by $\sqrt{2}$ and thus bring it to the form $\dfrac{\sqrt{10}+\sqrt{6}}{2}$; for, in this case, we shall only have to extract the square roots $\sqrt{10}$ and $\sqrt{6}$ to 3 places of decimals, and take half the sum for answer. Hence arises the importance of **rationalising** the denominators of surd fractions, *i.e.*, of clearing them of surds by choosing proper surd factors, wherewith to multiply both numerators and denominators of the fractions.

CAUTION.—In rationalising the denominator of any surd fraction, **the numerator must be multiplied by the rationalising factor as well as the denominator,** according to the ordinary rules for the transformation of fractions in algebra. Otherwise the value of the fraction would be altered. The new *numerator* is consequently often a more complicated surd expression than the original numerator, but the important point is that no surds occur in the new *denominator*.

30. Rationalising factors of monomial surds.

CASE I. Let each of the indices be less than unity.

Ex. Find a factor that will rationalise $a^{\frac{3}{4}} b^{\frac{7}{8}} c^{\frac{8}{9}}$.

A rationalising factor is evidently $a^{1-\frac{3}{4}} b^{1-\frac{7}{8}} c^{1-\frac{8}{9}}$ which is $= a^{\frac{1}{4}} b^{\frac{1}{8}} c^{\frac{1}{9}}$.

CASE II. Let the indices be any arithmetical numbers whatever.

It is clear that the surds, when simplified, will reduce to the form considered in Case I, when they may therefore be rationalised.

Ex. 1. Find a factor which will rationalise $\sqrt[5]{a^7} \sqrt[7]{b^{23}}$.

The expression $= a^{\frac{7}{5}} b^{\frac{23}{7}} = a^{1\frac{2}{5}} b^{3\frac{2}{7}} = (ab^3) \times (a^{\frac{2}{5}} b^{\frac{2}{7}})$.

Hence a rationalising factor is $a^{1-\frac{2}{5}} b^{1-\frac{2}{7}} = a^{\frac{3}{5}} b^{\frac{5}{7}}$.

20   SURDS.

Ex. 2. Required to calculate $\dfrac{5^{\frac{1}{2}}}{2^{\frac{1}{3}} \times 3^{\frac{3}{2}}}$ to 3 places of decimals.

A factor that rationalises the denominator is evidently $2^{\frac{2}{3}} \times 3^{\frac{1}{2}}$.
Hence the expression
$$= \frac{5^{\frac{1}{2}} \times 3^{\frac{1}{2}} \times 2^{\frac{2}{3}}}{2^{\frac{1}{3}} \times 2^{\frac{2}{3}} \times 3^{\frac{3}{2}} \times 3^{\frac{1}{2}}} = \frac{15^{\frac{1}{2}} \times 4^{\frac{1}{3}}}{2 \times 9}$$
$$= 3 \cdot 87298 \times 1 \cdot 58740 \times \tfrac{1}{18} = 6 \cdot 14796 \div 18 = \cdot 34155\ldots$$

31. Rationalising factors of binomial quadratic surds.

Binomial quadratic surds are of the forms
$$a\sqrt{b} \pm c\sqrt{d}, \text{ or } a \pm b\sqrt{c}.$$

DEFINITION.—When two binomial quadratic surds differ only in the sign which connects their terms, they are said to be **conjugate.** Thus $a\sqrt{b}+c\sqrt{d}$ and $a\sqrt{b}-c\sqrt{d}$ are conjugate surds.

Since $(a\sqrt{b}+c\sqrt{d})(a\sqrt{b}-c\sqrt{d}) = a^2b - c^2d$ which is rational, we conclude that **a rationalising factor of a binomial quadratic surd is its conjugate.**

Ex. 1. Find a rationalising factor of $5\sqrt{2} - \sqrt{3}$.
The required factor is $5\sqrt{2} + \sqrt{3}$
the rational result being
$$(5\sqrt{2})^2 - (\sqrt{3})^2 = 50 - 3 = 47.$$

Ex. 2. Find $\dfrac{\sqrt{6}}{\sqrt{2}+1}$ true to 3 places of decimals.
$$\frac{\sqrt{6}}{\sqrt{2}+1} = \frac{\sqrt{6}(\sqrt{2}-1)}{(\sqrt{2}+1)(\sqrt{2}-1)} = \frac{\sqrt{12}-\sqrt{6}}{2-1} = \sqrt{12}-\sqrt{6}.$$
$= 3 \cdot 4641\ldots - 2 \cdot 4494\ldots = 1 \cdot 0147\ldots = 1 \cdot 015$ (true to 3 places).

32. Rationalisation of trinomial quadratic surds.

The identity
$$(a+b+c)(b+c-a)(c+a-b)(a+b-c) = \\ 2b^2c^2 + 2c^2a^2 + 2a^2b^2 - a^4 - b^4 - c^4$$
shows that any trinomial quadratic surd can be rationalised by multiplying it by three other factors.

SURDS. 21

Ex. 1. The rationalising factor of $\sqrt{3}+\sqrt{2}-1$ is
$$(\sqrt{3}+\sqrt{2}+1)(\sqrt{3}-\sqrt{2}+1)(\sqrt{3}-\sqrt{2}-1).$$
Ex. 2. Find a rationalising factor of $\sqrt{2}+\sqrt{3}+\sqrt{10}$.
The rationalising factor is
$$(\sqrt{2}+\sqrt{3}-\sqrt{10})(\sqrt{2}-\sqrt{3}+\sqrt{10})(\sqrt{2}-\sqrt{3}-\sqrt{10}).$$
In practice it will be more convenient to effect the above by two operations as in the following method :—
$$(\sqrt{2}+\sqrt{3}+\sqrt{10})(\sqrt{2}+\sqrt{3}-\sqrt{10}) = (\sqrt{2}+\sqrt{3})^2 - (\sqrt{10})^2$$
$= 2+3+2\sqrt{6}-10 = 2\sqrt{6}-5$ which is a binomial surd.
Again, $(2\sqrt{6}-5)(2\sqrt{6}+5) = (2\sqrt{6})^2 - 5^2 = 24-25 = -1$.
Thus a factor that rationalises the given surd is
$$(\sqrt{2}+\sqrt{3}-\sqrt{10})(2\sqrt{6}+5),$$
which is, in fact, equal to
$$(\sqrt{2}+\sqrt{3}-\sqrt{10})(\sqrt{2}-\sqrt{3}+\sqrt{10})(-\sqrt{2}+\sqrt{3}+\sqrt{10})$$
as can be very easily verified.

33. Rationalisation of any binomial surd.

CASE I. First, let the binomial be $a^{\frac{1}{m}} - b^{\frac{1}{n}}$.

If $x = a^{\frac{1}{m}}, y = b^{\frac{1}{n}}$, and $p =$ the L.C.M. of m and n, then x^p and y^p are rational quantities.

Now take the identity
$$x^p - y^p = (x-y)(x^{p-1} + x^{p-2}y + x^{p-3}y^2 \ldots + xy^{p-2} + y^{p-1}).$$
This shows that $x^p - y^p$ is divisible by $x-y$; that is, that the difference of the pth powers of $a^{\frac{1}{m}}$ and $b^{\frac{1}{n}}$, which is rational, is divisible by $a^{\frac{1}{m}} - b^{\frac{1}{n}}$.

Hence a factor that rationalises $a^{\frac{1}{m}} - b^{\frac{1}{n}}$ is
$$x^{p-1} + x^{p-2}y + \ldots + xy^{p-2} + y^{p-1},$$
where x, y, p denote respectively $a^{\frac{1}{m}}, b^{\frac{1}{n}}$ and the L.C.M. of m and n.

CASE II. Next let the binomial be $a^{\frac{1}{m}} + b^{\frac{1}{n}}$.
Take x, y, p as in Case I.
If p is even, the identity that we take is
$$x^p - y^p = (x+y)(x^{p-1} - x^{p-2}y + x^{p-3}y^2 - \ldots + xy^{p-2} - y^{p-1}),$$
which shows that a factor which rationalises $x+y$ is the other factor on the right-hand side, and that the rational result is $x^p - y^p$.

SURDS.

If p be odd, we take the identity
$$x^p + y^p = (x+y)(x^{p-1} - x^{p-2}y + \ldots - xy^{p-2} + y^{p-1}),$$
which shows that the rational result in this case will be $x^p + y^p$.

Ex. 1. Find a factor that rationalises $5^{\frac{1}{2}} - 7^{\frac{1}{3}}$.

Since the L.C.M. of 2 and 3 is 6, we have, substituting x and y for $5^{\frac{1}{2}}$ and $7^{\frac{1}{3}}$ respectively,
$$x^6 - y^6 = (x-y)(x^5 + x^4y + x^3y^2 + x^2y^3 + xy^4 + y^5)$$
which shows that a factor which rationalises $5^{\frac{1}{2}} - 7^{\frac{1}{3}}$ is
$$(5^{\frac{1}{2}})^5 + (5^{\frac{1}{2}})^4(7^{\frac{1}{3}}) + (5^{\frac{1}{2}})^3(7^{\frac{1}{3}})^2 + (5^{\frac{1}{2}})^2(7^{\frac{1}{3}})^3 + (5^{\frac{1}{2}})(7^{\frac{1}{3}})^4 + (7^{\frac{1}{3}})^5$$
$$= 25.5^{\frac{1}{2}} + 25.7^{\frac{1}{3}} + 5.5^{\frac{1}{2}}.7^{\frac{2}{3}} + 35 + 7.5^{\frac{1}{2}}.7^{\frac{1}{3}} + 7.7^{\frac{2}{3}};$$
and that the rational result is $(5^{\frac{1}{2}})^6 - (7^{\frac{1}{3}})^6 = 5^3 - 7^2 = 125 - 49 = 76.$

Ex. 2. Find a factor which rationalises $a + b^{\frac{1}{3}}$.

Here p, the L.C.M. of 1 and 3 is odd. Hence the identity that we take is
$$x^3 + y^3 = (x+y)(x^2 - xy + y^2)$$
which shows that a factor which rationalises $a + b^{\frac{1}{3}}$ is $a^2 - ab^{\frac{1}{3}} + b^{\frac{2}{3}}$, and that the rational result is $a^3 + (b^{\frac{1}{3}})^3$, that is, $a^3 + b$.

NOTE. We may remark in passing that surds like
$$a^{\frac{2}{3}} - a^{\frac{1}{3}}b^{\frac{1}{3}} + b^{\frac{2}{3}} \text{ and } a + a^{\frac{1}{2}}b^{\frac{1}{2}} + b$$
may be rationalised by utilising identities like
$$(a^2 - ab + b^2)(a+b) = a^3 + b^3; \quad (a^2 + ab + b^2)(a^2 - ab + b^2) = a^4 + a^2b^2 + b^4.$$

Ex. Rationalise the denominator of $\dfrac{1}{\sqrt[3]{25} - \sqrt[3]{5} + 1}$.

$$\frac{1}{\sqrt[3]{25} - \sqrt[3]{5} + 1} = \frac{1}{(\sqrt[3]{5})^2 - \sqrt[3]{5} + 1} = \frac{\sqrt[3]{5}+1}{(\sqrt[3]{5}+1)\{(\sqrt[3]{5})^2 - \sqrt[3]{5} + 1\}}$$
$$= \frac{\sqrt[3]{5}+1}{(\sqrt[3]{5})^3 + 1} = \frac{\sqrt[3]{5}+1}{5+1} = \frac{\sqrt[3]{5}+1}{6}.$$

SURDS.

EXERCISES II.

64. Rationalise $a^{\frac{1}{5}}b^{\frac{3}{4}}c^{\frac{1}{3}}$ and $a^{\frac{7}{5}}b^{\frac{3}{2}}c^3$.

Calculate the values of the following to 3 places of decimals:

[Given $\sqrt{3} = 1\cdot 73201$; $\sqrt{5} = 2\cdot 2360$; $\sqrt{6} = 2\cdot 4494$; $\sqrt[3]{7} = 1\cdot 91291$; $\sqrt[3]{25} = 2\cdot 9240$.]

65. $\dfrac{15}{\sqrt{3}}$. **66.** $\dfrac{12\sqrt{2}}{\sqrt{3}}$. **67.** $\dfrac{6}{\sqrt{5}}$. **68.** $\dfrac{\sqrt{3}}{\sqrt{2}}$.

69. $\dfrac{1}{\sqrt{24}}$. **70.** $\dfrac{1}{\sqrt{300}}$. **71.** $\dfrac{3}{\sqrt[3]{49}}$. **72.** $\dfrac{4}{\sqrt[3]{5}}$.

Find the values of :

73. $\{\sqrt{(a+b)} - c\}\{\sqrt{(a+b)} + c\}$.

74. $(\sqrt{2} + 3\sqrt{\tfrac{1}{2}}) \div \tfrac{1}{2}\sqrt{\tfrac{1}{2}}$. **75.** $(\sqrt{3} + \sqrt{2}) \div (\sqrt{3} - \sqrt{2})$.

76. $4 \div (\sqrt{5} - 1)$.

Rationalise the denominators of, and simplify :

77. $\dfrac{15 + 7\sqrt{3}}{9 + 5\sqrt{3}}$. **78.** $\dfrac{1}{2\sqrt{2} - \sqrt{3}}$. **79.** $\dfrac{4\sqrt{7} + 3\sqrt{2}}{5\sqrt{2} + 2\sqrt{7}}$.

80. $\dfrac{1}{a - \sqrt{(a^2 - x^2)}} - \dfrac{1}{a + \sqrt{(a^2 - x^2)}}$.

81. $\dfrac{1}{\sqrt{2} - 1} + \dfrac{1}{\sqrt{3} - \sqrt{2}} + \dfrac{1}{2 - \sqrt{3}}$.

82. $\dfrac{1}{4(1 + \sqrt{x})} + \dfrac{1}{4(1 - \sqrt{x})} + \dfrac{1}{2(1 + x)}$.

83. $\dfrac{1}{\sqrt{2} + \sqrt{3} + \sqrt{5}}$.

84. $\dfrac{3 + 4\sqrt{3}}{\sqrt{6} + \sqrt{2} - \sqrt{5}}$. **85.** $\dfrac{1}{\sqrt{30} + 3\sqrt{2} + 2\sqrt{3}}$.

86. $\dfrac{3 + 2\sqrt{5}}{1 - \sqrt{3} + 5}$. **87.** $\dfrac{(15 + 5\sqrt{5})(\sqrt{5} - 2)}{5 - \sqrt{5}}$.

88. $\dfrac{13\sqrt{15} - 7\sqrt{21}}{13\sqrt{1\tfrac{2}{3}} - 7\sqrt{2\tfrac{1}{3}}}$. **89.** $\dfrac{x + 2 + \sqrt{(x^2 - 4)}}{x + 2 - \sqrt{(x^2 - 4)}}$.

SURDS.

90. Find the rationalising factors and the rationalised results of:
   (i) $3^{\frac{1}{3}}+4^{\frac{1}{3}}$.   (ii) $a^{\frac{1}{2}}-b^{\frac{1}{4}}$.   (iii) $3-4^{\frac{1}{5}}$.   (iv) $a^{\frac{1}{3}}-b^{\frac{1}{4}}$.

91. Simplify $\left(\dfrac{2-\sqrt{(2+\sqrt{2})}}{2+\sqrt{(2+\sqrt{2})}}\right)^{\frac{1}{2}}$.

92. Simplify $\dfrac{\sqrt{\frac{3}{2}}-\sqrt{\frac{8}{27}}}{\sqrt{2}\left(\sqrt{3}+\dfrac{1}{\sqrt{3}}\right)}$.

93. Simplify $\dfrac{\left(\dfrac{1}{\sqrt{3}}-\dfrac{1}{\sqrt{5}}\right)(\sqrt{5}+\sqrt{3})}{(\sqrt{5}+\sqrt{3})^2-(\sqrt{5}-\sqrt{3})^2}$.

94. Find to five decimal places the value of
$$\dfrac{\sqrt{7}-\sqrt{5}}{\sqrt{7}+\sqrt{5}}.$$

95. Show that
$$\dfrac{2\sqrt{2}+\sqrt{3}+\sqrt{5}}{2\sqrt{2}-\sqrt{3}-\sqrt{5}}=-\dfrac{1}{15}(15+6\sqrt{10}+10\sqrt{6}+8\sqrt{15}).$$

96. Show that
$$\dfrac{3}{1-\sqrt{2}+\sqrt{3}}+\dfrac{1}{1-\sqrt{2}-\sqrt{3}}-\dfrac{2}{1+\sqrt{2}-\sqrt{3}}+\dfrac{3}{\sqrt{2}}=1.$$

97. Show that $\dfrac{7}{2^{\frac{3}{4}}+2^{\frac{1}{4}}+1}=1+2\sqrt{2}-3\sqrt[4]{2}+\sqrt[4]{8}$.

98. Show that $\dfrac{5}{\sqrt[3]{36}+\sqrt[3]{6}+1}=\sqrt[3]{6}-1$.

99. If $x=\sqrt[3]{r+\sqrt{r^2+q^3}}+\sqrt[3]{r-\sqrt{r^2+q^3}}$,
show that $x^3+3qx-2r=0$.

Rationalise the denominators of

100. $\dfrac{3+\sqrt{2}}{\sqrt{2}-1+\sqrt{3}}$.   101. $\dfrac{1}{1+\sqrt{2}+\sqrt{3}+\sqrt{6}}$.

102. $\dfrac{1}{1-\sqrt{3}-\sqrt{5}+\sqrt{15}}$.

103. If $a^{\frac{1}{3}}+b^{\frac{1}{3}}+c^{\frac{1}{3}}=0$, then $(a+b+c)^3=27abc$.

34. The product and quotient of two similar quadratic surds are rational; *and, conversely, if the product or quotient of two quadratic surds is rational, the surds are similar.*

(i) Let $a\sqrt{m}$ and $b\sqrt{m}$ be two similar surds. Their product is abm which is rational. Their quotient $=a/b$ which is also rational.

(ii) Let the product of two surds $\sqrt{a}$ and $\sqrt{b}$ be m, a rational quantity. Then $\sqrt{b} = \dfrac{m}{\sqrt{a}} = \dfrac{m}{a}\sqrt{a} = $ a rational quantity $\times \sqrt{a}$.

Hence $\sqrt{b}$ and $\sqrt{a}$ are similar.

It may be proved indirectly by applying the above results that *the product and quotient of two dissimilar quadratic surds are irrational; and, conversely, if the product or quotient of two quadratic surds is irrational, the surds are dissimilar.*

35. A quadratic surd cannot be equal to the sum or difference of a rational quantity and a quadratic surd.

For if possible, let $\sqrt{a} = b \pm \sqrt{c}$.

Squaring, we have $a = b^2 + c \pm 2b\sqrt{c}$,

which gives $\sqrt{c} = \pm \dfrac{a - b^2 - c}{2b}$.

Thus a surd is equal to a rational quantity, which is absurd.

36. A quadratic surd cannot be equal to the sum or difference of two quadratic surds, unless all the three surds are similar.

We have to show that, if $\sqrt{a} \pm \sqrt{b} \pm \sqrt{c} = 0$, then $\sqrt{a}$, $\sqrt{b}$, and $\sqrt{c}$ are similar.

Now, $\sqrt{a} \pm \sqrt{b} = \mp \sqrt{c}$.

$\therefore a + b \pm 2\sqrt{(ab)} = c$.

$\therefore \pm 2\sqrt{(ab)} = c - a - b$, which is rational.

26   SURDS.

Hence $\sqrt{a}$ and $\sqrt{b}$ are similar surds [§ 34].
It can be similarly proved that $\sqrt{a}$ and $\sqrt{c}$ are similar.
Hence $\sqrt{a}$, $\sqrt{b}$ and $\sqrt{c}$ are all similar.

37. If $a+\sqrt{b} = c+\sqrt{d}$; then $a = c$ and $b = d$.

For, if not, let $a = c + m$.
Then $\qquad c + m + \sqrt{b} = c + \sqrt{d}$.
$\qquad\qquad \therefore m + \sqrt{b} = \sqrt{d}$.

That is, a quadratic surd is equal to the sum of a rational quantity and a quadratic surd, which is absurd. [§ 35.] Hence, &c.

NOTE. Thus we see that an equation like $a+\sqrt{b}=c+\sqrt{d}$ which contains both rational quantities and surds is equivalent to two independent equations.

38. If $\sqrt{(a+\sqrt{b})} = \sqrt{c}+\sqrt{d}$,
then $\qquad \sqrt{(a-\sqrt{b})} = \sqrt{c}-\sqrt{d}$.

For by squaring $\sqrt{(a+\sqrt{b})} = \sqrt{c}+\sqrt{d}$,
$\qquad a+\sqrt{b} = c + d + 2\sqrt{(cd)}$.
$\therefore a = c+d$; and $\sqrt{b} = 2\sqrt{(cd)}$. [§ 37.]
$\therefore a - \sqrt{b} = c + d - 2\sqrt{(cd)}$.

Hence supposing $c > d$
$$\sqrt{(a-\sqrt{b})} = \sqrt{c} - \sqrt{d}.$$

39. To find the square root of $a+\sqrt{b}$ by inspection.—If the square root of $a+\sqrt{b}$ is expressible in the form of $\sqrt{x}+\sqrt{y}$, the root may sometimes be obtained by inspection.

Ex. 1. Find the square root of $8 + 2\sqrt{15}$.
Let $\qquad \sqrt{(8+2\sqrt{15})} = \sqrt{x}+\sqrt{y}$.
Hence $\qquad 8 + 2\sqrt{15} = x + y + 2\sqrt{(xy)}$.

Thus, we have to find two numbers such that their sum is 8 and their product is 15.

The numbers are obviously 5 and 3;

and the required square root is therefore $\sqrt{5}+\sqrt{3}$.

Ex. 2. Simplify $\sqrt{\left(\dfrac{\sqrt{7}+\sqrt{3}}{\sqrt{7}-\sqrt{3}}\right)}$.

To rationalise the denominator of the fraction under the root, we must multiply above and below by $\sqrt{7}+\sqrt{3}$.

Hence the given expression $=\sqrt{\left(\dfrac{(\sqrt{7}+\sqrt{3})^2}{7-3}\right)}=\dfrac{1}{2}(\sqrt{7}+\sqrt{3})$.

40. To find the square root of $a+\sqrt{b}$.

Let $\sqrt{(a+\sqrt{b})}=\sqrt{x}+\sqrt{y}$.

Then, squaring, we have $a+\sqrt{b}=x+y+2\sqrt{(xy)}$ which gives by § 37

$x+y=a$ and $2\sqrt{(xy)}=\sqrt{b}$.

To find x and y from these two equations, we have

$(x+y)^2-4xy=a^2-b$.

$\therefore (x-y)^2=a^2-b$;

Hence $\quad x-y=\sqrt{(a^2-b)}$; and $x+y=a$.

$\therefore 2x=a+\sqrt{(a^2-b)}$;

and $\quad\quad\quad\quad 2y=a-\sqrt{(a^2-b)}$.

Hence, $x=\tfrac{1}{2}\{a+\sqrt{(a^2-b)}\}$; and $y=\tfrac{1}{2}\{a-\sqrt{(a^2-b)}\}$.

Thus the required root

$=\sqrt{\{\tfrac{1}{2}a+\tfrac{1}{2}\sqrt{(a^2-b)}\}}+\sqrt{\{(\tfrac{1}{2}a-\tfrac{1}{2}\sqrt{(a^2-b)}\}}$,

also $\sqrt{(a-\sqrt{b})}=\sqrt{\{\tfrac{1}{2}a+\tfrac{1}{2}\sqrt{(a^2-b)}\}}-\sqrt{\{\tfrac{1}{2}a-\tfrac{1}{2}\sqrt{(a^2-b)}\}}$.

In order that a quadratic surd may be obtained as the square root of $a\pm\sqrt{b}$, it is evident from the values we have obtained for x and y in the previous work, (i) that a^2-b must be a perfect square and (ii) that a must be positive.

28   SURDS.

When these conditions are not satisfied*, a biquadratic surd of the form $\sqrt[4]{x} + \sqrt[4]{y}$ may sometimes be obtained as the square root. [See Ex. 2 below.]

If we are required to find the square root of an expression like $3 - 2\sqrt{2}$ in which the terms are of opposite signs, it will be well to begin by assuming the square root to be equal to $\sqrt{x} - \sqrt{y}$.

Ex. 1. Find the square root of $14 + 6\sqrt{5}$.

[Here $14^2 = 196$, and $(6\sqrt{5})^2 = 180$; and $(14)^2 - (6\sqrt{5})^2 = 4^2$.]

Both the conditions are satisfied, and a binomial surd can be obtained as the square root.

Let $\quad\quad \sqrt{(14 + 6\sqrt{5})} = \sqrt{x} + \sqrt{y}$ .... (i).

Then by § 38 $\quad \sqrt{(14 - 6\sqrt{5})} = \sqrt{x} - \sqrt{y}$ .... (ii).

$\therefore \sqrt{\{(14 + 6\sqrt{5})(14 - 6\sqrt{5})\}} = (\sqrt{x} + \sqrt{y})(\sqrt{x} - \sqrt{y})$.

$\therefore x - y = \sqrt{(196 - 180)} = 4$.

But since $\quad \sqrt{(14 + 6\sqrt{5})} = \sqrt{x} + \sqrt{y}$,

$14 + 6\sqrt{5} = x + y + 2\sqrt{(xy)}$; and $\therefore x + y = 14$.

$\therefore x = 9$; and $y = 5$.

$\therefore$ the required root $= \sqrt{9} + \sqrt{5} = 3 + \sqrt{5}$.

[NOTE. If we take $x - y = -4$ in the extraction of the square root, we get $y = 9$ and $x = 5$, which give us the same solution.]

Ex. 2. Find the square root of $5\sqrt{6} + 12$.

Here $(5\sqrt{6})^2 = 150$, $(12)^2 = 144$; and $(5\sqrt{6})^2 - (12)^2 = (\sqrt{6})^2$.

Here the difference of the squares of $5\sqrt{6}$ and 12 is not a perfect square. The required square root is not therefore a quadratic surd.

We therefore express $5\sqrt{6} + 12$ in the form $\sqrt{6}(5 + 2\sqrt{6})$.

If, of the two factors, $5 + 2\sqrt{6}$ satisfies the two conditions laid down already, its square root can be expressed in the form $\sqrt{m} + \sqrt{n}$.

By proceeding as before, the square root of $5 + 2\sqrt{6}$ is found to be $\sqrt{3} + \sqrt{2}$; and the square root of $\sqrt{6}$ is $\sqrt[4]{6}$.

Hence the square root of the given expression is $\sqrt[4]{6}(\sqrt{3} + \sqrt{2})$

$\quad\quad\quad\quad = \sqrt[4]{54} + \sqrt[4]{24}$.

This may be verified by squaring $\sqrt[4]{54} + \sqrt[4]{24}$.

---

* When $a^2 - b$ is a perfect square, but a is not positive, an imaginary quantity is obtained as the square root; for example, $-3 + 2\sqrt{2}$ gives $\sqrt{(-1)} - \sqrt{(-2)}$ as the square root.

SURDS. 29

41. The square root of a trinomial quadratic surd.

Sometimes the square root of a quantity of the form
$$a + \sqrt{b} + \sqrt{c} + \sqrt{d}$$
may be found by assuming
$$\sqrt{a + \sqrt{b} + \sqrt{c} + \sqrt{d}} = \sqrt{x} + \sqrt{y} + \sqrt{z}.$$
Then, $a + \sqrt{b} + \sqrt{c} + \sqrt{d} = x + y + z + 2\sqrt{(xy)} + 2\sqrt{(yz)} + 2\sqrt{(zx)}$.

We now suppose that $2\sqrt{(yz)} = \sqrt{b}$, $2\sqrt{(zx)} = \sqrt{c}$, $2\sqrt{(xy)} = \sqrt{d}$; if the values of x, y, z found from these three equations also satisfy $x + y + z = a$, the required square root has been found; but not otherwise.

Hence $x + y + z = a$ is sometimes called the *equation of condition*.

Ex. Find the square root of $6 + 2\sqrt{2} + 2\sqrt{3} + 2\sqrt{6}$.

Let $6 + 2\sqrt{2} + 2\sqrt{3} + 2\sqrt{6} = (\sqrt{x} + \sqrt{y} + \sqrt{z})^2$
$$= x + y + z + 2\sqrt{(xy)} + 2\sqrt{(xz)} + 2\sqrt{(yz)}.$$
Put $\quad 2\sqrt{(xy)} = 2\sqrt{2}\,;\ 2\sqrt{(xz)} = 2\sqrt{3}\,;\ 2\sqrt{(yz)} = 2\sqrt{6}$.
Thus $\quad xy = 2\,;\ xz = 3\,;\ yz = 6$.
$\therefore x^2 yz = 6$; and as $yz = 6$, $\therefore x^2 = 1$.
$\therefore x = 1$; and hence $y = 2$, and $z = 3$.

And since the equation $x + y + z = 6$ is satisfied by the values 1, 2, 3 of x, y, z respectively, the required root $= 1 + \sqrt{2} + \sqrt{3}$.

This may be verified by squaring $1 + \sqrt{2} + \sqrt{3}$.

42. The cube root of a binomial quadratic surd may sometimes be obtained by the following method.

Ex. Find the cube root of $45 - 29\sqrt{2}$.

Let $\qquad \sqrt[3]{(45 - 29\sqrt{2})} = x - \sqrt{y}$ .... (1).

Then, cubing, we have $45 - 29\sqrt{2} = x^3 - 3x^2 y^{\frac{1}{2}} + 3xy - y^{\frac{3}{2}}$.
$\therefore x^3 + 3xy = 45$; and $3x^2 y^{\frac{1}{2}} + y^{\frac{3}{2}} = 29\sqrt{2}$. [§ 37.]
$\therefore x^3 + 3x^2 y^{\frac{1}{2}} + 3xy + y^{\frac{3}{2}} = 45 + 29\sqrt{2}$.
$\therefore (x + y^{\frac{1}{2}})^3 = 45 + 29\sqrt{2}$.
$\therefore \sqrt[3]{(45 + 29\sqrt{2})} = x + \sqrt{y}$ ....(2).

Multiplying (1) and (2) we have,
$$x^2 - y = \sqrt[3]{\{(45)^2 - (29\sqrt{2})^2\}} = \sqrt[3]{343} = 7.$$

Thus, $x^2 - y = 7$; which gives $y = x^2 - 7$.
Substitute this value of y in $x^3 + 3xy = 45$.
We get $\quad\quad\quad x^3 + 3x(x^2 - 7) = 45$.
That is, $\quad\quad\quad 4x^3 - 21x = 45$.
We find, *by trial*, that $x = 3$; which gives $y = 2$.
$\therefore$ the cube root required $= 3 - \sqrt{2}$.

EXERCISES II.

104. If $\sqrt[3]{(a+\sqrt{b})} = x + \sqrt{y}$, prove that $\sqrt[3]{(a - \sqrt{b})} = x - \sqrt{y}$.

105. Prove that the product of two conjugate expressions is rational.

106. If the sum and the product of $a + \sqrt{b}$ and $c + \sqrt{d}$ are both rational, show that $c = a$ and $\sqrt{d} = -\sqrt{b}$; and hence show that $a + \sqrt{b}$ and $c + \sqrt{d}$ are conjugate surds.

107. Prove that the sum or difference of two (three or more) similar surds vanishes or is a surd similar to the original ones.

108. Prove that the sum or difference of two (three or more) dissimilar surds cannot be rational.

Extract the square roots of:

109. $7 + 4\sqrt{3}$. **110.** $36 + 10\sqrt{11}$. **111.** $14 - 6\sqrt{5}$.

112. $8 + 2\sqrt{15}$. **113.** $2\frac{1}{4} + \sqrt{5}$. **114.** $61 - 28\sqrt{3}$.

115. $30 + 12\sqrt{6}$. **116.** $19 - 4\sqrt{21}$. **117.** $4\frac{1}{2} + 2\sqrt{2}$.

118. $1\frac{7}{9} - 2\frac{2}{3}\sqrt{\frac{1}{3}}$. **119.** $3\sqrt{6} - 4\sqrt{3}$. **120.** $6 + 4\sqrt{3}$.

121. $\frac{35}{36} - \sqrt{\frac{2}{9}}$. **122.** $\sqrt{27} + \sqrt{24}$. **123.** $2 + \frac{1}{2}\sqrt{18}$.

124. $1 + \sqrt{(1 - m^2)}$. **125.** $2a + 2\sqrt{(a^2 - x^2)}$.

126. $ax - 2a\sqrt{(ax - a^2)}$. **127.** $2 + 2(1 - x)\sqrt{(1 + 2x - x^2)}$.

128. $10 - 2\sqrt{3} - 2\sqrt{6} + 6\sqrt{2}$.

129. $9 - 2\sqrt{6} - 4\sqrt{2} + 4\sqrt{3}$.

130. $5 - \sqrt{10} - \sqrt{6} + \sqrt{15}$.

Find, by inspection, the square roots of:

131. $32 - 10\sqrt{7}$. **132.** $8 - 2\sqrt{7}$. **133.** $10 - 4\sqrt{6}$.

134. $5 + \sqrt{24}$.

SURDS. 31

Find the cube roots of :
   135. $10 + 6\sqrt{3}$.   136. $16 + 8\sqrt{5}$.   137. $100 - 51\sqrt{3}$.
   138. $2\frac{1}{2} - 1\frac{3}{4}\sqrt{2}$.

Find the fourth roots of :
   139. $17 + 12\sqrt{2}$.   140. $\frac{3}{2}\sqrt{5} + 3\frac{1}{2}$.   141. $48\frac{1}{16} + 11\frac{1}{2}\sqrt{15}$.

142. Show that $(2 + \sqrt{3})^{\frac{3}{2}} + (2 - \sqrt{3})^{\frac{3}{2}} = 3\sqrt{6}$.

143. Show that $\dfrac{2+\sqrt{3}}{\sqrt{2+\sqrt{(2+\sqrt{3})}}} - \dfrac{2-\sqrt{3}}{\sqrt{2-\sqrt{(2-\sqrt{3})}}} = \sqrt{\dfrac{2}{3}}$.

144. Obtain the square root of $1 + (1-c^2)^{-\frac{1}{2}}$ in the form $m^{\frac{1}{4}} + n^{\frac{1}{4}}$.

145. Find the square root of $11 - 2\sqrt{30}$, and prove that
$$\dfrac{1}{\sqrt{(11-2\sqrt{30})}} = \dfrac{3}{\sqrt{5}-\sqrt{2}} + \dfrac{2\sqrt{2}}{\sqrt{3}+1}.$$

146. Show that $\sqrt{\{3\sqrt{5} - \sqrt{2} + \sqrt{(7 + 2\sqrt{10})}\}} = 2\sqrt[4]{5}$.

147. Prove that $\sqrt{[\,-\sqrt{3} + \sqrt{\{3 + 8\sqrt{(7 + 4\sqrt{3})}\}}\,]} = 2$.

148. Show that $\dfrac{1}{\sqrt{(11-2\sqrt{30})}} - \dfrac{3}{\sqrt{(7-2\sqrt{10})}} - \dfrac{4}{\sqrt{(8+4\sqrt{3})}} = 0$.

149. Show that the square root of $\dfrac{\sqrt{5}+1}{\sqrt{5}-1} = 1\cdot618$ to three places of decimals.

150. Show that the value of $\dfrac{\sqrt{(3-\sqrt{5})}}{\sqrt{2}+\sqrt{(7-3\sqrt{5})}} = \cdot 447$ to four places of decimals.

151. Simplify $\sqrt{\left(\dfrac{\frac{1}{2}\sqrt{3} - \frac{1}{\sqrt{2}}}{\frac{1}{2}\sqrt{3} + \frac{1}{\sqrt{2}}}\right)}$.

Find the values of :
   152. $\sqrt{[m\sqrt{\{m\sqrt{(m\ldots)}\}}]}$ to infinity.
[Let x be the required value. Then, on squaring, $x^2 = mx$.]
   153. $\sqrt{[p^2 - \frac{1}{4} + \sqrt{\{p^2 - \frac{1}{4} + \sqrt{(p^2 - \frac{1}{4}\ldots)}\}}]}$ to infinity.
[Let x be the required value. Then $x^2 = p^2 - \frac{1}{4} + x$.]
   154. $\sqrt{[7 + \sqrt{\{7 + \sqrt{(7 + \ldots)}\}}]}$ to infinity.

155. When the square root of $a + \sqrt{b} + \sqrt{c} + \sqrt{d}$ can be exhibited in the form $\sqrt{p} + \sqrt{q} + \sqrt{r}$, find the relation between a, b, c and d.

43. To find the square root of an algebraic expression by Indeterminate Coefficients.

If the expression whose square root is required is an *exact square*, its square root can always be determined by applying the Principle of Indeterminate Coefficients.

Ex. 1. Find $\sqrt{(4x^2 + 36x + 81)}$.

Let $4x^2 + 36x + 81 = (2x + m)^2 = 4x^2 + 4mx + m^2$.

Then $36 = 4m$ and $m = 9$. Thus the square root is $2x + 9$.

Ex. 2. When is $ax^2 + bx + c$ a square in x?

$$ax^2 + bx + c = \left\{ \sqrt{a}\sqrt{\left(x^2 + \frac{b}{a}x + \frac{c}{a}\right)} \right\}^2$$
$$= \{\sqrt{a}(x+m)\}^2, \text{ suppose.}$$

Squaring, we have

$$x^2 + \frac{b}{a}x + \frac{c}{a} = x^2 + 2mx + m^2. \quad \therefore \frac{b}{a} = 2m, \text{ and } \frac{c}{a} = m^2.$$

Hence $\dfrac{b^2}{a^2} = \dfrac{4c}{a}$; $\therefore b^2 = 4ac$ is the condition required.

Ex. 3. Find $\sqrt{(1 - 2x + 5x^2 - 4x^3 + 4x^4)}$.

If the square root be assumed to be $1 + mx + nx^2$, squaring this and equating it to the given expression, we have

$1 - 2x + 5x^2 - 4x^3 + 4x^4 = 1 + 2mx + (m^2 + 2n)x^2 + 2mnx^3 + n^2x^4$.

Hence $-2 = 2m$; $5 = m^2 + 2n$; $-4 = 2mn$; $4 = n^2$.

[Of these equations the first two are sufficient to determine the values of m and n. The other two will be satisfied by the values of m and n thus found, if the given quantity is a perfect square.]

Thus $m = -1$ and $n = 2$. Hence the required square root

$$= 1 - x + 2x^2.$$

44. Cube root by Indeterminate Coefficients.

The student may proceed as in the corresponding investigation for the square root.

Ex. If $ax^3 + bx^2 + cx + d$ is a perfect cube, show that $b^2 = 3ac$, and $c^2 = 3bd$.

SURDS. 33

Let $ax^3 + bx^2 + cx + d = (mx + n)^3 = m^3x^3 + 3m^2nx^2 + 3mn^2x + n^3$.
Equating the coefficients of like powers of x, we have
$$a = m^3\ ;\ b = 3m^2n\ ;\ c = 3mn^2\ ;\ \text{and}\ d = n^3.$$
Hence $b^2 = 9m^4n^2 = 3m^3 \cdot 3mn^2 = 3ac\ ;$
and $c^2 = 9m^2n^4 = 3n^3 \cdot 3m^2n = 3db.$

EXERCISES II.

Find, by the method of Indeterminate Coefficients, the square roots of the following expressions :—

156. $x^4 + 2x^3 + 3x^2 + 2x + 1$.
157. $9 - 12x + 10x^2 - 4x^3 + x^4$.
158. $4a^4 - 4a^3 - 7a^2 + 4a + 4$.
159. $1 - 4ax + 2a^2x^2 + 4a^3x^3 + a^4x^4$.
160. $1 - 12x + 60x^2 - 160x^3 + 240x^4 - 192x^5 + 64x^6$.
161. $a^6 - 4a^5b + 8a^4b^2 - 10a^3b^3 + 8a^2b^4 - 4ab^5 + b^6$.
162. Prove that $x^4 + ax^3 + bx^2 + cx + d$ is a perfect square, if
$$c^2 = a^2d \text{ and } 4ab = a^3 + 8c.$$
163. If $mx^4 + nx^3 + 22x^2 + 12x + 9$ is a square, find m and n.
164. If $4x^4 + 8x^3 + mx + n$ is a square, find m and n.
165. If $x^6 - 4x^5 + 10x^4 - 20x^3$ are the first four terms of an exact square, find the next three terms.
166. Show that if both $x^4 + ax^3 + bx^2 + cx + 1$ and
$$x^4 + 2ax^3 + 2bx^2 + 2cx + 1$$
are perfect squares, then $b = 3$ and $a = c = \pm 2$.
167. If $m^6 + 6m^5 - 40m^3 + pm - q$ is a perfect cube, find p and q.
168. If $x^6 - 6x^5 + 15x^4 + nx^3 + px^2 + qx + r$ is a perfect cube, find p, q, r, n.
169. Find the fourth root of
$$16a^4 - 96a^3x + 216a^2x^2 - 216ax^3 + 81x^4.$$
170. If $1 + px + qx^2 + rx^3$ is a perfect cube, then
$$p^2 = 3q \text{ and } q^2 = 3pr.$$

TUT. ALG. II. 3

CHAPTER III.

INEQUALITIES.

45. Laws of Inequalities.—As the student will often meet with the relations of unequal quantities to one another, we shall now draw attention to the main laws which govern inequalities.

DEFINITION. a is said to be **greater** than b when $a - b$ is positive; and a is said to be **less** than b when $a - b$ is negative.

Ex. To prove that $x^2 + 1 > 2x$, we have to prove that $x^2 + 1 - 2x > 0$, that is, to prove that the sign of the expression $x^2 + 1 - 2x$, that is $(x-1)^2$, is positive, which is obvious by the rule of signs unless $x = 1$.

We infer from the definition that *a positive quantity increases algebraically as it increases numerically, while a negative quantity decreases algebraically as it increases numerically.*

The letters in the remaining articles of this chapter stand for *real positive quantities*. The theorems of this chapter are almost axiomatic, and are mostly left to be proved by the student.

46. If $a > b$, then $a + c > b + c$,
and $a - c > b - c$,
but $c - a < c - b$.

For $(a + c) - (b + c) = a - b$ and is positive, and so on.

Thus, *an inequality still holds if each side be increased or decreased by the same quantity.*

Moreover *any term of an inequality can be transformed from one side to the other, if its sign be changed.*

Hence every inequality can be reduced to one or other of the forms $n > 0$, or $n < 0$.

47. If $a > b$, then $ma > mb$, and $\dfrac{a}{m} > \dfrac{b}{m}$;

but $\quad -ma < -mb$, **and** $\dfrac{a}{-m} < \dfrac{b}{-m}$.

Also $\qquad \dfrac{c}{a} < \dfrac{c}{b}$.

The last result is easiest seen by writing it in the form
$$\frac{bc}{ab} < \frac{ac}{ab}.$$

Thus *an inequality still holds good if both sides be multiplied or divided by the same positive quantity; but if both sides be multiplied or divided by the same negative quantity, the sign of the inequality must be reversed.*

Hence an inequality may be cleared of fractions.

Ex. 1. Thus $7 \times 5 > 3 \times 5$ but $7 \times (-5) < 3 \times (-5)$.

Ex. 2. If $\qquad a > b, \dfrac{1}{a} < \dfrac{1}{b}$.

48. If $a > b$, then $a^m > b^m$, but $a^{-m} < b^{-m}$.

This is obvious when m is an integer. If $m = p/q$ put $a \times x^q$, $b = y^q$ so that $a^m = x^p$, $b^m = y^p$. By the results for an integral index, $x^p >$ or $< y^p$ according as $x >$ or $< y$, or again according as $x^q >$ or $< y^q$; that is $a^m >$ or $< b^m$ according as $a >$ or $< b$.

49. The square of every real quantity is positive, and therefore greater than zero.

This proposition follows from the rule of signs.

INEQUALITIES.

50. We shall now give a few examples of simple inequalities.

Ex. 1. If x and y are unequal, prove that $x^2 + y^2 > 2xy$.

For since $(x-y)^2 > 0$,

we have $x^2 - 2xy + y^2 > 0$ or $x^2 + y^2 > 2xy$.

When $x=y$, the inequality becomes an equality.

From this example we see that $\dfrac{x}{y} + \dfrac{y}{x}$ is never less than 2.

Hence *the sum of any positive quantity and its reciprocal is never less than* 2.

Ex. 2. Prove that $a^2 + b^2 + c^2 > ab + bc + ca$, unless $a = b = c$.

Since $a^2 + b^2 > 2ab$, $b^2 + c^2 > 2bc$, $c^2 + a^2 > 2ca$,

therefore $2(a^2 + b^2 + c^2) > 2(ab + bc + ca)$.

Ex. 3. Prove that $x^{m+n} + y^{m+n} > x^m y^n + y^m x^n$, except when $x = y$.

We have to establish that

$$x^{m+n} - x^m y^n + y^{m+n} - y^m x^n \text{ is positive.}$$

The expression

$$= x^m(x^n - y^n) - y^m(x^n - y^n) = (x^m - y^m)(x^n - y^n).$$

Here the factors are both positive or both negative, and the product is therefore positive.

Ex. 4. Prove that $x^{2n} + 1 > x^{2n-1} + x$, unless $x = 1$.

Since $(x^{2n-1} - 1)(x - 1)$ is always positive, the result follows that
$$x^{2n} + 1 > x^{2n-1} + x.$$

Ex. 5. Prove that $x^n + \dfrac{1}{x^n} > x^{n-1} + \dfrac{1}{x^{n-1}}$.

Dividing both sides of the inequality in Ex. 4 by x^n, the required result follows.

The inequality shows that, as n increases, $x^n + \dfrac{1}{x^n}$ increases; thus

$$x^2 + \frac{1}{x^2} > x + \frac{1}{x}; \quad x^3 + \frac{1}{x^3} > x^2 + \frac{1}{x^2}; \text{ and so on.}$$

INEQUALITIES.

Ex. 6. Find which is greater, $\sqrt{3}+\sqrt{17}$, or $\sqrt{13}+\sqrt{6}$.

$$\sqrt{3}+\sqrt{17} > \text{or} < \sqrt{13}+\sqrt{6},$$

according as $\quad 3+17+2\sqrt{51} >$ or $< 13+6+2\sqrt{78}$,

that is, $\quad\quad\quad 1+2\sqrt{51} >$ or $< 2\sqrt{78}$,

that is, $\quad\quad 1+204+4\sqrt{51} >$ or < 312,

that is, $\quad\quad\quad\quad 4\sqrt{51} >$ or < 107.

But $4\sqrt{51} < 4\sqrt{64}$ or 32, and is therefore < 107.

Thus $\sqrt{13}+\sqrt{6}$ is the greater.

Ex. 7. If $7x-5 > 3x+4$, find a limit to the value of x.

Since $\quad\quad 7x-5 > 3x+4, \quad \therefore\ 7x-3x > 4+5$.

That is $4x > 9$, or $x > 2\frac{1}{4}$.

EXERCISES III.

1. If $a > b$, and $c > d$, prove that $a-c$ may be greater than, equal to, or less than $b-d$.

2. Generalise the following case :
$-20 < -4$; and $-10 < -5$; but $(-20) \times (-10) > (-4)(-5)$.

3. If $5x+7 > 3-2x$, find a limit to x.

4. If $5x+8 > 3x+2$, and $3x+7 < 2(1-x)$, find the integral value of x.

5. If $\dfrac{4x}{x+1} > 3$, is $4x$ necessarily greater than $3(x+1)$?

Prove (with the exceptions stated in each case) that the following inequalities are generally true. [The letters represent positive quantities.]

6. $m^3+1 > m^2+m$, unless $m=1$.

7. $x^5+y^5 > x^4y+xy^4$, unless $x=y$.

8. $(b+c-a)^2+(c+a-b)^2+(a+b-c)^2 > bc+ca+ab$,
   unless $a=b=c$.

9. $(a+b)^2 > 4ab$, unless $a=b$.

10. $(a+b)(b+c)(c+a) > 8abc$, unless $a=b=c$.

11. $a^2b+b^2c+c^2a+ab^2+bc^2+ca^2 > 6abc$, unless $a=b=c$.

INEQUALITIES.

12. $\dfrac{a+b}{c} + \dfrac{b+c}{a} + \dfrac{c+a}{b} > 6$, unless $a = b = c$.

13. $\left(\dfrac{x}{a} + \dfrac{y}{b}\right)\left(\dfrac{a}{x} + \dfrac{b}{y}\right) > 4$, unless $\dfrac{x}{a} = \dfrac{y}{b}$.

14. $(a+b+c)\left(\dfrac{1}{a} + \dfrac{1}{b} + \dfrac{1}{c}\right) > 9$, unless $a = b = c$.

15. $\left(\dfrac{l}{a} + \dfrac{m}{b} + \dfrac{n}{c}\right)\left(\dfrac{a}{l} + \dfrac{b}{m} + \dfrac{c}{n}\right) > 9$, unless $\dfrac{l}{a} = \dfrac{m}{b} = \dfrac{n}{c}$.

16. $x^7 + y^7 > x^4y^3 + x^3y^4$, unless $x = y$.
17. $x^7 + 1 > x^6 + x$, unless $x = 1$.
18. $x^5 + x^{-5} > x^2 + x^{-2}$, unless $x = 1$.
19. $(a+b+c)^3 > 27abc < 9(a^3+b^3+c^3)$, unless $a = b = c$.
20. $xyz > (y+z-x)(z+x-y)(z+y-z)$, unless $x = y = z$.
21. Prove that $\sqrt{10} + \sqrt{7} < \sqrt{5} + \sqrt{13}$.
22. Prove that $\sqrt{15} + \sqrt{7} > \sqrt{20} + \sqrt{3}$.
23. If $x^2 = a^2 + b^2$, and $y^2 = c^2 + d^2$, show that xy will be greater than $ad + bc$ if x and y are of the same sign, and ac is not equal to bd.

51. Number of digits in products and powers of integers.—The truth of the following statements will be obvious to the student:

If an integer contains p *digits, it is less than* 10^p *and not less than* 10^{p-1}.

Ex. Of the numbers of 4 digits, the lowest is $1000 \,(=10^3)$ and the highest is $9999 = (10^4 - 1)$; similarly, of the numbers of p digits the lowest is 10^{p-1}, and the highest is $10^p - 1$.

Again *if an integer is not less than* 10^p *and is less than* 10^{p+q}; p *and* q *being positive integers, the number of digits which it contains may be* $p+1, p+2, p+3, \ldots$ *or* $p+q$.

Ex. An integer which is not less than 10^2 and less than 10^5 contains 3, 4 or 5 digits. The reasoning is general, and will apply to the statement above.

52. *If* a *contains* m *digits, and* b, n *digits, then* ab *contains* $m + n$ *or* $m + n - 1$ *digits.*

Since a contains m digits, it is less than 10^m and not less* than 10^{m-1}. Similarly $b < 10^n$ and not $< 10^{n-1}$.

---

* The statement that a is not less than b is sometimes written thus: $a \not< b$.

INEQUALITIES. 39

Hence $ab < 10^{m+n}$ and not $< 10^{m+n-2}$.

Thus ab contains either $m+n$ or $m+n-1$ digits.

Cor. *If a number contains* n *digits, its square contains* 2n *or* 2n -1 *digits.*

This is obvious from putting $a = b$, and $\therefore m = n$ in the preceding result.

53. *If* a, b, c *contain* p, q, r *digits respectively, then* abc *contains* $p+q+r, p+q+r-1$ *or* $p+q+r-2$ *digits.*

The property may be proved independently as in § 52, or may be deducted therefrom by finding the number of digits first in $a \times b$, and then in the product of ab and c.

Cor. *If a number contains* m *digits, then its cube contains* 3m, $3m-1$, *or* $3m-2$ *digits.*

This is the peculiar case of the above result, in which $a = b = c$, and $\therefore p = q = r = m$.

Ex. 1. If a contains m digits, how many digits are there in $2a$?

$2a$ is the product of 2 and a of which 2 is a number of one digit and a is a number of m digits. Hence the product $2a$ contains $m+1$ or m digits.

EXERCISES III.

24. If a contains 5 digits, how many digits are there (i) in a^2 ? (ii) in $3a^2$?

28. If a contains 3 digits, how many digits are there (i) in $3a$? (ii) in $3a^3$?

26. How many digits are there in the square of the square of a number of 3 digits ?

27. Find the number of digits in $(99,999)^3$.

25. Explain the consideration that leads you to find the exact number of digits in $(21,387)^2$; and in $(40,343)^2$.

29. If, of two integers a and b, a has m digits and b n digits, show that the number of digits in the integral portion of a/b is $m-n$ or $m-n+1$, a being greater than b.

30. If x contains p digits and n is any positive integer, prove that x^n will contain not more than np and not less than $np-n+1$ digits.

CHAPTER IV.

RATIO.

54. Two kinds of comparison.—When we begin to compare two magnitudes of the same kind, we find that the comparison may be made and expressed in two different ways.

(i) By considering *by what quantity* the one exceeds the other. The comparison, in this case, is of the nature of *subtraction*.

Thus taking two sticks of 7 yards and 9 yards in length, we state that the one is longer than the other by the *absolute* difference of measure of 2 yds.

(ii) By observing *how many times* the one is greater than the other. The comparison, in this case, is of the nature of *division*.

Thus take two persons, A and B, of whom A is worth, say, two thousands and B twenty thousands. Which of these two is the richer? and by how much? Of course B, who is ten times as rich as A. In such a case, it is only rarely that we think of stating the actual difference of fortune.

The quantities must be *of the same kind*. Thus there is no meaning in comparing 5 tons with 7 miles.

The second mode of comparison leads up to the definition of ratio which we give in the next article.

55. Definitions.—Ratio is the relation of two quantities of the same kind to each other in respect of magnitude, the relation being determined by considering what multiple, part or parts the one is of the other,

RATIO.

The ratio of a to b is written thus :—$a:b$.

a and b are called the **terms** of the ratio, and of these a is called the **antecedent** and b the **consequent** of the ratio.

56. Ratio, Division, and Fractions.—These three subjects although apparently different are identical in principle.

Thus, taking $a \div b$, $\dfrac{a}{b}$, a/b, and $a:b$ we find that the four symbols of connection represent the same algebraical operation of division*. Hence **the value of the ratio $a:b$ is that of the corresponding fraction $\dfrac{a}{b}$,** the antecedent of the ratio standing in the place of the numerator, and the consequent in that of the denominator.

Two ratios are said to be *equal* if the fractions which they represent are equal. Thus $6:4 = 3:2$ and $10:5 = 2:1$. It is *not* however usual to write the latter relation as $10:5 = 2$.

57. Composition of Ratios.—DEFINITIONS.—

(i) When two ratios are multiplied together, they are said to be **compounded**.

Thus $ab:cd$ is said to be compounded of $a:c$ and $b:d$.

The ratio compounded of $a:b$, $b:c$, $c:d$, and $d:e$ is $abcd:bcde$ which is equal to $a:e$.

(ii) The ratios $a:b$ and $b:a$ are said to be **reciprocal** to each other.

When $a:b$ and $b:a$ are compounded, $ab:ba$ or unity is the result.

---

* The symbol of division / in a/b is called the *solidus*. It has come into wide use on account of the economy of space it effects. The symbol of ratio : has been explained to be nothing but the symbol of division $\div$ with the line between the points removed.

(iii) The ratio compounded of two equal ratios $a:b$ and $a:b$, that is, $a^2:b^2$, is called the **duplicate ratio** of $a:b$.

(iv) The ratio compounded of three equal ratios $a:b$, $a:b$, $a:b$, that is, $a^3:b^3$, is called the **triplicate ratio** of $a:b$.

(v) Similarly the ratio $a^{\frac{1}{2}}:b^{\frac{1}{2}}$ is called the **subduplicate** of $a:b$, $a^{\frac{1}{3}}:b^{\frac{1}{3}}$ the **subtriplicate** of $a:b$, and $a^{\frac{3}{2}}:b^{\frac{3}{2}}$ the **sesquiplicate** of $a:b$.

Ex. Compound the duplicate ratio of 2 : 3, the triplicate ratio of 3 : 4, and the subduplicate ratio of 64 : 36.

The duplicate of 2 : 3 is 4 : 9;
the triplicate of 3 : 4 is 27 : 64;
and the subduplicate of 64 : 36 is 8 : 6.

Hence we have to compound 4 : 9, 27 : 64 and 8 : 6.

Changing these ratios into their equivalent fractional forms, we have for the result $\frac{4}{9} \times \frac{27}{64} \times \frac{8}{6} = \frac{1}{4}$, or 1 : 4.

EXERCISES IV.

1. If $a:b = 6:7$, find $7a + 10b : 18b - 7a$.

2. If $x:y = 3:4$, find $4x + 3y : 12x - 3y$.

3. If $\dfrac{5a^2 + 7b^2}{3a^2 - b^2} = 1\frac{49}{47}$, find $a:b$.

4. Compound the ratios
$a + x : a - x$; $a^3 - x^3 : a^2 - ax + x^2$; and $a^3 + x^3 : a^2 + ax + x^2$.

5. Compound the reciprocal of 5 : 6, the subduplicate of 100 : 36, and the subtriplicate of 8 : 1.

6. Show that the ratio compounded of the duplicate of 4 : 5, the subduplicate of 25 : 16, and the reciprocal of 4 : 5 is unity.

7. Find the ratio compounded of 5 : 8, the reciprocal of 15 : 4, the triplicate of 2 : 3, and the subduplicate of 9 : 16.

8. From a cask which can contain a gallons of wine and is full, b gallons are drawn off and the vessel is filled up with water. If this process is repeated, show that the quantity of wine remaining in the cask $= (a-b)^2/a$ gallons.

58. Comparison of ratios.—Since the value of a ratio is that of the corresponding fraction, to compare ratios we have simply to compare the equivalent fractions.

Ex. Which is greater, $5:7$ or $5:10$? The question only means, "Which of the two fractions $\frac{5}{7}$, $\frac{5}{10}$ is greater than the other?"

Since $7 < 10$ it follows from § 47 that $\frac{5}{7}$ or $5:7$ is the greater.

We shall now establish a few theorems relating to the comparison of ratios, taking the terms of these ratios to be positive.

We require the following definitions.

59. Definitions.—A ratio of equality is one of which the antecedent is equal to the consequent.

Thus $5:5$, $a:a$ are ratios of equality, and are equal to unity.

A **ratio of greater inequality** is one of which the antecedent is greater than the consequent.

Thus $5:3$ and $25:17$ are ratios of greater inequality, and correspond to improper fractions.

A ratio of less inequality is one of which the antecedent is less than the consequent.

Thus $3:5$ and $17:25$ are ratios of less inequality, and correspond to proper fractions.

60. A ratio is not altered in value, if both its terms are multiplied or divided by the same quantity.

Thus, $a:b$ is of the same value as $ma:mb$, or $\dfrac{a}{m}:\dfrac{b}{m}$.

This property enables us to arrange two or more given ratios in the order of their magnitude.

61. A ratio of equality is unaltered, a ratio of greater inequality is diminished, and a ratio of less inequality is increased, *if the same positive quantity be added to both its terms.*

Let $\dfrac{a}{b}$ be any ratio, and let $\dfrac{a+x}{b+x}$ be the ratio formed by adding the same positive quantity x to both its terms.

(i) If $a = b$, then evidently $a + x = b + x$.

(ii) If a and b are unequal, then
$$\frac{a+x}{b+x} > = \text{ or } < \frac{a}{b},$$
according as $(a+x)b > =$ or $< (b+x)a$; [§ 47.]
that is, $ab + bx > =$ or $< ab + ax$;
that is, $bx > =$ or $< ax$;
that is, $b > =$ or $< a$. [§ 47.]

If $b > a$, that is, if the given ratio is a ratio of less inequality, then $\dfrac{a+x}{b+x} > \dfrac{a}{b}$ which means that the ratio is increased.

If $b < a$, then $\dfrac{a+x}{b+x} < \dfrac{a}{b}$. This means that a ratio of greater inequality is diminished.

Ex. 1. To find the least value to which a ratio of greater inequality can be diminished by the addition of the same positive quantity to both its terms.

The form $\dfrac{a+x}{b+x} = 1 + \dfrac{a-b}{b+x}$ shows that the ratio can be brought as near to unity as we please by increasing x, though it can never become absolutely equal to unity. Similarly a ratio of less inequality which is increased by the addition of the same positive quantity to both its terms can never become a ratio of equality, but if x be sufficiently increased, the value of the resulting ratio can be made to approach as near unity as we please.

Hence the results of this article may be enunciated as follows :—

While a ratio of equality is unaltered, a ratio of inequality is made more nearly equal to a ratio of equality by the addition of the same positive quantity to both its terms.

RATIO.

62. A ratio of equality is unaltered, a ratio of greater inequality is increased, and a ratio of less inequality is diminished *if the same positive quantity not greater than the less term be subtracted from each of its terms.*

Let $\dfrac{a}{b}$ be any ratio, and let $\dfrac{a-x}{b-x}$ be the ratio formed by subtracting the same positive quantity x which is less than the less of two quantities a and b from each of its terms.

(i) If $a = b$, then evidently $a - x = b - x$.

(ii) If a and b are unequal, then

$$\frac{a-x}{b-x} > = \text{ or } < \frac{a}{b},$$

according as $b(a-x)$ is $> =$ or $< a(b-x)$; [§ 47.]
that is, $\qquad\qquad -b > =$ or $< -a$.
that is, $\qquad\qquad b < =$ or $> a$. [§ 47.]

If $b < a$, then $\dfrac{a-x}{b-x} > \dfrac{a}{b}$, which means that a ratio of greater inequality is increased.

If $b > a$, then $\dfrac{a-x}{b-x} < \dfrac{a}{b}$, which means that a ratio of less inequality is diminished.

NOTE. A ratio of greater inequality will become infinite when $x = b$, and a ratio of less inequality will become zero when $x = a$.

Ex. What is the least integer which when subtracted from both the terms of 5 : 6 will give a ratio less than 13 : 18 ?

Let $\qquad\qquad \dfrac{5-x}{6-x} = \dfrac{13}{18}.$

Solving this equation, we get $x = 2\frac{2}{5}$.

Hence 3 is the integer required; for if we subtract 2 which is less than $2\frac{2}{5}$, the resulting ratio will be greater than 13 : 18.

EXERCISES IV.

9. (i) Compare the ratios 15 : 28, and 16 : 29.

(ii) Arrange in order of magnitude, 5 : 6, 7 : 8 and 13 : 16.

10. Find the least integer that should be added to the terms of 15 : 28 to make the resulting ratio greater than 15 : 16.

11. What is the least integer which when added to both the terms of 71 : 6 will give a ratio less than 3 ?

12. Show that no integer under 12 will, when subtracted from both the terms of 31 : 13, give a ratio greater than 31 : 2.

13. Find the least integer that should be subtracted from the terms of 5 : 7 to make the resulting ratio not greater than 1 : 2.

14. Which is greater, $\dfrac{a+5b}{a+6b}$ or $\dfrac{a+6b}{a+7b}$?

15. Show that $a-b : a+b$ is greater or less than $a^2-b^2 : a^2+b^2$ according as $a : b$ is a ratio of less or greater inequality.

16. What quantity must be added to each term of the ratio $a : b$ that it may be equal to $c : d$?

17. What quantity must be added to each term of the triplicate of $a : b$ to make it equal to $a : b$?

18. What quantity must be added to each term of the duplicate of $c : d$ to make it equal to $c : d$?

19. If 1 : 2 is duplicated by subtracting x from its terms, find x.

63. If $\dfrac{a}{b} = \dfrac{c}{d} = \dfrac{e}{f} = \ldots$, then each of these ratios

$$= \left(\dfrac{pa^n + qc^n + re^n + \ldots}{pb^n + qd^n + rf^n + \ldots} \right)^{\tfrac{1}{n}}.$$

Let $\quad \dfrac{a}{b} = \dfrac{c}{d} = \dfrac{e}{f} = \ldots = k.$

[*This method of expressing the antecedents of equal ratios in terms of their consequents is extremely useful in solving questions in proportion.*]

Then $a = bk$; $c = dk$; $e = fk$;
and so on.

Hence $pa^n = pb^n k^n$; $qc^n = qd^n k^n$; $re^n = rf^n k^n$; ...
By addition,
$$pa^n + qc^n + re^n + ... = k^n(pb^n + qd^n + rf^n + ...).$$
$$\therefore (pa^n + qc^n + re^n + ...)^{\frac{1}{n}} = k(pb^n + qd^n + rf^n + ...)^{\frac{1}{n}}.$$

That is, $\left(\dfrac{pa^n + qc^n + re^n + ...}{pb^n + qd^n + rf^n + ...}\right)^{\frac{1}{n}} = k = \dfrac{a}{b} = \dfrac{c}{d} = \dfrac{e}{f} = ...$

The theorem of this article is an extremely important one, and the following particular cases of it may be proved by the same method.

(i) If $\dfrac{a}{b} = \dfrac{c}{d}$, each fraction is equal to $\dfrac{a+c}{b+d} = \dfrac{a-c}{b-d}$.

(ii) If $\dfrac{a}{b} = \dfrac{c}{d} = \dfrac{e}{f}$, then each ratio $= \dfrac{a+c+e}{b+d+f}$

$= \dfrac{\text{the sum of all the antecedents}}{\text{the sum of all the consequents}}$.

Ex. 1. If $\dfrac{a-b}{a+b} = \dfrac{2b-c}{2b+c} = \dfrac{c-2a}{c+2a}$,

then prove that each of the ratios $= 0$, unless $2a + 2b + c = 0$.

Let $\dfrac{2a-2b}{2a+2b} = \dfrac{2b-c}{2b+c} = \dfrac{c-2a}{c+2a} = k$.

Then

$(2a - 2b) + (2b - c) + (c - 2a) = [(2a+2b) + (2b+c) + (c+2a)]k$,

or $(4a + 4b + 2c)k = 0$ whence $k = 0$ or $2a + 2b + c = 0$.

Ex. 2. If $\dfrac{a}{b} = \dfrac{c}{d} = \dfrac{e}{f}$, then $\dfrac{a^2 + c^2 + e^2}{ab + cd + ef} = \dfrac{ab + cd + ef}{b^2 + d^2 + f^2}$.

Let each ratio $\dfrac{a}{b} = \dfrac{c}{d} = \dfrac{e}{f} = k$,

$$\therefore \frac{a^2+c^2+e^2}{ab+cd+ef} = \frac{b^2k^2+d^2k^2+f^2k^2}{b^2k+d^2k+f^2k} = k,$$

and similarly $\frac{ab+cd+ef}{b^2+d^2+f^2} = k$, which proves the result.

Ex. 3. If $\frac{a}{b} = \frac{c}{d} = \frac{e}{f}$, then each is equal to the ratio

$$\left(a+\frac{1}{b}+c+\frac{1}{d}+e+\frac{1}{f}\right) \div \left(b+\frac{1}{a}+d+\frac{1}{c}+f+\frac{1}{e}\right).$$

Let $\frac{a}{b} = \frac{c}{d} = \frac{e}{f} = k$. Then the given ratio

$$= \left(bk+\frac{1}{b}+dk+\frac{1}{d}+fk+\frac{1}{f}\right) \div \left(b+\frac{1}{bk}+d+\frac{1}{dk}+f+\frac{1}{fk}\right) = k.$$

Ex. 4. If $ad = bc$, prove that $abcd = \dfrac{a^2+b^2+c^2+d^2}{a^{-2}+b^{-2}+c^{-2}+d^{-2}}$.

Let $ad = bc = k$. Then $a = kd^{-1}$, $b = kc^{-1}$, $abcd = k^2$, and so on whence the required result follows.

Ex. 5. If $\dfrac{bz-cy}{b-c} = \dfrac{cx-az}{c-a}$, each $= \dfrac{ay-bx}{ab}$.

Equating each of the given quantities to k, we have
$$bz - cy = k(b-c); \text{ and } cx - az = k(c-a).$$
Multiply the first equation by a, the second by b, and add.

We get $bcx - acy = (bc - ac)k$, which gives the required result.

64. The following very important generalisation, of which the above theorem is a particular case, may be proved similarly.

If $\dfrac{a}{b} = \dfrac{c}{d} = \dfrac{e}{f} = \ldots$, *then each fraction is equal to the nth root of a fraction, of which the numerator is a homogeneous expression of the nth degree in* a, c, e, …, *and the denominator is the same homogeneous expression with* b *substituted for* a, d *for* c, f *for* e, *etc.*

Ex. 1. If $\dfrac{a}{b} = \dfrac{c}{d}$, then each $= \left(\dfrac{a^6+c^6+2ac^5}{b^6+d^6+2bd^5}\right)^{\frac{1}{6}}$.

For put $a/b = k$, then $a = bk$, $c = dk$. Substituting for a, c the result follows at once.

Ex. 2. If $\dfrac{a}{b} = \dfrac{c}{d} = \dfrac{e}{f}$,

each $= \left(\dfrac{a^3 + ace + c^3 + e^3}{b^3 + bdf + d^3 + f^3}\right)^{\frac{1}{3}}$

$= \left(\dfrac{a^4 + 5ace^2}{b^4 + 5bdf^2}\right)^{\frac{1}{4}} = \left(\dfrac{pa^5 + qc^5 + re^5}{pb^5 + qd^5 + rf^5}\right)^{\frac{1}{5}}$.

Ex. 3. If $\dfrac{a^2p}{x} = \dfrac{b^2q}{y} = \dfrac{c^2r}{z}$, and $\dfrac{x^2}{a^2} + \dfrac{y^2}{b^2} + \dfrac{z^2}{c^2} = 1$,

find the values of x, y, z.

$\dfrac{a^4p^2}{x^2} = \dfrac{a^2p^2}{x^2/a^2} = \dfrac{b^2q^2}{y^2/b^2} = \dfrac{c^2r^2}{z^2/c^2} = \dfrac{a^2p^2 + b^2q^2 + c^2r^2}{x^2/a^2 + y^2/b^2 + z^2/c^2} = a^2p^2 + b^2q^2 + c^2r^2$.

$$\therefore x = \pm \dfrac{a^2p}{\sqrt{(a^2p^2 + b^2q^2 + c^2r^2)}}\,;$$

hence $\qquad y = \pm \dfrac{b^2q}{\sqrt{(a^2p^2 + b^2q^2 + c^2r^2)}}$, etc.

65. If $a:b$, $c:d$, $e:f$... be a number of unequal ratios of which the consequents are all positive, then

$$a + c + e + \ldots : b + d + f + \ldots$$

lies between the greatest and the least of them.

[For example, the ratio $5 + 6 : 7 + 8$ lies in value between $5:7$ and $6:8$; and the ratio $5 - 6 : 7 + 8$ lies in value between $5:7$ and $-6:8$.]

The property may be proved as follows:

Let $a:b$ be the greatest of the ratios, and be denoted by k; then the other ratios $c:d$, $e:f$,... are less than k.

Thus, $a = bk$; $c < dk$; $e < fk$; and so on, since the consequents of the ratios $c:d$, $e:f$,... are all positive. [See § 47.]

Hence $\qquad a + c + e + \ldots < k(b + d + f + \ldots)$.

That is, $a + c + e + \ldots : b + d + f + \ldots$ is less than k, where k stands for the greatest of the ratios.

It can be similarly proved that the new ratio is greater than the least of the given ratios.

Hence it lies in value between the greatest and the least of the given ratios.

EXERCISES IV.

20. Prove that, if a ratio of greater inequality and another of less inequality be compounded together the compound ratio is intermediate in magnitude between them.

21. If $\dfrac{a}{b} = \dfrac{b}{c} = \dfrac{c}{d}$, then $\dfrac{a^2 + b^2 + c^2}{ab + bc + cd} = \dfrac{ab + bc + cd}{b^2 + c^2 + d^2}$.

22. If $ad = bc$, show that $(abcd)^2 = \dfrac{a^4 + b^4 + c^4 + d^4}{a^{-4} + b^{-4} + c^{-4} + d^{-4}}$.

23. If $\dfrac{a}{b} = \dfrac{c}{d} = \dfrac{e}{f}$, then each of these ratios $= \dfrac{2a + 7c + 3e}{2b + 7d - 3f}$.

24. If $\dfrac{a}{b} = \dfrac{c}{d} = \dfrac{e}{f} = \dfrac{g}{h}$; then each of these ratios
$$= \sqrt{\dfrac{a^2 + ce + g^2}{b^2 + df + h^2}} = \sqrt{\dfrac{ac + eg}{bd + fh}} = \sqrt[3]{\dfrac{a^3 + ceg}{b^3 + dfh}}.$$

25. If $\dfrac{a}{b} = \dfrac{c}{d}$, then $\dfrac{1}{ma} + \dfrac{1}{nb} + \dfrac{1}{pc} + \dfrac{1}{qd} = \dfrac{1}{bc}\left(\dfrac{a}{q} + \dfrac{b}{p} + \dfrac{c}{n} + \dfrac{d}{m}\right)$.

26. If $\dfrac{a}{b} > \dfrac{c}{d}$, then $\sqrt{\dfrac{a^2 + c^2}{b^2 + d^2}} < \dfrac{a}{b}$ and $> \dfrac{c}{d}$.

27. If $\dfrac{a}{b} > \dfrac{c}{d} > \dfrac{e}{f}$, then $\sqrt[3]{\dfrac{a^3 + c^3 + e^3}{b^3 + d^3 + f^3}} < \dfrac{a}{b}$ and $> \dfrac{e}{f}$.

28. If $\dfrac{a}{b + c - a} = \dfrac{b}{c + a - b} = \dfrac{c}{a + b - c}$, then each ratio $= 1$.

29. If $\dfrac{bx - ay}{cy - az} = \dfrac{cx - az}{by - ax} = \dfrac{z + y}{x + z}$, then each of these fractions $= \dfrac{x}{y}$, unless $b + c = 0$.

30. If $\dfrac{y + z}{ay + bz} = \dfrac{z + x}{az + bx} = \dfrac{x + y}{ax + by}$, each of these fractions $= \dfrac{2}{a + b}$, unless $x + y + z = 0$.

31. If $\dfrac{a - b}{ay + bx} = \dfrac{b - c}{bz + cx} = \dfrac{c - a}{cy + az} = \dfrac{a + b + c}{ax + by + cz}$, then each of these ratios $= \dfrac{1}{x + y + z}$, if $a + b + c$ be not equal to zero.

32. If $\dfrac{x}{a^2} = \dfrac{y}{b^2} = \dfrac{z}{c^2}$, then $\dfrac{x + y + z}{a^2 + b^2 + c^2} = \dfrac{\dfrac{x}{a} + \dfrac{y}{b} + \dfrac{z}{c}}{a + b + c}$.

RATIO.   51

33. If $\dfrac{x^2-yz}{a}=\dfrac{y^2-zx}{b}=\dfrac{z^2-xy}{c}$,

then $(x+y+z)(a+b+c)=ax+by+cz$.

34. If $ap=bq=cr$, then prove that $\dfrac{p^2}{qr}+\dfrac{q^2}{rp}+\dfrac{r^2}{pq}=\dfrac{bc}{a^2}+\dfrac{ca}{b^2}+\dfrac{ab}{c^2}$.

35. If $\dfrac{a+bx}{b+cy}=\dfrac{b+cx}{c+ay}=\dfrac{c+ax}{a+by}$, then each $=\dfrac{1+x}{1+y}$,

unless $a+b+c=0$.

36. If $\dfrac{b^2-ac}{a+c-2b}=\dfrac{c^2-bd}{b+d-2c}$, then each $=\dfrac{bc-ad}{a+d-b-c}$,

b and c being unequal.

37. If $\dfrac{x}{b+c-a}=\dfrac{y}{c+a-b}=\dfrac{z}{a+b-c}$, show that

$(b-c)x+(c-a)y+(a-b)z=0$.

38. If $\dfrac{bz+cy}{b-c}=\dfrac{cx+az}{c-a}=\dfrac{ay+bx}{a-b}$, show that

$(a+b+c)(x+y+z)=ax+by+cz$.

39. If $\dfrac{a^{n+1}}{x^n}=\dfrac{b^{n+1}}{y^n}=\dfrac{c^{n+1}}{z^n}=x+y+z$,

then each of these fractions $=\left(a^{\frac{n+1}{n}}+b^{\frac{n+1}{n}}+c^{\frac{n+1}{n}}\right)^{\frac{n}{n+1}}$.

66. Homogeneous equations. If $\dfrac{a}{b}=\dfrac{c}{d}=\dfrac{e}{f}=\ldots$,
and $a, c, e\ldots$ occur homogeneously in any equation, the equation is unaltered if $b, d, f,\ldots$ are substituted for $a, c, e\ldots$ respectively*.

---

* This result is of great use in solving questions on the properties of triangles. Thus if $\sin^2 A+\sin^2 B=\sin^2 C$, the equation $a^2+b^2=c^2$ readily follows, showing that the triangle is right-angled.

Ex. If $pa^3 + qc^2e + race = 0$ be the homogeneous equation in a, c, e, to show that $pb^3 + qd^2f + rbdf = 0$.

Let $$\frac{a}{b} = \frac{c}{d} = \frac{e}{f} = \ldots = k.$$

Then $a = bk$; $c = dk$; $e = fk$;...

Therefore $pa^3 + qc^2e + race = 0$

becomes $pb^3k^3 + qd^2fk^3 + rbdfk^3 = 0$, that is,

$$k^3(pb^3 + qd^2f + rbdf) = 0.$$

67. Ratios from homogeneous equations.

When the number of given homogeneous equations is one less than the number of unknown quantities, the ratios which these unknown quantities bear to one another can be determined.

Ex. 1. Find the ratio of x to y from $ax + by = cx + dy$.

We have $x(a - c) = y(d - b)$;

whence $x : y = d - b : a - c$.

Ex. 2. If $x^2 + 4y^2 = 5xy$, find $x : y$.

The equation $x^2 - 5xy + 4y^2 = 0$ gives us $\dfrac{x^2}{y^2} - 5\dfrac{x}{y} + 4 = 0$.

Solving this quadratic equation in $\dfrac{x}{y}$, we get $\dfrac{x}{y} = 1$, or 4.

Thus the ratio of x to y is either $1 : 1$ or $4 : 1$.

Ex. 3. If $acx^2 - bcxy = adxy - bdy^2$, find $x : y$.

Here, $acx^2 - bcxy - adxy + bdy^2 = 0$.

Hence $(ax - by)(cx - dy) = 0$. $\therefore ax = by$; or $cx = dy$.

Thus the ratio of $x : y = b : a$; or $d : c$.

Ex. 4. Find $x : y : z$ from

$$ax + by + cz = 0 \quad\ldots\ldots\ldots\ldots\ldots\ldots(1).$$
$$a'x + b'y + c'z = 0 \quad\ldots\ldots\ldots\ldots\ldots\ldots(2).$$

To obtain $x : y$, we have to eliminate z from the two equations.

Multiplying (1) by c', we have $ac'x + bc'y + cc'z = 0$.

Multiplying (2) by c, we have $a'cx + b'cy + cc'z = 0$.

RATIO. 53

By subtraction and transposition, $x(ca' - c'a) = y(bc' - b'c)$.

Thus the ratio of x to y is $\dfrac{bc' - b'c}{ca' - c'a}$.

Similarly the ratio of y to z is $\dfrac{ca' - c'a}{ab' - a'b}$.

Hence $x : y : z = bc' - b'c : ca' - c'a : ab' - a'b$.

That is, $\dfrac{x}{bc' - b'c} = \dfrac{y}{ca' - c'a} = \dfrac{z}{ab' - a'b}$.

68. The Rule of Cross-Multiplication. The relations of the last example can be easily obtained by the following rule:—

RULE. *Write the coefficients in order in two lines, beginning from the second, and repeat the second coefficients as in the following diagram* :—

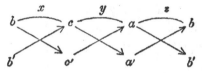

Then, make the six multiplications as indicated by the arrows, noting that the descending products are positive, while the ascending ones are negative. From these six products, form the three expressions $bc' - b'c$, $ca' - c'a$, *and* $ab' - a'b$. *Then*

$$\dfrac{x}{\text{first expression}} = \dfrac{y}{\text{second expression}} = \dfrac{z}{\text{third expression}}.$$

This rule is called the **Rule of Cross-Multiplication.**

Ex. Find the ratios of $x : z$ and $y : z$ from

$$4x + 3y = 2z \;;\; 5x = 5y - z.$$

The equations being

$$4x + 3y - 2z = 0, \text{ and } 5x - 5y + z = 0,$$

the coefficients, when arranged as in the foregoing diagram, stand thus :

$$\begin{array}{cccc} 3 & -2 & 4 & 3 \\ -5 & 1 & 5 & -5. \end{array}$$

We therefore get
$$\frac{x}{3 \times 1 - (-5) \times (-2)} = \frac{y}{(-2) \times 5 - 1 \times 4} = \frac{z}{4 \times (-5) - 5 \times 3}.$$

Hence $\dfrac{x}{3-10} = \dfrac{y}{-10-4} = \dfrac{z}{-20-15}$.

$\therefore x : z = 1 : 5$; and $y : z = 2 : 5$.

69. Solution of simultaneous equations.

We may apply the rule to the solution of two simultaneous equations of two unknown quantities of the first degree.

Since the equations $ax + by + cz = 0$,
and $a'x + b'y + c'z = 0$,
give us the ratios of $x : z$ and $y : z$,
therefore the equations $ax + by + c = 0$,
and $a'x + b'y + c' = 0$
which are obtained from the first pair by the substitution of unity for z must give the ratios of $x : 1$ and $y : 1$, that is, the values of x and y.

Ex. 1. Solve $\quad x + 2y = 9$; $3x - y = 13$.

By transposition, $\quad x + 2y - 9 = 0$,
$\quad\quad\quad\quad\quad\quad\quad 3x - y - 13 = 0$.

Hence, arranging the coefficients as shown before,

$$\begin{array}{cccc} 2 & -9 & 1 & 2 \\ -1 & -13 & 3 & -1, \end{array}$$

we get $\dfrac{x}{2 \times (-13) - (-1) \times (-9)} = \dfrac{y}{(-9) \times 3 - (-13) \times 1}$
$$= \dfrac{1}{1 \times (-1) - 3 \times 2}.$$

Hence $\dfrac{x}{-35} = \dfrac{y}{-14} = \dfrac{1}{-7}$; which give $x = 5$; and $y = 2$.

Ex. 2. Solve $9x + 2y - 7z = 0.$
$x + y - 2z = 0.$
$3x^2 + 2yz = 181.$

From the first two equations, by the method of cross-multiplication,

$$\frac{x}{2\times(-2) - 1\times(-7)} = \frac{y}{(-7)\times 1 - (-2)\times 9} = \frac{z}{9\times 1 - 1\times 2}.$$

$$\therefore \frac{x}{3} = \frac{y}{11} = \frac{z}{7} = k \text{ (suppose)}.$$

$$\therefore x = 3k, y = 11k, \text{ and } z = 7k.$$

Thus from the third equation we have

$$3x^2 + 2yz = 27k^2 + 154k^2 = 181.$$

Hence $k^2 = 1$ and $k = \pm 1.$

$\therefore x = 3, y = 11,$ and $z = 7;$ or $x = -3, y = -11,$ and $z = -7.$

Ex. 3. Eliminate x, y, z from

$$ax + by + cz = 0;$$
$$bx + cy + az = 0;$$
$$cx + ay + bz = 0.$$

From the first two equations by the method of cross-multiplication,

$$\frac{x}{ab - c^2} = \frac{y}{bc - a^2} = \frac{z}{ac - b^2} = k \text{ (suppose)}.$$

$\therefore x = k(ab - c^2), y = k(bc - a^2),$ and $z = k(ac - b^2).$

Thus the third equation becomes

$$k(abc - c^3 + abc - a^3 + abc - b^3) = 0.$$

Hence the required result is $3abc - a^3 - b^3 - c^3 = 0.$

This new equation or the expression on its left-hand side is called the **resultant** or **eliminant** of the given equations.

We have thus eliminated three unknown quantities from a system of three homogeneous equations. Similarly four unknown quantities may be eliminated from a system of four homogeneous equations, and so on.

EXERCISES IV.

40. If $2x^2 + 2y^2 = 5xy$, find the ratio of $x : y$.

41. If $5x^2 + 12y^2 = 23xy$, find the ratio of $x : y$.

42. Find $x : y$ from $7xy - 2y^2 = 3x^2$.

Use the rule of cross-multiplication to solve the following questions:

43. Find the ratios of x, y, z from
$$5x + 4y = 9z \text{ and } 7z - 4y = 3x.$$

44. Determine the ratios of x, y, z from
$$2x + 3y = 2z \text{ and } 5z - 3x = 10y.$$

45. Show that the resultant is $bc - ad = 0$, when x, y are eliminated from $ax + by = 0$, and $cx + dy = 0$.

46. Find $x : y$ and $x : z$ from $x = 2y + 3z$ and $x - 5y = z$.

Solve the following simultaneous equations:

47. $6x + 7y = 46$, and $5x + 3y = 27$.

48. $13x - 17y + 54 = 0$, and $7x + 28 - 9y = 0$.

49. $\dfrac{x}{a} + \dfrac{y}{b} = \dfrac{c}{d}$, and $\dfrac{x}{m} + \dfrac{y}{n} = \dfrac{p}{q}$.

50. $2x + y = z$.
$15x - 3y = 2\tfrac{1}{4}z$.
$4x + 3y + 2z = 18$.

51. $x^2 + xy + yz = 12$.
$17x + 4y = 21z$.
$43x - 21y = 22z$.

52. $x + y + z = a + b + c$.
$(b - c)x + (c - a)y + (a - b)z = 0$.
$c(x - a) + a(y - b) + b(z - c) = 0$.

Eliminate x, y, and z from:

53. $x + y + z = 0$.
$(b - c)x + (c - a)y + (a - b)z = 0$.
$(b + c)x + (c + a)y + (a + b)z = 0$.

54. $bx = ay - z$; $cy = bz - x$; and $az = cx - y$.

55. $\quad ax + by + cz = 0$.
$\quad a'x + b'y + c'z = 0$.
$\quad a''x + b''y + c''z = 0$.

70. Approximate ratios.

It is sometimes of use to be able to find the approximate values of expressions which are too complicated for easy calculation.

When a is very large when compared with x, the ratio between the powers of $a \pm x$ and of a can be expressed approximately as in the following examples.

Ex. 1. Let $a \pm x : a$ be a ratio, where a is very large when compared with x.

Then $(a \pm x)^2 : a^2 = a^2 \pm 2ax + x^2 : a^2$

$$= a \pm 2x + \frac{x^2}{a} : a = a \pm 2x : a \text{ nearly,}$$

since x^2/a must be a very small fraction and may therefore be neglected.

Ex. 2. Similarly

$$(a \pm x)^3 : a^3 = a^3 \pm 3a^2x + 3ax^2 \pm x^3 : a^3$$

$$= a \pm 3x + \frac{3x^2}{a} \pm \frac{x^3}{a^2} : a$$

$$= a \pm 3x : a \text{ nearly.}$$

NOTE.—In these two examples the same sign must be taken throughout.

EXERCISES IV.

56. Show that $(10{,}001)^2 : (10{,}000)^2 = 10{,}002 : 10{,}000$ nearly.

57. Show that $9875^3 : 9876^3 = 9875 : 9878$ nearly, $= 9873 : 9876$ nearly.

Of the two approximate fractions which is the greater? Express the difference as a vulgar fraction, and show that it is less than a ten millionth.

58. Show that when a is very small compared with m,

$(m+a)^4 : m^4$ is very nearly $m+4a : m$.

59. The length of each side of a cube is increased by ·00004 of itself. Find approximately the increase in the area of each face, and in the volume of the cube.

CHAPTER V.

PROPORTION.

71. Definitions. When two ratios are equal, the four terms taken in order are called **proportionals,** and are said to be **in proportion.**

Thus, if $a:b$ and $c:d$ are equal, a, b, c, d are proportionals. If $a:b=c:d=e:f$, then a, b, c, d, e, f are called proportionals. The proportion is written in two ways, namely

$$a:b=c:d\,;$$

or $$a:b::c:d\,;$$

of which the first line is read " the ratio of a to b equals the ratio of c to d," or more briefly, " a to b equals c to d," and the second line* is read " a is to b as c is to d."

If $a:b=c:d$, then a and d, the extreme terms, are called the **extremes,** and b and c, the middle terms, the **means;** and d is called the **fourth proportional** to a, b and c.

72. Properties of Proportionals.—The product of the extremes of a proportion is equal to the product of the means.

Let $a:b=c:d$. Then $\dfrac{a}{b} = \dfrac{c}{d}$.

Hence $\dfrac{a}{b} \times bd = \dfrac{c}{d} \times bd$; that is $ad = bc$.

---

* The four points which constitute the symbol :: have been explained to be the extremities of = the sign of equality.

PROPORTION. 59

Cor. (The Rule of Three.) When three terms of a proportion are given, the fourth can be found by solving a simple equation.

73. Conversely if the product of two quantities be equal to the product of two others, the four form a proportion, *the extremes being the two factors of one of the products, and the means the two factors of the other.*

Let $ad = bc$. Then $\dfrac{ad}{bd} = \dfrac{bc}{bd}$, whence $\dfrac{a}{b} = \dfrac{c}{d}$, or $a : b = c : d$.

It may be similarly proved that $b : a = d : c$, &c.

74. If $a : b = c : d$, then $a : c = b : d$.
[Alternando.]
[Note. *Alternando* is applicable to concrete quantities, only when all the terms of the proportion are of the same kind.]

75. If $a : b = c : d$, then $b : a = d : c$.
[Invertendo.]

76. If $a : b = c : d$, then $a + b : b = c + d : d$.
[Componendo.]

Since $\dfrac{a}{b} = \dfrac{c}{d}$, we have $\dfrac{a}{b} + 1 = \dfrac{c}{d} + 1$. $\therefore \dfrac{a+b}{b} = \dfrac{c+d}{d}$.

That is, $a + b : b = c + d : d$.

77. If $a : b = c : d$, then $a - b : b = c - d : d$.
[Dividendo.]

Since $\dfrac{a}{b} = \dfrac{c}{d}$, we have $\dfrac{a}{b} - 1 = \dfrac{c}{d} - 1$.

$\therefore a - b : b = c - d : d$.

[Note. The names *Invertendo, Alternando, Componendo,* and *Dividendo* are the Latin names of the corresponding propositions in Euclid, Book V.]

78. If $a : b = c : d$, then
$a + b : a - b = c + d : c - d$.

Since $\dfrac{a}{b} = \dfrac{c}{d}$, $\dfrac{a+b}{b} = \dfrac{c+d}{d}$ [§ 76.]

60   PROPORTION.

and $$\frac{a-b}{b} = \frac{c-d}{d}$$ [§ 77.]

Hence $\dfrac{a+b}{b} \times \dfrac{b}{a-b} = \dfrac{c+d}{d} \times \dfrac{d}{c-d}$.

Therefore $a+b : a-b = c+d : c-d$.

This inference is called *Composition and Division*, or sometimes *Mixing*.

79. If $a : b = c : d$,
then $ma^r + nb^r : pa^r + qb^r = mc^r + nd^r : pc^r + qd^r$.

Let $$\frac{a}{b} = \frac{c}{d} = k.$$

Then $a = bk$; and $c = dk$.

$$\therefore \frac{ma^r + nb^r}{pa^r + qb^r} = \frac{mb^r k^r + nb^r}{pb^r k^r + qb^r} = \frac{b^r(mk^r + n)}{b^r(pk^r + q)} = \frac{mk^r + n}{pk^r + q}.$$

Similarly, $\dfrac{mc^r + nd^r}{pc^r + qd^r} = \dfrac{d^r(mk^r + n)}{d^r(pk^r + q)} = \dfrac{mk^r + n}{pk^r + q}.$

Hence the required result follows.

COR. The following properties as well as §§ 76—78 will be found to be particular cases of this result.

If $a : b = c : d$, then

(i) $ma : pb = mc : pd$.

(ii) $\dfrac{a}{m} : \dfrac{b}{p} = \dfrac{c}{m} : \dfrac{d}{p}$.

(iii) $a^n : b^n = c^n : d^n$.

(iv) $ma + nb : pa + qb = mc + nd : pc + qd$.

80. Generalization.—*If* a, b, c, d *be proportionals, then the ratio of any two homogeneous expressions of the same degree in* a *and* b *is equal to that of the same two expressions with* c *and* d *substituted for* a *and* b *respectively.*

The proof is similar to that of § 79 and consists in putting $a/b = c/d = k$ and simplifying each of the ratios that have to be proved equal.

PROPORTION. 61

Ex. $\qquad \dfrac{2a^2 + 3ab - b^2}{a^2 - 5b^2} = \dfrac{2c^2 + 3cd - d^2}{c^2 - 5d^2}$

and $\qquad \dfrac{ma^3 + na^2b - pb^3}{qb^3 - ra^3} = \dfrac{mc^3 + nc^2d - pd^3}{qd^3 - rc^3}.$

NOTE. All the four expressions are, in each case, homogeneous and of the same degree.

81. If four quantities are proportionals, the sum of the greatest and the least is greater than the sum of the other two. [Euclid, V. 25.]

We divide the proof into two parts.

(i) To show that if a be the greatest of the four proportionals a, b, c, d, then d is the least.

Since $a : b = c : d$, and $a > b$, $\therefore c > d$.

Again, since $a : c = b : d$ (*Alternando*), and $a > c$, $\therefore b > d$.

And, by hypothesis, $a > d$.

Hence d is the least of the four quantities.

(ii) We have next to prove that $a + d > b + c$.

Since $\qquad a : b = c : d,$
$\qquad\qquad a - b : b = c - d : d.$
$\qquad \therefore\ a - b : c - d = b : d.$

And since $\qquad b > d,\ \therefore\ a - b > c - d.$
$\qquad \therefore\ a - b - c + d > 0.$
$\qquad \therefore\ a + d > b + c.$

This proposition is a particular case of the following example.

Ex. If a is the greatest of the four proportionals, a, b, c, d, then $a^n + d^n > b^n + c^n$.

For by § 79 Cor. (iii), $a^n : b^n = c^n : d^n$.

82. Illustrative Examples.

Ex. 1. If $a : b = c : d$, then $a + b : a - b = c + d : c - d$.

[We now give another proof of the important property already proved in § 78.]

Let $\qquad \dfrac{a}{b} = \dfrac{c}{d} = k.$

PROPORTION.

Then
$$\frac{a+b}{a-b} = \frac{bk+b}{bk-b} = \frac{k+1}{k-1}.$$

Similarly,
$$\frac{c+d}{c-d} = \frac{dk+d}{dk-d} = \frac{k+1}{k-1}.$$

Hence the result required follows.

Ex. 2. State the equation $x^3 + 1 = a^3 - b^3$ as a proportion.

Since $(x+1)(x^2-x+1) = (a-b)(a^2+ab+b^2)$,

we have $x+1 : a-b = a^2+ab+b^2 : x^2-x+1.$ [§ 73.]

Ex. 3. What quantity should be added to each of the four quantities a, b, c, d to obtain four proportionals?

If the required quantity be x, we have
$$a+x : b+x = c+x : d+x.$$
$$\therefore (a+x)(d+x) = (b+x)(c+x).$$

Solving the equation, we have $x = \dfrac{bc-ad}{a-b-c+d}.$

Ex. 4. If the ratio of $a : b$ is unaltered, when x and y are added to its terms in order, show that $x : y = a : b$.

Since
$$a+x : b+y = a : b.$$
$$\therefore a+x : a = b+y : b. \qquad [\textit{Alternando}.]$$
$$\therefore x : a = y : b. \qquad [\textit{Dividendo}.]$$
$$\therefore x : y = a : b. \qquad (\textit{Alternando}.)$$

Ex. 5. If $ax+by : cx+dy$ has the same value for all values of x and y, then a, b, c, d are proportionals.

For by hypothesis
$$\frac{ax+by}{cx+dy} = \frac{a.1+b.0}{c.1+d.0} = \frac{a.0+b.1}{c.0+d.1},$$

or
$$\frac{a}{c} = \frac{b}{d} \text{ whence } \frac{a}{b} = \frac{c}{d}.$$

Ex. 6. Find $x : y$ from
$$x+2y : x-2y = x^2+2(x-2y)^2 : x^2-2(x-2y)^2.$$

By Composition and Division, we have
$$\frac{x+2y+x-2y}{x+2y-x+2y} = \frac{x^2+2(x-2y)^2+x^2-2(x-2y)^2}{x^2+2(x-2y)^2-x^2+2(x-2y)^2}.$$

PROPORTION. 63

By simplification, we get $\dfrac{x}{2y} = \dfrac{x^2}{2(x-2y)^2}$.

Hence $x^2 - 5xy + 4y^2 = 0$, or $(x-y)(x-4y) = 0$.

∴ the ratio of x to y is either $1:1$ or $4:1$ [See § 67, Ex. 2.]

EXERCISES V.

1. If there be two or more sets of proportionals of the same number of terms, the products of the corresponding terms are proportionals.

2. Find a fourth proportional to 16, 15 and 8.
   If 4, x, 7 and 8 are proportionals, find x.

3. State the following equations as proportions :
   (i) $a^2 - 1 = b^3 + c^3$.   (ii) $xy = x^2 + 3x + 2$.
   (iii) $x^2 + 7x + 12 = y^2 - 20y + 96$.

NOTE. Questions of this class admit of any number of different answers.

4. Find x, if $5 + x : 7 + x = 5 : 7$.

5. For what value of x will $4 + x : 5 + x$ be the duplicate ratio of $2:3$?

6. Find x, if $a + x : b + x$ be in the subduplicate ratio of $a : b$.

7. If the ratio of $5 : 6$ is unaltered when x and y are added to its terms in order, show that $x : y = 5 : 6$.

8. What number must be subtracted from 4, 5, 6, 8 to obtain four proportionals?

9. What number must be added to 2, 1, 5, 3 to get four proportionals?

10. Show that no finite number added to or subtracted from 4, 5, 6, 7 will give four proportionals.

11. Show that there is no finite quantity which when added to each of four proportionals will make the resulting quantities proportionals.

12. If the six quantities a, b, c, d, e, f are such that $a:b = c:d$ and $e:b = f:d$, then $a + e : b = c + f : d$.

13. With the conditions of the 12th question, prove that
$$a^n + e^n : c^n + f^n = b^n : d^n.$$

PROPORTION.

14. If $x:y$ in the duplicate ratio of $a:b$, and $a:b$ in the sub-duplicate ratio of $a+x:a-y$, then $2x:a=x-y:y$.

15. If $a-b$ be the least of the four proportionals $a+b$, $c+d$, $c-d$, and $a-b$, prove that b is greater than d.

16. Prove that, if all the quantities be positive, $(pa+qb)/(pc+qd)$ lies between a/c and b/d if these are unequal, and is equal to them if they are equal.

17. If $a:b=c:d$ prove the following equalities:
  (i) $a^2bd+b^2c+bc=ab^2c+abd+ad$.
  (ii) $a+b+c+d:b+d=c+d:d$.
  (iii) $a^3+b^3:c^3+d^3 = \dfrac{a^4}{a+b} : \dfrac{c^4}{c+d}$.
  (iv) $a+mb:a-nb=c+md:c-nd$.

18. If $x+y:x-y=a:b$, prove that
$$2x^2(ab-b^2)+2y^2(ab+b^2)=(a^2-b^2)(x^2-y^2).$$

19. If $x:y=a:b$, then each of these ratios $= \dfrac{\sqrt{(x^2+a^2p^2)}}{\sqrt{(y^2+b^2p^2)}}$.

20. If $a^2-c^2:b^2-d^2=a^2:b^2$, prove that $a:b=\pm c:d$.

Also find the value of the ratio
$$(a^2d^2+b^2c^2)(ad+bc)^2 : a^2b^2c^2d^2.$$

If $a:b=c:d$, prove that

21. $a(a+b+c+d)=(a+b)(a+c)$.

22. $a^4+b^4:a^2b^2=c^4+d^4:c^2d^2$.

23. $\sqrt{(a-b)}:\sqrt{(c-d)}=\sqrt{a}-\sqrt{b}:\sqrt{c}-\sqrt{d}$.

24. If $(a+b+c+d)(a-b-c+d)=(a-b+c-d)(a+b-c-d)$, prove that a, b, c, d are proportionals.

25. Solve $\dfrac{\sqrt{(x+5)}+\sqrt{(x-16)}}{\sqrt{(x+5)}-\sqrt{(x-16)}}=2\tfrac{1}{8}$.

26. If $a:b=c:d$, then $(a^2+c^2)(b^2+d^2)=(ab+cd)^2$.

27. Prove that if
$$\frac{bx+cy+dz}{b+c+d-a}=\frac{cx+dy+az}{c+d+a-b}=\frac{dx+ay+bz}{d+a+b-c}=\frac{ax+by+cz}{a+b+c-d},$$
each $=\tfrac{1}{2}(x+y+z)$, unless $a+b+c+d=0$.

28. If $(a+b+c)x = (-a+b+c)y = (a-b+c)z = (a+b-c)w$, prove that $\dfrac{1}{y}+\dfrac{1}{z}+\dfrac{1}{w}=\dfrac{1}{x}$.

29. Solve the simultaneous equations
$$\frac{\sqrt{(y+x)}+\sqrt{(y-x)}}{\sqrt{(1+x)}+\sqrt{(1-x)}}=\frac{\sqrt{(y+x)}-\sqrt{(y-x)}}{\sqrt{(1+x)}-\sqrt{(1-x)}};$$
$$x^2+xy=2.$$

30. The times in which two boatmen row the same distance up a stream are as $a:b$; and their times of rowing equal distances down the stream are as $c:d$. Find the ratio of their rates of rowing in still water.

31. Find four proportionals such that the sum of the extremes is 21, the sum of the means 19, and the sum of the squares of all the four, 442.

32. If m shillings in a row reach as far as n sovereigns, and a pile of p shillings is as high as a pile of q sovereigns, compare the weights of a sovereign and a shilling, taking their specific gravities to be r and s respectively.

83. Continued Proportion. DEFINITIONS.—

When the ratio of the first to the second of any number of quantities is equal to the ratio of the second to the third, this again to the ratio of the third to the fourth, and so on, the quantities are said to be in **continued proportion**.

Thus if $\dfrac{a}{b}=\dfrac{b}{c}=\dfrac{c}{d}=\dfrac{d}{e}$, a, b, c, d, e are said to be in continued proportion.

NOTE. A continued proportion of three terms is in reality a full proportion of four terms of which the two middle terms are equal.

When a, b, c are in continued proportion, b is said to be a **mean proportional** between a and c, the extremes.

When a, b, c, d are in continued proportion, b and c are said to be **two mean proportionals** between a and d; and so on.

84. Properties of Continued Proportionals.—

If a, b, c be in continued proportion, then $ac = b^2$.

Conversely if $ac = b^2$, then a, b, c are in continued proportion.

These are easy inferences from §§ 72, 73. A few other inferences will be found among the exercises that follow.

85. Euclid's definitions of Duplicate and Triplicate Ratio.

When three quantities are in continued proportion, the first has to the third the duplicate ratio of the first to the second (or of the second to the third).

Let a, b, c be three quantities in continued proportion.

Then $\quad \dfrac{a}{b} = \dfrac{b}{c}. \quad \therefore \dfrac{a}{b} \times \dfrac{b}{c} = \dfrac{a}{b} \times \dfrac{a}{b}.$

Thus, $a : c = a^2 : b^2 = b^2 : c^2$, which means that a has to c the duplicate ratio of $a : b$ or of $b : c$.

Similarly *when four quantities are in continued proportion, the first has to the fourth the triplicate ratio of any term to the next.*

Let the proportionals be a, b, c, d.

Then $\dfrac{a}{b} = \dfrac{b}{c} = \dfrac{c}{d}. \quad \therefore \dfrac{a}{b} \times \dfrac{b}{c} \times \dfrac{c}{d} = \dfrac{a}{b} \times \dfrac{a}{b} \times \dfrac{a}{b}.$

Thus, $a : d = a^3 : b^3 = b^3 : c^3 = c^3 : d^3$.

86. To find a mean proportional between two given quantities. (Euclid VI. 13.)

Let a and b be two given quantities, and x the required mean proportional. Then $a : x = x : b$. Thus $x^2 = ab$, and $x = \sqrt{ab}$.

87. To find two mean proportionals between two given quantities.

Let a and b be two given quantities, and x and y the required mean proportionals.

Then $\dfrac{a}{x} = \dfrac{x}{y} = \dfrac{y}{b}$. $\therefore \dfrac{a}{b} = \dfrac{a^3}{x^3} = \dfrac{y^3}{b^3}$.

Thus $x^3 = a^2 b$, and $y^3 = b^2 a$.

$\therefore x = \sqrt[3]{a^2 b}$, and $y = \sqrt[3]{ab^2}$.

Ex. Find two mean proportionals between a and $2a$*.

If x and y be the two quantities required,

$$\dfrac{a}{x} = \dfrac{x}{y} = \dfrac{y}{2a}.$$

$$\therefore \dfrac{a}{2a}\left(=\dfrac{1}{2}\right) = \dfrac{a^3}{x^3} = \dfrac{y^3}{8a^3}.$$

Hence $x = a\sqrt[3]{2}$ and $y = a\sqrt[3]{4}$.

EXERCISES V.

33. Find a mean proportional to 75 and 12.
34. Find a third proportional to 4 and 5.
35. Find a mean proportional to $\sqrt{27} - 3\sqrt{2}$ and $\sqrt{27} + 3\sqrt{2}$.
36. If a, x and $a-x$ are continued proportionals, find x.
[The problem corresponds to Euclid II. 11, or VI. 30.]
37. Find a third proportional to $5 + 2\sqrt{3}$, and $37 + 20\sqrt{3}$.
38. If 2, y, 4 and x are continued proportionals, find x and y.
39. Find two mean proportionals between 3 and 5.

---

* The corresponding geometrical problem (the *Delian Problem*), "To find two mean proportionals between two given straight lines," is often referred to as the *Duplication of the Cube*, since the side of a cube which is double the volume of another cube can be determined as soon as two mean proportionals can be found between the side of the given cube and twice its length.

PROPORTION.

40. Find two mean proportionals between 2 and 54.

41. Find two mean proportionals between 2 and 5.

42. Find three mean proportionals between (i) 1 and 16, (ii) between 2 and 8.

43. If, of three continued proportionals, the first is a square, the third is also a square.

If, of four continued proportionals, the first is a cube, the fourth is also a cube.

44. Find x so that $a+x$, $b+x$, and $c+x$ may be continued proportionals.

If a, b, c are continued proportionals, prove that

45. $a+b : a-b = b+c : b-c$.

46. $a-b : b = b-c : c$.

47. $a+c > 2b$.

48. $ma+nb : pa+qb = mb+nc : pb+qc$.

49. $ab+bc$ is a mean proportional between a^2+b^2, and b^2+c^2.

50. $a+b+c : a-b+c = (a+b+c)^2 : a^2+b^2+c^2$.

51. $(a-b+c)(a+b+c)(a^2-b^2+c^2) = a^4+b^4+c^4$.

52. $a+b : b+c = a^2(b-c) : b^2(a-b)$.

53. If $a : b$ be in the duplicate ratio of $a-c : b-c$, a and b being unequal, then c is a mean proportional between a and b.

54. If four quantities are proportionals, and the second of them is a mean proportional between the third and the fourth, the third will be a mean proportional between the first and the second.

55. If four positive quantities are in continued proportion, show that the difference between the first and the last is at least three times as great as the difference of the other two.

56. If $a : b = b : c = c : d$, show that $b+c$ is a mean proportional between $a+b$ and $c+d$.

57. If $a : b = c : d$, then show that the mean proportional between b and c is a mean proportional between the mean proportional between a and b, and that between c and d.

58. If the n quantities $a_1, a_2, a_3 \ldots a_n$ be in continued proportion, then $a_1 : a_n = a_1{}^{n-1} : a_2{}^{n-1} = a_2{}^{n-1} : a_3{}^{n-1}$; and so on. Hence find the values of a_2 and a_3 in terms of a_1 and a_n.

88. Incommensurable quantities.

DEFINITIONS.—Two quantities are said to be **commensurable** when their ratio can be expressed as the ratio of two integers.

Thus, any two arithmetical numbers such as $\frac{3}{16}$ and $\frac{7}{17}$ are commensurable, because their ratio can be expressed as the ratio of the two integers 51 : 112.

Two quantities are said to be **incommensurable** when their ratio cannot be expressed as the ratio of two integers.

Thus $\sqrt{2}$ and 1 which are proportional to the lengths of the diagonal and the side of a square are incommensurable, since $\sqrt{2}$ cannot be exactly determined.

A **single** numerical quantity is said to be **incommensurable** when it is incommensurable with unity.

Thus $\sqrt{2}$, $\sqrt{3}$ and all other irrational numbers are incommensurable. It has been mentioned that *irrational* literally means "not expressible by an arithmetical ratio."

Though all irrational numbers are incommensurable, every incommensurable number is not irrational. Thus π which the student meets with in Trigonometry is incommensurable, though it is not irrational in the current sense of the word.

89. Any root of an integer either is an integer or is incommensurable.

Let N be the given integer. Then if $\sqrt[n]{N}$ were equal to the arithmetical fraction a/b we should have $N = a^n/b^n$; also if a/b were expressed in its lowest terms, a, b would contain no common factor and a^n/b^n would evidently be expressed in its lowest terms. Therefore the integer N would be equal to a fraction which is impossible. Hence $\sqrt[n]{N}$ must be either an integer or an incommensurable number.

COR. *A root of an integer can never be expressed as a recurring decimal,* for a recurring decimal can always be reduced to a vulgar fraction.

EXERCISES V.

59. Prove that the square root of a fraction is incommensurable unless, when reduced to its lowest terms, both its numerator and denominator are exact squares.

60. Find the integral part of $28\sqrt{2}$, and hence find two fractions differing by $\frac{1}{28}$, between which $\sqrt{2}$ lies.

90. Euclid's test of proportionality applies to commensurable and incommensurable quantities alike.

The geometrical test of proportion is 'The first of four magnitudes has to the second the same ratio which the third has to the fourth, when, any equimultiples whatsoever of the first and the third being taken, and any equimultiples whatsoever of the second and the fourth, according as the multiple of the first is greater than, equal to, or less than the multiple of the second, so is the multiple of the third greater than, equal to, or less than the multiple of the fourth.' (Euc. V., Def. 4.)

The test may be symbolically given as follows :

(1) The four magnitudes A, B, C, D, will be proportional, if $mC >, =$ or $< nD$, according as $mA >, =$ or $< nB$, m and n being any integers.

(2) That is, A, B, C, D will be proportional, if, m and n being any integers whatsoever, $mC = nD$ when $mA = nB$; or mC lies between nD and $(n+1) D$, when mA lies between nB and $(n+1) B$.

It is evident that the test applies equally well whether A and B be commensurable or incommensurable. If A and B are commensurable, suitable values of m and n can be found, which will make mA equal to nB. If, in this case, mC is also equal to nD, one part of the test is satisfied, and the other parts are fulfilled as a consequence of this. But when A and B are incommensurable as also C and D, mA will never be equal to nB and mC to nD, since $\frac{A}{B}$ or $\frac{C}{D}$ cannot be equal to $\frac{n}{m}$, a fraction with integral numerator and denominator.

Now, what the definition states is that, in such a case, A, B, C, D may be asserted to be proportional, if, in *all* cases, mC lies between nD and $(n+1) D$ when mA lies between nB and $(n+1) B$.

91. Extension of the algebraical definition to incommensurable quantities.

Since, by the algebraical definition, the ratio or the relative magnitude of two quantities is measured by the *number of times* and *parts of times* the one is contained in the other, the terms of an algebraical ratio, and therefore of an algebraical proportion, cannot but be commensurable quantities. To take an example, no multiple whatever of the side of a square will make an exact multiple of its diagonal.

We shall show in this article that, though two incommensurable quantities have not an assignable numerical ratio to each other, we can always find a ratio which is as near as we please to the exact ratio of the two quantities.

PROPORTION. 71

DEFINITION. Two magnitudes A, B are said to have a **greater** or **less** ratio than two whole numbers n, m according as m times A is greater or less than n times B.

Let A and B be two incommensurable quantities. It is obvious that mA can never be equal to nB, m and n being any integers whatever. Suppose n is the quotient with a remainder, when mA is divided by B. Hence mA is greater than nB and less than $(n+1)B$.

$\therefore \dfrac{A}{B}$ is greater than $\dfrac{n}{m}$, and $\dfrac{A}{B}$ is less than $\dfrac{n+1}{m}$.

Thus, $\dfrac{A}{B}$ lies between $\dfrac{n}{m}$ and $\dfrac{n+1}{m}$.

Since the difference between $\dfrac{n}{m}$ and $\dfrac{n+1}{m}$ is $\dfrac{1}{m}$, and $\dfrac{A}{B}$ lies between these two fractions, $\dfrac{A}{B}$ differs from $\dfrac{n}{m}$ by a quantity which is less than $\dfrac{1}{m}$. Since m may be any integer whatever, it may be taken to be as great as we please. Hence $\dfrac{1}{m}$ can be made as small as we please. Thus the difference between the ratio of the two incommensurable quantities A and B, and that of the two commensurable quantities m and n can be made as small as we please. Hence the ratio of two incommensurable quantities can be expressed arithmetically to any required degree of accuracy.

Now, since proportion is the equality of ratios, and since ratios of incommensurable quantities can thus be estimated arithmetically to any degree of accuracy, the algebraical definition of proportion may be extended to incommensurable quantities.

92. To deduce Euclid's test of proportion from the algebraical test.

Four quantities A, B, C, D are algebraically proportional when $\dfrac{A}{B} = \dfrac{C}{D}$.

Since $\dfrac{A}{B} = \dfrac{C}{D}$, therefore $\dfrac{mA}{nB} = \dfrac{mC}{nD}$.

Therefore $mA >, =$ or $< nB$, according as $mC >, =$ or $< nD$, which is the geometrical test of proportionality.

93. To deduce the algebraical test of proportion from Euclid's test.

If $mC >, =$ or $< nD$, according as $mA >, =$ or $< nB$, m and n being any positive integers whatever, then shall

$$\frac{A}{B} = \frac{C}{D}.$$

If $\dfrac{A}{B}$ be not $= \dfrac{C}{D}$, then one of them must be greater than the other.

Let $\dfrac{A}{B}$ be the greater. Now take a fraction with integral numerator and denominator between $\dfrac{A}{B}$ and $\dfrac{C}{D}$, and call it $\dfrac{n}{m}$.

Here, $\dfrac{A}{B} > \dfrac{n}{m}$ and $\dfrac{n}{m} > \dfrac{C}{D}.$

$$\therefore mA > nB, \text{ and } nD > mC.$$

[That is, equimultiples being taken of the first and the third, and of the second and the fourth, we find that while the multiple of the first is greater than that of the second, the multiple of the third is less than that of the fourth.]

But this is contrary to the hypothesis. Therefore, $\dfrac{A}{B}$ is not unequal to $\dfrac{C}{D}$, as was to be proved.

Alternative Proof. Since
$$mA > \text{ or } < nB \text{ as } mC > \text{ or } < nD$$
$$\therefore \frac{A}{B} > \text{ or } < \frac{n}{m} \text{ as } \frac{C}{D} > \text{ or } < \frac{n}{m}.$$

Therefore any arithmetic ratio $m : n$ is either greater than both the ratios $A : B$ and $C : D$ or is less than both. Hence the ratios $A : B$ and $C : D$ lie within the same limits however close these limits may be, and therefore these ratios cannot be unequal.

CHAPTER VI.

VARIATION.

94. Variation and Proportion. Questions in proportion are sometimes proposed and considered in a form which is different from that discussed in the preceding chapters and which will be found to involve a new notation and phraseology.

DEFINITIONS.—**Constants** are quantities which retain the same fixed values in an expression or equation, and which are independent of the values of all other quantities.

Variables are quantities that admit of an infinite number of values in the same expression or equation.

Ex. $2x = 3y$. Here 2 and 3 are constants, and x and y are variables. If the values of x are derived from those of y, y to which we may give any value that we like is called an *independent variable*, and x that becomes changed in consequence is called a *dependent variable*. Similarly in $by = cx$, b and c may be constants, x an independent, and y a dependent variable.

95. The simplest relations of two variables.

Two variable quantities x and y may be so connected that
$$x = ky \quad \ldots\ldots\ldots\ldots(\text{i}),$$
or
$$x = \frac{k}{y} \quad \ldots\ldots\ldots\ldots(\text{ii}),$$
where k is a constant.

VARIATION.

[The letter k is used to denote a constant quantity in the articles that follow.]

Direct Variation.—In $x = ky$, where k is a constant, the value of x depends on that of y. If $y = 5$, $x = 5k$, and if $y = \frac{1}{2}$, $x = \frac{1}{2}k$.

If now, when y is changed to y', x becomes changed to x', $x' = ky'$, and

$$\therefore \frac{x}{x'} = \frac{ky}{ky'} = \frac{y}{y'}.$$

Thus we see that, when y is increased or decreased, x is not arbitrarily increased or decreased, but its value is increased or decreased proportionally to the value of y.

In such a case x is said to *vary directly* as y, or more shortly to *vary* as y.

DEFINITION.—One variable is said to **vary directly**, (or more shortly to **vary**) **as** another quantity, when the two are so related that any change in the value of the second produces a *proportionate* change in the value of the first.

Arithmetical illustration. The interest on a sum of money varies directly as the principal, other things remaining unchanged.

Geometrical illustration. If the altitude of a rectangle be a fixed length, the area varies as the base.

For, if A be the number of square feet in a rectangle and h and b the number of linear feet in the altitude and the base respectively, $A = hb$. Since h is a constant in this equation, any change in b leads to a proportionate change in A. Hence the area varies as the base when the height is constant.

Inverse Variation. Taking the other equation $x = k/y$ where k is a constant, we have $x = k\,(1/y)$, which gives us, as in the preceding part of the article, the proportion

$$x : x' = 1/y : 1/y'.$$

Thus x varies as the reciprocal of y, an increase in the value of y leading to a proportionate decrease in the value of x, and a decrease in y to a proportionate increase in x.

In this case x is said to *vary inversely* as y.

DEFINITION.—One variable is said **to vary inversely as** another when it varies as its reciprocal, *i.e.*, when the two are so related that any increase or decrease in the

VARIATION. 75

latter leads to a *proportionate* decrease or increase in the former.

Arithmetical illustration. If a given sum of money be distributed equally among a number of beggars, the share of each person varies inversely as the number of beggars relieved.

Geometrical illustration. If the area of a rectangle be constant, the altitude varies inversely as the base.

For since $A = bh$, $h = A/b = $ a constant quantity $\times 1/b$. Hence the altitude varies inversely as the base.

96. The simplest relations of three variables.

Joint Variation. Of the relations of three dependent quantities to one another, it is usual to discuss the following two cases:

$$x = kyz \ldots\ldots(\text{i}). \qquad x = k\frac{y}{z} \ldots\ldots(\text{ii}).$$

If $x = kyz$, then $x : x' = yz : y'z'$, it being assumed that x', y', and z' are three simultaneous values of x, y and z.

In this case x is said to *vary jointly* as y and z.

DEFINITION.—One variable is said to **vary jointly as** two or more others, when the first varies as the product of the other two.

Arithmetical illustration. The quantity of work that is turned out varies jointly as the number of men and the number of days.

Geometrical illustration. The area of a triangle varies jointly as the base and the altitude.

For $A = \frac{1}{2}hb = $ a constant quantity $\times hb$.

If $y = z$, then $x = kyz$ becomes turned into $x = ky^2$. In this case x varies as the square of y.

If $x = kyzuv$, where x, y, z, u, v are variable quantities, then x varies jointly as y, z, u, v.

If $x = ky/z$, then $x : x' = y/z : y'/z'$; x', y', z' being three simultaneous values of x, y, z. Here x is said to *vary directly* as y and *inversely* as z.

Arithmetical illustration. The time of performing a journey varies directly as the distance to be travelled, and inversely as the rate of moving.

Geometrical illustration. The altitude of a triangle varies directly as the area and inversely as the base.

VARIATION.

97. Equations of variation and absolute equations.

The statement that x varies as y is known as an **equation of variation,** and is written $x \propto y$, the symbol $\propto$ standing for *varies as*.

The equation $x = ky$ may be called an *absolute equation* as distinguished from an equation of variation.

We have seen that the absolute equations $x = ky$ and $x = kyz$, where k is a constant, are identical with the equations of variation $x \propto y$ and $x \propto yz$. We shall now show how, conversely, equations of variation can be written as absolute equations.

98. If one variable varies as another, then any two simultaneous values of the two quantities are in a constant ratio.

Let A and B be two variables, a, b two simultaneous values of them, and let the value of A change to a' when that of B is changed into b'. Then, by definition of variation, the two values of A are proportional to the corresponding values of B or $a' : a = b' : b$. Therefore $a : b = a' : b'$.

Putting k for each of the equal ratios $a : b$ or $a' : b'$, this result may be stated in the following form:—

If $A \propto B$, then $A = kB$, where k is a constant.

To find the constant k in any case, it is clearly enough to know a single pair of corresponding values.

Thus, if $A \propto B$, and A is 3 when B is 7, $k = \frac{3}{7}$. Hence $A = \frac{3}{7}B$.

99. *Hence the three following equations are but different modes of expressing the same fact:—*

(i) $A \propto B$.

(ii) $A = kB$, where k is a constant.

(iii) $a : a' = b : b'$ where a, b, and a', b' are pairs of simultaneous values of A and B.

NOTE. The side and the area of a square are mutually dependent on each other, and increase or diminish together. But the area of a

VARIATION. 77

square does not vary as the side, as the changes in the two are not proportional.

100. Points to be remembered.—Before proceeding further, we shall do well to draw the attention of the student to the following important points :—

The word 'quantity' is used in this chapter in a general, indefinite sense, to denote variables, without any reference to particular values. Thus length, weight, cost, area, time, &c., are quantities.

We say, for example, that the cost of sugar varies as its weight, while nobody will think of stating that the cost of 1 shilling's worth of sugar varies as 3 lbs., since it is impossible for a fixed quantity to vary.

NOTE. To avoid any confusion that might result, we shall, in the *propositions* that follow, use *capital* letters such as A, B, C to denote *variables* regarded as general, indefinite quantities, and the *small* letters a, b, c to indicate *particular values* of these variable quantities. In *examples*, the lower-case end letters of the alphabet, such as x, y, z, will also be used to denote variables.

When $A \propto B$, we have to remember that A and B are not necessarily *of the same kind*.

When we state that the time of performing a journey varies as the distance, we mean that the *number* of hours spent in the journey equals the *number* of miles multiplied by a constant factor.

Similarly, the angle at the centre of a circle varies as the arc on which it stands, though it has no ratio to it. [Euc. VI. 33.]

When two or more variations occur in the same problem, the *same* constant should not be used more than once in turning equations of variation into absolute equations.

Thus, if $A \propto B$, $A/B = k$ and $A = kB$ where k is a constant. But if, in the same problem, C is given to vary as D, then C/D is likewise a constant, but it may not have the same constant value k as A/B. Hence $C \propto D$ is written $C = mD$, where m is another constant.

Ex. 1. If x varies as y, and $x = 4$ when $y = 3$, find x when $y = 10$.

Since $x \propto y$, $x = ky$, where k is constant. Hence $4 = 3k$, and $k = \frac{4}{3}$. Thus $x = \frac{4}{3}y$; and when $y = 10$, $x = \frac{4}{3} \times 10 = 13\frac{1}{3}$.

Ex. 2. If y is equal to the sum of three terms of which the first $\propto x^2$, the second $\propto x$, and the third is constant; and if $y = 6$, 11, 18 when $x = 1, 2, 3$ respectively; find the relation between x and y.

VARIATION.

Of the three terms the sum of which is equal to y, we may write the first $= kx^2$, the second $= mx$, and the third $= n$, where k, m, n are constants.

Hence $\qquad y = kx^2 + mx + n.$

Substituting successively in this equation the values 1 and 6, 2 and 11, and 3 and 18 for x and y respectively, we have

$$6 = k + m + n;$$
$$11 = 4k + 2m + n;$$
and $\qquad 18 = 9k + 3m + n.$

Solving these simultaneous equations, we have

$$k = 1, \ m = 2 \text{ and } n = 3.$$

Hence the required relation is

$$y = x^2 + 2x + 3.$$

Ex. 3. If the volume of a sphere varies as the cube of its radius, find the radius of a sphere whose volume equals the sum of the volumes of three spheres whose radii are 3 ft., 4 ft., and 5 ft., respectively.

If v denotes the volume of a sphere and r its radius, then $v \propto r^3$; that is, $v = k \cdot r^3$ where k is a constant.

Thus the sum of the volumes of the three spheres whose radii are 3 ft., 4 ft. and 5 ft. $= k (3^3 + 4^3 + 5^3)$ cub. ft. $= 216\,k$ cub. feet.

Therefore, if the required radius of the sphere whose volume equals the sum of the volumes of the three spheres be r, its volume

$$kr^3 \text{ cub. feet} = 216\,k \text{ cub. feet.}$$

Hence $r^3 = 216$, whence $r = 6$, or the radius $= 6$ feet.

EXERCISES VI.

1. Examine whether the following examples are instances of direct or inverse variation and obtain the required results in this way.

(i) If a train moving with a uniform velocity of 1 foot per second traverses a certain distance in 1 minute, how long would it have taken to traverse the same distance had the velocity been $1\frac{1}{2}$ feet per second?

(ii) If 2 lbs. of sugar cost 5*d.*, what will 5 lbs. cost?

(iii) If 10 men reap a field in 6 hours, in how many hours will 18 men reap the same field?

VARIATION.

2. If $7x+5y=4x+3y$, show that $x \propto y$.

If $x^2y^2+1=2xy$, show that x varies inversely as y.

3. Prove that, when one quantity varies as another, the sum or arithmetical difference of two values of the one is proportional to the sum or arithmetical difference of the corresponding values of the other.

4. If $y^2 \propto a^2-x^2$, and when $x=\sqrt{(a^2-b^2)}$, $y=b^2/a$, find the relation between x and y.

5. If $x/y \propto x-y$, and $y/x \propto x^2+xy+y^2$, show that x^3-y^3 is invariable.

6. If x varies inversely as y^2, and $x=4a$ when $y=3a$, find the relation between x and y.

7. If x varies as y and $x=5a$ when $y=3b$, find y when $x=\dfrac{a}{b}$.

8. If x varies jointly as y and z, and if $x=1$ when $y=2$ and $z=3$, find x when $y=4$ and $z=5$.

9. If x varies directly as y and inversely as z, and if $x=1$ when $y=2$ and $z=3$, find x when $y=4$ and $z=5$.

10. $x=p+q$ where $p \propto y^2$ and $q \propto \dfrac{1}{y}$. Find the equation between x and y, if $y=1$ when $x=7$, and $y=2$ when $x=14$.

11. If $a \propto b^2c$, and 1, 2, 3 be simultaneous values of a, b, c, express a in terms of b and c.

12. If y equals the sum of two quantities of which one is constant and the other varies as xy, and if $y=-2\frac{1}{3}$, when $x=2$, and $y=1$, when $x=-2$, express y in terms of x.

13. If $x \propto y-z$, $y \propto q^2$ and $z \propto q$; and if $x=0$ when $q=5$; and $x=6$ when $q=6$; find q when $x=-6$.

14. If $a^x \propto b^y$, and if $x=7$ when $y=5$; then $a^{x-7}=b^{y-5}$.

15. If x equals the sum of three quantities of which the first varies as y, the second as y^2 and the third as y^3, and if $x=3$, 14 and 39, when $y=1$, 2, 3 respectively, find x when $y=4$.

16. If $x=p+q+r$, where p is constant, q varies as y, and r as y^2; and if $x=6$, 17, 34 when $y=1$, 2, 3 respectively; find the equation connecting x and y.

17. If x varies inversely as the mth power of y, and y varies inversely as the nth power of z, and $z=c$ when $x=a$, find the equation between x and z.

18. If $a \propto bc$ and $b \propto ac$, show that c is constant.

VARIATION.

19. Two globes whose radii are r and r' are melted and formed into a single globe. Find its radius, having given that the volume of a globe varies as the cube of its radius.

20. If the area of a circle varies as the square of its radius, show that the area of a circle whose radius is $6\frac{1}{2}$ feet is equal to the sum of the areas of two circles whose radii are 6 feet and $2\frac{1}{2}$ feet.

21. A point moves with a speed which is different in different miles but is invariable in the same mile, and its speed in any mile varies inversely as the number of miles travelled before it commences this mile. If the second mile be described in 2 hours, find the time taken to describe the nth mile.

[NOTE. Its speed in the nth mile $=k/(n-1)$ miles per hour, where k is a constant. Hence it describes the nth mile in $(n-1)/k$ hrs. Hence $(2-1)/k=2$, which gives $k=\frac{1}{2}$.]

22. A railway engine without a train can go 40 miles an hour, and its speed is diminished by a quantity which varies as the square root of the number of carriages attached. With 9 waggons its speed is 32 miles an hour. Find the greatest number of carriages which the engine can move.

23. The illumination from a source of light varies inversely as the square of the distance; how much further from a candle must a book which is now 8 inches off be removed, so as to receive just half as much light?

24. The value of diamonds varies as the square of their weight, and the square of the value of rubies varies as the cube of their weight. A diamond of a carats is worth m times the value of a ruby of b carats, and both together are worth £c. Required the values of a diamond and a ruby, each weighing n carats.

25. The expenses of a charitable institution are partly constant, and partly vary as the number of the inmates. When the inmates are 960, and 3000, the expenses are respectively £112. 5s. 4d., and £180. Find the outlay for a thousand inmates.

26. The square of a planet's time of revolution varies as the cube of its distance from the sun; the distances of the earth and Mercury from the sun being 91 and 35 millions of miles, find, in years, the time of Mercury's revolution.

27. If the quantity of water which flows through pipes in a given time varies as the squares of their diameters, and two vessels whose contents are in the ratio of 26 to 9 be filled by two pipes respectively in 2 and 13 minutes, compare the diameters of the pipes.

28. A 'compensated' chronometer goes accurately at 60° F., and its rate of error varies as the square of the deviation from this temperature; if it gains 2·4 seconds a day when the temperature is 80° F., how many seconds does it gain in a week when the temperature is 45° F.?

We shall now establish a number of elementary propositions in variation.

101. *If $A \propto B$, then $B \propto A$.*
For if $A = kB$ then $B = (1/k)A$.

102. *If $A \propto B$, and $B \propto C$, then $A \propto C$.*
For if $A = mB$ and $B = nC$ where m and n are constants, then $A = mnC$, whence $A \propto C$.

103. *If $A \propto BC$, then $B \propto A/C$, and $C \propto A/B$.*
For if $A = kBC$ where k is a constant, then $B = 1/k \times A/C$. Hence $B \propto A/C$.
It follows similarly that $C \propto A/B$.

104. *If $A \propto C$ and $B \propto C$, then $A \pm B \propto C$, and $AB \propto C^2$.*
Let $A = mC$, and $B = nC$, where m and n are constants.
Then $A \pm B = (m \pm n) C$, and $AB = mnC^2$.
Hence $(A \pm B) \propto C$, and $AB \propto C^2$.

105. *If $A \propto B$, then $AC \propto BC$ where C is any quantity constant or variable.*
Let $A = kB$ where k is a constant.
Then $AC = kBC$, $\therefore AC \propto BC$.

106. *If $A \propto B$, and $C \propto D$, then $AC \propto BD$ and $A/C \propto B/D$.*
For if $A = mB$, and $C = nD$, where m and n are constants, then $AC = mnBD$ and $A/C = m/n \times B/D$.
Hence $AC \propto BD$ and $A/C \propto B/D$.

107. *If $A \propto B$, then any homogeneous expression in A and B varies as any other homogeneous expression of the same degree in A and B.*
Similarly if $A \propto B \propto C \propto ...$, any homogeneous expression in $A, B, C...$ varies as any other homogeneous expression of the same degree in $A, B, C...$.
We shall illustrate this proposition by the following examples.

VARIATION.

Ex. 1. If $x + y \propto x - y$ then $x^2 + y^2 \propto xy$.

Since $x + y = k(x - y)$ where k is constant.
$$x^2 + y^2 + 2xy = k^2(x^2 + y^2 - 2xy).$$
$$\therefore (x^2 + y^2)(1 - k^2) = 2xy(-1 - k^2).$$
Thus $\qquad x^2 + y^2 = \{2(k^2 + 1)/(k^2 - 1)\}xy.$

The coefficient of xy on the right-hand side of this equation is constant, since k is constant.
$$\therefore x^2 + y^2 \propto xy.$$

Ex. 2. If $a \propto b \propto c$, then $a^3 + 2a^2c + 3b^3 + 2c^3 \propto 5ac^2 + 6a^3$.

Let $a = mc$, and $b = nc$, where m and n are constant.

Then
$$\frac{a^3 + 2a^2c + 3b^3 + 2c^3}{5ac^2 + 6a^3} = \frac{m^3c^3 + 2m^2c^3 + 3n^3c^3 + 2c^3}{5mc^3 + 6m^3c^3}$$
$$= \frac{m^3 + 2m^2 + 3n^3 + 2}{5m + 6m^3} = \text{a constant.}$$
$$\therefore a^3 + 2a^2c + 3b^3 + 2c^3 \propto 5ac^2 + 6a^3.$$

Ex. 3. If $a + 2b \propto b - 3a$, then $2a^8 + 7a^5b^3 + 5b^8 \propto a^2b(a^5 + 4b^5)$.

It will not be easy in all such cases to deduce the second equation directly from the first. Hence we divide our proof into two parts.

(i) $a + 2b = k(b - 3a)$ where k is constant.
$$\therefore a(1 + 3k) = b(k - 2). \text{ Hence } a \propto b.$$

(ii) Since $a \propto b$, let $a = mb$, where m is constant. Then
$$\frac{2a^8 + 7a^5b^3 + 5b^8}{a^2b(a^5 + 4b^5)} = \frac{2m^8b^8 + 7m^5b^8 + 5b^8}{m^2b^3(m^5b^5 + 4b^5)}$$
$$= \frac{b^8(2m^8 + 7m^5 + 5)}{b^8 \times m^2 \times (m^5 + 4)} = \frac{2m^8 + 7m^5 + 5}{m^2(m^5 + 4)};$$

which is a constant, thus establishing the required relation.

EXERCISES VI.

29. If $ax + by \propto cx + dy$, then $x \propto y$; and if $y^{-1} - x^{-1} \propto x - y$, then x varies inversely as y.

30. If $a \propto b$, then $a^n + 2b^n \propto 3a^n - 4b^n$.

31. If $a + b \propto a - 2b$, then $a^6 + 5b^6 \propto 2a^6 + 3a^2b^4 + b^6$.

32. If $a \propto b$, and $b \propto c$, then $a^5 + a^2c^3 + bc^4 \propto a^5 + b^5 + c^5$.

VARIATION.

33. If $a \propto b$, and $b \propto c$, then $a^n + b^n + c^n \propto a^n - 2b^n$.

34. If $a \propto b$, and $b \propto c$,

then $\qquad pa + qb + rc \propto p'\sqrt{ab} + q'\sqrt{bc} + r'\sqrt{ca}.$

35. If $a \propto b + c$, and $b \propto c + a$, then $a \propto b$.

[Turn the two equations of variation into absolute equations, and eliminate c.]

36. If $a \propto b + 5c$, and $b \propto 7c + 3a$, then $a \propto b$.

37. If $2a - 3b \propto b + 5a$, then $\dfrac{1}{3a^2} + \dfrac{1}{4b^2} \propto \dfrac{1}{2a^2} - \dfrac{1}{3b^2}.$

38. If $a \propto b^2$, $b^3 \propto c^4$, $c^5 \propto d^6$, and $d^7 \propto e^4$, show that $\dfrac{a}{e} \times \dfrac{b}{e} \times \dfrac{c}{e} \times \dfrac{d}{e}$ is constant.

39. There are three quantities; the first varies as x^2, the second inversely as x, and the third jointly as the first and the second. Show that the first varies jointly as the three.

40. If x, y, and z are variable but their sum is constant, and if $(x - y + z)(x + y - z) \propto yz$, prove that $yz \propto y + z - x$.

41. If $a^2 + b^2 \propto c^2 + d^2$, and $ab \propto cd$, does $a + b \propto c + d$?

42. If $z \propto x + y$, $u \propto x - y$, and $x \propto u + z$, then show that $y \propto x$ generally. What exception is there to the conclusion?

108. If A, B, C be variables such that $A \propto B$ when C is kept constant, and $A \propto C$ when B is constant, then $A \propto BC$ when both B and C vary.

Let A have the value a when B has the value b and C the value c.

Let A change from a to a_1, when B changes from b to b_1 and C remains constant and equal to c.

Let A change from a_1 to a_1' when B remains b_1 and C changes from c to c'.

Then by the definition of variation

$$\dfrac{a}{a_1} = \dfrac{b}{b_1} \ldots \text{(i)} \quad \text{and} \quad \dfrac{a_1}{a_1'} = \dfrac{c}{c'} \ldots \text{(ii)}.$$

84   VARIATION.

Multiplying (i) and (ii), we have $\dfrac{a}{a_1'} = \dfrac{bc}{b_1c'}$.

This shows that when BC changes from bc to b_1c', A changes from a to a quantity a_1', such that a, a_1', bc and b_1c' are proportionals.

Hence A varies as BC, that is A varies as B and C jointly (§ 96, Def.).

COR. 1. *If A varies separately as either of the quantities $B, C, D,\ldots$ when the rest are kept constant, it varies as their product when all are allowed to vary.*

For when of the variables $B, C, D\ldots$ only B varies, $A \propto B$; and when C alone varies, $A \propto C$.

∴ when B and C alone vary, $A \propto BC$.

And when D alone varies, $A \propto D$.

Hence when B, C, D alone vary, $A \propto BCD$. And so on.

COR. 2. If $A \propto B$ when C is constant, and $A \propto 1/C$ when B is constant, $A \propto B/C$ when B and C are variable.

For when C is constant, $1/C$ is constant, and therefore
$$A \propto B \cdot 1/C \text{ or } B/C.$$

Arithmetical illustration. The length of a journey varies as the rate of travelling when the time of performing the journey is constant, and varies as the time when the rate is constant. Hence the length varies jointly as the rate and the time, when both of them vary.

Geometrical illustration. The area of a triangle varies as the base when the altitude is constant, and varies as the altitude when the base is constant. Hence it varies jointly as the base and altitude, when both of them vary.

Ex. If 4 men earn £2 in 5 days, find what 16 men will earn in 30 days and verify from first principles that the amount earned varies jointly as the number of men and the number of days.

Here,   4 men in 5 days earn £2.
∴   16 men in 5 days earn £8.
∴   16 men in 30 days earn £48,
and since $2 : 48 = 4 \times 5 : 16 \times 30$,
the number of £s earned varies jointly as the number of men and the number of days.

VARIATION. 85

109. Problems in Variation. The preceding example affords an instructive illustration of the method of proof employed in § 108. Conversely the general theorem can be employed to solve questions on "compound proportion." In all such questions the simultaneous values of certain variables are given, and we are required to find what value one of them assumes when the rest have other given values.

Ex. 1. If 18 compositors of equal ability can set up 24 sheets in 8 days, how many sheets can 45 compositors of the same ability set up in 14 days?

Here, when the number of compositors is constant, the number of sheets $\propto$ the number of days.

And when the number of days is constant, the number of sheets $\propto$ the number of compositors.

Hence the number of sheets varies jointly as the number of compositors and the number of days, when both are variable.

Or calling these variables x, y, z, we have $x \propto yz$; and $\therefore x = kyz$, where k is constant.

Since 24, 18 and 8 are simultaneous values of x, y and z respectively, $24 = k \times 18 \times 8$.

Hence $k = \frac{1}{6}$, and $\therefore x = \frac{1}{6}yz$.

$\therefore$ when $y = 45$ and $z = 14$, $x = \frac{1}{6} \times 45 \times 14 = 105$.

Therefore the compositors can set up 105 sheets.

Ex. 2. If 10 men take 18 days to mow 60 acres of grass, how long will 15 men take to mow 80 acres?

The number of days varies directly as the number of acres when the number of men is constant.

And the number of days varies inversely as the number of men when the number of acres is constant.

Calling the three last-named variables x, y, z respectively we have therefore $x \propto z/y$; that is $x = kz/y$ where k is a constant.

Since 18, 10 and 60 are simultaneous values of x, y, z respectively,

$$18 = 60k/10 = 6k.$$

$\therefore k = 3$, which gives $x = 3z/y$.

Hence when $y = 15$ and $z = 80$, $x = 3 \times 80/15 = 16$.

Therefore the men will take 16 days.

EXERCISES VI.

43. If $x \propto 1/y$ when z is constant, and $x \propto 1/z$ when y is constant, prove that $x \propto 1/yz$ when both y and z are variable.

44. Given that $s \propto t^2$ when f is constant, and $s \propto f$ when t is constant, and also that $2s = f$ when $t = 1$, find the equation connecting f, s, and t.

45. If the volume of a pyramid varies as its base when its altitude is constant, and varies as its altitude when its base is constant, and if a pyramid the base of which is 4 ft. square and the height of which is 6 ft. contains 32 cubic feet, find the base of another pyramid of 30 cubic feet, the height of which is 10 ft.

46. If a garrison of 700 men have provisions for 6 weeks at 12 oz. for each man *per diem*, how many can the same quantity maintain for 12 weeks at 10 oz. per man *per diem*?

47. Ten English labourers can do as much in 6 days as 9 French labourers can do in 7 days; a Frenchman receives one franc per cubic metre excavated; how many pence must an Englishman receive per cubic yard excavated that his daily earnings may be 5 per cent. more than a Frenchman's? [A metre may be taken as equal to 399 inches and a franc to 10 pence.]

48. The value of a silver coin varies as the square of its diameter while its thickness remains the same, and varies as its thickness, while its diameter remains the same. Two silver coins have their diameters in the ratio of 4 : 3; find the ratio of their thicknesses if the value of the first be four times the value of the second.

49. If $A \propto B$ when C is constant, and $A \propto C$ when B is constant, prove that when $B \propto C$, then will $A \propto B^2$ or C^2.

Apply this property to compare the areas of two similar triangles.

50. The electric resistance of a wire of given material varies directly as the length and inversely as the square of the diameter. What must be the length and diameter of a wire which is to have double the resistance but only two-thirds the weight of a wire of the same material 100 feet long, and ·018 inch diameter?

CHAPTER VII.

EQUATIONS REDUCIBLE TO QUADRATICS.

ONE UNKNOWN QUANTITY.

101. FIRST METHOD. **Reduction by a change of the unknown quantity.**—Occasionally equations of a higher degree than the second happen to be reducible to quadratics owing to their peculiar forms.

Equations of the form $ax^{2n}+bx^n+c=0$ (in which there are only two powers of an unknown quantity, and the index of one of the powers is double that of the other) may always be solved by regarding them as quadratics in which the unknown quantity is x^n.

Ex. 1. Solve the equation $ax^{2n} + bx^n + c = 0$.
This may be treated as a quadratic equation in x^n.
$$\therefore x^n = \frac{-b \pm \sqrt{(b^2 - 4ac)}}{2a}.$$
$$\therefore x = \sqrt[n]{\frac{-b \pm \sqrt{(b^2 - 4ac)}}{2a}}.$$

Ex. 2. Solve $x^{-1} + x^{-\frac{1}{2}} = \frac{3}{4}$.
The equation being treated as a quadratic in $x^{-\frac{1}{2}}$,
we have $x^{-\frac{1}{2}} = \frac{-1 \pm \sqrt{(1+3)}}{2} = -\frac{3}{2}$ or $\frac{1}{2}$.
$$\therefore x^{\frac{1}{2}} = -\frac{2}{3} \text{ or } 2; \text{ and } x = \frac{4}{9} \text{ or } 4.$$

On substitution it will be found that the value $x = 4$ satisfies the equation $1/x + 1/\sqrt{x} = \frac{3}{4}$. The value $x = \frac{4}{9}$ does not do so but is the solution of the equation $1/x - 1/\sqrt{x} = \frac{3}{4}$ obtained by taking $x^{-\frac{1}{2}}$ with the negative sign.

EQUATIONS REDUCIBLE TO QUADRATICS.

Ex. 3. Solve $x^{\frac{1}{4}} + x^{\frac{1}{2}} = 6$.

The equation, by arrangement and transposition, becomes
$$x^{\frac{1}{2}} + x^{\frac{1}{4}} - 6 = 0,$$
and is a quadratic in $x^{\frac{1}{4}}$.

$$\therefore x^{\frac{1}{4}} = \frac{-1 \pm \sqrt{(1+24)}}{2} = -3 \text{ or } 2, \text{ and } x = 81 \text{ or } 16.$$

The value $x = 16$ satisfies $\sqrt{x} + \sqrt[4]{x} = 6$ but the value $x = 81$ is the solution of $\sqrt{x} - \sqrt[4]{x} = 6$.

Ex. 4. Solve $x^2 + \dfrac{1}{x^2} + x + \dfrac{1}{x} = 4$.

Adding 2 to both sides, we have $\left(x + \dfrac{1}{x}\right)^2 + \left(x + \dfrac{1}{x}\right) = 6$.

Solving as a quadratic, $x + \dfrac{1}{x} = \dfrac{-1 \pm \sqrt{(1+24)}}{2} = -3 \text{ or } 2$.

Hence the original equation reduces to the two quadratics

$$(a) \quad x + \frac{1}{x} = -3, \text{ and } (b) \quad x + \frac{1}{x} = 2.$$

Solving these quadratics we have

$$(a) \quad x = \frac{-3 \pm \sqrt{5}}{2} \text{ or } (b) \quad x = 1 \pm 0.$$

Ex. 5. Solve $\dfrac{5}{x^2 + 6x + 2} = \dfrac{3}{x^2 + 6x + 1} - \dfrac{4}{x^2 + 6x + 8}$.

Suppose $x^2 + 6x + 2 = y$.

Then $\dfrac{5}{y} = \dfrac{3}{y-1} - \dfrac{4}{y+6} = \dfrac{22-y}{y^2 + 5y - 6}$.

This is a quadratic in y, and gives $y = 2$ or $-2\frac{1}{2}$.

Hence the original equation reduces to the two quadratics

$$(a) \quad x^2 + 6x + 2 = 2, \text{ and } (b) \quad x^2 + 6x + 2 = -2\frac{1}{2}.$$

Thus we obtain 4 values for x, viz.,

$$(a) \quad 0 \text{ or} -6, \quad (b) \quad -3 \pm \tfrac{3}{2}\sqrt{2}.$$

NOTE. Another convenient method would be to assume
$$y = x^2 + 6x + 9 = (x+3)^2.$$

EQUATIONS REDUCIBLE TO QUADRATICS.

EXERCISES VIII.

Solve the following equations:

1. $x^4 - 12x^2 + 27 = 0$.
2. $x^{\frac{1}{4}} - x^{-\frac{1}{4}} = 1\frac{1}{2}$.
3. $x + 3\sqrt{(5x)} = 50$.
4. $x + \sqrt{x} = 6$.
5. $\dfrac{1}{y^4} - \dfrac{11}{5y^2} = -\dfrac{6}{5}$.
6. $8(x^{-3} + x^3) = 65$.
7. $x^2 + x^{-2} = a^2 + a^{-2}$.
8. $3\{(x+7)^{\frac{1}{2}} + (x+7)^{-\frac{1}{2}}\} = 10$.
9. $x^2 = 21 + \sqrt{(x^2 - 9)}$.
10. $\dfrac{x^2}{x^2 + 3x + 2} + \dfrac{2(x^2 + 3x + 2)}{x^2} = 12\frac{1}{6}$.
11. $\dfrac{5}{x^2 + 6x + 8} = \dfrac{1}{x^2 + 6x + 5} + \dfrac{4}{x^2 + 6x + 9}$.
12. $\left(x - \dfrac{1}{x}\right)^2 + 7\left(x - \dfrac{1}{x}\right) = 12\frac{3}{4}$.

111. The preceding equations may all be regarded as **quadratics in some function of** x, that is quadratics in which the unknown quantity is replaced by some expression involving x (such as a power of x). If this "function" or expression is put equal to y the equation transforms into a quadratic in y. To discover the proper transformation, the *form* in which x enters into the equation must be carefully noted.

Thus in p. 88, Ex. 5, x only occurs in the form $x^2 + 6x$ and this *should at once suggest* the transformation $y = x^2 + 6x$ or for greater convenience $y = x^2 + 6x + 2$.

112. Equations of the form
$$x^2 + bx + c\sqrt{\{a(x^2 + bx) + d\}} = e$$
may be solved by the substitution $y = \sqrt{\{a(x^2 + bx) + d\}}$.

For multiplying by a and adding d to both sides we have
$$ax^2 + abx + d + ac\sqrt{(ax^2 + abx + d)} = ae + d.$$

This being a quadratic in $\sqrt{(ax^2 + abx + d)}$, it will give us two values for that expression, and when the resulting quadratic equations are solved, we obtain four values for x. Only those values which make y positive satisfy the original equation, the others satisfy the equation $x^2 + bx - c\sqrt{\{a(x^2 + bx) + d\}} = e$.

EQUATIONS REDUCIBLE TO QUADRATICS.

Ex. 1. Solve $x^2 + 3\tfrac{1}{2}x - 4 = \sqrt{(16 - 7x - 2x^2)}$.

We shall reduce this to a quadratic equation in $\sqrt{(16 - 7x - 2x^2)}$.

By transposition, $x^2 + 3\tfrac{1}{2}x - 4 - \sqrt{(16 - 7x - 2x^2)} = 0$.

Multiplying by -2, we have
$$-2x^2 - 7x + 8 + 2\sqrt{(16 - 7x - 2x^2)} = 0.$$

Adding 8 to both sides we have
$$16 - 7x - 2x^2 + 2\sqrt{(16 - 7x - 2x^2)} = 8.$$

Let $\qquad y = \sqrt{(16 - 7x - 2x^2)}$,

then $\qquad\quad y^2 + 2y = 8$.

$\qquad\qquad \therefore y = -4$, or 2.

Hence the original equation is equal to the two quadratics

(*a*) $16 - 7x - 2x^2 = 16$, $\qquad$ (*b*) $16 - 7x - 2x^2 = 4$.

Hence $x = 0$, $-3\tfrac{1}{2}$, or $\tfrac{1}{4}\{-7 \pm \sqrt{145}\}$.

On making trial it will be found that the values $x = 0$ and $x = -3\tfrac{1}{2}$ do not sitisfy the given equation but are solutions of the equation
$$x^2 + 3\tfrac{1}{2}x - 4 = -\sqrt{(16 - 7x - 2x^2)}.$$

EXERCISES VII.

Solve the following equations :

13. $(2x^2 - 7x)^2 + x^2 = 3\tfrac{1}{2}x$. $\qquad$ 14. $(2x^2 - x - 3)^2 + x^2 = \tfrac{1}{2}(x + 3)$.

15. $3x - \sqrt{(2x^2 + 6x + 1)} = 1 - x^2$.

16. $x^2 - x + \sqrt{(x^2 - 7x + 8)} = 6x + 4$.

17. $2x^2 - 2x + 2\sqrt{(2x^2 - 7x + 6)} = 5x - 6$.

18. $x^2 - x + 5\sqrt{(2x^2 - 5x + 6)} = \tfrac{1}{2}(3x + 33)$.

19. $x^2 + 5x - 10 = \sqrt{(x^2 + 5x + 2)}$. $\quad$ 20. $ax^2 + \sqrt{(ax^2 - bx + c)} = bx$.

21. $x + \sqrt{(x^2 - ax + b^2)} = \dfrac{x^2}{a} + b$. $\quad$ 22. $\sqrt{(10 - x^2 - x)} = x^2 + x - 4$.

23. $\left(5x + \dfrac{5}{x} + 1\right)^2 + 2\left(5x + \dfrac{5}{x}\right) = 141$.

24. $5x + \sqrt{5x - \dfrac{5}{x}} = \dfrac{5}{x}$. $\qquad$ 25. $\left(x + \dfrac{1}{x}\right)^2 + 3\left(x + \dfrac{1}{x}\right) = 10$.

26. $\left(x - \dfrac{1}{x}\right)^2 + 2\left(x + \dfrac{1}{x}\right) = 7\tfrac{1}{4}$.

EQUATIONS REDUCIBLE TO QUADRATICS. 91

113. Equations of the form
$$(x+a)(x+b)(x+c)(x+d) = e$$
may, by a change of the unknown quantity, be reduced to a quadratic and solved, if the sum of any two of the quantities a, b, c, d be equal to the sum of the other two.

For suppose $a+b = c+d$.

Then the given equation becomes
$$\{x^2 + x(a+b) + ab\}\{x^2 + x(c+d) + cd\} = e,$$
which may be solved by putting either
$$y = x^2 + x(a+b) = x^2 + x(c+d),$$
or $y = \{x + \tfrac{1}{2}(a+b)\}^2$ giving $\{y - \tfrac{1}{4}(a-b)^2\}\{y - \tfrac{1}{4}(c-d)^2\} = e$.

Ex. 1. Solve $(x+3)(x+5)(x-2)(x-4) = 120$.

Since $3 + (-2) = 5 + (-4)$, the equation can be solved by the present method.

Now, $\qquad (x^2 + x - 6)(x^2 + x - 20) = 120$.

If $y = x^2 + x - 6$, $\quad y(y - 14) = 120$. $\quad \therefore\; y = 20$ or -6.

Thus the given equation is equal to the two quadratics

$\qquad$ (a) $\quad x^2 + x - 6 = 20\quad$ and $\quad$ (b) $\quad x^2 + x - 6 = -6$.

Hence $\qquad x = \tfrac{1}{2}\{-1 \pm \sqrt{(105)}\}$, 0, or -1.

Ex. 2. Solve $\dfrac{5}{x^2 - 5x + 4} + \dfrac{5}{x^2 - 11x + 28} = x^2 - 8x + 15$.

Resolving the compound expressions into factors, we have
$$\frac{5}{(x-4)(x-1)} + \frac{5}{(x-7)(x-4)} = (x-5)(x-3).$$
$$\therefore\; \frac{5(2x-8)}{(x-4)(x-1)(x-7)} = (x-5)(x-3).$$
$$\therefore\; \frac{10}{(x-1)(x-7)} = (x-5)(x-3);$$
whence $\qquad (x^2 - 8x + 7)(x^2 - 8x + 15) = 10$.

Let $\qquad y = (x-4)^2 = x^2 - 8x + 16$.

Then $\qquad (y-9)(y-1) = 10$, that is $y^2 - 10y = 1$.

$\therefore\; y = 5 \pm \sqrt{26}\;$ and $\; x = 4 \pm \sqrt{y} = 4 \pm \sqrt{(5 \pm \sqrt{26})}$.

EXERCISES VII.

Solve the following equations:

27. $(x-2)(x+3)(x+6)(x+1)+56=0$.
28. $(x+1)(x+3)(x+5)(x+7)=9$.
29. $(2x+a)(2x+5a)(3x+8a)(3x+a)=30a^4$.
30. $\dfrac{1}{6x^2-7x+2}+\dfrac{1}{12x^2-17x+6}=8x^2-6x+1$.
31. $\dfrac{1}{x^2-7x+12}+\dfrac{1}{x^2-9x+20}=\dfrac{2}{9}(x^2-1)$.
32. $\dfrac{1}{3x^2-7x+4}+\dfrac{2}{3x^2-10x+8}=\dfrac{1}{4}(x^2+3x+2)$.

*114. Equations of the form

$$(x+a)^4+(x+b)^4=d$$

can be easily solved as in the following example.

Ex. Solve $(x+1)^4+(x+5)^4=82$.

Let y stand for the semi-sum of $x+1$ and $x+5$; that is, for $x+3$.
Then $\qquad (y-2)^4+(y+2)^4=82$.
By simplification, $\quad 2y^4+48y^2+32=82$.
Hence $\qquad y^4+24y^2-25=0$.

$$\therefore y^2=\frac{-24\pm\sqrt{(576+100)}}{2}=1 \text{ or } -25.$$

Thus $\qquad y=\pm 1 \text{ or } \pm 5\sqrt{-1}$.
$\therefore x=y-3=-2,\ -4,\ \text{or}\ -3\pm 5\sqrt{-1}$.

EXERCISES VII.

Solve the following equations:

33. $(x+1)^4+x^4=1$.
34. $(x+1)^4+(x-3)^4=256$.
35. $(1-x)^4+(1+x)^4=82$.
36. $(x+1)^5+(x+3)^5=0$.
37. $(x+1)^5+(x-3)^5=0$.
38. $(x+2)^6+(x+4)^6=2$.

EQUATIONS REDUCIBLE TO QUADRATICS.

115. Second Method. Reduction by Rationalisation.—This method is applicable to **equations involving surds,** i.e. equations where the unknown quantity occurs under a radical.

Equations of this class may generally be rationalised by transposition and squaring, which must be repeated until they are cleared of radicals.

Ex. Solve $\sqrt{(x+1)} + \sqrt{(x-2)} = \sqrt{(3x)}$.
On squaring, $(x+1) + (x-2) + 2\sqrt{\{(x+1)(x-2)\}} = 3x$.
By transposition, $2\sqrt{\{(x+1)(x-2)\}} = x+1$.
By squaring and simplification, $x^2 - 2x - 3 = 0$.
$$\therefore x = \frac{2 \pm \sqrt{(4+12)}}{2} = \frac{2 \pm 4}{2} = 3 \text{ or } -1.$$

Caution. *The result of the first operation of squaring the left-hand side is not* $(x+1) + (x-2)$ *but* $(x+1) + (x-2) + 2\sqrt{\{(x+1)(x-2)\}}$, since the square of $a+b$ is not $a^2 + b^2$ but $a^2 + 2ab + b^2$. The new equation contains one radical instead of three. To remove this we must *transpose the rational terms to the opposite side* before again squaring.

116. Extraneous Solutions. The values of x obtained by clearing of radicals are *not all necessarily solutions of the original equation,* and trial must be made to determine which of them actually do satisfy it. In making the trial, it must be remembered that square roots (unless otherwise specified) are considered to be positive. Hence any values of x which arise from negative values of *even* roots are to be rejected.

Ex. Consider the equations :
  (a) $\sqrt{(x-p)} + \sqrt{(x-q)} = \sqrt{(2x-p-q)}$.
  (b) $\sqrt{(x-p)} - \sqrt{(x-q)} = \sqrt{(2x-p-q)}$.
  (c) $\sqrt{(x-p)} - \sqrt{(x-q)} = -\sqrt{(2x-p-q)}$.
  (d) $\sqrt{(x-p)} + \sqrt{(x-q)} = -\sqrt{(2x-p-q)}$.

Each of these equations, on being solved as in § 115 Ex. above, gives p and q for its roots. But on verification, we find that whereas the first equation is satisfied by both p and q, the second is satisfied only by q, the third only by p, while the fourth is satisfied neither by p nor by q. But p and q are obtained as the roots of all the four equations, and we naturally inquire what the explanation of this discrepancy may be.

117. *If a rational equation be derived from a given irrational equation, then every solution of the irrational equation must satisfy the derived rational equation but no solution of the derived equation need satisfy the original equation.*

The student will remember the following identity,
$$(\sqrt{a}+\sqrt{b}+\sqrt{c})(\sqrt{a}+\sqrt{b}-\sqrt{c})(\sqrt{a}-\sqrt{b}+\sqrt{c})(\sqrt{a}-\sqrt{b}-\sqrt{c})$$
$$=a^2+b^2+c^2-2ab-2ac-2bc.$$

This shows that if either of the four equations
$$\sqrt{a}+\sqrt{b}+\sqrt{c}=0, \quad \sqrt{a}+\sqrt{b}-\sqrt{c}=0,$$
$$\sqrt{a}-\sqrt{b}+\sqrt{c}=0, \quad \sqrt{a}-\sqrt{b}-\sqrt{c}=0,$$
be rationalised, the resulting equation will be
$$a^2+b^2+c^2-2ab-2ac-2bc=0.$$

This is satisfied if *either* of its factors be equal to zero, **and it does not necessarily follow that this factor was the one that we** originally started with. In fact the first factor can never vanish.

EXERCISES VII.

Solve the following equations, and by verification point out in each case the extraneous solutions (*i.e.*, solutions not satisfying the equations).

39. $5\sqrt{(x-3)} + 2\sqrt{(x+1)} = \sqrt{(x+13)}$.
40. $4\sqrt{(x-1)} - \sqrt{(x+4)} = \sqrt{(x+20)}$.
41. $3\sqrt{x} + 2\sqrt{(5-x)} = 8$.
42. $\sqrt{(x-1)} + 7\sqrt{x} = \sqrt{(13x+36)}$.
43. $\sqrt{(x+1)} - 4\sqrt{(x-4)} + 5\sqrt{(x-7)} = 0$.
44. $4\sqrt{(x+2)} - \sqrt{(x+7)} - 5\sqrt{(x-1)} = 0$.
45. $\sqrt{(1+x^2)} + \sqrt{(4+x^2)} + \sqrt{(9+x^2)} = \sqrt{(14+x^2)}$.

118. Sometimes the rationalisation of an equation is simplified by the removal of a common factor.

Ex. 1. Solve $2\sqrt{(x^2-9x+18)} - \sqrt{(x^2-4x-12)} = x-6$.

Resolving the expressions within the roots into factors, we have
$$2\sqrt{\{(x-3)(x-6)\}} - \sqrt{\{(x+2)(x-6)\}} = \sqrt{\{(x-6)(x-6)\}}.$$

EQUATIONS REDUCIBLE TO QUADRATICS.

One solution is evidently $x - 6 = 0$ or $x = 6$.

Dividing by the common factor $\sqrt{(x-6)}$ we obtain
$$2\sqrt{(x-3)} - \sqrt{(x+2)} = \sqrt{(x-6)}.$$
By squaring, $4(x-3) + x + 2 - 4\sqrt{\{(x-3)(x+2)\}} = x - 6$.

By transposition and squaring, $16(x+2)(x-3) = 16(x-1)^2$.

This equation happens to be of the first degree, and gives 7 as another solution. $\therefore x = 6$ or 7, both of which satisfy the original equation.

119. In many instances an irrational equation may be rationalised by employing an identity.

Ex. $\qquad \sqrt{(x^2 - 5x + 20)} + \sqrt{(x^2 - 5x + 4)} = 4$ ....(i).

[N.B. The equation can also be solved by putting y for $x^2 - 5x$.]

Multiply both sides by the conjugate surd, viz.
$$\sqrt{(x^2 - 5x + 20)} - \sqrt{(x^2 - 5x + 4)}.$$
$\therefore (x^2 - 5x + 20) - (x^2 - 5x + 4)$
$\qquad = 4\{\sqrt{(x^2 - 5x + 20)} - \sqrt{(x^2 - 5x + 4)}\}$......(ii).

Now $\qquad (x^2 - 5x + 20) - (x^2 - 5x + 4) = 16$...............(iii).

[As 16 is the difference of $x^2 - 5x + 20$ and $x^2 - 5x + 4$, (iii) is true for all values of x and is therefore an identity.]

Substituting and dividing by 4, we have
$$\sqrt{(x^2 - 5x + 20)} - \sqrt{(x^2 - 5x + 4)} = 16/4 = 4 \text{ ...... (iii)}.$$
But $\qquad \sqrt{(x^2 - 5x + 20)} + \sqrt{(x^2 - 5x + 4)} = 4$ ....(i).

By subtraction, $2\sqrt{(x^2 - 5x + 4)} = 0$. $\therefore x^2 - 5x + 4 = 0$.

The roots of this equation are therefore 4 and 1, both of which satisfy (i).

EXERCISES VII.

Solve the following equations, and point out the extraneous roots.

46. $\sqrt{(2x^2 + 7x - 9)} - \sqrt{(x^2 - 5x + 4)} = \sqrt{(x^2 - 1)}$.

47. $5 - x + \sqrt{(30 - 11x + x^2)} = \sqrt{(55 - 21x + 2x^2)}$.

48. $\sqrt{(3x + x^2)} + \sqrt{(3x - x^2)} - 2x = 0$.

EQUATIONS REDUCIBLE TO QUADRATICS.

49. $\sqrt{(5x^2 - 9x + 4)} + \sqrt{(2x^2 - 2x)}$
$\qquad = \sqrt{(5x^2 - 8x + 3)} + \sqrt{(2x^2 - 3x + 1)}$.
50. $\sqrt{(2x^2 + 10x + 8)} - \sqrt{(x^2 + 6x + 5)} = \sqrt{(x + 1)}$.
51. $\sqrt{(x^2 - 5x + a)} + \sqrt{(x^2 - 5x + b)} = \sqrt{a} + \sqrt{b}$.
52. $\sqrt{(x^2 - ax + 49)} - \sqrt{(x^2 - ax + 16)} = 3$.
53. $\sqrt{\{a(a + b + x)\}} + \sqrt{\{a(a + b - x)\}} = x$.
54. $\sqrt{(x^2 + 5x + 2)} + \sqrt{(x^2 + 15x + 2)} = 10$.
55. $\sqrt{(x^2 + 5x + 2)} + \sqrt{(x^2 + 15x + 2)} = 5x$.
56. $\sqrt{(ax + b^2)} + \sqrt{(bx + a^2)} = a - b$.

120. Where surd expressions occur in the denominators of fractions the equations may be simplified by rationalising these denominators (§§ 29-32).

Ex. 1. Solve $\dfrac{1}{\sqrt{(x+4)} - \sqrt{x}} - \dfrac{1}{\sqrt{(x-4)} + \sqrt{x}} = \sqrt{(x-4)}$.

Rationalising the denominators, we have
$$\frac{\sqrt{(x+4)} + \sqrt{x}}{(x+4) - x} - \frac{\sqrt{(x-4)} - \sqrt{x}}{(x-4) - x} = \sqrt{(x-4)}.$$

Hence $\sqrt{(x+4)} + \sqrt{x} + \sqrt{(x-4)} - \sqrt{x} = 4\sqrt{(x-4)}$.
$\qquad \therefore \sqrt{(x+4)} = 3\sqrt{(x-4)}$.

By squaring, $\quad x + 4 = 9(x - 4), \quad \therefore 8x = 40$.

Hence $\qquad\qquad x = 5$.

Ex. 2. Solve $\dfrac{x-7}{\sqrt{(x-3)} - 2} + \dfrac{x-5}{\sqrt{(x-4)} - 1} = 4\sqrt{(x-3)}$.

We have $\dfrac{(x-3) - 4}{\sqrt{(x-3)} - 2} + \dfrac{(x-4) - 1}{\sqrt{(x-4)} - 1} = 4\sqrt{(x-3)}$.

Dividing the denominators into the numerators,
$$\sqrt{(x-3)} + 2 + \sqrt{(x-4)} + 1 = 4\sqrt{(x-3)}.$$

By transposition, squaring and simplification, we get
$$(x-4)(16x - 73) = 0.$$

Hence $x = 4$ or $4\tfrac{9}{16}$, both of which satisfy the equation.

EXERCISES VII.

Solve the following equations, and point out the extraneous roots.

57. $\dfrac{1}{\sqrt{(2+x)}-\sqrt{2}} + \dfrac{1}{\sqrt{(2-x)}+\sqrt{2}} = \dfrac{\sqrt{2}}{x}$.

58. $\dfrac{1}{\sqrt{(a-x)}+\sqrt{a}} + \dfrac{1}{\sqrt{(a+x)}-\sqrt{a}} = \dfrac{\sqrt{a}}{x}$.

59. $\dfrac{1}{\sqrt{(x^2-1)}-x} + \dfrac{1}{\sqrt{(x^2-1)}+x} = 3$.

60. $\dfrac{\sqrt{(x^2+5x)}+\sqrt{(x^2-3x)}}{\sqrt{(x^2+5x)}-\sqrt{(x^2-3x)}} + \dfrac{\sqrt{(x^2+5x)}-\sqrt{(x^2-3x)}}{\sqrt{(x^2+5x)}+\sqrt{(x^2-3x)}} = 2\tfrac{1}{2}$.

61. $\dfrac{x}{\sqrt{(x+p)}-\sqrt{p}} + \dfrac{x-2p}{\sqrt{(x-p)}+\sqrt{p}} = \sqrt{(2x+p)}$.

62. $\dfrac{5}{\sqrt{(x+5)}-\sqrt{x}} + \dfrac{12}{\sqrt{(x+12)}+\sqrt{x}} = \sqrt{(9x+13)}$.

63. $\dfrac{a}{\sqrt{(x+a)}-\sqrt{x}} + \dfrac{b}{\sqrt{(x+b)}+\sqrt{x}} = \sqrt{(x+a+b)}$.

121. THIRD METHOD. **Reduction by Factorisation.**

When one or more of the roots is given and the others are required the equation can always be reduced, as in the following examples.

Ex. 1. Solve $x^3 - 15x^2 + 74x = 120$, having given that $x = 4$ is one of the roots.

Since 4 is one of the roots of the equation

$$x^3 - 15x^2 + 74x - 120 = 0,$$

the left-hand side vanishes when $x = 4$; it must therefore be exactly divisible by $x - 4$.

$$\therefore\ x^2(x-4) - 11x(x-4) + 30(x-4) = 0.$$

That is, $(x-4)(x^2 - 11x + 30) = 0$.

Hence the two other roots are given by $x^2 - 11x + 30 = 0$.

Solving this quadratic we have

$$x = \dfrac{11 \pm \sqrt{(121-120)}}{2} = 6\ \text{or}\ 5.$$

The three roots are therefore 4, 5, and 6.

EQUATIONS REDUCIBLE TO QUADRATICS.

Ex. 2. Solve $x^4 - 5x^3 + 4x^2 + 8x - 8 = 0$,
having given that two* of the roots are $1 + \sqrt{5}$ and $1 - \sqrt{5}$.

We have $(x - 1 - \sqrt{5})(x - 1 + \sqrt{5}) = (x-1)^2 - (\sqrt{5})^2 = x^2 - 2x - 4$.

Dividing $x^4 - 5x^3 + 4x^2 + 8x - 8 = 0$ by $x^2 - 2x - 4$,
we have $x^2 - 3x + 2 = 0$. Hence the remaining roots are 1 and 2.

EXERCISES VII.

Solve completely the following equations having the given values for one or more of their roots :

64. $x^3 - x^2(5 + a) + x(6 + 5a) - 6a = 0.$ $x = 2.$
65. $2x^3 - 3x^2(a + 2) + ax(a + 9) = 3a^2.$ $x = \tfrac{1}{2}a.$
66. $x^2(x + 4) + (2x + 1)(x + 4) = 2.$ $x = -2.$
67. $(x - 4)^2 + 2(x - 4) = 2x^{-1} - 1.$ $x = 2 + \sqrt{3},$ or $2 - \sqrt{3}.$
68. $(x^2 - 5)^2 = 2(x^2 - 2x + 5).$ $x = \sqrt{6} - 1,$ or $-(\sqrt{6} + 1).$
69. $9x^3 - 6 = 13x.$ $x = \tfrac{1}{3}(1 \pm \sqrt{10}).$

122. *When one or more roots can be found by inspection* we proceed in the same way.

Ex. 1. Solve $x(x + 1)(x + 2) = m(m + 1)(m + 2)$.

Here $x = m$ is obviously a root. Hence $x - m$ is a factor of
$$x(x + 1)(x + 2) - m(m + 1)(m + 2),$$
and we find
$$x(x + 1)(x + 2) - m(m + 1)(m + 2)$$
$$= x^3 - m^3 + 3(x^2 - m^2) + 2(x - m)$$
$$= (x - m)\{x^2 + xm + m^2 + 3(x + m) + 2\} = 0.$$

Hence $x^2 + x(m + 3) + (m^2 + 3m + 2) = 0$, a quadratic equation whose roots are
$$x = \tfrac{1}{2}\{-(m + 3) \pm \sqrt{(1 - 6m - 3m^2)}\}.$$

---

* It can be proved that when $a + \sqrt{b}$ is a root of an equation with rational coefficients, $a - \sqrt{b}$ is also a root ; and that when $a + \sqrt{-b}$ is a root of an equation with real coefficients, $a - \sqrt{-b}$ is also a root.

EQUATIONS REDUCIBLE TO QUADRATICS.

Ex. 2. Solve $(a-x)^3 + (b-x)^3 = (a+b-2x)^3$.

By inspection we find a and b to be two of the roots of this equation. Hence $(x-a)(x-b)$ is a factor of
$$(a-x)^3 + (b-x)^3 - (a+b-2x)^3,$$
and we easily find
$$(a-x)^3 + (b-x)^3 - (a+b-2x)^3 = -3(a+b-2x)(a-x)(b-x).$$

Hence the third root is given by $a+b-2x = 0$ or $x = \tfrac{1}{2}(a+b)$.

The question can be solved otherwise by the following method :—
The equation is
$$(a-x)^3 + (b-x)^3 = (a+b-2x)^3 = \{(a-x)+(b-x)\}^3$$
$$= (a-x)^3 + 3(a-x)^2(b-x) + 3(a-x)(b-x)^2 + (b-x)^3.$$

Hence $\qquad 3(a-x)^2(b-x) + 3(a-x)(b-x)^2 = 0.$

Thus $\qquad 3(a-x)(b-x)(a-x+b-x) = 0.$

The roots of the equation are therefore a, b, and $\tfrac{1}{2}(a+b)$.

Ex. 3. Solve $x^3 - 15x^2 + 62x - 72 = 0$.

A root may sometimes be obtained by guessing, and $x = 2$ is thus found to be a solution of this equation. [The remainder-theorem and the theorem for the sum of the coefficients in a product will be found useful in this connection.]

Hence $(x-2)(x^2 - 13x + 36) = 0$, which gives $x = 2, 4,$ or 9.

EXERCISES VII.

Solve the following equations :

70. $x^2 + \dfrac{a}{x} = b^2 + \dfrac{a}{b}$.

71. $\dfrac{x^2}{10} + \dfrac{7}{x} = 7\tfrac{1}{10}$.

72. $(x-m)(x-n)(x-p) = (n+p)(p+m)(m+n)$.

73. $(x-p-q)(x-q-r)(x-r-p) = pqr$.

74. $(x+a)(x+b)(x+c) - abc = 0$.

75. $(x-a)^2 + (x-b)^2 = (2x-a-b)^2$.

76. $(x-3)^3 + (x-4)^3 = (2x-7)^3$.

77. $(x-a)^3 + (b-x)^3 = (b-a)^3$.

78. $\sqrt[3]{(7-x)} + \sqrt[3]{(x-5)} = \sqrt[3]{2}$.

79. $x^3 - 10x^2 + 31x = 30$.

80. $x^4 + 10x^3 + 33x^2 + 40x + 12 = 0$.

123. Fourth Method. Reduction of Reciprocal Equations.

DEFINITION.—**Reciprocal Equations** are equations in which the coefficients of terms equidistant, respectively, from the highest and lowest terms are equal.

Such equations are called *reciprocal* since they are not altered when $1/x$ is written for x.

Hence the roots of reciprocal equations occur in pairs, those of any pair being reciprocals of each other.

Thus, $ax^3 + bx^2 + bx + a = 0$, $ax^3 - bx^2 - bx + a = 0$,
$ax^4 + bx^3 + cx^2 + bx + a = 0$, $ax^4 - bx^3 + cx^2 - bx + a = 0$,
are reciprocal equations.

Substituting $1/x$ for x in $ax^4 - bx^3 + cx^2 - bx + a = 0$,
$$\frac{a}{x^4} - \frac{b}{x^3} + \frac{c}{x^2} - \frac{b}{x} + a = 0;$$
this gives us, by simplification, $a - bx + cx^2 - bx^3 + ax^4 = 0$, which is identical with the original equation.

Ex. 1. Solve $2x^3 - 3x^2 - 3x + 2 = 0$.

We have $\quad 2(x^3 + 1) - 3x(x + 1) = 0$.
$\therefore (x + 1)\{2(x^2 - x + 1) - 3x\} = 0$.
$\therefore x = -1, 2$ and $\tfrac{1}{2}$. Of the three roots, 2 and $\tfrac{1}{2}$ are reciprocals, and -1 is its own reciprocal.

Ex. 2. Solve $ax^4 - bx^3 + cx^2 - bx + a = 0$.

Dividing the equation by x^2, we have $ax^2 - bx + c - \dfrac{b}{x} + \dfrac{a}{x^2} = 0$.
$$\therefore a\left(x^2 + \frac{1}{x^2}\right) - b\left(x + \frac{1}{x}\right) + c = 0.$$

To bring this to the form of a quadratic, add and subtract $2a$.
$$\therefore a\left(x^2 + \frac{1}{x^2} + 2\right) - b\left(x + \frac{1}{x}\right) + c - 2a = 0.$$

Putting $y = x + \dfrac{1}{x}$, we have $ay^2 - by + c - 2a = 0$.
$$\therefore y = \frac{b \pm \sqrt{(b^2 - 4ac + 8a^2)}}{2a} = m \text{ or } n \text{ (suppose)}.$$

EQUATIONS REDUCIBLE TO QUADRATICS. 101

Thus the given equation has four roots, which are those of the two quadratics, $x + \dfrac{1}{x} = m$, and $x + \dfrac{1}{x} = n$.

Ex. 3. Solve $6x^5 + 41x^4 + 97x^3 + 97x^2 + 41x + 6 = 0$.
Here $\quad 6(x^5 + 1) + 41x(x^3 + 1) + 97x^2(x + 1) = 0$.
Hence $(x + 1)\{6(x^4 - x^3 + x^2 - x + 1) + 41x(x^2 - x + 1) + 97x^2\} = 0$.
Hence either $x = -1$ or $6x^4 + 35x^3 + 62x^2 + 35x + 6 = 0$, which is another reciprocal equation.

Solving this equation, as in Ex. 2 we get
$$6x^2 + 35x + 62 + \frac{35}{x} + \frac{6}{x^2} = 0.$$

Hence $\quad 6\left(x^2 + \dfrac{1}{x^2}\right) + 35\left(x + \dfrac{1}{x}\right) + 62 = 0$.

Putting $\quad y = x + \dfrac{1}{x}, \quad x^2 + \dfrac{1}{x^2} = y^2 - 2$.

$\therefore\ 6y^2 + 35y + 50 = 0$ which gives $y = -2\frac{1}{2}$ or $-3\frac{1}{3}$.

$\therefore$ either $\quad x + \dfrac{1}{x} = -2\frac{1}{2}$ giving $x = -2$ or $-\frac{1}{2}$;

or $\quad x + \dfrac{1}{x} = -3\frac{1}{3}$ giving $x = -3$ or $-\frac{1}{3}$.

x has thus five values $-1, -2, -\frac{1}{2}, -3, -\frac{1}{3}$.

*124. Equations of such forms as
$$ax^4 + bx^3 + cx^2 - bx + a = 0,$$
in which coefficients of corresponding *odd* powers of x are equal but of opposite sign, may be regarded as reciprocal and solved by putting $y = x - 1/x$. The roots occur in pairs, each pair being reciprocals of one another but of opposite signs.

Again equations of *odd* degree such as
$$ax^3 - bx^2 + bx - a = 0, \quad ax^3 + bx^2 - bx - a = 0,$$
$$ax^5 - bx^4 + cx^3 - cx^2 + bx - a = 0$$
are satisfied by $x = 1$ and they reduce to ordinary reciprocal equations on dividing by $x - 1$.

EQUATIONS REDUCIBLE TO QUADRATICS.

Ex. 1. Solve $6x^4 + 7x^3 - 36x^2 - 7x + 6 = 0$.

Here $\quad 6\left(x^2 + \dfrac{1}{x^2}\right) + 7\left(x - \dfrac{1}{x}\right) - 36 = 0$.

Putting $y = x - \dfrac{1}{x}$ we get $6y^2 + 7y - 24 = 0$ whence $y = -\dfrac{8}{3}$ or $\dfrac{3}{2}$,
and the corresponding values of x will be found to be $-3, \tfrac{1}{3}, 2, -\tfrac{1}{2}$.

*125. Equations of such forms as
$$ax^4 + bkx^3 + ck^2x^2 + bk^3x + ak^4 = 0$$
reduce at once to reciprocal equations in x/k on being divided throughout by k^4.

EXERCISES VII.

Solve the following equations:

81. $ax^3 + bx^2 + bx + a = 0.$ 82. $5x^3 + 31x^2 + 31x + 5 = 0.$
83. $x^5 - 4x^4 + 3x^3 + 3x^2 - 4x + 1 = 0.$
84. $x^4 + x^3 - 4x^2 + x + 1 = 0.$
85. $14x^4 - 135x^3 + 278x^2 - 135x + 14 = 0.$
86. $3x^4 - 20x^3 - 94x^2 - 20x + 3 = 0.$
87. $x^5 + x^4 - 2x^3 - 2x^2 + x + 1 = 0.$
88. $8x^4 - 42x^3 + 29x^2 + 42x + 8 = 0.$
89. $12x^4 + 77x^3 + 96x^2 - 77x + 12 = 0.$

126. Fifth Method. Reduction by Proportionals. In this article we shall give a few examples showing how the properties of proportionals may be sometimes utilised to shorten the work of solving an equation.

Ex. Solve $\sqrt{\dfrac{1+bx}{1-bx} \times \dfrac{1-ax}{1+ax}} = 1$.

We have $\sqrt{\dfrac{1+bx}{1-bx}} = \dfrac{1+ax}{1-ax}$. $\therefore \dfrac{1+bx}{1-bx} = \dfrac{1+2ax+a^2x^2}{1-2ax+a^2x^2}$.

$\therefore \dfrac{(1+bx)-(1-bx)}{(1+bx)+(1-bx)} = \dfrac{(1+2ax+a^2x^2)-(1-2ax+a^2x^2)}{(1+2ax+a^2x^2)+(1-2ax+a^2x^2)}$.

EQUATIONS REDUCIBLE TO QUADRATICS.

$$\therefore \frac{bx}{1} = \frac{2ax}{1+a^2x^2}.$$

$\therefore x = 0$; or $b(1+a^2x^2) = 2a$, which gives $x = \pm\sqrt{\left(\dfrac{2}{ab} - \dfrac{1}{a^2}\right)}$.

Ex. 2. Solve $(1+x)^{\frac{2}{n}} - (1-x)^{\frac{2}{n}} = (1-x^2)^{\frac{1}{n}}$.

Dividing all the terms by $(1-x)^{\frac{2}{n}}$, we have

$$\left(\frac{1+x}{1-x}\right)^{\frac{2}{n}} - 1 = \frac{(1-x^2)^{\frac{1}{n}}}{(1-x)^{\frac{2}{n}}} = \left(\frac{1+x}{1-x}\right)^{\frac{1}{n}} = y \text{ (suppose)}.$$

Thus $y^2 - 1 = y$; which gives $y = \dfrac{1 \pm \sqrt{5}}{2}$.

$$\therefore \frac{1+x}{1-x} = \frac{(1 \pm \sqrt{5})^n}{2^n}. \quad \therefore x = \frac{(1 \pm \sqrt{5})^n - 2^n}{(1 \pm \sqrt{5})^n + 2^n}.$$

EXERCISES VII.

Solve the following equations:

90. $(1+2x)^{\frac{1}{2}} - 2(1-2x)^{\frac{1}{2}} = (1-4x^2)^{\frac{1}{4}}$.

91. $\dfrac{\sqrt{(2x^2+3x+2)} - \sqrt{(x^2-3x+6)}}{\sqrt{(2x^2+3x+2)} + \sqrt{(x^2-3x+6)}} = \dfrac{\sqrt{(2x+12)} - \sqrt{(x+2)}}{\sqrt{(2x+12)} + \sqrt{(x+2)}}$.

92. $x^4 - 5x^2 = 2x - 3$.

93. $x\sqrt{\left(\dfrac{6}{x} - x\right)} = \dfrac{1+x^2}{\sqrt{x}}$.

94. $(x+4)(x+1) + \sqrt{\{(x+5)(x-3)\}} = 3x + 31$.

95. $\sqrt{(1-x+x^2)} - \sqrt{(1+x+x^2)} = m$.

96. $2^{x+1} + 2^{2x} = 8$.*   97. $4^{x+1} + 2^{4x+2} = 8$.

98. $4^{x+1} + 2^{4x+2} = 80$.   99. $2^{x^2} : 2^{3x} = 16 : 1$.

---

* Equations in which the unknown quantity appears in the index are instances of what are called *transcendental equations*.
In Ex. 96 put $2^x = y$ then $2y + y^2 = 8$.

CHAPTER VIII.

SIMULTANEOUS QUADRATIC EQUATIONS.

TWO UNKNOWN QUANTITIES.

127. CASE I. When one of the equations is linear.

In dealing with equations involving two unknown quantities, we first take the case in which, of the two given equations, one is of the first degree and the other is quadratic. Such equations can always be solved by the following

RULE. *From the simple equation, express* x *in terms of* y *and the known quantities; and substitute this value in the other equation. Thus we obtain a quadratic equation in* y.

In this way any two equations can be solved which are of the forms

$$ax + by = c\ ;\ mx^2 + nxy + py^2 + qx + ry + s = 0.$$

Ex. 1. Solve $5x + 6y = 11$; $3x^2 + 4y^2 = 7$.

Since $5x + 6y = 11$, therefore $x = \frac{1}{5}(11 - 6y)$.

Hence, $3x^2 + 4y^2 = \frac{3}{25}(11 - 6y)^2 + 4y^2 = 7$.

By simplification, $3(11 - 6y)^2 + 100y^2 = 175$.

$\therefore 208y^2 - 396y + 363 = 175$.

From this quadratic equation we get $y = 1$ or $\frac{47}{52}$.

And since $x = \frac{1}{5}(11 - 6y)$, the corresponding values of x are 1 and $1\frac{3}{26}$ respectively.

Thus the roots of the given equations are
$$x = 1,\ y = 1\ ;\ \text{or}\ x = 1\tfrac{3}{26},\ y = \tfrac{47}{52}.$$

SIMULTANEOUS QUADRATIC EQUATIONS.

[NOTE. In stating the results the student must be careful *to arrange the values of* x, y *in pairs*. It should be noted that $x = 1$, $y = \frac{47}{52}$ will satisfy neither of the given equations.]

Ex. 2. Solve $x + 2y = 1$; $x^2 + xy + 2y^2 + 3x + 4y = 4$.

From the first equation $x = 1 - 2y$.

Substituting this value in the second equation, we have
$$(1 - 2y)^2 + (1 - 2y)y + 2y^2 + 3(1 - 2y) + 4y = 4.$$
Hence $4y^2 - 5y + 4 = 4$, which gives $y = 0$, or $1\frac{1}{4}$.

And, since $x = 1 - 2y$, its corresponding values are $x = 1$ or $-1\frac{1}{2}$.

Hence the roots of the given equations are
$$x = 1,\ y = 0;\ \text{or}\ x = -1\frac{1}{2},\ y = 1\frac{1}{4}.$$

Ex. 3. Solve $x^2 - y^2 = 1$; $x^4 - 2y^4 = 1$.

If $x^2 = m$, and $y^2 = n$, the equations become
$$m - n = 1,\ \text{and}\ m^2 - 2n^2 = 1.$$

From the first equation, we have $m = 1 + n$.

Hence $\qquad (1 + n)^2 - 2n^2 = 1.$

$\qquad\qquad\qquad \therefore\ 2n - n^2 = 0.$

That is, $\qquad n = 0$ or 2,

and $\qquad\qquad \therefore\ m = 1$ or 3.

Therefore either $\qquad x = \pm 1,\quad y = 0,$

or $\qquad\qquad\qquad x = \pm\sqrt{3},\ y = \pm\sqrt{2},$

the signs of x, y being independent of each other.

EXERCISES VIII.

Solve the following equations :

1. $x + y = 5$; $x^2 + y^2 = 13$.   2. $2x + 3y = 7$; $2x^2 + x + 3y^2 = 13$.
3. $2x - y = 0$; $x^2 + 2x + 3y + 4xy = 17$.
4. $x + y = 2$; $x^2 + xy + y^2 + x + y = 5$.
5. $x - y = a$; $y^2 + ay + bx = 0$.   6. $x - y = 2$; $xy = 15$.
7. $x - y = 1$; $x^3 - y^3 = 19$.
   [Note that $x^2 + xy + y^2 = (x^3 - y^3)/(x - y) = 19$.]
8. $x^2 + 2y^2 = 5$; $x^4 + 4y^4 = 17$.
9. $x^2 + y^2 = 13$; $x^6 + y^6 = 793$.

128. When two equations involving two unknown quantities are *both* quadratic they cannot *always* be solved. We shall however consider some special cases where the equations can be so transformed as to make the solution depend on quadratics in one unknown quantity.

129. Case II. **When one of the equations contains only terms of the second degree.**

Any two simultaneous quadratic equations involving two unknown quantities can always be solved if one is of the form $ax^2 + bxy + cy^2 = 0$, and the other of the form $px^2 + qxy + ry^2 + sx + ty = d$.

Rule. *Divide the first equation by* y^2 *and solve as a quadratic in* x/y. *From the results substitute for* x *in terms of* y *in the other equation.*

[What we have here to note is that the ratio of $x : y$ is readily given by the first equation. See § 67.)]

Ex. Solve $\quad 3x^2 + 2xy - y^2 = 0$;
$$x^2 + y^2 + 2x = 12.$$
Dividing both sides of the first equation by y^2, we have
$$3\left(\frac{x}{y}\right)^2 + 2\left(\frac{x}{y}\right) - 1 = 0, \text{ which gives } \frac{x}{y} = \frac{1}{3} \text{ or} - 1.$$
Thus $\quad x = \tfrac{1}{3}y \text{ or } - y.$

(*a*) If $x = \dfrac{y}{3}$, $x^2 + y^2 + 2x = \dfrac{y^2}{9} + y^2 + \dfrac{2y}{3} = 12.$

Thus $\quad 10y^2 + 6y = 108$, which gives $y = 3 \text{ or } - 3\tfrac{3}{5}.$
Corresponding to these values of y, $x = 1 \text{ or } - 1\tfrac{1}{5}$.

(*b*) If $x = -y$, $x^2 + y^2 + 2x = y^2 + y^2 - 2y = 12.$
Thus $\quad y^2 - y = 6$; and $y = 3 \text{ or } - 2.$
Corresponding to these values of y, $x = -3, \text{ or } 2.$

Hence the four solutions are

$$\left.\begin{array}{l} x = 1 \\ y = 3 \end{array}\right\} \quad \left.\begin{array}{l} x = -1\tfrac{1}{5} \\ y = -3\tfrac{3}{5} \end{array}\right\} \quad \left.\begin{array}{l} x = -3 \\ y = 3 \end{array}\right\} \quad \left.\begin{array}{l} x = 2 \\ y = -2 \end{array}\right\}.$$

SIMULTANEOUS QUADRATIC EQUATIONS.

EXERCISES VIII.

Find the ratio of x to y from the following equations:

10. $x^2 - 2xy = 35y^2$.
11. $2x^2 - 13xy = 7y^2$.
12. $3x^2 + 4xy + y^2 = 0$.

Solve the following equations:

13. $x^2 - 4xy + 3y^2 = 0$; $2x^2 - 2x + y = 13$.
14. $x^2 + 2xy + 2y^2 = 3x^2 + xy + y^2 = 5$.
15. $8x^2 - 6xy + y^2 = 0$; $x^2 + y^2 + x = 6$.
16. $x^2 + xy = 2y^2$; $x^2 + 2xy + 3y^2 + 4x + 5y = 15$.

130. CASE III. **When terms of the first degree are absent from both equations. [Homogeneous Equations.]**

Equations can always be solved if they are of the form

$$ax^2 + bxy + cy^2 = d \dots \dots \dots \dots \text{(i)},$$
$$mx^2 + nxy + py^2 = q \dots \dots \dots \dots \text{(ii)}.$$

FIRST METHOD.—RULE. **Assume $x = vy$.** *Substitute in both equations, and by division form an equation involving* v *only. Having found* v *substitute* vy *for* x *and proceed as in Case I.*

Ex. Solve $3xy - 4y^2 = 2$; $x^2 + 3xy - 5y^2 = 7$.

To obtain the ratio of $x : y$, suppose $x = vy$, and substitute this value of x in both the equations.

Then $\qquad y^2(3v - 4) = 2$;
and $\qquad y^2(v^2 + 3v - 5) = 7$.

By cross-multiplication*,

$\qquad 7(3v - 4) = 2(v^2 + 3v - 5)$. $\therefore$ $2v^2 - 15v + 18 = 0$.

Hence $v = 1\frac{1}{2}$ or 6, which gives $x = 1\frac{1}{2}y$ or $6y$.

---

* $y = 0$ evidently does not satisfy the given equations.

108   SIMULTANEOUS QUADRATIC EQUATIONS.

(i) Taking $x = 1\frac{1}{2}y$, we have $3xy - 4y^2 = 4\frac{1}{2}y^2 - 4y^2 = 2$.

$\therefore \frac{1}{2}y^2 = 2$ which gives $y = \pm 2$.

The corresponding values of $x (= 1\frac{1}{2}y) = \pm 3$.

(ii) Taking $x = 6y$, we have $3xy - 4y^2 = 14y^2 = 2$.

Hence $y = \pm \sqrt{\frac{1}{7}}$, and $\therefore x = \pm 6\sqrt{\frac{1}{7}}$.

We thus obtain four pairs of roots which are

$\left.\begin{array}{c} x = 3 \\ y = 2 \end{array}\right\}, \left.\begin{array}{c} x = -3 \\ y = -2 \end{array}\right\}, \left.\begin{array}{c} x = \frac{6}{7}\sqrt{7} \\ y = \frac{1}{7}\sqrt{7} \end{array}\right\}, \left.\begin{array}{c} x = -\frac{6}{7}\sqrt{7} \\ y = -\frac{1}{7}\sqrt{7} \end{array}\right\}.$

131. SECOND METHOD.—RULE. *Multiply Equation* (ii) (§ 130) *by* d *and* (i) *by* q *and subtract so as to eliminate the constant terms.* *We thus obtain an equation of the form*

$$(md - qa)x^2 + (nd - qb)xy + (pd - cq)y^2 = 0.$$

From this equation the ratio of $x : y$ may be easily determined as in Case II (§ 129).

Ex. Solve $3xy - 4y^2 = 2$; $x^2 + 3xy - 5y^2 = 7$.

Dividing the first equation by the second, we have

$$\frac{3xy - 4y^2}{x^2 + 3xy - 5y^2} = \frac{2}{7}.$$

By cross-multiplication and transposition,

$$2x^2 - 15xy + 18y^2 = 0,$$

which gives $\qquad x = 1\frac{1}{2}y$ or $6y$.

The remainder of the solution is the same as in § 130, Ex.

EXERCISES VIII.

Solve the following equations:

17. $x^2 - xy = 2$; $2x^2 + y^2 = 9$.   18. $x^2 + 3y^2 = 43$; $x^2 + xy = 28$.
19. $2x^2 + 3xy = 26$; $3y^2 + 2xy = 39$.
20. $x^2 + xy + 4y^2 = 6$; $3x^2 + 8y^2 = 14$.
21. $x^2 - xy = 6$; $x^2 + y^2 = 61$.   22. $x^2 + 2xy = 45$; $xy + 3y^2 = 22$.

SIMULTANEOUS QUADRATIC EQUATIONS.

By putting vy for x (or otherwise), solve the following:

23. $x^2 + y^2 = 2y$; $2xy - y^2 = y$.
24. $x^2 + 5y^2 = 21x$; $3x^2 + 2xy + y^2 = 11x$.
25. $x^2 + 2y^2 - 18y = 0$; $2x^2 + 3xy - 24y^2 = 20y$.
26. $x^2 - x - y = 0$; $2x^2 + xy + 2y^2 = 5(x + y)$.
27. $2x^2 = x + y$; $6x^2 - 3y^2 = 3x$.
28. $xy + y^2 = 4x + y$; $5xy + 2y^2 = 8x + 5y$.
29. $2x^2 = x + y$; $2xy + y^2 = 3x$.

[The last equation becomes a cubic on the substitution of vy for x. By inspection $v = 1$ is one solution, the others are found as in § 122.]

132. CASE IV. When the equations are symmetrical.

Equations in which the two unknown quantities enter *symmetrically* and can be interchanged without altering the equations can always be solved if they are quadratic by the following

RULE. *Put* $x = u + v$, $y = u - v$ *and substitute in the equations. The equations then reduce to the forms*
$$au^2 + bv^2 + cu = d, \quad mu^2 + nv^2 + pu = q,$$
and v *may be eliminated by cross-multiplying by* b, n *respectively.*

[In the following example the equations are not quadratic.]
Ex. Solve $x^5 + y^5 = 32$; $x + y = 2$.
Let $\qquad x = u + v$; and $y = u - v$.
Since $\quad x + y = 2$, $u + v + u - v = 2$, and $\therefore u = 1$.
$\qquad \therefore x^5 + y^5 = (1 + v)^5 + (1 - v)^5 = 32$.
By simplification, $\quad 10v^4 + 20v^2 + 2 = 32$.
$\qquad \therefore v^4 + 2v^2 - 3 = 0$.
Thus $\quad v^2 = 1$ or -3, and $v = \pm 1$ or $\pm\sqrt{-3}$.
Hence $x = 1 \pm 1$ or $1 \pm \sqrt{-3}$, $y = 1 \mp 1$ or $1 \mp \sqrt{-3}$,
that is, $\quad \begin{matrix} x = 2 \\ y = 0 \end{matrix}\}$, $\quad \begin{matrix} x = 0 \\ y = 2 \end{matrix}\}$, $\quad \begin{matrix} x = 1 \pm \sqrt{-3} \\ y = 1 \mp \sqrt{-3} \end{matrix}\}$.

NOTE. Since the expressions that occur are symmetrical, two of the solutions can be obtained from the other two by interchanging x and y.

EXERCISES VIII.

Solve the following by putting $x = u + v$ and $y = u - v$ or otherwise:

30. $x^3 + y^3 = 8$; $x + y = 2$.

31. $x + y = 4$; $x^4 + y^4 = 9x^2y^2 + 1$.

32. $3x^2 + 4xy + 3y^2 = 3$; $x^2 + 3x + 3y = 4$.

33. $\dfrac{x}{y+1} + \dfrac{y}{x+1} = \dfrac{5}{3}$; $x^2 + y^2 = 2$.

34. $\dfrac{x}{5y+1} + \dfrac{y}{3x+1} = \dfrac{4}{15}$; $3x + 5y = 2$.

35. $x^2 = 5x + 6y$; $y^2 = 6x + 5y$.

36. $x^3 = 2x + 7y$; $y^3 = 2y + 7x$.

133. Solutions depending on known identities.

The solution of the following questions mainly depending on identities familiar to the student will indicate some other expedients, whereby simultaneous equations of two unknown quantities can often be solved.

Ex. 1. Solve $x + y = 7$; $xy = 12$.

By squaring, $(x + y)^2 = x^2 + y^2 + 2xy = 49$;
and since $4xy = 48$, by subtraction,
$$x^2 + y^2 - 2xy = (x - y)^2 = 1. \quad \therefore x - y = \pm 1.$$
Thus from $x + y = 7$ and $x - y = \pm 1$, we get the two solutions, viz.
$$x = 4, y = 3 \quad \text{and} \quad x = 3, y = 4.$$

Ex. 2. Solve $x^4 + x^2y^2 + y^4 = 21$; $x^2 + xy + y^2 = 7$.

Since $x^4 + x^2y^2 + y^4 = (x^2 + xy + y^2)(x^2 - xy + y^2)$, identically
we have $\quad x^2 - xy + y^2 = 21 \div 7 = 3$.
But $\quad x^2 + xy + y^2 = 7$.

By addition, $x^2 + y^2 = 5$, and therefore $xy = 2$.

Hence, as in Ex. 1, $x + y = \pm 3$, $x - y = \pm 1$, where $x = \pm 2$, $y = \pm 1$ or $x = \pm 1$, $y = \pm 2$.]

Ex. 3. Solve $\quad x^3 + y^3 = (a + b)(x - y)$ .... (i).
$\quad\quad\quad\quad\quad\quad\quad x^2 - xy + y^2 = a - b$ .... (ii).

SIMULTANEOUS QUADRATIC EQUATIONS.

Dividing (i) by (ii), $x + y = (a + b)(x - y) \div (a - b)$.

Hence $\dfrac{x+y}{x-y} = \dfrac{a+b}{a-b}$; $\therefore \dfrac{x}{y} = \dfrac{a}{b}$.

Thus x being equal to ay/b, we can substitute for x in the second equation, and thus find

$$y = \pm b \sqrt{\dfrac{a-b}{a^2 - ab + b^2}}, \quad x = \pm a \sqrt{\dfrac{a-b}{a^2 - ab + b^2}}.$$

EXERCISES VIII.

Solve :

37. $x + y = 5$; $x^3 + y^3 = 35$.　　38. $x - y = 3$; $x^3 - y^3 = 117$.

39. $x + y = 10$; $\sqrt{\dfrac{x}{y}} + \sqrt{\dfrac{y}{x}} = \dfrac{10}{3}$.　　40. $\dfrac{x^2}{y} + \dfrac{y^2}{x} = 2$; $x + y = 2$.

41. $\sqrt[3]{x} + \sqrt[3]{y} = 3$; $x + y = 9$.　　42. $x^3 + y^3 = 28$; $x^2y + xy^2 = 12$.

43. $x^4 + x^2y^2 + y^4 = 133$; $x^2 + xy + y^2 = 19$.　　44. $x^2 + y^2 + xy = 84$; $x + y + \sqrt{(xy)} = 14$.

45. $x + \dfrac{1}{y} = 4\tfrac{1}{2}$; $y + \dfrac{1}{x} = 2\tfrac{1}{4}$.　　46. $x + \dfrac{2}{y} = 2$; $y + \dfrac{2}{x} = 4$.

47. $\dfrac{1}{x^2} + \dfrac{1}{xy} = \dfrac{1}{16}$; $\dfrac{1}{y^2} + \dfrac{1}{xy} = \dfrac{1}{9}$.　　48. $\dfrac{1}{x} + \dfrac{1}{y} = \dfrac{3}{4}$; $\dfrac{1}{x^2} + \dfrac{1}{y^2} = \dfrac{5}{16}$.

49. $\dfrac{1}{x} + \dfrac{1}{y} = \dfrac{13}{36}$;

$\dfrac{1}{\sqrt{x}} + \dfrac{1}{\sqrt{y}} = \dfrac{5}{6}$.

50. $xy - \dfrac{x}{y} = 5$;

$xy - \dfrac{y}{x} = \dfrac{1}{5}$.

51. $x^2 + xy + 3x = 6$; $y^2 + xy + 3y = 12$.　　52. $x^2 + y^2 + x^2y^2 = 49$; $x + y - xy = -1$.

134. Turning one of the equations into a simpler form.—Sometimes one of two simultaneous equations can be very readily turned into a simpler form by solving it as an ordinary quadratic equation.

Ex. 1. Solve $4x^2y^2 + 1 = 4xy$; $x^2 + y^2 = 3$.

From the first equation which is a quadratic in xy, we get $xy = \tfrac{1}{2}$. This coupled with the second gives $x + y = \pm 2$ and $x - y = \pm\sqrt{2}$, whence $x = 1 \pm \tfrac{1}{2}\sqrt{2}$, $y = 1 \mp \tfrac{1}{2}\sqrt{2}$ or $x = -1 \pm \tfrac{1}{2}\sqrt{2}$, $y = -1 \mp \tfrac{1}{2}\sqrt{2}$.

SIMULTANEOUS QUADRATIC EQUATIONS.

Ex. 2. Solve $\dfrac{x+y}{x-y}+\dfrac{x-y}{x+y}=\dfrac{5}{2}$; $x^2+xy=12$.

Let $\dfrac{x+y}{x-y}=z$; then $z+\dfrac{1}{z}=\dfrac{5}{2}$.

This quadratic equation gives $\dfrac{x+y}{x-y}=z=2$ or $\tfrac{1}{2}$.

Thus $x=3y$ or $-3y$.

Substituting $\pm 3y$ for x in $x^2+xy=12$ we shall obtain
$x=\pm 3,\ y=\pm 1$ or $x=\pm 3\sqrt{2},\ y=\mp\sqrt{2}$.

Ex. 3. Solve $x^2+y^2+x+y=18$; $xy=6$.

Since $2xy=12,\ x^2+y^2+2xy+x+y=30$.
$\therefore (x+y)^2+(x+y)=30$.

Thus $x+y=5$ or -6; and $xy=6$;
whence $x=3,\ y=2$, or $x=2,\ y=3$, or $x=-3\pm\sqrt{3},\ y=-3\mp\sqrt{3}$.

EXERCISES VIII.

Solve :

53. $\dfrac{x^2}{4y^2}+\dfrac{x}{y}=3$; $3x-2y=4$.

54. $(x+y)^2+(x+y)=6$; $x-y=1$.

55. $\dfrac{x+y}{x-y}+\dfrac{x-y}{x+y}=3\tfrac{1}{3}$; $x^2+xy=6$.

56. $\dfrac{\sqrt{(x^2+3y^2)}+\sqrt{(x^2-3y^2)}}{\sqrt{(x^2+3y^2)}-\sqrt{(x^2-3y^2)}}=3$; $x^2+4y^2=9$.

57. $(x+y)^{\frac{2}{3}}+6(x-y)^{\frac{2}{3}}=5(x^2-y^2)^{\frac{1}{3}}$; $x-y=2$.

58. $x+y=5$; $(x^2+y^2)(x^3+y^3)=455$.

59. $\dfrac{(ac+1)(x^2+1)}{x+1}=\dfrac{(a^2+1)(xy+1)}{y+1}$;
$\dfrac{(ac+1)(y^2+1)}{y+1}=\dfrac{(c^2+1)(xy+1)}{x+1}$.

60. $4^x=2^y$; $27^{xy}=9^{y+1}$. 61. $9^x=27^y$; $64^{xy}=512^{x+1}$.

CHAPTER IX.

SIMULTANEOUS QUADRATIC EQUATIONS.

THREE UNKNOWN QUANTITIES.

135. In this chapter, we solve some typical equations of the second degree involving three unknown quantities.

CASE I. **Equations involving products.** In the equations considered in this article, we obtain, or are given, the products, two by two, of the unknown quantities themselves, or expressions containing them.

Ex. 1. Solve
$$3y + 2z = 2yz \quad \text{(i)},$$
$$4z + x = zx \quad \text{(ii)},$$
$$3x + 2y = xy \quad \text{(iii)}.$$

Dividing (i) by yz, (ii) by xz, (iii) by xy, we have

$$\frac{3}{z} + \frac{2}{y} = 2 \quad \text{(iv)};$$

$$\frac{4}{x} + \frac{1}{z} = 1 \quad \text{(v)};$$

and
$$\frac{3}{y} + \frac{2}{x} = 1 \quad \text{(vi)}.$$

Multiplying equation (v) by 3, and subtracting equation (iv), we have

$$\frac{12}{x} - \frac{2}{y} = 1. \quad \text{But by (vi)} \; 6\left(\frac{3}{y} + \frac{2}{x}\right) = 6.$$

$$\therefore \frac{12}{x} - \frac{2}{y} - \frac{18}{y} - \frac{12}{x} = 1 - 6 = -5.$$

Thus $y = 4$, and therefore from the given equations $x = 8$ and $z = 2$. $x = y = z = 0$ is obviously another solution.

TUT. ALG. II.   8

Ex. 2. Solve
$$x(px + qy + rz) = a \quad \text{(i)},$$
$$y(px + qy + rz) = b \quad \text{(ii)},$$
$$z(px + qy + rz) = c \quad \text{(iii)}.$$

Multiplying (i) by p, (ii) by q, (iii) by r, and adding, we have
$$(px + qy + rz)^2 = ap + bq + cr.$$
Thus $px + qy + rz = \pm\sqrt{(ap + bq + cr)}$;

$\therefore x = \dfrac{a}{\pm\sqrt{(ap + bq + cr)}}$; and similarly $y = \dfrac{b}{\pm\sqrt{(ap + bq + cr)}}$; &c.

EXERCISES IX.

Solve the following equations:

1. $yz = 4z - 3y$;
   $zx = 2z - 3x$;
   $2xy = 2x + y$.

2. $yz = 3(y + z)$;
   $zx = 2(x + z)$;
   $xy = 4(x + y)$.

3. $yz = py + qz$;
   $zx = qz + rx$;
   $xy = rx + py$.

4. $4x(x + 2y + 4z) = 21$;
   $y(x + 2y + 4z) = 10\frac{1}{2}$;
   $z(x + 2y + 4z) = 21$.

5. $x(x + y + z) = 6$;
   $y(x + y + z) = 12$;
   $z(x + y + z) = 18$.

6. $x(y + z - x) = 39 - 2x^2$;
   $y(z + x - y) = 52 - 2y^2$;
   $z(x + y - z) = 78 - 2z^2$.

7. $(y + z)(x + y + z) = 1$;
   $(z + x)(z + y + z) = 3$;
   $(x + y)(x + y + z) = 4$.

Ex. 3. Solve $xy = 6$; $xz = 8$; $yz = 12$.

We have $\dfrac{xy \times xz}{yz} = \dfrac{6 \times 8}{12}$.

Thus $x^2 = 4$, and $x = \pm 2$. $\therefore y = \pm 3$, and $z = \pm 4$.

Ex. 4. Solve
$$x(y + z) = 11 \quad \text{(i)},$$
$$y(z + x) = 35 \quad \text{(ii)},$$
$$z(x + y) = 36 \quad \text{(iii)}.$$

Adding (i) and (ii) and subtracting (iii), we have
$$xy + xz + xy + yz - xz - yz = 11 + 35 - 36.$$

Thus $2xy = 10$, and therefore $xy = 5$.

Hence from equations (i) and (ii) $xz = 6$ and $yz = 30$; whence as in Ex. 3, $x = \pm 1$, $y = \pm 5$, $z = \pm 6$.

SIMULTANEOUS QUADRATIC EQUATIONS. 115

Ex. 5. Solve $yz + zx + xy = 12 - x^2 = 15 - y^2 = 20 - z^2$.
We have $x^2 + xy + xz + yz = 12$, $\therefore (x+y)(x+z) = 12$.
Similarly, $(y+x)(y+z) = 15$, and $(z+x)(z+y) = 20$.
Now proceeding as in Ex. 3 we get,
$$x + y = \pm 3, \; x + z = \pm 4, \text{ and } y + z = \pm 5.$$
$$\therefore x = \pm 1, \; y = \pm 2, \text{ and } z = \pm 3.$$

Ex. 6. Solve
$$xy + 5(x+y) = 47 \ldots\ldots\ldots\ldots\ldots\ldots (i)$$
$$yz + 5(y+z) = 65 \ldots\ldots\ldots\ldots\ldots\ldots (ii),$$
$$zx + 5(z+x) = 55 \ldots\ldots\ldots\ldots\ldots (iii).$$

Adding 25 to both the sides of equation (i), we have
$$xy + 5(x+y) + 25 = 47 + 25.$$
[The student should observe why 25 has been added to both sides.]
Hence $\qquad (x+5)(y+5) = 72.$
Similarly $\qquad (y+5)(z+5) = 90,$
and $\qquad (z+5)(x+5) = 80.$

Hence as in Ex. 3, $x + 5 = \pm 8$, $y + 5 = \pm 9$, $z + 5 = \pm 10$, giving either $x = 3$, $y = 4$, $z = 5$ or $x = -13$, $y = -14$, $z = -15$.

EXERCISES IX.

Solve the following equations:

8. $yz = a^2$; $zx = b^2$; $xy = c^2$.
9. $xyz = p^2x = q^2y = r^2z$.
10. $x = y = z = xyz$.
11. $42x = 35y = 30z = xyz$.
12. $x(y+z) = 36$;
    $y(z+x) = 50$;
    $z(x+y) = 56$.
13. $xy + xz + yz = 26$;
    $xy + xz = 14$;
    $xy + yz = 18$.
14. $xyz = 2(x+y) = \frac{3}{2}(x+z) = \frac{6}{5}(y+z)$.
15. $xy + xz + yz = 30 - x^2 = 35 - y^2 = 42 - z^2$.
16. $x^2 - (y-z)^2 = 15$;
    $y^2 - (z-x)^2 = 5$;
    $z^2 - (x-y)^2 = 3$.
17. $(x+3)(y+5) = 24$
    $(y+5)(z+7) = 48$
    $(z+7)(x+3) = 32$.
18. $xy + 3y + 5x = 9$;
    $yz + 5z + 7y = 13$;
    $zx + 7x + 3z = 11$.

19. $xy + x + y = 41$;
$xz + x + z = 55$;
$yz + y + z = 47$.

20. $xy = x + y + 5$;
$yz = y + z + 11$;
$zx = z + x + 7$.

21. $xy + 3(x+y) = 11$;
$yz + 3(y+z) = 21$;
$zx + 3(z+x) = 15$.

22. $xy = 2y + 3x - 8$;
$yz = 3z + 4y - 9$;
$zx = 4x + 2z - 14$.

23. $6yz = 2y + 3z + 9$;
$3zx = 6z + x + 4$;
$2xy = x + 4y + 13$.

24. $yz = cy + bz + p^2$;
$zx = az + cx + q^2$;
$xy = bx + ay + r^2$.

136. Case II. **When two of the equations are linear.** In this article, the first two equations are simple, while the third is a quadratic.

Rule. *The values of* x *and* y *in the first two equations having been expressed in terms of* z, *substitute these values in the third equation, which thus becomes a quadratic in* z.

Ex. Solve
$$2x + y - z = 3 \quad \text{(i)},$$
$$3x + y + 2z = 7 \quad \text{(ii)},$$
$$xy + y^2 + z = 7 \quad \text{(iii)}.$$

Subtracting (i) from (ii), we have
$x = 4 - 3z$; and therefore $y = 7z - 5$.
$$\therefore xy + y^2 + z = (4 - 3z)(7z - 5) + (7z - 5)^2 + z = 7.$$
By simplification, $14z^2 - 13z = 1$.
Thus $z = 1$, or $-\frac{1}{14}$.
$\therefore y = 7z - 5 = 2$ or $-5\frac{1}{2}$; and $x = 4 - 3z = 1$ or $4\frac{3}{14}$.

EXERCISES IX.

Solve:

25. $x + y + z = 1$;
$3x - 2y + 2z = 2$;
$x^2 + 2y^2 + x + z = 1$.

26. $x - y + 2z = 0$;
$3x - 2y + z = 4$;
$x^2 + x + y + z = 24$.

SIMULTANEOUS QUADRATIC EQUATIONS. 117

27. $x - y = 1$;
$2x + 3y + z = 8$;
$x^2 + y + z = 6$.

28. $x - y = 1$;
$x - z = 2$;
$xy + xz + yz = x + y + z + 5$.

137. Case III. Application of cross-multiplication.

The first step in solving such examples as the following is to find the ratios of x, y and z to one another; and these we obtain from the first two equations by the *rule of cross-multiplication* explained in z 68.

Ex. 1. Solve $\quad 2x + 4y - 7z = 0$ .... (i),
$\qquad\qquad\qquad 4x - 5y - z = 0$ .... (ii),
$\qquad\qquad\qquad x^2 + y^2 + z^2 = 17$ .... (iii).

From the rule, we obtain from (i) and (ii)

$$\frac{x}{(4 \times -1) - (-5 \times -7)} = \frac{y}{(-7 \times 4) - (-1 \times 2)} = \frac{z}{(2 \times -5) - (4 \times 4)}.$$

Thus $\qquad\qquad \dfrac{x}{-39} = \dfrac{y}{-26} = \dfrac{z}{-26}$;

or $\qquad\qquad \dfrac{x}{3} = \dfrac{y}{2} = \dfrac{z}{2} = k$ (suppose).

∴ by (iii) $\quad x^2 + y^2 + z^2 = 9k^2 + 4k^2 + 4k^2 = 17$.
Hence $\qquad\qquad k^2 = 1$, and $k = \pm 1$.
Thus $\qquad\qquad x = \pm 3$, $y = \pm 2$, and $z = \pm 2$.

Ex. 2. Solve $\qquad x + y = 3z$ .... (i),
$\qquad\qquad\qquad 4x + 3y = 10z$ .... (ii),
$\qquad\qquad\qquad x^2 + x + yz + y = z + 5$ .... (iii).

We obtain from the first two equations the ratios of x, y and z to one another by the rule of cross-multiplication. We have

$$\frac{x}{(1 \times -10) - (3 \times -3)} = \frac{y}{(-3 \times 4) - (-10 \times 1)} = \frac{z}{(1 \times 3) - (4 \times 1)}.$$

∴ $\dfrac{x}{1} = \dfrac{y}{2} = \dfrac{z}{1} = k$ (suppose).

∴ $x = k$; $y = 2k$; $z = k$.

Now substitute these values in (iii).
$$\therefore k^2 + k + 2k^2 + 2k = k + 5.$$
Thus $3k^2 + 2k = 5$, which gives $k = 1$ or $-\tfrac{5}{3}$.
$x = 1,\ y = 2,\ z = 1,$ or $x = -\tfrac{5}{3},\ y = -\tfrac{10}{3},\ z = -\tfrac{5}{3}.$

EXERCISES IX.

Solve the following equations:

29. $3x + 4y = 5z$;
    $2x + y = 2z$;
    $x^2 + x + y^2 = z^2 + 3.$

30. $3x + 4y = 5z$;
    $2x + y = 2z$;
    $2x^2 + x + y^2 = z^2 + 12.$

31. $x + 5y = 6z$;
    $3x - 2y = z$;
    $x^2 + y^2 + z = 10.$

32. $x^2 + 2y^2 = z^2$;
    $7x^2 + 5y^2 = 3z^2$;
    $xy + 2yz = 3xz + 5.$

138. Miscellaneous Examples.

Ex. 1. Solve $\quad x - y + z = 5$ .... (i),
$\qquad\qquad\qquad\quad x^2 + y^2 - z^2 = 41$ ....(ii),
$\qquad\qquad\qquad\quad xy = 18$ ....(iii).

From (ii) and (iii), $x^2 + y^2 - z^2 - 2xy = (x - y)^2 - z^2 = 5.$
$$\therefore (x - y + z)(x - y - z) = 5.$$
But by (i) $\quad x - y + z = 5. \therefore x - y - z = 1.$
Hence $\quad (x - y + z) - (x - y - z) = 5 - 1 = 4.$
Thus $\quad z = 2. \therefore x - y = 3,$ and $xy = 18.$
Hence the student will easily find
$$x = -3,\ y = -6\ \text{or}\ x = 6,\ y = 3.$$

Ex. 2. Solve $\dfrac{yz}{cy + bz} = \dfrac{zx}{az + cx} = \dfrac{xy}{bx + ay}$; $x + y + z = m.$

Multiplying the first expression above and below by x, the second by y, and the third by z, we have
$$\frac{xyz}{cyx + bxz} = \frac{xyz}{ayz + cxy} = \frac{xyz}{bxz + ayz}.$$
The numerators being equal, we get
$$ayz = bxz = cxy.$$
Hence, $\qquad \dfrac{x}{a} = \dfrac{y}{b} = \dfrac{z}{c} = k$ (suppose).
$$\therefore x + y + z = k(a + b + c) = m.$$

Hence $\quad k = \dfrac{x}{a} + \dfrac{m}{a+b+c}.$

Thus $\quad x = \dfrac{ma}{a+b+c}; \; y = \dfrac{mb}{a+b+c}; \; z = \dfrac{mc}{a+b+c}.$

EXERCISES IX.

Solve the following equations

33. $x + 2y + 3z = 0$;
    $6x + 3y + 2z = 0$;
    $xyz + 6(x + 8y + 27z) = 0$.

34. $x + y - z = 4$;
    $x^2 + y^2 - z^2 = 12$;
    $xy = 6$.

35. $x - 2y + 3z = 2$;
    $x^2 + 4y^2 - 9z^2 = -4$;
    $xy = 1$.

36. $\dfrac{y+z}{3} = \dfrac{z+x}{5} = \dfrac{x+y}{6}$;
    $x^2 + y^2 + z^2 = 21$.

37. $\dfrac{y}{x-z} = \dfrac{y+x}{z} = \dfrac{x}{y}$;
    $x + y + z = 9$.

38. $x + y + z = 14$;
    $x^2 + y^2 + z^2 = 84$;
    $xz = y^2$.

39. $x^2 + y^2 + z^2 = 50$;
    $y^2 + 14 = 2xz$;
    $y^2 = xz$.

40. $x^2 + y^2 + z^2 = 14$;
    $xy + xz + yz = 11$;
    $x + y + 2z = 9$.

41. $x^2 - yz = 3$; $y^2 - xz = 5$; $z^2 - xy = -1$.

42. $x+y+z = 1$; $x^2+y^2+z^2+6xy = 0$; $\dfrac{x}{y+z} + \dfrac{y}{z+x} + \dfrac{z}{x+y} = 0$.

43. $x^2 - y^2 - z^2 + 2yz = 3$; $y^2 - z^2 - x^2 + 2zx = -9$;
    $z^2 - x^2 - y^2 + 2xy = -3$.

44. $\dfrac{yz}{y+z} = 2$; $\dfrac{zx}{z+x} = 3$; $\dfrac{xy}{x+y} = 1$.

45. $x + y + z = 1$; $x^2 + y^2 + z^2 = 13$; $x^3 + y^3 + z^3 = 19$.

CHAPTER X.

PROBLEMS LEADING TO QUADRATICS.

139. Method of solution.—In applying Algebra to the solution of arithmetical problems, there are three distinct operations to be performed.

(i) Frame an equation or equations by expressing the conditions of the problem in algebraical language.

(ii) Solve the equation or equations by the methods given in the previous chapters; and

(iii) Ascertain by trial whether the values found are true arithmetical solutions of the problem.

Ex. By selling a horse for £24, I lose as much per cent. as it cost me in pounds. What was the original price of the horse?

(i) Let x pounds be the original price of the horse.

Then I lose x per cent. of £x, i.e. £$\frac{x^2}{100}$.

Hence $$x - \frac{x^2}{100} = 24.$$

(ii) Now $x^2 - 100x = -2400$. Hence $(x-50)^2 = 10^2$;

$$\therefore x = 40 \text{ or } 60.$$

(iii) Both roots of the equation satisfy the conditions of the problem, and are therefore true solutions of the question.

140. Negative and fractional solutions.

Sometimes the solution of a problem turns out to be inapplicable to the given question as a problem of arithmetic, from a negative or fractional result being obtained in a question requiring a positive integral answer.

Ex. 1. A workman earns £1 10*s*. in a certain time. If he had worked for 5 days more, and had been allowed 6*d*. less for each day that he worked, he would still have earned the same amount. How long did he work?

(i) Let x denote the number of days he worked.

Then $\dfrac{30}{x}$ is the number of shillings he earned on each day.

Hence $(x+5)\left(\dfrac{30}{x} - \dfrac{1}{2}\right) = 30.$

(ii) By simplification and transposition, we get $x^2 + 5x - 300 = 0$, which gives $x = 15$ or -20.

(iii) While of the two values 15 satisfies the conditions of the problem, the negative value -20 cannot be accepted as a solution.

Hence the time required is 15 days.

The negative root suggests the following modified form of the problem:—

Ex. 2. A workman earns £1 10*s*. in a certain time. If he had worked for 5 days *less* and earned 6*d*. *more* for each day that he worked he would still have earned the same amount. How long did he work? *Ans.* 20 days.

Ex. 3. Twice the square of the number of my cows is 18 less than 13 times the number. How many cows have I?

If x be the number of cows, the problem leads to the equation

$$2x^2 - 13x + 18 = 0,$$

the roots of which will be found to be 2 and $4\frac{1}{2}$.

Hence the number of cows must be 2 since it cannot be fractional.

141. Surd solutions.—Irrational results happen sometimes to be wholly inapplicable. Even where they can be admitted as proper solutions, they furnish us, of course, with endless decimals or quantities that cannot be determined exactly.

PROBLEMS LEADING TO QUADRATICS.

Ex. 1. A person bought a certain number of oxen for £40, and after losing 2, sold the rest for £1 a head more than they cost him, thus gaining £6 by the transaction. How many did he buy?

If x denotes the number of oxen, we have

$$(x-2)\left(\frac{40}{x}+1\right) = 46,$$

giving $x = 4 \pm \sqrt{96}.$

Now $\sqrt{96}$ lies between 9 and 10, ∴ the positive root lies between 13 and 14. The interpretation will be found to be that if he bought 13 he would gain *less* than £6, if he bought 14 he would gain *more* than £6.

Ex. 2. Divide a straight line into two parts so that the rectangle contained by the whole and one of the parts shall be equal to the square of the other part.

If the whole line be a and the greater part be x, we have

$$a(a-x) = x^2, \text{ giving } x = \frac{a}{2}(-1 \pm \sqrt{5}).$$

Neither solution enables us to state the ratio of $a:x$ as an exact arithmetical fraction.

The geometrical construction given in Euclid, II. 11, is based on the positive solution. The negative result leads to the geometrical solution of the corresponding proposition connected with *external* section.

142. Impossible solutions.

Ex. Find a number such that its square shall be equal to the number decreased by 1.

Denoting the number by x, we have

$$x^2 = x - 1, \text{ which gives } x = \frac{+1 \pm \sqrt{(-3)}}{2}.$$

The values of x indicate that there are no numbers satisfying the conditions of the problem.

143. All solutions fail.

Ex. If from the square of a number 7 is subtracted, and to the square root of the difference, twice the number be added, the sum obtained is 5. Find the number.

Denoting the number by x, we have

$$2x + \sqrt{(x^2-7)} = 5, \text{ whence } x = 4 \text{ or } 2\tfrac{2}{3}.$$

Upon trial it will be found that neither value satisfies the conditions of the problem. For explanation of this irregularity, refer to §§ 116, 117.

EXERCISES X.

1. A party at an hotel had a bill of £4 to pay; but two having absconded, those who remained had each 2s. more to pay; how many were there at first?

2. What are ashes per 100 loads when 8 more loads for a sovereign lowers the price a penny per load? Of what problem is the negative answer a solution?

3. A person sells a horse for £144 and gains as much per cent. as the horse cost him. Find what he paid for the horse.

4. A person is about to distribute £6 among some beggars when two others come in, and thus each person's share is diminished by 2s. Find the number of persons relieved. Also explain the negative result.

5. I have to walk a distance of 144 miles, and I find that if I increase my speed by $1\frac{1}{2}$ miles per hour, I can do the journey in 16 hours less than if I walk at my usual rate. Find my usual rate of walking. Explain the negative result.

6. One man can reap a field in five days less than another man, and if they work together they can reap it in 6 days; find in what time each can do it alone. Explain the double result.

7. A number consists of three digits, and it is greater by one than 36 times the sum of its digits; also the middle digit is twice the product of the other two, and is greater by two than their sum. Find the number.

8. The area of a rectangular field is $7\frac{1}{2}$ acres, and the sum of the lengths of the two sides exceeds the length of either diagonal by 110 yards. Find the lengths of the sides.

9. If 32 men can be arranged in the form of a hollow square 2 deep, find the number of men presented in the front.

10. A number of men are first formed into a solid square, and afterwards into a hollow square three deep; the front presented in the latter formation has 75 men more than the front in the solid square. Determine the number of men.

11. The sum of the squares of two numbers is 650, and the product of the numbers is 323. What are they?

12. Find two numbers such that their sum may be 24 and the sum of their cubes 5256.

PROBLEMS LEADING TO QUADRATICS.

13. There is a number consisting of two digits, such that the difference of the cubes of the digits is 109 times the difference of the digits. Also the number exceeds twice the product of its digits by the digit in the unit's place. Find the number.

14. Find two numbers, such that their sum is 9 and the sum of their fourth powers 2417.

15. A man walks a certain distance in a certain time. He calculates that if he had walked a mile per hour slower than he did, he would have taken 6 hours more than three-fourths of the time he actually took: but if he had walked a mile per hour faster he would have taken 2 hours longer than half the time he actually took. Find the distance walked, and the rate of walking.

16. Two numerical quantities are such that their sum, their product, and the difference of their squares, are all equal; find them.

17. Divide an integer a into two parts, such that one of them is the square of the other.

Show that the parts cannot be rational, unless a is the product of two consecutive integers.

18. In a lake the head of a bulrush was observed 9 inches above the water; and when moved by the gentle breeze it sank in the water at 36 inches distance; required the depth of the water. [The stem originally stood up straight.]

19. At the bottom of a post is a serpent's hole; on the top of the pole is perched a peacock; the post in nine yards high, and the peacock descries the serpent moving to its hole when distant from it three times the height of the post; the peacock flies down in the line of the hypotenuse and seizes the serpent when both have passed over an equal distance. At what distance from the serpent's hole did they meet?

*20. The sum of £19,950 invested in the 4% gives an income of £8 8s. more than in the 3%. If each stock were to rise 1, the difference would be £9 17s. 11d. What is the price of each stock?

21. Two rectangular fields each contain one acre; one of the fields is four poles shorter and two poles broader than the other; find the length and breadth of each field.

22. A number consisting of three digits is equal to 76 times the sum of its digits. The third digit is twice the second, and if the second digit be increased by 2, and then squared, the result will be the first digit. Find the number.

23. Find two equal rectangles the sum of whose bases is a; and which are such that if the first had the altitude of the second, its area would be b; and if the second had the altitude of the first, its area would be c.

PROBLEMS LEADING TO QUADRATICS.

24. A and B are towns beside a river which runs at the rate of 4 miles an hour. A waterman rows from A to B and back again, and takes 39 minutes more to do it than if there had been no stream. The next day he does the same with another waterman, with whose help he can row half as fast again, and they are now eight minutes longer than if there had been no stream. At what rate does the waterman row by himself without any stream?

25. A fradulent tradesman contrives to employ his false balance both in buying and selling a certain article; thereby gaining at the rate of 11 per cent. more on his outlay that he would gain were the balance true. If however the articles were placed in the opposite scale-pan to be weighed, he would neither gain nor lose by the transaction. Determine the legitimate gain per cent.

26. A boat goes along a straight reach of a canal at 6 miles an hour. A person living 4 miles from the canal sets out to catch the boat three-quarters of an hour before it passes the point of the canal nearest to his residence. If his rate of moving be 4 miles an hour, find what point of the canal he should make for, in order that he may reach it just with the boat? Draw a diagram and interpret fully the results of your investigation.

27. Obtain an equation to find the sides of a rectangle of given perimeter b inscribed in a circle of given radius a. When is the problem impossible? Work out the solution for the case of

$$a = 97, \ b = 548.$$

CHAPTER XI.

ZERO AND INFINITY.

144. Two meanings of Zero.

(i) The ordinary meaning of *zero* is the total absence of any magnitude—that is, the result obtained by subtracting a number from another that is absolutely **equal** to it.

When 2 is subtracted from 2, or a from a, we obtain zero, or what is sometimes called *absolute zero*, to distinguish the zero thus defined from that explained below.

(ii) There is another meaning of *zero*, that the student often meets with in Algebra.

If a finite number, say unity, is divided by a billion, what is the result? The quotient is a very small number. If this very small number is raised to the billionth power, what is the result? The result, it will be said, is an inconceivably small number, which is very nearly nothing. If this number, which we shall suppose we can conceive, is again raised to the billionth power, we get a number still more minute which is still more nearly equal to absolute zero, though it is not actually nothing. Continuing this process the results approach nearer and nearer to absolute zero, which value is, however, never actually attained by any number thus obtained, however small.

DEFINITION. **Zero** in this sense, may be defined as a number which is smaller than any assignable fraction of unity.

145. Meaning of Infinity.—If *zero* is, in this sense, less than any assignable number whatever, *infinity*, on the other hand, is a number which is greater than any that can be imagined.

It is, of course, impossible to conceive the greatest possible number; for were it possible to think of this number distinctly, it would be as easy to think of a greater number. It therefore follows that if we take ever so large a number, it will still be less than infinity. Hence the following

DEFINITIONS. **Infinity** may be defined as a number larger than any that can be named or conceived. The symbol that stands for infinity is ∞.

Thus the statement that $x = \infty$ does not mean that infinity has a precise value, and that x is equal to it. It must not therefore be inferred from the equations $x = \infty$ and $y = \infty$ that x is necessarily equal to y.

A quantity is said to be **finite** when it is neither infinitely large nor infinitely small.

146. Vanishing Fractions.—DEFINITIONS.

A fraction is called a **vanishing fraction** when it assumes the form $\frac{0}{0}$, *i.e.* when the numerator and denominator simultaneously become zero for certain values of the variables which they contain.

When a variable x continually approaches nearer and nearer to a given constant value a so that the difference between x and a becomes less than any assignable finite quantity however small; then a is called the **limiting value** or **limit** of x.

Ex. Find the limit of the value of $\frac{x^2 + 3x - 4}{x(x-1)}$ when $x = 1$.

The substitution of 1 for x in the fraction produces $\frac{0}{0}$.

To ascertain if the fraction has any finite limit, we must consider what becomes of the fraction, when x is *very nearly* equal to 1. And since both numerator and denominator vanish when $x = 1$, they must both be divisible by $x - 1$. Simplifying, we have

$$\frac{x^2 + 3x - 4}{x(x-1)} = \frac{x+4}{x}$$ for all values of x except 1.

This form shows that the fraction may be made as nearly equal to 5 as we please, by taking x sufficiently near to the given value, unity. Hence we conclude that the *limit* required is 5; by which we mean that, by bringing x sufficiently near to 1, we can cause the given expression to differ as little as we please from 5.

147. Properties of zero and infinity.

In this article we shall lay down (without proof) certain statements regarding 0 and ∞, taking the former symbol in the sense of a quantity which is infinitely small. In the following relations a denotes a finite quantity and (?) signifies that the value is indeterminate.

(i) ADDITION OF ZERO
$$0 + 0 = 0; \qquad a + 0 = a; \qquad \infty + 0 = \infty.$$

(ii) SUBTRACTION OF ZERO
$$0 - 0 = 0; \qquad a - 0 = a; \qquad \infty - 0 = \infty.$$

(iii) MULTIPLICATION BY ZERO
$$0 \times 0 = 0; \qquad a \times 0 = 0; \qquad \infty \times 0 = (?).$$

(iv) DIVISION BY ZERO
$$0 \div 0 = (?); \qquad a \div 0 = \infty; \qquad \infty \div 0 = \infty.$$

(v) DIVISION OF ZERO
$$0 \div 0 = (?); \qquad 0 \div a = 0; \qquad 0 \div \infty = 0.$$

(vi) ADDITION OF INFINITY
$$0 + \infty = \infty; \qquad a + \infty = \infty; \qquad \infty + \infty = \infty.$$

(vii) SUBTRACTION OF INFINITY
$$0 - \infty = -\infty; \qquad a - \infty = -\infty; \qquad \infty - \infty = (?).$$

(viii) MULTIPLICATION BY INFINITY
$$0 \times \infty = (?); \qquad a \times \infty = \infty; \qquad \infty \times \infty = \infty.$$

(ix) DIVISION BY INFINITY
$$0 \div \infty = 0; \qquad a \div \infty = 0; \qquad \infty \div \infty = (?).$$

(x) DIVISION OF INFINITY
$$\infty \div 0 = \infty; \qquad \infty \div a = \infty; \qquad \infty \div \infty = (?).$$

ZERO AND INFINITY.

148. Involution and Evolution. The limiting forms assumed by x^n when either x or n vanishes or becomes infinite may be stated as follows.

(i) POWERS OF ZERO.

$0^n = 0$ if n is positive; $\quad 0^0 = (?);\quad 0^n = \infty$ if n is negative.

(ii) POWERS OF INFINITY.

$\infty^n = \infty$ if n is positive; $\infty^0 = (?);\quad \infty^n = 0$ if n is negative.

(iii) POWERS WITH INDEX ZERO.
$\quad 0^0 = (?);\qquad\qquad a^0 = 1;\qquad \infty^0 = (?).$

(iv) POWERS WITH POSITIVE INFINITE INDEX.
$\quad a^\infty = \infty$ if $a > 1;\qquad 1^\infty = (?);\qquad a^\infty = 0$ if $a < 1.$

(v) POWERS WITH NEGATIVE INFINITE INDEX.
$\quad a^{-\infty} = 0$ if $a > 1;\qquad 1^{-\infty} = (?);\qquad a^{-\infty} = \infty$ if $a < 1.$

We now proceed to prove these results and to state them in words.

149. Powers of zero and infinity. From the nature of involution and evolution it is easy to see that
$$x^n = 0 \text{ when } x = 0 \text{ if } n \text{ is positive,}$$
or *any positive power of zero is itself zero.*

Again $\quad x^{-n} = 1/x^n = 1/0 = \infty,$

or *any negative power of zero is infinite.*

Put $y = 1/x$ and let $y = \infty$, then $x = 0$, and we have
$$y^n = 1/x^n = 1/0 = \infty \text{ if } n \text{ is positive,}$$
or *any positive power of infinity is infinite.*

Again $\quad y^{-n} = x^n = 0,$

or *any negative power of infinity is zero.*

150. Powers with index zero or infinity. In Chap. I. we proved that $a^0 = 1$ if a is finite. We shall now consider the value of a^x when x becomes infinite.

TUT. ALG. II.

Suppose a to be any quantity numerically greater than unity. Then since $a>1 : a^2>a$ numerically, $a^3>a^2$ and so on, thus a^x increases without limit as x increases without limit. Hence

$$a^x = \infty \text{ when } x = +\infty \text{ if } a>1,$$

or *any quantity greater than unity with positive infinite index is infinite.*

Again $\quad a^{-x} = 1/a^x = 1/\infty = 0,$

or *any quantity greater than unity with negative infinite index vanishes.*

Put $a = 1/b$. Then b is numerically less than unity, and we have

$$b^x = 1/a^x = 1/\infty = 0 \text{ when } x = +\infty \text{ and } b<1,$$

or *any quantity less than unity with positive infinite index vanishes**.

Again $\quad b^{-x} = 1/b^x = 1/0 = \infty,$

or *any quantity less than unity with negative infinite index is infinite.*

151. Change of sign in passing through infinity.

Consider the fraction

$$y = \frac{1}{x-a}.$$

When $x=a$, $y=\infty$. When $x<a$, y is negative, and when $x>a$, y is positive. Hence y changes from negative to positive in passing through the value ∞ so that its value must instantaneously change from $-\infty$ to $+\infty$ as x passes through the value a.

This shows that *a variable may change sign when it becomes infinite.*

---

* This property has an important application to the summation of an infinite geometrical progression.

ZERO AND INFINITY. 131

This property has very important applications to Trigonometry. Thus both sec A and tan A become infinite, and at the same time change sign from positive to negative when the angle A passes through the value 90°.

We now give two further examples illustrative of certain types of vanishing fractions.

Ex. 1. Find the limit when $x = 1$ of
$$\frac{\sqrt{(x-1)}}{\sqrt{x}-1}.$$

It is often convenient to replace the variable in a vanishing fraction by a new variable so chosen as to vanish at the required limit. In the present instance we put $\sqrt{x} = 1 + h$, so that h vanishes when $x = 1$. Let the given expression $= y$. Then on squaring,
$$y^2 = \frac{x-1}{(\sqrt{x}-1)^2} = \frac{(1+h)^2 - 1}{h^2} = \frac{2h + h^2}{h^2} = \frac{2}{h} + 1.$$

But $2/h = \infty$ when $h = 0$. Therefore $y^2 = \infty$ and consequently the given expression y becomes infinite when $x = 1$.

If $x > 1$, y^2 is positive and y is real. But if x is slightly less than unity so that h is a small negative quantity, $2/h + 1$ is negative and cannot be equal to the square of a real quantity, hence the given fraction y no longer has a real value.

Ex. 2. Find the limit of $\dfrac{ax^2 + bx + c}{dx^2 + ex + f}$ when $x = \infty$.

Put $x = 1/y$, then when $x = \infty$, $y = 0$, and we have to find the limit of
$$\frac{ay^{-2} + by^{-1} + c}{dy^{-2} + ey^{-1} + f}.$$
Multiplying numerator and denominator by y^2 we have
$$\text{Required limit} = \frac{a + by + cy^2}{d + ey + fy^2} = \frac{a + b.0 + c.0^2}{d + e.0 + f.0^2} = \frac{a}{d}.$$

When $x = 0$ the value of the fraction is evidently c/f, and this example illustrates the following general rule:

The limit of an expression when x $= \infty$ *depends only on the terms containing the* **highest** *powers of* x, *while its limit when* x $= 0$ *depends only on the terms containing the* **lowest** *powers of* x.

152. Infinite, zero, and indeterminate solutions of equations.

We shall conclude this chapter with a few particular cases, leading to infinite, zero and indeterminate solutions, of what has been called the "*Problem of the Couriers*"—a question that has been often solved and discussed.

ZERO AND INFINITY.

Problem. A *and* B *travel in the same direction at* u *and* v *miles an hour respectively, and* A *reaches* L *at a time* t *hours before* B *reaches a place* M, *which is* a *miles beyond* L. *When will they be together?*

$$\overline{\qquad \text{L} \qquad \text{M} \qquad \text{N} \qquad}$$

Let us suppose them to be together at N, x hours after A reaches L.

Then $LM = a$, $LN = ux$ hence

$$MN = ux - a = (x - t)v\,; \quad \therefore\ x = \frac{a - vt}{u - v}.$$

If $u = v$ and a is unequal to vt, then $x = \infty$; which means that if their rates be equal, the one can never overtake the other.

[If $u = v = 0$; then $x = \infty$, which means that if they do not move at all, the one cannot overtake the other.]

If $a = vt$, then $x = 0$; which means that in this case A and B are together at L.

If $a = 0$, $t = 0$, then $x = 0$; which means that they meet at L, and in this case L is the same place as M.

If $a = vt$, and $v = u$, $x = \frac{0}{0}$; an indefinite result which means that they will always be together. This must be the case since they are together at L (a being equal to vt), and their rates are equal.

Many other inferences can be drawn by equating one or more of the quantities a, u, v, t to zero. These are left to the student.

EXERCISES XI.

Show that the following expressions become vanishing fractions under the conditions given, and find the limits of their values:

1. $\dfrac{x^2 - 7x + 6}{x^2 - x}$ when $x = 1$. 2. $\dfrac{x^2 + 2x - 8}{x^2 + 7x - 18}$ when $x = 2$.

3. $\dfrac{x^2 + x}{x + \sqrt{x}}$ when $x = 0$. 4. $\dfrac{5x + 3\sqrt{x}}{x^2 + x}$ when $x = 0$.

5. $\dfrac{x^3 - 4x^2 + 5x - 2}{x^3 - 5x^2 + 7x - 3}$ when $x = 1$. 6. $\dfrac{x^2 + x^3 + x^4}{x^2 + x^5 + x^8}$ when $x = 0$.

7. Show that $(x + x^2) \div x^2 = 1$, when $x = \infty$.

8. Show that $(2x^4 + 3x^3) \div (4x^2 + 5x) = \infty$, when $x = \infty$.

9. Solve $x = x + a$, where a is finite.

10. Solve $x + a = x + b$, where a and b are finite.

CHAPTER XII.

THE THEORY OF QUADRATIC EQUATIONS.

153. In this chapter we shall deduce some important properties relating to the roots and coefficients of quadratic equations. We begin the chapter with a theorem relating to simple equations.

154. A linear equation in one unknown quantity has always one root, and only one.

Every simple equation can, after suitable reduction and transposition, be made to assume the form $ax + b = 0$. And of this equation, $-\dfrac{b}{a}$ is the root.

[NOTE. Another proof will be suggested by § 155.]

COR. A simple equation in x that is satisfied by more than one value of x is an identity.

155. A quadratic equation in one unknown quantity has always two and only two roots.

After suitable reduction and transposition, every quadratic equation can be brought to the form
$$ax^2 + bx + c = 0.$$
The roots of this equation being
$$\frac{-b + \sqrt{(b^2 - 4ac)}}{2a} \text{ and } \frac{-b - \sqrt{(b^2 - 4ac)}}{2a}$$
the value of x *must* be one or the other of these two quantities.

ALTERNATIVE PROOF. That *a quadratic equation cannot have more than two different roots* is sometimes proved as follows:—

If possible, let $ax^2 + bx + c = 0$ have three different roots α, β and γ. Since the equation is satisfied by each of these roots, we have

$$a\alpha^2 + b\alpha + c = 0 \quad\quad\quad\text{(i)},$$
$$a\beta^2 + b\beta + c = 0 \quad\quad\quad\text{(ii)},$$
and
$$a\gamma^2 + b\gamma + c = 0 \quad\quad\quad\text{(iii)}.$$

Subtracting (ii) from (i), we have
$$a(\alpha^2 - \beta^2) + b(\alpha - \beta) = 0 \quad\quad\quad\text{(iv)}.$$

Dividing both sides of (iv) by $(\alpha - \beta)$ which is not zero by supposition, we have
$$a(\alpha + \beta) + b = 0 \quad\quad\quad\text{(v)}.$$

Similarly from (ii) and (iii),
$$a(\beta + \gamma) + b = 0 \quad\quad\quad\text{(vi)}.$$

Subtracting (vi) from (v), we get
$$a(\alpha - \gamma) = 0.$$

Since the equation is quadratic, a is not zero, therefore $\alpha - \gamma = 0$; that is two of the three roots are identical, contrary to supposition.

Hence a quadratic equation cannot have more than two different roots.

COR. If a were $= 0$ we should readily find $b = c = 0$.

Since a quadratic equation $ax^2 + bx + c = 0$ cannot therefore be satisfied by more than two values of x unless $a = b = c = 0$, it follows that a quadratic equation in x is an identity if it is satisfied by more than two values of x.

NOTE. It can be similarly proved than an equation of the third degree in one unknown quantity cannot have more than three roots, an equation of the fourth degree in one unknown quantity more than four roots, &c.

THE THEORY OF QUADRATIC EQUATIONS.

156. To find the relations between the roots and coefficients of a quadratic equation.

Solving the equation $ax^2 + bx + c = 0$, we obtain as the roots

$$\frac{-b+\sqrt{(b^2-4ac)}}{2a} \text{ and } \frac{-b-\sqrt{(b^2-4ac)}}{2a}.$$

Call the roots α and β respectively.

Then we have, by actual addition and multiplication,

(i) $\quad \alpha + \beta = \dfrac{-b+\sqrt{(b^2-4ac)}}{2a} + \dfrac{-b-\sqrt{(b^2-4ac)}}{2a} = -\dfrac{b}{a}.$

(ii) $\quad \alpha\beta = \left(\dfrac{-b}{2a} + \dfrac{\sqrt{(b^2-4ac)}}{2a}\right)\left(\dfrac{-b}{2a} - \dfrac{\sqrt{(b^2-4ac)}}{2a}\right)$

$\qquad = \dfrac{b^2}{4a^2} - \dfrac{b^2-4ac}{4a^2} = \dfrac{4ac}{4a^2} = \dfrac{c}{a}.$

That is,

(i) **The sum of the roots**
$$= -\frac{b}{a} = -\frac{\text{coefficient of } x}{\text{coefficient of } x^2},$$

(ii) **The product of the roots**
$$= \frac{c}{a} = \frac{\text{absolute term}}{\text{coefficient of } x^2}.$$

These two relations are very important, and should be carefully remembered.

COR. By writing the equation $ax^2 + bx + c = 0$ in the form $x^2 + \dfrac{b}{a}x + \dfrac{c}{a} = 0$, the property may be stated thus :—

If, in a quadratic equation, the coefficient of the first term is unity, then

(i) **the sum of the roots is equal to the coefficient of the second term with the sign changed,**

(ii) **the product of the roots is equal to the third term.**

Ex. Show that the roots of $ax^2 + bx + a = 0$ are the reciprocals of each other.

Let the roots be a and β. Then $a\beta = a/a = 1$, which proves the statement.

157. If a and β are the roots of $ax^2+bx+c=0$, the equation may be written in the form
$$a(x-a)(x-\beta) = 0.$$

The given equation may be written in the form
$$a\left\{x^2 + \frac{b}{a}x + \frac{c}{a}\right\} = 0.$$

$\therefore a\{x^2 - \text{(sum of roots)}\, x + \text{product of roots}\} = 0.$

Hence, $\quad a\{x^2 - (a+\beta)x + a\beta\} = 0.$

That is, $\quad a(x-a)(x-\beta) = 0.$

This result is very important as it readily enables us **to factorise any quadratic expression** such as $ax^2 + bx + c$. All that we have to do is to *put the given expression equal to zero and solve the quadratic equation thus obtained*. If a, β be the roots, we have identically,
$$ax^2 + bx + c = a(x-a)(x-\beta).$$

EXERCISES XII.

Find the sum and product of the roots of the three following equations, without solving them :—

1. $\dfrac{1}{x+a+b} = \dfrac{1}{x} + \dfrac{1}{a} + \dfrac{1}{b}$.
2. $\dfrac{x}{x+1} + \dfrac{x+1}{x} = \dfrac{5}{2}$.
3. $x(x-1) + (x-1)(x-2) + x(x-2) = 2$.

4. Show that the roots of $ax^2 - ax + c = 0$ have unity for their sum.

5. Prove that $(x-1)^2(a^2+1) - (x^2+1)(a-1)^2 = 2(x-a)(ax-1)$ is an identity, by showing that the equality is satisfied by more than two values of x.

6. Show that the product of the roots of $acx^2 + b^2x + c^2 = 0$ is that of the roots of $ax^2 + bx + c = 0$.

THE THEORY OF QUADRATIC EQUATIONS. 137

Express the following equations in the form
$$a(x-\alpha)(x-\beta) = 0.$$

7. $\dfrac{1}{3}x^2 - \dfrac{1}{2}x = 9.$ 8. $\dfrac{x+4}{x-3} - \dfrac{2x-3}{x+4} = 7\tfrac{3}{8}.$

9. $\dfrac{x-1}{x+1} + \dfrac{x+3}{x-3} = 2\left(\dfrac{x+2}{x-2}\right).$

158. If one of the roots of a quadratic equation can be found by inspection, the other can be obtained from either

$$\alpha + \beta = -\dfrac{b}{a},$$

or
$$\alpha\beta = \dfrac{c}{a}.$$

The latter relation also enables us to write the roots in the altered forms

$$\alpha = \dfrac{c}{a\beta} = \dfrac{2c}{-b-\sqrt{(b^2-4ac)}}, \text{ and } \beta = \dfrac{c}{a\alpha} = \dfrac{2c}{-b+\sqrt{(b^2-4ac)}}.$$

Ex. Solve $(x-3)(x-4) = (a-3)(a-4).$
By inspection one of the roots $= a.$
Simplifying, we have, $x^2 - 7x - (a^2 - 7a) = 0.$
Here, the sum of the two roots $= 7.$
Hence the other root $= 7 - a.$

EXERCISES XII.

Solve by the method of this article:

10. $x - \dfrac{1}{x} = a - \dfrac{1}{a}.$

11. $x^2(a-b) + x(b-c) + (c-a) = 0.$

12. $a(x^2 - 1) = x(a^2 - 1).$

13. $a(x^2 + 1) = x(a + 1).$

14. $(x-5)(x-10) = (a-5)(a-10).$

15. Find, for what other value of x, the expression $(x-1)(x-2)$ has the same value as it has when $x = m.$

138 THE THEORY OF QUADRATIC EQUATIONS.

159. Discrimination of roots. We shall now show how, without solving a quadratic equation, we can learn the nature of its roots.

We take the typical equation $ax^2 + bx + c = 0$; the roots of which are

$$-\frac{b}{2a} + \frac{\sqrt{(b^2 - 4ac)}}{2a} \text{ and } -\frac{b}{2a} - \frac{\sqrt{(b^2 - 4ac)}}{2a}$$

and we suppose that a, b, and c are real rational quantities.

(i) **When $b^2 - 4ac$ is negative,** the roots of $ax^2 + bx + c = 0$ are **imaginary,** since an expression that contains the even root of a negative quantity is imaginary or impossible.

(ii) **When $b^2 - 4ac = 0$,** both the roots become $-\frac{b}{2a}$ and are therefore **real and equal.**

(iii) **When $b^2 - 4ac$ is positive,** the roots are **real,** but **unequal.**

When $b^2 - 4ac$ is a perfect square, the roots are **rational.**

When $b^2 - 4ac$ is not a perfect square, the roots are **irrational.**

NOTE. Since $b^2 - 4ac$ is of much use in discriminating the nature of the roots, it is sometimes called the **discriminant.**

CAUTION. The statement, sometimes met with, that "a quadratic has only one root when $b^2 = 4ac$ and no roots when $b^2 < 4ac$" is algebraically incorrect. In accordance with the Principle of the Permanence of Equivalent Forms (§ 7) **every quadratic has two roots** whose symbolical expressions are those given above and in § 155. If both expressions lead to the same arithmetical value, the quadratic is said to have two equal roots, and if the expressions have no arithmetical value, positive or negative, commensurable or incommensurable, the quadratic is said to have two imaginary or impossible roots.

THE THEORY OF QUADRATIC EQUATIONS.

160. To find when both roots are positive.

If α and β be the roots, both the roots will be positive only if $\alpha + \beta$ and $\alpha\beta$ be positive.

Therefore $-b/a$ and c/a are positive.

That is, b must be of the opposite sign to a and c of the same sign as a.

For the roots to be real, $b^2 - 4ac$ must not be negative.

161. To find when both roots are negative.

If α and β be the roots, both the roots will be negative, only if $\alpha + \beta$ be negative and $\alpha\beta$ be positive.

Hence $-b/a$ must be negative and c/a positive or b/a and c/a both positive.

That is, a, b, c must all be of the same sign.

For the roots to be real, $b^2 - 4ac$ must not be negative.

162. To find when one root is positive and the other root negative.

If α and β be the roots, it is necessary and sufficient that $\alpha\beta$ be negative.

Hence c and a must be of opposite sign.

The numerically greater root has the sign of $\alpha + \beta$, that is, of $-b/a$.

The condition for real roots is necessarily satisfied, for since $4ac$ is negative $b^2 - 4ac$ is essentially positive.

163. To find when the roots are numerically equal, but opposite in sign.

Since α has to be equal and opposite to β,
$$\alpha + \beta = 0.$$
Hence $\qquad -b/a = 0.$

Therefore the condition required is $b = 0$.

For the roots to be real, a and c must be of opposite signs.

140 THE THEORY OF QUADRATIC EQUATIONS.

164. To find when one root = 0.

If a and β be the roots, one root will be zero if $a\beta = 0$; that is, if $\dfrac{c}{a} = 0$, or $c = 0$.

That is, the absolute term must vanish.

165. To find when both roots = 0.

$$a + \beta = -\frac{b}{a} = 0 \text{; and } a\beta = \frac{c}{a} = 0.$$
$$\therefore b = 0 \text{; and } c = 0.$$

That is, the last two terms must vanish.

*166. To find when one root = ∞.

If $a = \infty$, then $1/a = 0$. Now
$$\frac{1}{a\beta} = \frac{a}{c} \text{ and } \frac{1}{a} + \frac{1}{\beta} = \frac{\beta + a}{a\beta} = -\frac{b}{a}\frac{a}{c} = -\frac{b}{c}.$$

The condition is that $1/a\beta = 0$, and $\therefore a = 0$.

That is, the first term must vanish.

*167. To find when both roots = ∞.

If a and β are both $= \infty$, $1/a$ and $1/\beta$ are both $= 0$ and therefore
$$\frac{1}{a\beta} = \frac{a}{c} = 0, \quad \frac{1}{a} + \frac{1}{\beta} = -\frac{b}{c} = 0.$$
$$\therefore a = 0 \text{ and } b = 0.$$

That is the first two terms must vanish.

*168. To find when the roots are 0 and ∞.

By § 164 one root will be zero if $c = 0$, and by § 166 one root will be infinite if $a = 0$.
$$\therefore a = 0 \text{ and } c = 0.$$

That is, the first and last terms must vanish.

*169. To find when the roots are indeterminate.

This will be the case when the equation is an identity which is satisfied by all values of x; therefore
$$a = 0, b = 0, c = 0.$$

That is all the coefficients must vanish identically.

THE THEORY OF QUADRATIC EQUATIONS.

170. Table of results. We tabulate the results of the last eleven articles for purposes of reference, taking the standard form of equation.

	Result.	Condition.
i	Both roots are imaginary,	if $b^2 - 4ac$ be negative.
ii	Both roots real and equal,	if $b^2 - 4ac = 0$.
iii	Both roots real and unequal	if $b^2 - 4ac$ is positive.
iv	Both roots rational,	if $b^2 - 4ac$ be a square.
v	Both roots irrational,	if $b^2 - 4ac$ be not a square.
vi	Both roots positive,	if a and c differ in sign from b.
vii	Both roots negative,	if a, b and c are of the same sign.
viii	Roots of opposite signs,	if c and a differ in sign.
ix	Roots numerically equal but opposite in sign,	if $b = 0$.
x	One root zero,	if $c = 0$.
xi	One root infinite,	if $a = 0$.
xii	Both roots zero,	if $c = 0$, $b = 0$.
xiii	Both roots infinite,	if $a = 0$, $b = 0$.
xiv	Roots ∞ and 0,	if $a = 0$, $c = 0$.
xv	Roots indeterminate,	if $a = 0$, $b = 0$, $c = 0$.

*171. The statement that $ax^2 + bx + c = 0$ has one of its roots $= \infty$ if $a = 0$, and both its roots $= \infty$ if $a = b = 0$, requires explanation. The symbol 0 does not here indicate the total absence of magnitude; for, then, the equation would become reduced in degree, and we should get as quadratics such curious forms as $bx + c = 0$, and $c = 0$.

THE THEORY OF QUADRATIC EQUATIONS.

When we state that if $a = 0$, one of the roots $= \infty$, we mean that if a becomes infinitely small, b and c being any finite magnitudes whatever, one of the roots will become infinitely great. The student can easily convince himself of the truth of the statement by writing the roots in the equivalent forms (see § 158)

$$\frac{-b - \sqrt{(b^2 - 4ac)}}{2a} \text{ and } \frac{2c}{-b - \sqrt{(b^2 - 4ac)}}.$$

It is obvious, from these expressions, that as a is infinitely decreased, one of the roots becomes infinite, while the other root approximates to $-c/b$.

When $a = 0$, and $b = 0$, we find with the roots in their altered forms, $\dfrac{2c}{-b + \sqrt{(b^2 - 4ac)}}$ and $\dfrac{2c}{-b - \sqrt{(b^2 - 4ac)}}$, that, as a and b become smaller and smaller, the roots of the equation increase without limit.

We also note that when $a = 0$, $b = 0$, and $c = 0$, x may have any value whatever, and that $ax^2 + bx + c = 0$ is thus an identity.

172. Illustrative Examples.

Ex. 1. Show that $5ax^2 - (5a + 3)x + 3 = 0$ has rational roots.

The discriminant of the equation $= (5a+3)^2 - 4 \times 5a \times 3 = (5a-3)^2$ which is a perfect square.

Hence the roots are rational.

Ex. 2. Find m if $x^2 + 4mx + m^2 + m + 2 = 0$ has equal roots.
Since the roots are equal, $(4m)^2 - 4(m^2 + m + 2) = 0$.
Simplifying, we have $3m^2 - m - 2 = 0$.
Solving the equation, we have $m = 1$ or $-\frac{2}{3}$.

Ex. 3. Prove that $x^2 + 7x - 5 = 0$ has one of its roots positive and the other negative, and that the negative root is numerically greater than the positive root.

If a and β be the roots, $a\beta = -5$, which shows that one root is positive and the other negative.

And since the sum of the roots is -7 and one of them is positive, the other root which is negative must be numerically greater than the positive root by 7.

Ex. 4. Prove that $(b - x)^2 - 4(a - x)(c - x) = 0$ has real roots.
Since the given equation becomes
$$3x^2 + 2x(b - 2a - 2c) + (4ac - b^2) = 0,$$
its roots will be real if $4(b - 2a - 2c)^2 - 12(4ac - b^2)$ be positive, or zero.

THE THEORY OF QUADRATIC EQUATIONS. 143

Now, $4(b-2a-2c)^2 - 12(4ac-b^2)$
$= 4\{b^2 + 4a^2 + 4c^2 - 4ab - 4bc + 8ac - 12ac + 3b^2\}$
$= 8\{2a^2 + 2b^2 + 2c^2 - 2ab - 2bc - 2ca\}$
$= 8\{(a-b)^2 + (b-c)^2 + (c-a)^2\} =$ a positive quantity.

The discriminant of the equation is thus positive, and the roots are therefore real.

Ex. 5. Find when $\dfrac{a}{x-a} + \dfrac{b}{x-b} = 5$ has roots equal in magnitude, but opposite in sign.

$$\dfrac{a}{x-a} + \dfrac{b}{x-b} = \dfrac{x(a+b) - 2ab}{x^2 - x(a+b) + ab} = 5.$$

Hence the equation is $5x^2 - 6x(a+b) + 7ab = 0$.
Thus the required condition is $a + b = 0$.

Ex. 6. Prove that the roots of $x^2 + mx + 2m = 1$ are real and different, if those of $x^2 + 2x = m - 1$ are imaginary.

Since the roots of the second equation are imaginary $4 - 4(1-m)$ is negative.

Hence m is negative.

Now the roots of the first equation will be real if $m^2 - 4(2m-1)$, that is, $m^2 - 8m + 4$ be positive or zero; here $m^2 + 4$ is obviously positive, and as m is negative $-8m$ is also positive. The roots are therefore real.

Ex. 7. Show that the values of x obtained from the equations $ax^2 + by^2 = 1$, and $ax + by = 1$, will be equal, if $a + b = 1$.

Since $ax + by = 1$, $y^2 = \dfrac{1 + a^2x^2 - 2ax}{b^2}$.

Substituting this value of y^2 in $ax^2 + by^2 = 1$, we have
$$ax^2 + \dfrac{1 + a^2x^2 - 2ax}{b} = 1.$$

Thus, $x^2(a^2 + ab) - 2ax + (1-b) = 0$.

The roots of this equation will be equal, if $4a^2 = 4(a^2 + ab)(1-b)$.

$\therefore 4ab(a+b)$ must be equal to $4ab$, which will be the case if $a + b = 1$.

EXERCISES XII.

Discriminate the roots of the following equations without actual solution:

16. $5x^2 + 6x - 7 = 0$.   **17.** $3x^2 - 16x + 21 = 0$.

18. $(x-1)(x-7) = 2(x-3)(x-4)$.   **19.** $5x^2 + 26x + 5 = 0$.

20. For what value of m will one of the roots of
$$x^2 + (5m+4)x + (5m+4) = 0$$
be five times the other root?

21. When will the roots of $x^2 + px + q = 0$ be in the ratio of $2:5$?

22. For what values of p will $75x^2 + 7px + 3 = 0$ have equal roots?

23. Show that the roots of $25x^2 + 5(p+3)x + 3p = 0$ are rational.

24. For what value of n will $2x^2 + nx + 5 = 0$ have equal roots?

25. For what value of a will $x^2 - 4mx + 4x + 3m^2 - 2m + 4a = 0$ have rational roots?

26. Prove that the roots of $(x-p)(x-q) = 5$ are always **real**.

27. Show that the roots of
$$(a^2 - bc)x^2 + (a+b)(a-c)x + a(b-c) = 0$$
are real.

28. When will $\dfrac{1}{x} + \dfrac{1}{x+a} = \dfrac{1}{m} + \dfrac{1}{m+a}$ have its roots equal in magnitude, but opposite in sign?

29. Show that the roots of
$$\dfrac{1}{a(a+b)} + \dfrac{a}{(x-a)(x-b)(a+b)} = \dfrac{1}{a(x-a)} \text{ are equal.}$$

30. When will $\dfrac{a}{bx+c} + \dfrac{b}{cx+a} + \dfrac{c}{ax+b} = 0$ have only one finite root?

31. Show that, in the equation $x^2 + mx = n$, if n be positive the roots are of opposite signs, but if n be negative, the roots are of the same sign.

32. Show that the roots of $4a\{x^2 - (a+c)x + ac\} = b^2(a-x)$ are real.

33. Show that the roots of $(b+c)x^2 - x(a+b+c) + a = 0$ are rational.

34. Show that the roots of $ax^2 + bx + c = 0$ are rational, if $b = a + c$.

35. Show that the roots of $(a^2 + b^2) x^2 - 2b(a+c) x + (b^2 + c^2) = 0$ will be real, only if b be a mean proportional between a and c.

36. If the roots of $x^2 - px + q = 0$ are two consecutive integers, prove that $p^2 = 4q + 1$.

37. The two values of x obtained from $ax^2 + by^2 = max + nby = 1$ will be equal, if $m^2 a + n^2 b = 1$.

38. The two values of x obtained from $x^2 + y^2 = mx + ny = 2mx$ are real.

39. Show that the roots of $bx^2 + (b-c) x = (c + a - b)$ are real, if those of $ax^2 + b(2x + 1) = 0$ are imaginary.

40. If $x^2 + 2(a - b) x + ab = 0$ has imaginary roots, then
$$4x^2 + 4(a - b) x + (4a^2 + 4b^2 - 11ab) = 0$$
has real roots.

41. If the roots of $ax^2 + 2bx + c = 0$ be real, then those of
$$\frac{ax^2 + 2bx + c}{x^2 + 1} = \frac{2(ac - b^2)}{a + c}$$ are equal or imaginary.

173. To construct a quadratic whose roots are m and n.

First Method. Since m, n are roots,
$$\therefore \text{ either } x - m = 0, \text{ or } x - n = 0,$$
and in both cases,
$$\therefore (x - m)(x - n) = 0.$$
That is, $\qquad x^2 - x(m + n) + mn = 0,$
the required quadratic.

Second Method. Since the roots are given to be m and n, therefore taking the coefficient of $x^2 = 1$,

(i) the coefficient of $x =$ the sum of the roots with the sign changed $= -(m + n)$.

(ii) the absolute term = the product of the roots = mn.

Hence the equation is $x^2 - x(m + n) + mn = 0$, as before.

THE THEORY OF QUADRATIC EQUATIONS.

Ex. 1. Form an equation with 2 and $1\frac{1}{2}$ as the roots.

If the coefficient of x^2 is 1,
the coefficient of $x\quad = -(2+1\frac{1}{2}) = -3\frac{1}{2}$,
and the absolute term $= 2 \times 1\frac{1}{2} = +3$.

Hence the equation required is
$$x^2 - 3\frac{1}{2}x + 3 = 0.$$

Clearing the equation of fractions we have
$$2x^2 - 7x + 6 = 0.$$

Ex. 2. Form an equation whose roots are 1, 2 and 4.

An equation whose roots are 1, 2 and 4 is
$$(x-1)(x-2)(x-4) = 0.$$

Simplifying, we have
$$x^3 - 7x^2 + 14x - 8 = 0.$$

Ex. 3. Obtain the quadratic equation whose roots are $1 + \sqrt{5}$ and $1 - \sqrt{5}$.

The coefficient of x^2 being unity,
the coefficient of $x = -(1-\sqrt{5}+1+\sqrt{5}) = -2$;
and the third term $= (1-\sqrt{5})(1+\sqrt{5}) = 1 - 5 = -4$.

Hence the equation is $\quad x^2 - 2x - 4 = 0$.

Ex. 4. Find the quadratic equation with rational coefficients one of whose roots is $5 + \sqrt{6}$.

It is clear from § 156 that when one of the roots is $5 + \sqrt{6}$, the other root *must* be $5 - \sqrt{6}$, for the coefficients to be rational. [See Exercise 106, Chapter II.]

Hence the equation is $\{x - (5+\sqrt{6})\}\{x - (5-\sqrt{6})\} = 0$.

That is, $\quad\quad\quad x^2 - 10x + 19 = 0$.

Alternative Method. Let $x = 5 + \sqrt{6}$.
Then $\quad\quad x - 5 = \sqrt{6}. \quad \therefore (x-5)^2 = 6$.
$\therefore x^2 - 10x + 19 = 0$, as before.

The equation has $5 - \sqrt{6}$ also for a root, since $(x-5)^2 = 6$ can be derived from $x - 5 = -\sqrt{6}$ also.

THE THEORY OF QUADRATIC EQUATIONS. 147

Ex. 5. Find the quadratic equation with real coefficients one of whose roots is $5-\sqrt{(-2)}$.

Let $x=5-\sqrt{(-2)}$.

$\therefore x-5=-\sqrt{(-2)}$. $\therefore (x-5)^2=-2$.

Thus the required equation is $x^2-10x+27=0$.

As in Ex. 4 the other root must be $5+\sqrt{(-2)}$.

The following properties may now be enunciated:

If one root of a quadratic equation with rational coefficients be a + $\sqrt{b}$, *the other root is* a − $\sqrt{b}$.

Similarly *if one of the roots of a quadratic equation with real coefficients be* a + $\sqrt{(-b)}$, *the other root is* a − $\sqrt{(-b)}$.

EXERCISES XII.

Construct a quadratic equation

42. With 3 and $-2\frac{1}{2}$ as roots.
43. With 2 and $-3\frac{3}{5}$ as roots.
44. With $3+\sqrt{5}$ and $3-\sqrt{5}$ as roots.
45. With $\frac{p}{q}$ and $-\frac{q}{p}$ as roots.
46. With $2-\sqrt{(-4)}$ and $2+\sqrt{(-4)}$ as roots.
47. With real coefficients, and with $1-\sqrt{-1}$ for a root.
48. With $\frac{1}{3}(4+2\sqrt{5})$ and $\frac{1}{3}(4-2\sqrt{5})$ as roots.
49. With rational coefficients, and with $2+\sqrt{5}$ for a root.
50. With rational coefficients, and with $12+\sqrt{7}$ for a root.
51. Form an equation with 7, 2 and 1 for roots.
52. Form an equation with $1+\sqrt{2}$, $1-\sqrt{2}$, and 4 as roots.
53. Form an equation with $\pm 5\sqrt{6}-7$ as roots.

174. Symmetric Functions.—*To express in terms of the coefficients of a quadratic equation, any function involving the roots symmetrically.*

We have to express in terms of a, b, c, the values of such expressions as $\alpha^2+\beta^2$, $\alpha^4+\beta^4+4\alpha^2\beta^2$, where α and β are the roots of

$$ax^2+bx+c=0.$$

THE THEORY OF QUADRATIC EQUATIONS.

RULE. *Express the given expression in terms of $\alpha + \beta$ and $\alpha\beta$, and then employ the formulæ*

$$\alpha + \beta = -\frac{b}{a} \quad and \quad \alpha\beta = \frac{c}{a}.$$

Ex. 1. α and β being the roots of $ax^2 + bx + c = 0$, express $\alpha^3\beta + \alpha\beta^3$ in terms of a, b, c.

$$\alpha^3\beta + \alpha\beta^3 = \alpha\beta\,(\alpha^2 + \beta^2) = \alpha\beta\,\{(\alpha + \beta)^2 - 2\alpha\beta\}$$
$$= \frac{c}{a}\left\{\left(-\frac{b}{a}\right)^2 - 2\,\frac{c}{a}\right\} = \frac{c}{a^3}(b^2 - 2ac).$$

Ex. 2. If a and b are the roots of $x^2 - px + q = 0$, express $a^4 + b^4$ in terms of p and q.

Since a and b are the roots of $x^2 - px + q = 0$, $a + b = p$, and $ab = q$.

$$\therefore\ a^4 + b^4 = (a^2 + b^2)^2 - 2a^2b^2 = \{(a+b)^2 - 2ab\}^2 - 2a^2b^2$$
$$= (p^2 - 2q)^2 - 2q^2 = p^4 - 4p^2q + 2q^2.$$

Ex. 3. If m and n are the roots of $x^2 - 5x + 8 = 0$, find the value of $m^4n + mn^4$.

$$m^4n + mn^4 = mn\,(m^3 + n^3) = mn\,(m+n)\,(m^2 - mn + n^2)$$
$$= mn\,(m+n)\,\{(m+n)^2 - 3mn\}.$$

And since $m + n = 5$, and $mn = 8$, the expression
$$= 8 \times 5 \times (5^2 - 3 \cdot 8) = 40.$$

EXERCISES XII.

54. If α and β be the roots of $x^2 + mx + n = 0$, express in terms of m and n the expressions

(i) $\alpha^2 + \beta^2$, (ii) $\dfrac{\alpha}{\beta} + \dfrac{\beta}{\alpha}$, (iii) $\alpha^3 + \beta^3$, (iv) $\dfrac{\alpha^2}{\beta} + \dfrac{\beta^2}{\alpha}$, (v) $\dfrac{\alpha^2}{\beta^2} + \dfrac{\beta^2}{\alpha^2}$.

55. Find the sum of the cubes of the roots of $x^2 - 5x + 7 = 0$.

56. Find the sum of the squares of the roots of $x^2 + mx + n = 0$.

57. If α and β be the roots of $5x^2 + 7x + 3 = 0$,

show that (i) $\dfrac{\alpha^2 + \beta^2}{\alpha^{-2} + \beta^{-2}} = \dfrac{9}{25}$; (ii) $\dfrac{\alpha^3 + \beta^3}{\alpha^{-1} + \beta^{-1}} = \dfrac{12}{125}$.

THE THEORY OF QUADRATIC EQUATIONS.

58. If x_1 and x_2 be the roots of $x^2+mx+n=0$, find the values of

(i) $\dfrac{x_1^2}{x_2+m}+\dfrac{x_2^2}{x_1+m}$. (ii) $(x_2+m)^{-1}+(x_1+m)^{-1}$.

(iii) $(x_2+m)^{-2}+(x_1+m)^{-2}$. (iv) $(x_2+m)^{-3}+(x_1+m)^{-3}$.

59. If a and β be the roots of $px^2+qx+q=0$, prove that

$$\sqrt{\frac{a}{\beta}}+\sqrt{\frac{\beta}{a}}+\sqrt{\frac{q}{p}}=0.$$

60. If a and b are the roots of $x^2+px+q=0$, and c, d the roots of $x^2+rx+s=0$, express the following in terms of p, q, r, s:

(i) $a(b+c+d)+b(c+d)+cd$.

(ii) $(a-c)^2+(b-d)^2+(b-c)^2+(a-d)^2$.

175. We shall now give several examples showing how to form a quadratic equation, whose roots bear some symmetrical relation to those of a given quadratic.

From Ex. 1 below we see that the roots of $cx^2 + bx + a = 0$ are the reciprocals of those of $ax^2 + bx + c = 0$.

Ex. 1. Find the equation the roots of which are the reciprocals of the roots of $mx^2+nx+r=0$.

If a and β be the roots of this equation, the roots of the required equation are $\dfrac{1}{a}$ and $\dfrac{1}{\beta}$.

Hence the required equation is $\left(x-\dfrac{1}{a}\right)\left(x-\dfrac{1}{\beta}\right)=0$.

That is, $\qquad x^2-x\left(\dfrac{a+\beta}{a\beta}\right)+\dfrac{1}{a\beta}=0.$

And since $a+\beta=-\dfrac{n}{m}$ and $a\beta=\dfrac{r}{m}$; the required equation becomes

$$x^2+\frac{n}{r}x+\frac{m}{r}=0.$$

That is, $\qquad rx^2+nx+m=0.$

THE THEORY OF QUADRATIC EQUATIONS.

Alternative Method. Substituting $\dfrac{1}{x}$ for x in the given equation, the required equation is

$$m\left(\dfrac{1}{x}\right)^2 + n\left(\dfrac{1}{x}\right) + r = 0.$$

[For the values of $\dfrac{1}{x}$ in this equation are those of x in the quadratic $mx^2 + nx + r = 0$.]

On simplification, the required equation becomes

$$rx^2 + nx + m = 0,$$

as before.

Ex. 2. If α and β are the roots of $x^2 + px + q = 0$, find the equation whose roots are $\alpha\beta + \alpha + \beta$ and $\alpha\beta - \alpha - \beta$.

Since α and β are the roots of $x^2 + px + q = 0$, $\alpha + \beta = -p$ and $\alpha\beta = q$.

∴ the roots of the required equation are $q - p$ and $q + p$.

Hence the required equation is $(x - q + p)(x - q - p) = 0$,
that is $\qquad x^2 - 2qx + q^2 - p^2 = 0$.

Ex. 3. If α and β be the roots of $x^2 - 2x + 5 = 0$, find the equation whose roots are $\alpha + 7$ and $\beta + 7$.

The new equation can be readily obtained by substituting $x - 7$ for x in the given equation $x^2 - 2x + 5 = 0$.

Hence the required equation is

$$(x - 7)^2 - 2(x - 7) + 5 = 0$$

of which the solutions are evidently $x - 7 = \alpha$ or β, as required.

Simplifying, the equation is $x^2 - 16x + 68 = 0$.

Thus an equation can always be found by this method the roots of which differ from those of a given quadratic equation by a given quantity.

Ex. 4. If α and β be the roots of $x^2 + px + q = 0$, find the equation whose roots are $m\alpha$ and $m\beta$.

Let $y = mx$. Substituting y/m for x in the given quadratic we have

$$(y/m)^2 + p(y/m) + q = 0,$$

or $\qquad y^2 + pmy + qm^2 = 0$.

But since the values of x are α and β, the values of y are $m\alpha$ and $m\beta$. Writing x instead of y for the unknown quantity,

∴ the required equation is $x^2 + mpx + m^2q = 0$.

NOTE. The coefficients of the new equation are those of the given equation multiplied by 1, m, and m^2 respectively in order.

EXERCISES XII.

61. If a and β be the roots of $x^2 - 5x + a = 0$, find the equation whose roots are $(a+2)^2$, and $(\beta+2)^2$.

62. If a and β are the roots of $ax^2 + bx + c = 0$, find the equation whose roots are

(i) $a^2 + \beta^2$, and $(a+\beta)^2$.   (ii) $\dfrac{a}{\beta}$ and $\dfrac{\beta}{a}$.

(iii) $ma + \beta$, and $m\beta + a$.

63. If m and n be the roots of $x^2 + px + q = 0$, and m^2 and n^2 those of $x^2 + rx + s = 0$, find r and s in terms of p and q.

64. If the roots of $ax^2 + bx + c = 0$ are the reciprocals of those of $a'x^2 + b'x + c' = 0$, show that $ab' = bc'$, and $aa' = cc'$.

65. Form the equations whose roots are the reciprocals of those of

(i) $5x^2 - 20x + 17 = 0$,   (ii) $qx - r = px^2$,   (iii) $ax^2 + bx + c = 0$.

66. Find the equation whose roots are the squares of the roots of $x^2 + px + q = 0$.

67. Find the equation whose roots are numerically equal to those of $x^2 + 7x + 8 = 0$, but of opposite signs.

68. Find the equation the sum of whose roots is m, and the sum of the squares of whose roots is n^2.

69. If a and b are the roots of $x^2 + mx + n = 0$, and c and d those of $x^2 + px + q = 0$, find the equation whose roots are $ac + bd$ and $ad + bc$.

70. The coefficient of x in a quadratic equation of the form $x^2 + mx + n = 0$ is copied as 13 instead of 11, and the roots thus obtained are -10 and -3. Find the roots of the original equation in the book.

71. The absolute term in a quadratic equation of the form $x^2 + mx + n = 0$ is misprinted 32 for 36, and the roots are therefore obtained as -8 and -4. Find the roots of the equation correctly printed.

72. If in copying a quadratic equation of the form $x^2 + mx + n = 0$, a mistake be made in writing out the coefficient of x, the roots are found to be 1 and 6; but if the absolute term is incorrectly taken down, the roots are 4 and 1; what is the correct equation?

176. Factors of Quadratic Expressions.

Some of the following properties are already familiar to the student. They are collected here for convenience.

(I) IF a IS A ROOT OF THE EQUATION $ax^2 + bx + c = 0$, THEN $x - a$ IS A FACTOR OF THE EXPRESSION $ax^2 + bx + c$.

Dividing $ax^2 + bx + c$ by $x - a$, we find that $aa^2 + ba + c$ is the remainder.

And since a is a root of $ax^2 + bx + c = 0$, this remainder $= 0$, and $ax^2 + bx + c$ is therefore divisible by $x - a$.

(II) IF $x - a$ IS A FACTOR OF $ax^2 + bx + c$, THEN a IS A ROOT OF $ax^2 + bx + c = 0$. (Converse of I.)

The remainder, as in the previous case, is $aa^2 + ba + c$ which $= 0$, since $x - a$ is a factor of $ax^2 + bx + c$.

Thus a is a root of $ax^2 + bx + c = 0$.

(III) THE EXPRESSION $ax^2 + bx + c = a(x - a)(x - \beta)$ WHERE a AND β ARE THE ROOTS OF THE EQUATION
$$ax^2 + bx + c = 0.$$

For
$$ax^2 + bx + c = a\left\{\left(x + \frac{b}{2a}\right)^2 - \frac{b^2 - 4ac}{4a^2}\right\}$$
$$= a\left\{x + \frac{b}{2a} + \frac{\sqrt{(b^2 - 4ac)}}{2a}\right\}\left\{x + \frac{b}{2a} - \frac{\sqrt{(b^2 - 4ac)}}{2a}\right\}$$
$$= a\left\{x - \frac{-b - \sqrt{(b^2 - 4ac)}}{2a}\right\}\left\{x - \frac{-b + \sqrt{(b^2 - 4ac)}}{2a}\right\}$$
$$= a\{x - a\}\{x - \beta\},$$

where $a = \dfrac{-b + \sqrt{(b^2 - 4ac)}}{2a}$, $\beta = \dfrac{-b - \sqrt{(b^2 - 4ac)}}{2a}$.

[See § 157 for alternative proof. A third proof is suggested by (I) (II).]

Cor. Hence the factors of $ax^2 + bx + c$ are imaginary, real and equal, or real and different, according as $b^2 - 4ac$ is negative, zero, or positive.

Similarly the factors of $ax^2 + bx + c$ are rational or irrational according as $b^2 - 4ac$ is or is not a perfect square.

THE THEORY OF QUADRATIC EQUATIONS. 153

(IV) ax^2+bx+c IS A PERFECT SQUARE IN x IF $b^2-4ac=0$.

Since $ax^2+bx+c=a(x-\alpha)(x-\beta)$,

and since the two factors $(x-\alpha)$ and $(x-\beta)$ become equal when $b^2-4ac=0$, we get $ax^2+bx+c=a(x-\alpha)^2$. The given expression is thus a complete square so far as x is concerned. (Compare § 43, Ex. 2.)

(V) THE FACTORS OF

$$ax^2 + bxy + cy^2 \text{ ARE } a(x-\alpha y)(x-\beta y)$$

WHERE α AND β ARE THE ROOTS OF $ax^2+bx+c=0$.

$$ax^2 + bxy + cy^2 = y^2\left(a\frac{x^2}{y^2}+\frac{bx}{y}+c\right)$$
$$= ay^2\left(\frac{x}{y}-\alpha\right)\left(\frac{x}{y}-\beta\right) = ay\left(\frac{x}{y}-\alpha\right)y\left(\frac{x}{y}-\beta\right)$$
$$= a(x-\alpha y)(x-\beta y).$$

COR. $ax^2+bxy+cy^2$ *is a perfect square in* x *and* y, *if* $b^2-4ac=0$.

*177. $ax^2 + 2hxy + by^2 + 2gx + 2fy + c$

is resolvable into two rational factors, if

$$abc + 2fgh - af^2 - bg^2 - ch^2 = 0.$$

For $ax^2 + 2x(hy+g) + (by^2+2fy+c)$ can be resolved into rational factors, if $4(hy+g)^2 - 4a(by^2+2fy+c)$ is a perfect square.

Simplifying and arranging this in descending powers of y, we find (supposing a does not vanish) that

$$y^2(h^2-ab) + 2y(hg-af) + (g^2-ac)$$

must be a square.

Hence the discriminant of this expression $=0$. [§ 176 IV.]

That is, $(hg-af)^2 - (h^2-ab)(g^2-ac) = 0$

or $a\{abc+2fgh-af^2-bg^2-ch^2\} = 0.$

178. Illustrative Examples.

Ex. 1. For what values of m will $x^2-(m+1)x+m+12\tfrac{1}{4}$ be a perfect square?

The expression will be a perfect square if $(m+1)^2 - 4m - 49 = 0$.
That is, $m^2 - 2m - 48 = 0$. ∴ $m = 8$ or -6.

Ex. 2. If $ax^2+by^2+cz^2+2ayz+2bzx+2cxy$ is resolvable into rational factors, then $a^3+b^3+c^3=3abc$.

The factors of $ax^2+2x(bz+cy)+(by^2+cz^2+2ayz)=0$ will be rational, if $4(bz+cy)^2-4a(by^2+cz^2+2ayz)$ be a perfect square.

Arranging this expression in the descending powers of y, we must have
$$y^2(c^2-ab) + 2yz(bc-a^2) + z^2(b^2-ac), \text{ a perfect square.}$$
Hence $\quad 4(bc-a^2)^2 - 4(c^2-ab)(b^2-ac) = 0.$
$$\therefore (bc-a^2)^2 - (c^2-ab)(b^2-ac) = 0.$$
On simplification and transposition, we have $a^3 + b^3 + c^3 = 3abc$.

Ex. 3. Resolve $2x^2 - 3xy - 7xz - 2y^2 + 4yz + 6z^2$ into factors.

Equating the expression to 0, and solving for x, we have
$$x = \frac{3y + 7z \pm \sqrt{\{(3y+7z)^2 + 16(y^2 - 2yz - 3z^2)\}}}{4}$$
$$= \tfrac{1}{4}\{(3y+7z) \pm (5y+z)\}$$
$$= 2y + 2z \text{ or } -\tfrac{1}{2}y + \tfrac{3}{2}z.$$

Hence the given expression
$$= 2(x - 2y - 2z)(x + \tfrac{1}{2}y - \tfrac{3}{2}z) \qquad [\S\ 176\ \text{iii.}]$$
$$= (x - 2y - 2z)(2x + y - 3z).$$

EXERCISES XII.

73. For what values of m will $x^2 + (m+1)x + 5$ be resolvable into rational factors?

74. Show that $2x^2 + 3xy + y^2 + 2y + 3x + 1$ is resolvable into rational factors, and find them.

75. What value of m will make $2x^2 + 3y^2 + 7xy + 4x + my + 2$ resolvable into rational factors?

76. Show that $mxy - 4(x^2 - y^2)$ is resolvable into two real factors.

77. If $x^2 + y(2x+3) + 4(x+2) + (3y-5)$ is a perfect square, find y.

78. If $x^2 + 5x + 4$ and $x^2 + mx + n$ have a common factor, find the relation between m and n.

79. If $x^2 + 3xy + 2y^2$ and $x^2 + 4xy + my^2$ have a common factor, find m.

80. Find when $ax^2 + bxy + cy^2$ and $a'x^2 + b'xy + c'y^2$ have a common factor.

Resolve into factors:

81. $2x^2 - 5xy - x - 25y - 3y^2 - 28$.

82. $x^2 + xy - 2y^2 - 11yz + 2xz - 15z^2$.

83. $2x^2 + xz - 2y^2 - 3yz - z^2$.

THE THEORY OF QUADRATIC EQUATIONS.

179. To find the condition that two quadratics may have two roots in common.

Let a, β be the roots of $ax^2 + bx + c = 0$, and a' and β' the roots of $a'x^2 + b'x + c' = 0$.

The equations will have both of their roots in common, only if $a + \beta = a' + \beta'$ and $a\beta = a'\beta'$.

Since $a + \beta$ is to be equal to $a' + \beta'$, $-\dfrac{b}{a} = -\dfrac{b'}{a'}$.

And since $a\beta$ is to be equal to $a'\beta'$, $\dfrac{c}{a} = \dfrac{c'}{a'}$.

$$\therefore \frac{a}{a'} = \frac{b}{b'} = \frac{c}{c'}.$$

Hence *two quadratic equations will have both of their roots common if the ratios of their corresponding coefficients are equal.*

NOTE. The two equations are equivalent; that is, the one is derivable from the other by multiplying by a constant.

180. To find the condition that two quadratics may have one root in common.

Let a be the common root of the two equations

$$ax^2 + bx + c = 0 \text{ and } a'x^2 + b'x + c' = 0.$$

Since a is a root of both the equations,

$$aa^2 + ba + c = 0 \dots\dots\dots\dots\dots \text{(i)},$$
and
$$a'a^2 + b'a + c' = 0 \dots\dots\dots\dots\dots \text{(ii)}.$$

By the rule of cross-multiplication [§ 68],

$$\frac{a^2}{bc' - b'c} = \frac{a}{ca' - c'a} = \frac{1}{ab' - a'b}.$$

$$\therefore \frac{a^2}{(bc' - b'c)} \times \frac{1}{(ab' - a'b)} = \left(\frac{a}{ca' - c'a}\right)^2.$$

Hence the condition required is

$$(ca' - c'a)^2 = (ab' - a'b)(bc' - b'c).$$

THE THEORY OF QUADRATIC EQUATIONS.

Cor. When this condition is satisfied, $ax^2 + bx + c$ and $a'x^2 + b'x + c'$ have a common linear factor, viz. $x - a$ (see § 176, 1).

Note. The common root (which is a) $= \dfrac{ca' - c'a}{ab' - a'b} = \dfrac{bc' - b'c}{ca' - c'a}$.

The other roots can be found by applying the formula for $a + \beta$, or for $a\beta$.

Ex. 1. Find the condition that $ax^3 + bx^2 + c = 0$ and $cx^3 + ax^2 + b = 0$ may have a common root.

Let a be the common root. Proceeding as in the foregoing proof, we get

$$\frac{a^3}{b^2 - ac} = \frac{a^2}{c^2 - ab} = \frac{1}{a^2 - bc}.$$

$\therefore a^3 = \dfrac{b^2 - ac}{a^2 - bc}$; and $a^2 = \dfrac{c^2 - ab}{a^2 - bc}$.

Eliminating a, we have $\left(\dfrac{b^2 - ac}{a^2 - bc}\right)^2 = \left(\dfrac{c^2 - ab}{a^2 - bc}\right)^3$.

Hence $(c^2 - ab)^3 = (b^2 - ac)^2 (a^2 - bc)$.

Ex. 2. If the ratios of the roots of $ax^2 + bx + c = 0$, and $px^2 + qx + r = 0$ be equal, then $b^2/ac = q^2/pr$.

Let the roots of the first equation be a and β, and those of the second a' and β'; and let $a/\beta = a'/\beta' = k$ (suppose).

Then $a = k\beta$; and $a' = k\beta'$.

Taking the first equation, we have

$$a + \beta = k\beta + \beta = \beta(k + 1) = -b/a \quad \dots\dots\dots\dots\dots\text{(i)};$$
and $$a\beta = k\beta^2 = c/a \quad \dots\dots\dots\dots\dots\dots\dots\dots\text{(ii)}.$$

To eliminate β, we divide the square of (i) by (ii).

Hence $\dfrac{(k+1)^2}{k} = \dfrac{b^2}{a^2} \div \dfrac{c}{a} = \dfrac{b^2}{ac}$.

Similarly from the second equation we get $\dfrac{(k+1)^2}{k} = \dfrac{q^2}{pr}$.

Hence $b^2/ac = q^2/pr$.

Ex. 3. The two equations $x^2 - cx + d = 0$, and
$$x^2 - ax + b = 0;$$
have one common root and the second equation has equal roots. Show that $2(b + d) = ac$.

If a and β be the roots of the first equation, and a and a those of the second, then $a + \beta = c$; $a\beta = d$; $2a = a$; and $a^2 = b$.

$\therefore 2(b + d) = 2(a^2 + a\beta) = 2a(a + \beta) = ac$.

EXERCISES XII.

84. Show that the roots of $ax^2+bx+c=0$ are the reciprocals of the roots of $px^2+qx+r=0$, if $a:b:c::r:q:p$.

85. Show that the roots of $5x^2+6x+m=0$ are the reciprocals of the roots of $mx^2+6x+5=0$.

86. When will the roots of $ax^2+bx+c=0$ be treble the roots of $bx^2+cx+a=0$?

87. Show that the roots of $px^2+qx+r=0$ are r/p times the roots of $rx^2+qx+p=0$.

88. Show that the ratio of the roots of
$$(a-b)x^2+(b-c)x+(c-a)=0$$
is equal to that of the roots of $(c-a)x^2+(b-c)x+(a-b)=0$.

89. If $x^2+7x+12=0$ and $x^2+21x+a=0$ are such that the sums of the squares of their roots are equal, show that $a=208$.

90. Show that the roots of one of the equations in Question 88 are the reciprocals of those of the other.

91. When are the roots of $x^2+px+q=0$ the reciprocals of those of
$$x^2+rx+s=0?$$

92. Show that the equations $x^2+px+r=0$ and $x^2+rx+p=0$ will have a common root if $p=r$, or if $1+p+r=0$.

93. Prove that the ratios of the roots of $x^2+px+r=0$ and $x^2-px+r=0$ are equal.

94. Show (by inspection) that $(a-b)x^2+(b-c)x+(c-a)=0$, and $(c-a)x^2+(a-b)x+(b-c)=0$ have a common root.

95. If $p+q+r=0$, show that each pair of the equations $x^2+px+qr=0$; $x^2+qx+rp=0$, and $x^2+rx+pq=0$ will have a common root.

96. Prove that if $ax^2+bx+c=0$ and $bx^2+cx+a=0$ have a common root, either $a=0$, or $a^3+b^3+c^3=3abc$.

97. If $ax^2+bx+c=0$ has a common root with $bx^2+cx+a=0$, then its other root is a root of $a^2bx^2+ax(2b^2-ac)+a^3+b^3-abc=0$.

98. If a, β are the roots of $ax^2+bx+c=0$, and γ, δ those of $a'x^2+b'x+c'=0$, prove that
$$a^2a'^2(a-\gamma)(a-\delta)(\beta-\gamma)(\beta-\delta)=(ca'-c'a)^2-(ab'-a'b)(bc'-b'c).$$
Hence deduce the condition of § 180.

CHAPTER XIII.

MAXIMA AND MINIMA.

181. THE property that **the square of a real quantity can never be negative** introduces us at once to the notion of maximum and minimum values.

Thus when x is real, x^2 is either positive or zero, hence if a, b are constants $a^2 + x^2$ is never less than a^2 and $b^2 - x^2$ is never greater than b^2. We therefore say that the *minimum* value of $a^2 + x^2$ is a^2, and that the *maximum* value of $b^2 - x^2$ is b^2. In deducing similar results for what are sometimes called *adfected* quadratic expressions (i.e. expressions of the form $ax^2 + bx + c$ where b is different from zero), the examples that follow will show that the best rule in most cases is to *make the variable part of such an expression into a complete square or sum of squares*.

DEFINITION. An expression (e.g. $ax^2 + bx + c$) whose value changes when the variable x changes in value is often called a **function** of x, and is sometimes written thus: $f(x)$. The general type of a **rational integral function** of x is

$$ax^n + bx^{n-1} + cx^{n-2} + \ldots\ldots + px + q,$$

where n is a positive integer. The number of terms in this function is evidently $n + 1$, the function being of the nth degree; but some of the terms may, of course, be wanting, the coefficients of these terms being equal to zero.

If, in any investigation, $f(x) = 2x^3 + 5x^2 + 7$,
then $f(y) = 2y^3 + 5y^2 + 7$;
and $f(5) = 2 \cdot 5^3 + 5 \cdot 5^2 + 7 = 382.$

NOTE. The quadratic equation $ax^2 + bx + c = 0$ is nothing but the quadratic expression $ax^2 + bx + c$ equated to zero. The question, "Solve $ax^2 + bx + c = 0$" means "find those particular values of x for which the expression $ax^2 + bx + c$ assumes the value zero." Similarly to find the values of x which make $ax^2 + bx + c$ equal to a given value k we should only have to solve the quadratic equation $ax^2 + bx + (c - k) = 0$.

MAXIMA AND MINIMA.

182. Illustrative Examples.

In the examples that follow, every given quadratic expression should first be represented in the form

$$a\left\{\left(x+\frac{b}{2a}\right)^2 - \frac{b^2-4ac}{4a^2}\right\},$$

and should then be resolved, when $b^2 - 4ac$ is positive, into real and linear factors, $a(x-\alpha)(x-\beta)$.

Ex. 1. For what values of x is $x^2 - 2x + 10$ least?

$$x^2 - 2x + 10 = (x-1)^2 + 9.$$

As x is real, the least value of $(x-1)^2$ is 0, and therefore that of $x^2 - 2x + 10$ is 9.

Ex. 2. Find the sign of $4x - x^2 - 4$ for real values of x.

$$4x - x^2 - 4 = -(x^2 - 4x + 4) = -(x-2)^2.$$

Since $(x-2)^2$ is always positive for real values of x except when $x = 2$, the given expression is always negative except when $x = 2$, in which case it vanishes.

Ex. 3. Find when $2x^2 - 7x + 5$ is negative.

Call the expression $= f(x)$.

Then $f(x) = 2x^2 - 7x + 5 = 2(x^2 - 3\frac{1}{2}x + 2\frac{1}{2}) = 2\{(x - 1\frac{3}{4})^2 - \frac{9}{16}\}$
$= 2(x - 1\frac{3}{4} + \frac{3}{4})(x - 1\frac{3}{4} - \frac{3}{4}) = 2(x-1)(x-2\frac{1}{2}).$

When $x > 2\frac{1}{2}$, $x - 2\frac{1}{2}$ and $x - 1$ are both positive and $(x-1)(x-2\frac{1}{2})$ is therefore positive. Thus $f(x)$ is positive.

When $x = 2\frac{1}{2}$, $f(x) = 0$.

When $x < 2\frac{1}{2}$ and > 1, $x - 1$ is positive and $x - 2\frac{1}{2}$ is negative, and $f(x)$ is therefore negative.

When $x = 1$, $f(x) = 0$.

When $x < 1$, $x - 1$ is negative and $x - 2\frac{1}{2}$ is negative, and $f(x)$ is therefore positive.

Hence $f(x)$ is negative only for values of x between $2\frac{1}{2}$ and 1.

183. To investigate the sign of $ax^2 + bx + c$ for real values of x. We have

$$ax^2 + bx + c = a\left\{x^2 + \frac{bx}{a} + \frac{c}{a}\right\} = a\left\{x^2 + \frac{b}{a}x + \frac{b^2}{4a^2} - \frac{b^2}{4a^2} + \frac{c}{a}\right\}$$

$$= a\left\{\left(x + \frac{b}{2a}\right)^2 - \frac{b^2 - 4ac}{4a^2}\right\}.$$

CASE I. LET $b^2 - 4ac$ BE NEGATIVE.

Since $b^2 - 4ac$ is negative, $(b^2 - 4ac)/4a^2$ is negative, and therefore $-(b^2 - 4ac)/4a^2$ is positive. And $(x + b/2a)^2$, being a square, is necessarily positive or zero for real values of x.

Hence the expression within the large brackets is positive, and $ax^2 + bx + c$ is therefore always of the same sign as a.

CASE II. LET $b^2 - 4ac = 0$.

Then $ax^2 + bx + c = a\left(x + \frac{b}{2a}\right)^2$.

Here $(x + b/2a)^2$ is always positive, except when $x = -b/2a$ in which case $(x + b/2a)^2 = 0$.

Hence $ax^2 + bx + c$ has always the same sign as a, except when $x = -b/2a$ in which case $ax^2 + bx + c = 0$.

CASE III. LET $b^2 - 4ac$ BE POSITIVE.

Then $ax^2 + bx + c$ and a will have the same sign if and only if

$$\left(x + \frac{b}{2a}\right)^2 - \frac{b^2 - 4ac}{4a^2} \text{ be positive; i.e. if } \left(x + \frac{b}{2a}\right)^2 > \frac{b^2 - 4ac}{4a^2},$$

i.e. if $x + \dfrac{b}{2a}$ be numerically greater than $\dfrac{\sqrt{(b^2 - 4ac)}}{2a}$,

i.e. if $x + \dfrac{b}{2a}$ be either positive and $> \dfrac{+\sqrt{(b^2 - 4ac)}}{2a}$ or negative and algebraically $< \dfrac{-\sqrt{(b^2 - 4ac)}}{2a}$,

i.e. if either $x > \dfrac{-b + \sqrt{(b^2 - 4ac)}}{2a}$ or $x < \dfrac{-b - \sqrt{(b^2 - 4ac)}}{2a}$.

In like manner $ax^2 + bx + c$ and a will differ in sign if x lies between the limits $\dfrac{-b + \sqrt{(b^2 - 4ac)}}{2a}$ and $\dfrac{-b - \sqrt{(b^2 - 4ac)}}{2a}$.

MAXIMA AND MINIMA. 161

Hence the general result of our investigation is as follows :

ax² + bx + c *and a have the same sign for real values of* x, *except when* b² − 4ac *is positive and* x *is taken so as to lie between* $\dfrac{-b+\sqrt{(b^2-4ac)}}{2a}$ *and* $\dfrac{-b-\sqrt{(b^2-4ac)}}{2a}$.

184. Connection between the sign of $ax^2 + bx + c$ and the roots of $ax^2 + bx + c = 0$.

Since the roots of $ax^2 + bx + c = 0$ are imaginary, real and equal, or real and unequal, according as $b^2 - 4ac$ is negative, zero, or positive, the results of the last article may be stated as follows :

$ax^2 + bx + c$ and a agree in sign except when the roots of
$$ax^2 + bx + c = 0$$
are real and different, and x lies between the roots.

Ex. 1. Find the sign of $2x^2 + x + 8$ for real values of x.
$$2x^2 + x + 8 = 2(x^2 + \tfrac{1}{2}x + 4)$$
$$= 2(x^2 + \tfrac{1}{2}x + \tfrac{1}{16} + 3\tfrac{15}{16}) = 2\{(x + \tfrac{1}{4})^2 + 3\tfrac{15}{16}\}.$$

Here the expression within the double brackets is necessarily positive for real values of x, since $(x+\tfrac{1}{4})^2$ can never be negative. The given expression is therefore positive.

Ex 2. Find the sign of $3x^2-10x+3$ for different real values of x.
$$3x^2 - 10x + 3 = 3\{x^2 - \tfrac{10}{3}x + 1\} = 3\{(x - \tfrac{5}{3})^2 - \tfrac{16}{9}\}$$
$$= 3(x - \tfrac{5}{3} + \tfrac{4}{3})(x - \tfrac{5}{3} - \tfrac{4}{3}) = 3(x - \tfrac{1}{3})(x - 3).$$

Hence the expression will be positive if x lies outside the limits $\tfrac{1}{3}$ and 3, negative if x lies between $\tfrac{1}{3}$ and 3, and zero if $x = \tfrac{1}{3}$ or 3.

Ex. 3. To investigate the sign of $\dfrac{4-x}{x^2-7x+15}$.

The expression $= \dfrac{4-x}{(x-3\tfrac{1}{2})^2 + 2\tfrac{3}{4}}$.

It is evident that the denominator must be positive for all real values of x. Hence we have only to attend to the sign of the numerator.

Now for all values of x from $+\infty$ to $+4$, the numerator is negative, and therefore the given fraction is also negative. If $x = 4$, the given expression $= 0$. If x lies between $+4$ and $-\infty$, the numerator is positive, and the fraction is also therefore positive.

Ex. 4. For what values of x is
$(2x - 7 - 5x^2)(x^2 - 7x + 12)(x + 3)$ positive?

The expression $= -5(x^2 - \tfrac{2}{5}x + \tfrac{7}{5})(x^2 - 7x + 12)(x + 3)$
$= -5\{(x - \tfrac{1}{5})^2 + \tfrac{34}{25}\}\{(x - 3\tfrac{1}{2})^2 - (\tfrac{1}{2})^2\}(x + 3)$
$= -5\{(x - \tfrac{1}{5})^2 + \tfrac{34}{25}\}(x - 4)(x - 3)\{x - (-3)\}$
$=$ a negative quantity $\times (x - 4)(x - 3)\{x - (-3)\}$.

Here when $x > 4$, the expression is negative.

$x = 4$,, ,, zero.
$x < 4$ and > 3 ,, ,, positive.
$x = 3$,, ,, zero.
$x < 3$ and > -3 ,, ,, negative.
$x = -3$,, ,, zero.
$x < -3$,, ,, positive.

Hence the expression is positive, if x lies between 3 and 4, or x is less than -3.

Ex. 5. Show that $a^2x^2 + (b - k)x + 1$ will have the same sign for all real values of x, if k lies within certain limits.

The expression will always have the same sign (that of a^2) only if $(b - k)^2 - 4a^2$ is negative or zero.

That is, if $(b - k + 2a)(b - k - 2a)$ is negative or zero.

That is, if $\{k - (b + 2a)\}\{k - (b - 2a)\}$ is negative or zero.

That is, if k does not lie without the limits $b + 2a$ and $b - 2a$.

EXERCISES XIII.

Investigate the sign of the following expressions for real values of x:

1. $x + 1$. 2. $a^2 - x$. 3. $-3a^2/x$. 4. $1/(4 - x)$.
5. $x^2 - 20x + 99$. 6. $x^2 + 16x + 64$.
7. $20x - x^2 - 99$. 8. $2x - 3 - 4x^2$.
9. $7x^2 + 6x + 7$. 10. $5x - x^2 - 4$.
11. $(x^2 - 3x + 2)(x^2 - x + 7)$. 12. $(x^2 - 3x + 17)(3x - x^2 - 2)$.
13. $\dfrac{x^2 - 2x + 4}{x - 3}$. 14. $(x^2 + 6x + 9)(x + 4)$.

15. $\dfrac{x^2-3x+2}{x-3}$. 16. $\dfrac{x^2-3x+2}{12x-x^2-32}$.

17. $(x^2-5x+6\tfrac{1}{4})(3x+2-2x^2)$. 18. $a^2x^2+4ax+7$.

19. If the roots of $ax^2+bx+c=0$ are imaginary, then
$a^2x^2+abx+ac$ is positive.

20. If a and b have opposite signs, then $(b-a-cx)^2-4ax(c+bx)$ is always positive.

21. Show that ax^2+bx+a will always have the same sign, if b lies between $2a$ and $-2a$.

22. Show that $mx^2+(b-m)x+m$ will always have the same sign, if m lies beyond the limits $-b$ and $b/3$.

23. Show that $x^2+(2b-m)x+b^2$ will always have the same sign if m lies between 0 and $4b$.

185. To trace the changes in the value of ax^2+bx+c as x passes through all real values from $-\infty$ to $+\infty$.

We have $ax^2+bx+c = a\left(x^2+\dfrac{b}{a}x+\dfrac{c}{a}\right)$

$$= a\left\{\left(x+\dfrac{b}{2a}\right)^2 - \dfrac{b^2-4ac}{4a^2}\right\}.$$

As x increases from $-\infty$ to $-b/2a$

$(x+b/2a)^2$ decreases from $+\infty$ to 0,

$\therefore x^2+bx/a+c/a$ decreases from $+\infty$ to $-(b^2-4ac)/4a^2$.

As x increases from $-b/2a$ to $+\infty$

$(x+b/2a)^2$ increases from 0 to $+\infty$,

$\therefore x^2+bx/a+c/a$ increases from $-(b^2-4ac)/4a^2$ to $+\infty$.

Hence the *minimum* value of

$x^2+bx/a+c/a$ is $-(b^2-4ac)/4a^2$

and this value corresponds to $x=-b/2a$.

Hence (i) if a is *positive*, the *minimum* value of

ax^2+bx+c is $-(b^2-4ac)/4a$

and the expression is capable of assuming all values between this and $+\infty$;

(ii) if a is *negative*, the *maximum* value of
$$ax^2 + bx + c \text{ is } -(b^2 - 4ac)/4a$$
and the expression is capable of assuming all values between this and $-\infty$.

COR. 1. If $b^2 - 4ac$ is negative the minimum value of
$$x^2 + bx/a + c/a$$
is positive and therefore the expression itself is always positive since it cannot be less than its minimum value. (§ 183, Case I.)

COR. 2. If $b^2 - 4ac$ is positive, then $x^2 + bx/a + c/a$ decreases from $+\infty$ to the *negative* value $-(b^2 - 4ac)/4a^2$ and then increases to $+\infty$. In passing from positive to negative values and *vice versâ* the expression must pass through the value zero.

Therefore $\qquad x^2 + bx/a + c/a$

is negative when x lies between the values for which it vanishes, i.e. when x lies between the roots of $ax^2 + bx + c = 0$. (§ 183, Case III.)

186. Useful method for finding maxima and minima.

That $-(b^2 - 4ac)/4a$ is the maximum or minimum value of $ax^2 + bx + c$, according to the sign of a, and that this value corresponds to $x = -b/2a$ can also be proved as follows :

Let $\qquad ax^2 + bx + c = y.$
Then $\qquad ax^2 + bx + c - y = 0.$
$$\therefore x = \frac{-b \pm \sqrt{\{b^2 - 4a(c+y)\}}}{2a}.$$

For x to be real, the least value of $b^2 - 4ac + 4ay$ is zero.

$\therefore$ the least value of $4ay$ is $-(b^2 - 4ac)$, giving $x = -b/2a$.

Hence the limit to the value of the expression
$$ax^2 + bx + c \text{ is } -(b^2 - 4ac)/4a.$$

187. Maxima and Minima of fractional expressions.

It will be found that equations similar to those of the previous article are obtained immediately on equating the following expressions to y.

Ex. 1. Is $\dfrac{2x-3}{x^2}$ capable of assuming all real values ?

Let the given expression $= y$, $\therefore 2x - 3 = yx^2$.
Hence $\qquad yx^2 - 2x + 3 = 0.$

MAXIMA AND MINIMA.

As x is real, $2^2 - 4 \cdot y \cdot 3$ is positive, or zero.

That is, $12y - 4$ is negative or zero, $\therefore$ the greatest value of $12y$ is 4.

$\therefore$ the greatest value of y is $\frac{1}{3}$.

Hence y may have any value between $-\infty$ and $\frac{1}{3}$ (inclusive), but cannot have any value between $\frac{1}{3}$ and $+\infty$.

Ex. 2. Is $\dfrac{1}{2x - x^2 - 7}$ capable of assuming all real values?

Let the expression $= y$, $\therefore yx^2 - 2yx + (7y + 1) = 0$.

The discriminant $= 4y^2 - 4y(7y+1)$
$= -24y^2 - 4y = -24y(y + \frac{1}{6}) = -24(y - 0)\{y - (-\frac{1}{6})\}$.

Since this must be zero or positive if x is real, $y - 0$ and $y - (-\frac{1}{6})$ must be of opposite signs.

Hence y cannot lie beyond the limits 0 and $-\frac{1}{6}$.

Alternative Method. The denominator
$$2x - x^2 - 7 = -6 - (x-1)^2.$$

For real values of x this lies between $-\infty$ and -6,

$$\therefore \frac{1}{2x - x^2 - 7} \text{ must lie between } -\frac{1}{\infty} \text{ and } -\frac{1}{6},$$

that is, between 0 and $-\frac{1}{6}$.

188. If the sum of two quantities is constant, their product is greatest when they are equal.

This theorem is identical with the following geometrical theorem.

Ex. Divide a straight line so that the rectangle contained by the two parts may be the greatest possible.

First Method. If the two parts be x and y, so that the length of the given line is $x + y$, the identity
$$4xy = (x+y)^2 - (x-y)^2$$
shows that the product xy will be greatest when $(x-y)^2 = 0$; that is, when $x = y$, or the two parts are equal.

Hence the line must be bisected.

Second Method. If the given line be taken to be $2a$, and one of the parts be $a + x$, the other part is $a - x$, and the rectangle is therefore $a^2 - x^2$. It is evident that $a^2 - x^2$ is greatest when $x = 0$, in which case the two parts are equal.

166   MAXIMA AND MINIMA.

Third Method. Let a be the whole line, and x one of the parts. Then the other part is $a-x$. And we have to find for what value of x the value of $x(a-x)$ is greatest.

Now $\quad x(a-x) = xa - x^2 = \tfrac{1}{4}a^2 - (\tfrac{1}{4}a^2 - xa + x^2)$
$\qquad\qquad\qquad = \tfrac{1}{4}a^2 - (\tfrac{1}{2}a - x)^2.$

This identity shows that $x(a-x)$ is greatest when $(\tfrac{1}{2}a-x)^2 = 0$, that is, when $x = \tfrac{1}{2}a$. Hence the line has to be bisected as before.

Fourth Method. Let $x(a-x) = y$. $\therefore x^2 - ax + y = 0$.

Hence $\quad\quad\quad\quad x = \tfrac{1}{2}\{a \pm \sqrt{(a^2 - 4y)}\}.$

Since x is to be real, $a^2 - 4y$ must not be negative.

Hence the greatest value of $4y = a^2$, which shows that the greatest value of $y = \tfrac{1}{4}a^2$.

In this case $x = \tfrac{1}{2}a$, showing that the line is bisected.

Fifth Method. A geometrical solution can be readily inferred from Euc. II. 5.

189. If the product of two quantities is constant, their sum is numerically least when they are equal.

This is proved in Ex. 1 below. In like manner the identities

$$2(x^2+y^2) = (x+y)^2 + (x-y)^2 \text{ and } x^2 + y^2 = 2xy + (x-y)^2$$

show that if either $x+y$ or xy is given, x^2+y^2 is least when $x=y$, and if x^2+y^2 is given both xy and $(x+y)^2$ are greatest when $x=y$.

Ex. 1. Of all rectangles having the same area, the rectangle which has the least perimeter is a square.

Let x and y be two adjacent sides of a rectangle, so that the given fixed area may be represented by xy; the identity

$$(x+y)^2 = 4xy + (x-y)^2$$

shows that $(x+y)^2$ is least when $(x-y)^2 = 0$. Hence the perimeter, $2(x+y)$, is least when the sides of the rectangle, x and y, are equal.

Ex. 2. Find the maximum and minimum values of $\dfrac{x+1}{x^2+x+1}$.

Let $\qquad\qquad x+1 = k. \quad \therefore x = k-1.$

$\therefore$ the expression $= \dfrac{k}{(k-1)^2+(k-1)+1} = \dfrac{k}{k^2-k+1} = \dfrac{1}{k+k^{-1}-1}.$

Since $\left(k+\dfrac{1}{k}\right)^2 = \left(k-\dfrac{1}{k}\right)^2 + 4$, its minimum value is 4.

Hence if $k+k^{-1}$ is positive it is not less than 2, so that the minimum value of $k+k^{-1}-1$ is 1 and the maximum value of the given expression $= 1/1 = 1$.

Similarly if $k+k^{-1}$ is negative its maximum value is -2 and the minimum value of the given expression $= -\frac{1}{3}$.

190. Maxima and minima of two variables connected by a quadratic equation.

Ex. 1. For what limits of y is x real in the equation

$$yx^2 - (2y+1)x + y - 3 = 0?$$

Since x is to be real, the discriminant of this equation must be zero or positive.

This discriminant $= (2y+1)^2 - 4y(y-3) = 16y + 1$.

Hence $16y + 1$ must be zero or positive.

Hence y must not be less than $-\frac{1}{16}$.

Ex. 2. Find the maximum and minimum values of y for which x is real in

$$x^2 - 2xy - (y+3) = x(2xy + 3).$$

Transpose all the terms to one side, and arrange them in descending powers of x, thus:

$$x^2(1-2y) - x(2y+3) - (y+3) = 0.$$

Since x is real, the discriminant of this equation must be zero, or positive. This discriminant

$$= (2y+3)^2 + 4(1-2y)(y+3)$$
$$= 21 - 4y^2 - 8y = -4\left(y - \tfrac{3}{2}\right)\left(y + \tfrac{7}{2}\right).$$

$\therefore -4\left(y - \tfrac{3}{2}\right)\left(y + \tfrac{7}{2}\right)$ must be zero or positive*.

$\therefore \left\{y - \tfrac{3}{2}\right\}\left\{y - \left(-\tfrac{7}{2}\right)\right\}$ must be zero or negative.

$\therefore y$ must not lie beyond the limits $\tfrac{3}{2}$ and $-\tfrac{7}{2}$.

Hence if x be real, the maximum value of y is $\tfrac{3}{2}$, and its minimum value $-\tfrac{7}{2}$.

---

* When the discriminant is a quadratic expression it should be resolved into simple *real* factors whenever this is possible—which will be the case when the discriminant of this discriminant is positive.

168 MAXIMA AND MINIMA.

Ex. 3. If x and y are real in $x^2 + 3y^2 = 2x + 12y + 3$, find the maximum and minimum values of x and y.

The question may be solved as in the first two examples, but the following method will also be found instructive. Writing the equation
$$x^2 - 2x + 3y^2 - 12y - 3 = 0,$$
then
$$(x-1)^2 - 1 + 3(y-2)^2 - 12 - 3 = 0.$$
$$\therefore (x-1)^2 + 3(y-2)^2 = 16.$$

Since $y - 2$ is real the greatest value of $(x-1)^2$ is 16.

Hence $x - 1$ lies between $+4$ and -4.

$\therefore x$ lies between 5 and -3.

Similarly the greatest value of $3(y-2)^2$ is 16.

Hence $(y-2)^2$ is less than $1\tfrac{1}{3}$.

$\therefore y - 2$ lies between $+\sqrt{1\tfrac{1}{3}}$ and $-\sqrt{1\tfrac{1}{3}}$.

$\therefore y$ lies between $2 + \tfrac{1}{3}\sqrt{3}$ and $2 - \tfrac{1}{3}\sqrt{3}$.

Ex. 4. Under what conditions is $(x-a)(x-c)/(x-b)$ capable of assuming all real values?

Let the given expression $= y$.

Then $\qquad x^2 - x(a+c+y) + (ac+by) = 0.$

Since x is real, $(a+c+y)^2 - 4(ac+by)$ must not be negative.

Hence $y^2 + 2y(a+c-2b) + (a-c)^2$ must not be negative, that is, must not have a different sign from the coefficient of y^2.

Hence the discriminant of this expression, viz.,
$$4(a+c-2b)^2 - 4(a-c)^2 = 16(b-a)(b-c),$$
must be negative or zero*.

Hence $b - a$ and $b - c$ must be of opposite signs, and the required condition is that b must not lie beyond the limits a and c.

Ex. 5. If $\dfrac{x^2 - x}{1 - px}$ is capable of assuming all real values for real values of x, find the limits of p.

Let the expression $= y$. Then
$$x^2 - x = y - pxy, \therefore x^2 - x(1-py) - y = 0. \quad\dots\dots\dots\dots(i)$$
The discriminant $= (1-py)^2 + 4y$
$$= p^2y^2 - 2y(p-2) + 1. \quad\dots\dots\dots\dots\dots\dots\dots(ii)$$

---

* If this discriminant be positive, the expression in y will be sometimes positive and sometimes negative, and therefore x will not be real for *all* values of y.

This last expression must be positive for all values of y.

Now the coefficient of y^2, which is p^2, is clearly positive. Hence the discriminant of the expression (ii), viz.
$$4(p-2)^2 - 4p^2 = 16(1-p)$$
must be negative or zero.

Hence $1-p$ must be negative or zero. That is, p is not less than 1.

Ex. 6. Find an expression in x that cannot lie between 1 and 2 for any real value of x.

Let the expression be denoted by y.

Since y cannot lie between 1 and 2, $(y-1)(y-2)$ is not negative.

That is, $y^2 - 3y + 2$ is positive or zero.

We must now find an equation in x, having $y^2 - 3y + 2$, *i.e.* $y^2 - 4.1.\frac{1}{4}(3y-2)$ as its discriminant. Comparing this with $b^2 - 4ac$ we see that *one* such equation is $x^2 + yx + \frac{1}{4}(3y-2) = 0$.

Hence $y = \dfrac{2-4x^2}{4x+3}$, and this is an expression which can never lie between 1 and 2.

EXERCISES XIII.

Find whether the following expressions are capable of assuming all values for real values of x.

24. $1-x$.
25. $x^2 + 4x + 5$.
26. $6x - 2x^2 - 13$.

27. $x^2 + 2x + \frac{1}{2}$.
28. $3x - x^2 - 1$.
29. $3x - 2x^2 - 7$.

30. $3x - 2x^2 - 1$.
31. $x^2 + 6x + 9$.
32. $1/(x^2 + 2x + 7)$.

33. $\dfrac{1}{2x - x^2 - 1}$.
34. $\dfrac{x^2}{x^2 + 2x + 4}$.
35. $\dfrac{x^2 + 2x + 1}{x^2 + 2x + 7}$.

36. $\dfrac{x^2 - 4x + 4}{x^2 - 8x + 12}$.
37. $\dfrac{x^2 - 4x + 4}{x^2 - 6x + 9}$.
38. $\dfrac{x^2 - 2x + 21}{6x - 14}$.

39. If x is real in $x^2 + 25 = 3x \cdot \dfrac{2+3y}{y-1}$, find limits for y.

40. If x is real in $4x^2(1-3y) + 9 - y = 4x(2y-9)$, find limits for y.

41. Show that $1 + \dfrac{1}{x-2} - \dfrac{1}{x-1}$ cannot lie between 1 and -3.

42. Find the value of x for which $2x - x^2$ is greatest.

MAXIMA AND MINIMA.

If x, y are both real in the following equations, find the limits to their values :

43. $x^2 + 2y^2 = 2x + 6y + \tfrac{3}{4}$.   **44.** $x^2 + x + y^2 + y = 0$.

45. $x^2 = y^2 + 2y + 5$.   **46.** $x^2 + 2ay + 4y^2 - 3a^2 = 0$.

47. Show that $\dfrac{(x-a)(x-b)}{(x-c)(x-d)}$ can have any value (where x is real), provided that only one of the two quantities a and b lies between c and d.

48. Show that $\dfrac{x^2 - 3x + 2}{31x - x^2 - 30}$ can assume all real values.

49. Divide 12 into two parts so that the sum of their squares may be the least.

50. Divide 12 into two parts such that the square of one part, together with 3 times the square on the other part, may be the least possible.

51. Resolve 16 into 2 factors, such that the sum of their squares shall be a minimum.

52. Find that fraction which exceeds its square by the greatest possible amount.

53. Of all squares inscribed in a given square, find that which is the least.

54. Find an expression of the second degree in x which shall have the values of 9, 24, 21 when $x = -2, 1, 4$ respectively. Find also the greatest value of this expression.

55. What value of x makes $a^4 + b^3x - c^2x^2$ a maximum ?

56. What value of x makes $a + \sqrt[3]{(a^3 - 2a^2x + ax^2)}$ a minimum ?

57. In a given circle, inscribe the greatest rectangle possible.

58. Show that no real values of x and y other than 4 and 4 will satisfy $\qquad x^2 - xy + y^2 = 4(x + y - 4)$.

59. In $x^2 + y^2 = 2y(2y - 3)$, show that y must not lie between 0 and 2 if x is real, but that x may take any value if y is real.

60. Find an expression in x which can never lie between 4 and -4.

61. Find two expressions in x which lie between 3 and $\tfrac{1}{3}$.

62. A purchaser is to take a rectangular plot of land fronting a street; three times its frontage together with twice its depth are equal to 96 yards. What is the greatest number of square yards he may take ?

[Let x and y be the length and breadth in yards. We then have to find the maximum value of xy, it being given that $3x + 2y = 96$; that is, we have to find the greatest value of $\tfrac{1}{3}(96 - 2y)y$.]

CHAPTER XIV.

ARITHMETICAL AND GEOMETRICAL PROGRESSION.

191. The simpler properties of arithmetical progression (A.P.) and geometrical progression (G.P.) are dealt with fully in most elementary text books and we shall assume the student to be familiar with them. In this chapter we shall consider a few of the higher properties of these two progressions, treated as far as possible by general methods such as are applicable to other series as well.

192. DEFINITIONS. **A series** is any succession of quantities which are formed in order according to some definite rule called the **law** of the series.

Each of the quantities is called a **term** of the series, and the amount of all the terms added together is called the **sum** of the series.

Thus 5, 10, 15,...... and 3, 6, 12, 24,...... are series.

DEFINITION. The nth term *when written as a function of n* where n is literal*, is called the **general term** of the series.

If the general term is known every term of the series may be written down by giving different values to n.

The sum of a series is sometimes denoted by writing the Greek letter Σ (sigma) in front of the general term, so that Σ stands for "*the sum of such terms as*" that following it.

---

* A *literal* quantity in contradistinction to a numerical quantity is one which is represented by a letter instead of a number.

172   ARITHMETICAL AND GEOMETRICAL PROGRESSION.

Thus if the nth term is $5n$ we see by putting $n=1, 2, 3, 4$ in succession that the first four terms are 5, 10, 15, 20 respectively, and similarly the 100th term is 500.

Also $\Sigma (5n)$ denotes the sum $5+10+15+\ldots\ldots+5n$ and $\Sigma_{n=1}^{n=20} (5n)$ denotes the sum obtained by giving n all values from $n=1$ to $n=20$ in the general term, that is the sum of the first 20 terms.

DEFINITIONS. An **infinite series** is a series containing an infinitely large number of terms. It is sometimes called an *endless series*.

If by adding more and more terms of an infinite series the sum continually approximates to a finite limit, in such a manner that, by sufficiently increasing the number of terms, the difference between their sum and the finite limit can be made less than any assignable quantity, that limit is called the **sum of the infinite series.**

193. Arithmetical Progression. The general term of an A.P. whose first term is a and common difference d is known to be

$$a + (n-1)d = l \text{ suppose*} \ldots\ldots\ldots\ldots (1)$$

and the formula for the sum may now be written

$$s = \Sigma\{a + (n-1)d\} = na + \tfrac{1}{2}n(n-1)d = \tfrac{1}{2}n(a+l)\ldots\ldots(2).$$

The sum of an infinite series in A.P. is *never* finite as it sometimes is for a G.P.

The following properties are easily proved.

If the same quantity be added to or subtracted from all the terms of an A.P., *the resulting quantities will form an* A.P., *with the same common difference as before.*

If, however, all the terms of an A.P. *be multiplied or divided by the same quantity, the resulting quantities will form an* A.P. *with a new common difference.*

---

* CAUTION. Although l is equal to the nth term **it must not be inferred that the general term of the series is** l, for the single letter l gives no indication of the *form* of the nth term or the way in which the various successive terms are derived. For a similar reason it would be incorrect to write the sum Σl.

ARITHMETICAL AND GEOMETRICAL PROGRESSION. 173

Ex. If P, Q, and R be the pth, q^{th}, and r^{th} terms of an A.P., then
$$p(Q-R)+q(R-P)+r(P-Q)=0.$$

If a be the first term and d the common difference,
$P = a+(p-1)d\,;\; Q=a+(q-1)d\,;\; R=a+(r-1)d.$
$\therefore\; Q-R=(q-r)d\,;\; R-P=(r-p)d\,;\;\text{and}\; P-Q=(p-q)d.$
$\therefore\; p(Q-R)+q(R-P)+r(P-Q)=p(q-r)d+q(r-p)d+r(p-q)d$
$\qquad = d(pq-pr+qr-qp+rp-rq)=0.$

This example gives an interesting relation between the positions and values of any three terms of an A.P.

194. Converse Propositions. It will be observed that the nth term of an A.P. $[a+(n-1)d]$ is represented by an expression of the first degree in n; and the sum of n terms of an A.P. $[na+\tfrac{1}{2}n(n-1)d]$ is represented by an expression of the second degree in n with the absolute term wanting*.

We shall now show conversely that *a series is arithmetic if its nth term is given to be an expression of the first degree in* n; *or if the sum of* n *terms is given to be an expression of the second degree in* n *without the absolute term.*

(i) Let the nth term of a series be $en+f$.

Then the $(n-1)^{\text{th}}$ term of the series can be derived from this by writing $n-1$ for n, and it is therefore $= e(n-1)+f$.

Hence denoting the nth term and the $(n-1)^{\text{th}}$ term by u_n and u_{n-1}, we have
$$u_n - u_{n-1} = \{en+f\} - \{e(n-1)+f\} = e.$$

Thus the difference between any term and the immediately preceding term is constant.

Hence the series is arithmetic with constant difference e and first term $= e.1+f = e+f$.

(ii) Let the sum of the first n terms of a series be gn^2+hn.

Then the sum of the first $n-1$ terms $= g(n-1)^2+h(n-1)$.

Hence, denoting these by s_n and s_{n-1} respectively, we have
$$u_n = s_n - s_{n-1} = (gn^2+hn) - \{g(n-1)^2+h(n-1)\}$$
$$= 2gn-(g-h),$$
which is an expression in n of the first degree.

Hence from what has been shown above the series is arithmetic. The constant difference is easily seen to be $2g$, and the first term is $2g-(g-h)=g+h$.

---

*An expression like an^2+bn+2 which contains the absolute term can never represent the sum of n terms of a series.

174   ARITHMETICAL AND GEOMETRICAL PROGRESSION.

Ex. 1. If the nth term of a series is always $3n + 2$, find the sum of n terms.

The nth term being $3n + 2$, the first term is $3 + 2 = 5$.

Hence $\quad s_n = \tfrac{1}{2}n\{5 + (3n + 2)\} = \tfrac{1}{2}n\,(3n + 7)$.

Ex. 2. Find the series of which the sum of n terms is always n^2.
Here $s_n - s_{n-1} = n^2 - (n-1)^2 = 2n - 1$. Thus $u_n = 2n - 1$.

Hence giving n the values 1, 2, 3... in succession, we find the series to be $1 + 3 + 5 + 7 + \ldots\ldots$

*195. Negative values of n. An A.P. may be continued backwards by writing the terms $a-d, a-2d,\ldots\ldots$ before the first term a, thus :

$\ldots\ldots a - 3d,\ a - 2d,\ a - d,\ a,\ a + d,\ a + 2d,\ldots\ldots a + (n-1)\,d$.

The terms to the left of a are obtained by putting
$$n = 0,\ -1,\ -2,\ldots \text{ in } a + (n-1)\,d.$$

If it is required to find the number of terms whose sum amounts to a given value s, we have to solve the equation
$$s = \tfrac{1}{2}n\{2a + (n-1)\,d\}$$
as a quadratic in n, and it sometimes happens that one or both roots are negative. If $-n_1$ is a negative value of n which satisfies this equation we have

$$s = \tfrac{1}{2}(-n_1)\{2a + (-n_1 - 1)\,d\},$$
which gives $\quad -s = \tfrac{1}{2}n_1\{2(a-d) - (n_1 - 1)\,d\}$.

This shows that if in solving any example we get a negative integral value of n, and we count *backwards* to n_1 places beginning with $a - d$, the term immediately preceding a, the sum so obtained will be $-s$.

Ex. 1. How many terms of the series $3\tfrac{1}{2} + 5\tfrac{1}{2} + 7\tfrac{1}{2} + \ldots\ldots$ should be taken, so that the sum may be 21?

Taking the formula, $\quad s = \tfrac{1}{2}n\{2a + (n-1)\,d\}$,
we have $\quad\quad\quad\quad 21 = \tfrac{1}{2}n\{7 + (n-1) \times 2\}$;
$\therefore 2n^2 + 5n = 42$, which gives $n = 3\tfrac{1}{2}$ or -6.

Of these values the negative solution indicates that -21 is the sum of the six terms within the vertical lines below, where $1\tfrac{1}{2}$ is the first term and -2 is the common difference.

$\mid -8\tfrac{1}{2},\ -6\tfrac{1}{2},\ -4\tfrac{1}{2},\ -2\tfrac{1}{2},\ -\tfrac{1}{2},\ 1\tfrac{1}{2},\mid 3\tfrac{1}{2},\ 5\tfrac{1}{2},\ 7\tfrac{1}{2}$.

Moreover the positive fractional value $3\tfrac{1}{2}$ indicates that there is no exact number of terms of which the sum is 21, and that while the sum of 3 terms of the series $(=16\tfrac{1}{2})$ is less, the sum of four terms $(=26)$ is greater, than 21.

ARITHMETICAL AND GEOMETRICAL PROGRESSION. 175

Another fact is sometimes proved in connection with the negative integral value of n in the solution of $s=\frac{1}{2}n\{2a+(n-1)d\}$, when the other value is a positive integer.

Ex. 2. Let $s = 16$, $a = 7$, and $d = 2$.
We have $16 = \frac{1}{2}n\{14 + (n-1)\times 2\}$; which gives $n = 2$ or -8.
$|-5, -3, -1, 1, 3, 5, | 7, 9$.

If now we start with the last term given by the positive value of n (*i.e.* the second term) and count backwards to as many places as are indicated by the negative value (viz. 8 places), the result will be the given sum.

196. Illustrative Examples.

Ex. 1. The sums of n terms of two A.P.'s are in the ratio of $7n + 2 : n + 4$. Find the ratio of their 5th terms.

Let the first terms and the common differences of the two series be a_1, a_2, and d_1, d_2 respectively.

The sum of the first series $= \frac{1}{2}n\{2a_1 + (n-1)d_1\}$;
and the sum of the second series $= \frac{1}{2}n\{2a_2 + (n-1)d_2\}$.

$$\therefore \frac{2a_1 + (n-1)d_1}{2a_2 + (n-1)d_2} = \frac{7n+2}{n+4}.$$

The ratio of the fifth terms of the series

$$= \frac{a_1 + 4d_1}{a_2 + 4d_2} = \frac{2a_1 + 8d_1}{2a_2 + 8d_2} = \frac{2a_1 + (9-1)d_1}{2a_2 + (9-1)d_2} = \frac{7.9+2}{9+4} = 5.$$

Ex. 2. The A.P. 7, 9, 11, 13... is divided into groups as follows : 7 ; | 9, 11 ; | 13, 15, 17 ; | 19, 21, 23, 25 ; | and so on ; find the sum of the numbers in the nth group.

Since the first group contains *one* term, the second group *two* terms, the third group *three* terms, and so on ; the last term in the nth group is the $(1+2+3+...+n)^{\text{th}}$ term of the series 7, 9, 11... ; that is, the $\{\frac{1}{2}n(n+1)\}^{\text{th}}$ term of the series 7, 9, 11, 13....

Hence applying the formula $l = a + (n-1)d$, we see that the last term of the nth group $= 7 + \{\frac{1}{2}n(n+1) - 1\}2 = n^2 + n + 5$.

And the group contains n terms, therefore its first term
$$= n^2 + n + 5 - (n-1)d = n^2 + n + 5 - 2n + 2 = n^2 - n + 7.$$
And the number of terms is n.

Hence the sum required $= \frac{1}{2}n\{(n^2+n+5)+(n^2-n+7)\} = n(n^2+6)$.

[This result can be easily verified by putting $n = 1$, 2, or 3.]

176   ARITHMETICAL AND GEOMETRICAL PROGRESSION.

Ex. 3. Find the sum of $8 - 11 + 14 - 17 + \ldots$ to n terms.

First let n be even and $= 2m$.

Then $s_n = s_{2m} = (8-11) + (14-17) + (20-23) + \ldots$ to m terms
$= (-3) + (-3) + \ldots$ to m terms $= -3m = -1\frac{1}{2}n$.

Next let n be odd and $= 2m + 1$.

Then $s_n = 8 - (11 - 14) - (17 - 20) - \ldots$ to $m + 1$ terms
$= 8 + 3m = (3m + 1\frac{1}{2}) + 6\frac{1}{2} = 1\frac{1}{2}n + 6\frac{1}{2}$.

To find a *single expression* which gives the sum of the series whether n is odd or even, we take *half* the sum of the two sums $(=3\frac{1}{4})$, and *half* their difference $(=1\frac{1}{2}n+3\frac{1}{4})$, and connect these two results as under

$$3\frac{1}{4} - (-1)^n (1\frac{1}{2}n + 3\frac{1}{4}).$$

Ex. 4. Find the only A.P. beginning with 1 in which the sum of the first half of any even number of terms has to that of the second half the same constant ratio; and find that ratio.

$s_n : s_{2n} - s_n = \frac{1}{2}n \{2 + (n-1)d : \frac{1}{2}.2n \{2 + (2n-1)d\} - \frac{1}{2}n\{2 + (n-1)d\}$

$$\therefore \frac{s_n}{s_{2n} - s_n} = \frac{2 + (n-1)d}{2 + (3n-1)d} = 1 - \frac{2d}{3d + (2-d)/n}.$$

Since the ratio is constant for all values of n, $(2-d)/n = 0$, which makes $d = 2$, and the ratio $= \frac{1}{3}$.

Ex. 5. Show that the number of terms in a complete *homogeneous* expression of the nth degree in x, y, z is $\frac{1}{2}(n+1)(n+2)$; and verify the statement, taking $n = 2$.

The terms may be classified into (i) those that do not contain x; (ii) those that contain x; (iii) those that contain x^2; $\ldots$ lastly, those that contain x^n.

The first set consists of terms of the nth degree in y, z; and these are $n+1$ in number. The second set consists of terms of the $n-1$th degree in y, z which are n in number. The third set contains $n-1$ terms; and so on, the last set consisting of only one term.

Hence the required number
$$= 1 + 2 + 3 + \ldots + (n+1) = \frac{1}{2}(n+1)(n+2).$$

For $n = 2$ the result is 6, which accords with the fact that the homogeneous expression of the second degree consists of *six* terms of the form

$$ax^2 + by^2 + cz^2 + fyz + gzx + hxy.$$

EXERCISES XIV.

1. Show that the sum of the numbers in any one of the following rows is the square of an odd number:

$$\begin{array}{ccccc} 1 & & & & \\ 2 & 3 & 4 & & \\ 3 & 4 & 5 & 6 & 7 \\ \end{array}$$
....

2. Prove that the sum of $3n$ terms in A.P. is three times the sum of the $n+1$th to the $2n$th terms inclusive.

3. The first term of an A.P. is n^2-n+1, and the common difference is 2; prove that the sum of n terms is n^3; and thence show that $1^3 = 1$; $2^3 = 3 + 5$; $3^3 = 7 + 9 + 11$; and so on.

4. The first term of an A.P. is $n^{m-1}-n+1$, and the common difference is 2; show that the sum of n terms is n^m; and thence show that $1^4 = 1$; $2^4 = 7 + 9$; $3^4 = 25 + 27 + 29$; and so on.

5. Show that the number of terms in a general integral expression of the nth degree in x and y is $\frac{1}{2}(n+1)(n+2)$; and verify the statement by taking $n=2$ and $n=3$.

[The general integral expression of the second degree in x and y may be taken as $1+x+y+x^2+xy+y^2$, and of the third degree as
$$1 + x + y + x^2 + xy + y^2 + x^3 + x^2y + xy^2 + y^3.]$$

6. Show that $1 - 2 + 3 - 4 + \ldots\ldots$ to n terms
$$= \tfrac{1}{4}\{1 - (2n+1)(-1)^n\}.$$

7. Show that $1 - 3 + 5 - 7 + \ldots\ldots$ to n terms $= (-1)^{n+1}n$.

8. Find the sum of $a - (a+d) + (a+2d) - (a+3d) + \ldots\ldots$ to n terms, and verify your answer by taking $n=1$, and $n=2$.

9. Divide $\tfrac{1}{2}n(n+4)$ into n parts in A.P. such that the first term and common difference shall be independent of the value of n.

10. If the ratio of the sums of n terms of two arithmetical progressions be as $2n:n+1$, find the ratio of their 8th terms.

11. There are two series in A.P. the sums of which to n terms are as $13-7n:3n+1$; prove that their first terms are as $3:2$, and their second terms are as $-4:5$.

12. If the sum of m terms of a series is to the sum of n terms as $m^2:n^2$, show that the mth term is to the nth term as $2m-1$ is to $2n-1$. Hence show that the series is arithmetical, and find it.

TUT. ALG. II.

178   ARITHMETICAL AND GEOMETRICAL PROGRESSION.

13. If a, b, c are in A.P., turn $a^2+4ac+c^2$ into an expression involving a, b, c symmetrically.
[Assume $a^2+4ac+c^2=x(a^2+b^2+c^2)+y(bc+ca+ab)$ and find x, y.]

14. If the natural numbers are divided into groups as follows :—
1 ; | 2, 3 ; | 4, 5, 6 ; | 7, 8, 9, 10 ; | and so on ;
show that the sum of the numbers in the nth group is $\frac{1}{2}n(n^2+1)$.

15. Prove that the sum of the terms within the nth bracket of the series
$$(1)+(3+5)+(7+9+11)+(13+15+17+19)+\ldots$$
is n^3, and that the sum of the terms in the first n brackets is
$$\tfrac{1}{4}n^2(n+1)^2.$$

16. The series of natural numbers is written as follows :—

$$\begin{array}{ccccc} & & 1 & & \\ & 2, & 3, & 4 & \\ 5, & 6, & 7, & 8, & 9 \end{array}$$
....

Show that the sum of the numbers in the nth row is $n^3+(n-1)^3$.

[Verify the results in the last three examples by taking $n=1$, and $n=2$.]

17. If $s_1, s_2, s_3 \ldots s_p$ be the sums of n terms of p arithmetical progressions, the first terms of which are respectively $1, 2, 3, \ldots p$, and the common differences are $1, 3, 5, \ldots (2p-1)$; show that
$$s_1+s_2+s_3+\ldots+s_p = \tfrac{1}{2}np(np+1).$$

18. If there be p A.P.'s of which each begins with unity, and if the common differences are $1, 2, 3, \ldots p$; show that the sum of their n^{th} terms is $\tfrac{1}{2}\{(n-1)p^2+(n+1)p\}$.

19. If $s_1, s_2, s_3, \ldots s_{2n}$ be the sums of n terms of $2n$ A.P.'s whose first terms are the same, and common differences $d, 2d, 3d, \ldots 2nd$; then shall
$$(s_2+s_4+s_6+\ldots+s_{2n})-(s_1+s_3+s_5+\ldots+s_{2n-1})=\tfrac{1}{2}n^2(n-1)d.$$

20. There are a number of series in A.P. whose common differences are $1, 2, 3,\ldots$; show that if the sum of n terms of each of these be n^2, their first terms will form a decreasing A.P. whose first term is $\frac{1}{2}(n+1)$, and common difference $\frac{1}{2}(n-1)$.

197. Geometrical Progression.
The general term of a G.P. whose first term is a and common ratio r is ar^{n-1}, also the formula for the sum may be written
$$s_n = \Sigma ar^{n-1} = a\frac{r^n-1}{r-1} = a\frac{1-r^n}{1-r}.$$

ARITHMETICAL AND GEOMETRICAL PROGRESSION.

If r lies between $+1$ and -1, it follows from § 148 (iv) and § 150 that $r^\infty = 0$, hence the limit of the sum when taken to infinity is

$$S = \sum_{n=1}^{n=\infty} ar^{n-1} = \frac{a}{1-r}.$$

The reader will have little difficulty in proving that *if all the terms of a* G.P. *be multiplied or divided by the same quantity, the resulting quantities will form a* G.P. *with the same common ratio as before.*

Moreover *if all the terms of a* G.P. *be raised to the same power, the resulting quantities will be in* G.P. *with a new common ratio.*

Ex. If M, N, P be the mth, n^{th}, and p^{th} terms of a G.P. show that $M^{n-p} . N^{p-m} . P^{m-n} = 1$.

If a and r be the first term and the common ratio, we have

$$M = ar^{m-1}; \quad N = ar^{n-1}; \quad P = ar^{p-1};$$

$$\therefore M^{n-p} = a^{n-p} . r^{(m-1)(n-p)}; \quad N^{p-m} = a^{p-m} . r^{(n-1)(p-m)}; \text{ and}$$

$$P^{m-n} = a^{m-n} . r^{(p-1)(m-n)}.$$

$$\therefore M^{n-p} . N^{p-m} . P^{m-n}$$
$$= a^{n-p+p-m+m-n} . r^{(m-1)(n-p)+(n-1)(p-m)+(p-1)(m-n)}$$
$$= a^0 r^0 = 1.$$

This example furnishes an interesting relation between the positions and values of any three terms of a G.P.

198. *To show that if the sum of n terms of a series be a (r^n-1), the series is a* G.P.

Since* $s_n = a(r^n-1), \quad s_{n-1} = a(r^{n-1}-1).$

$$\therefore t_n = s_n - s_{n-1} = ar^n - ar^{n-1} = ar^{n-1}(r-1).$$
$$\therefore t_{n-1} = ar^{n-2}(r-1).$$
$$\therefore t_n : t_{n-1} = ar^{n-1}(r-1) : ar^{n-2}(r-1) = r : 1.$$

Thus the ratio of any term to the preceding one is constant, and the series is therefore geometrical.

199. Continued Product of a G.P.

If there are n *terms in* G.P. *of which the first and last are* a *and* l, *their product is* $(al)^{\frac{n}{2}}$.

If p be the continued product of n quantities in G.P., we have to prove that $p^2 = (al)^n$, where a and l are the first and last terms.

---

* t_n denotes the nth term of a series.

180 ARITHMETICAL AND GEOMETRICAL PROGRESSION.

If the series be $a, ar, ar^2, ar^3 \ldots \ldots \frac{l}{r^2}, \frac{l}{r}, l;$

$$p = a \times ar \times ar^2 \times \ldots \ldots \times \frac{l}{r^2} \times \frac{l}{r} \times l$$

and $\quad p = l \times \frac{l}{r} \times \frac{l}{r^2} \times \ldots \ldots \times ar^2 \times ar \times a,$

where the factors are written in the reverse order,

$\therefore p^2 = al \times al \times al \times \ldots \ldots$ to n factors $= (al)^n.$

COR. We have incidentally proved that *the product of any two terms equidistant from the beginning and the end of a* G.P. *is equal to the product of the extreme terms.*

200. To prove the arithmetical rule for finding the value of a recurring decimal.

Let the recurring decimal consist of a non-recurring part P, containing p digits, followed by a recurring part Q containing q digits; and let S denote the value of the recurring decimal.

Then $S = P(\frac{1}{10})^p + Q(\frac{1}{10})^{p+q} + Q(\frac{1}{10})^{p+2q} + Q(\frac{1}{10})^{p+3q} + \ldots$ to infinity.

$$\therefore S = P(\tfrac{1}{10})^p + \frac{Q(\tfrac{1}{10})^{p+q}}{1 - (\tfrac{1}{10})^q} = \frac{P}{10^p} + \frac{Q}{10^p(10^q - 1)}$$

$$= \frac{10^q P + Q - P}{(10^q - 1)10^p}.$$

Now, since $10^q - 1$ is a number consisting of q nines, the denominator consists of q nines followed by p ciphers. Also $10^p P + Q$ is the integral number formed of all the digits, and P the integral number containing the digits which do not recur. Hence we have the ordinary rule of arithmetic for reducing a recurring decimal to a vulgar fraction.

201. *To prove that in a* G.P. *continued to infinity, each term bears a constant ratio to the sum of all the terms which follow it, the common ratio being supposed to be less than unity.*

Let the G.P. be $a, ar, ar^2, \ldots \ldots$

We shall show that the ratio of the nth term to the sum of all the terms which follow it is constant, whatever n may be.

ARITHMETICAL AND GEOMETRICAL PROGRESSION.

Now, the nth term $= ar^{n-1}$; and the sum of all the terms which follow it is

$$ar^n + ar^{n+1} + ar^{n+2} \ldots\ldots \text{to infinity} = \frac{ar^n}{1-r}.$$

Hence the ratio of the nth term to the sum of all the terms which follow it $= ar^{n-1} \div \dfrac{ar^n}{1-r} = \dfrac{1-r}{r}$; which does not involve n, and is therefore constant whatever n may be.

Ex. 1. In an infinite G.P. whose terms are all positive, and whose common ratio is less than unity, find under what circumstances any term is greater than, equal to, or less than the sum of all the terms which follow it.

Let the G.P. be a, ar, ar^2,......to infinity.

Then the nth term $>$, $=$, or $<$ the sum of all the terms following it, according as $ar^{n-1} >$, $=$, or $< \dfrac{ar^n}{1-r}$;

that is, as $1-r >$, $=$, or $< r$;
that is, as $2r <$, $=$, or > 1.

Hence any term of an infinite G.P. will be greater than, equal to, or less than the sum of all the terms that succeed it, according as the common ratio is less than, equal to, or greater than $\frac{1}{2}$.

Ex. 2. Find the common ratio, if, in an infinite G.P. the ratio of any term to the sum of all the terms which follow it is $\frac{1}{3}$.

Let a, ar, ar^2 ...... be the G.P.

Then the nth term : the sum of all the terms following it

$$= ar^{n-1} : \frac{ar^n}{1-r} = 1-r : r.$$
$$\therefore (1-r) : r = 1 : 3; \text{ whence } r = \tfrac{3}{4}.$$

202. Illustrative Examples.

Ex. 1. Find the least number of terms of the series

$$\tfrac{1}{2} + \tfrac{1}{6} + \tfrac{1}{18} + \ldots\ldots$$

which shall differ from the sum to infinity by less than $\dfrac{1}{10,000}$.

$$s_n = \frac{a}{1-r} - \frac{ar^n}{1-r}; \text{ and } S = \frac{a}{1-r}.$$

Hence $S - s_n = \dfrac{ar^n}{1-r}$.

182 ARITHMETICAL AND GEOMETRICAL PROGRESSION.

We are here required to find the least value of n which will make

$$\frac{ar^n}{1-r} < \frac{1}{10,000}, \text{ where } a = \tfrac{1}{2}, \text{ and } r = \tfrac{1}{3}.$$

We have therefore to make

$$\frac{\tfrac{1}{2}(\tfrac{1}{3})^n}{1-\tfrac{1}{3}} < \frac{1}{10,000}\;;\; \text{that is,}\; \frac{1}{2}\cdot\frac{1}{3^n} < \frac{1-\tfrac{1}{3}}{10,000} \text{ or } \frac{1}{15,000}\;;$$

that is, $\dfrac{1}{2 \cdot 3^n} < \dfrac{1}{15,000}$; that is, $2 \cdot 3^n > 15,000$;

that is, $\qquad\qquad\qquad 3^n > 7,500.$

By raising 3 to positive integral powers beginning from 3^2, we find that $3^8 < 7,500$, while $3^9 > 7,500$.

Hence the required value of n is 9.

That is, the least number of terms is 9.

Ex. 2. If a, b, c, d be in G.P. then $a^2 + d^2 > b^2 + c^2$, unless the common ratio be ± 1.

If r be the common ratio, then $b = ar$; $c = ar^2$; and $d = ar^3$.

$$\therefore\; a^2 + d^2 = a^2 + a^2 r^6 = a^2(1 + r^6);$$
and $\qquad b^2 + c^2 = a^2 r^2 + a^2 r^4 = a^2(r^2 + r^4).$

Here, $r^6 + 1 > r^2 + r^4$, since $r^6 - r^4 - r^2 + 1$ or $(r^2-1)(r^4-1) > 0$, whether r is less or greater than 1.

And as a^2 is positive, $a^2(1 + r^6) > a^2(r^2 + r^4)$, which establishes the result.

Ex. 3. Sum $(x+y) + (x^2 + xy + y^2) + (x^3 + x^2 y + xy^2 + y^3) + \ldots$ to n groups.

The sum to n groups

$$= \frac{x^2 - y^2}{x-y} + \frac{x^3 - y^3}{x-y} + \frac{x^4 - y^4}{x-y} + \ldots \text{ to } n \text{ terms}$$

$$= \frac{1}{x-y}\{(x^2 - y^2) + (x^3 - y^3) + (x^4 - y^4) + \ldots \text{ to } n \text{ terms}\}$$

$$= \frac{1}{x-y}\{(x^2 + x^3 + x^4 + \ldots \text{ to } n \text{ terms})$$
$$\qquad\qquad - (y^2 + y^3 + y^4 + \ldots \text{ to } n \text{ terms})\}$$

$$= \frac{1}{x-y}\left\{x^2\frac{1-x^n}{1-x} - y^2\frac{1-y^n}{1-y}\right\}.$$

ARITHMETICAL AND GEOMETRICAL PROGRESSION.   183

Ex. 4. Sum to n terms : $5 + 55 + 555 + \ldots\ldots$

Let s_n stand for the sum of n terms. Then
$$s_n = 5 + 55 + 555 + \ldots\ldots \text{ to } n \text{ terms}$$
$$= 5(1 + 11 + 111 + \ldots\ldots \text{ to } n \text{ terms}).$$
$$\therefore \frac{9s_n}{5} = 9 + 99 + 999 + \ldots\ldots \text{ to } n \text{ terms}$$
$$= (10-1) + (10^2-1) + (10^3-1) + \ldots\ldots \text{ to } n \text{ terms}$$
$$= (10 + 10^2 + 10^3 + \ldots\ldots \text{ to } n \text{ terms}) - n$$
$$= 10 \cdot \frac{10^n - 1}{10 - 1} - n = \frac{10}{9}(10^n - 1) - n,$$
$$\therefore s_n = \frac{5}{9}\left\{\frac{10}{9}(10^n - 1) - n\right\} = \frac{50}{81}(10^n - 1) - \frac{5}{9}n.$$

It may be noted that this example is a particular case of Ex. 3.

Ex. 5. Sum the series
$$\frac{1}{2} + \frac{3}{2^2} + \frac{5}{2^3} + \frac{1}{2^4} + \frac{3}{2^5} + \frac{5}{2^6} + \frac{1}{2^7} + \ldots\ldots \text{ to } 12 \text{ terms}.$$

If s be the required sum,

$$s = \left(\frac{1}{2} + \frac{3}{2^2} + \frac{5}{2^3}\right) + \left(\frac{1}{2^4} + \frac{3}{2^5} + \frac{5}{2^6}\right) + \ldots\ldots \text{to four groups}$$
$$= \left(\frac{1}{2} + \frac{3}{2^2} + \frac{5}{2^3}\right) + \frac{1}{2^3}\left(\frac{1}{2} + \frac{3}{2^2} + \frac{5}{2^3}\right)$$
$$\qquad + \frac{1}{2^6}\left(\frac{1}{2} + \frac{3}{2^2} + \frac{5}{2^3}\right) + \frac{1}{2^9}\left(\frac{1}{2} + \frac{3}{2^2} + \frac{5}{2^3}\right)$$
$$= \left(\frac{1}{2} + \frac{1}{2^4} + \ldots\ldots \text{ to four terms}\right)\left(1 + \frac{3}{2} + \frac{5}{4}\right)$$
$$= \left(1 + \frac{3}{2} + \frac{5}{4}\right) \times \frac{1}{2}\frac{1 - (\frac{1}{2})^{12}}{1 - (\frac{1}{2})^3}$$
$$= \frac{15}{4} \times \frac{4(2^{12} - 1)}{7 \cdot 2^{12}} = \frac{8775}{4096}.$$

EXERCISES XIV.

21. Explain the apparent paradox that the sum of an infinite geometrical progression is a finite quantity.

22. Find a value of n which will make $\dfrac{a^n + b^n}{a^{n-1} + b^{n-1}}$ the G.M. of a and b.

ARITHMETICAL AND GEOMETRICAL PROGRESSION.

23. Show that the sum of n geometric means between a and b is
$$\frac{a^{1/(n+1)}b - b^{1/(n+1)}a}{b^{1/(n+1)} - a^{1/(n+1)}}.$$

24. Prove that the ratio of the sums of n terms of two geometrical series having the same common ratio is the ratio of their p^{th} terms.

25. Find the value of $\sqrt[3]{\{a\sqrt[4]{\{b\sqrt[3]{\{a\sqrt[4]{(b......)}\}}\}\}}}$ continued to infinity.

26. Find the ratio of any term to the sum of all the terms which follow it, in the infinite G.P. $3 + 2\frac{1}{4} + \ldots$

27. Construct an infinite G.P. of which the ratio of any term to the sum of all the terms which follow it is p.

28. Construct an infinite G.P. having 2 for the first term, and each term equal to the sum of all the terms following it.

29. Prove that $1 = (1-r) + (1-r)r + (1-r(r^2 + \ldots$ *ad inf.*, where r is less than unity, and show that by giving different values to r we may subdivide unity into an infinite number of parts forming a G.P. of which the first term may have any value whatever.

30. Find $1 + (1+r) + (1+r+r^2) + (1+r+r^2+r^3) + \ldots$ to n groups.

31. Sum to n terms $4 + 44 + 444 + \ldots$

32. Sum to n terms:
$$\left(1 + \frac{1}{r}\right)^2 + \left(1 + \frac{1}{r^3}\right)^2 + \left(1 + \frac{1}{r^5}\right)^2 + \ldots$$

33. Show that the middle term of a G.P. containing an odd number of terms is equal to the ratio of the continued product of its odd terms to that of its even terms.

34. If $a, b, c, d, \ldots$ be in G.P., show that
$$(a^3 + b^3)^{-1}, (b^3 + c^3)^{-1}, (c^3 + d^3)^{-1}, \ldots$$
are also in G.P., and find the sum of n terms of the latter series.

35. If each even term of an infinite series be $\frac{1}{2}$ the immediately preceding term, and each odd term be $\frac{1}{3}$ of the immediately preceding term, find the sum of all the terms, the first term being unity.

Sum the following series:

36. $\dfrac{5}{2} + \dfrac{3}{2^2} + \dfrac{5}{2^3} + \dfrac{3}{2^4} + \ldots$ to 12 terms.

37. $\dfrac{1}{3} + \dfrac{2}{3^2} + \dfrac{4}{3^3} + \dfrac{1}{3^4} + \dfrac{2}{3^5} + \dfrac{4}{3^6} + \ldots$ to 9 terms.

ARITHMETICAL AND GEOMETRICAL PROGRESSION. 185

38. Prove that, in any G.P., of which the terms are all positive, the sum of the first and the last terms is greater than the sum of any other two terms equidistant from the ends.

39. If s_1, s_2, s_3 are the sums to n terms of n geometrical progressions whose first terms are each unity, and common ratios are 1, 2, 3, ..., show that

$$s_1 + s_2 + 2s_3 + 3s_4 + \ldots + (n-1)s_n = 1^n + 2^n + 3^n \ldots + n^n.$$

40. Show that the sums of all the products two and two of the terms of the infinite G.P. 1, $r, r^2, r^3 \ldots = \dfrac{r}{(1-r)(1-r^2)}$, r being less than 1.

41. Show that the sum of the products of n terms of a G.P. taken two and two together is $\dfrac{r}{r+1} \cdot s_n \cdot s_{n-1}$ where r is the common ratio, and s_n and s_{n-1} are the sums of n and $(n-1)$ terms.

42. If a_1 be the first of the n geometric means between a and b, then $a : b = a^{n+1} : a_1^{n+1}$.

203. Definitions.—The arithmetic mean of n quantities is the sum of all the quantities divided by n and the **geometric mean of n quantities** is the nth root of their product.

Thus the A.M. of 3, 6, and 12
$$= \frac{3+6+12}{3} = 7$$
and their G.M. $+ (3 \times 6 \times 12)^{\frac{1}{3}} = 6$.

It is evident that both the A.M. and G.M. of any number of positive quantities are intermediate in value between the greatest and least of the quantities.

204. The A.M. of two unequal positive quantities is greater than their G.M.

Let a, b be the quantities, A their A.M., G their G.M.

Then $\qquad A = \dfrac{a+b}{2}$ and $G = \sqrt{(ab)}$.

$\therefore A - G = \tfrac{1}{2}\{a+b-2\sqrt{(ab)}\} = \tfrac{1}{2}(\sqrt{a}-\sqrt{b})^2.$

Now $(\sqrt{a}-\sqrt{b})^2$ is positive unless $a = b$, for since a and b are positive, $\sqrt{a}$ and $\sqrt{b}$ are both real.

Hence $A > G$ if a and b are unequal.

If $a=b$, then $A=G$ and both are equal to a and b.

205. The A.M. of any number of unequal positive quantities is greater than their G.M.

Let A be the A.M. and G the G.M. of the n positive quantities $a, b, c, d, \ldots\ldots$ Then by definition

$$A = \frac{a+b+c+d+\ldots\ldots}{n}, \quad G = (abcd\ldots\ldots)^{\frac{1}{n}}.$$

The quantities being unequal, suppose a, b to be the greatest and least of them respectively, so that G lies between a and b (§ 203). Let k be a quantity such that
$$ab = kG,$$
then the G.M. of the n quantities $a, b, c, d\ldots\ldots$ is evidently equal to the G.M. of the n quantities $G, k, c, d\ldots\ldots$.

But $\qquad a : G = k : b$,

and by § 81, the sum of the greatest and least terms of the proportion is greater than the sum of the other two. Hence
$$a + b > G + k.$$
$$\therefore \frac{G+k+c+d+\ldots\ldots}{n} < \frac{a+b+c+d+\ldots\ldots}{n},$$

that is, the A.M. of $G, k, c, d\ldots\ldots$ is less than the A.M. of $a, b, c, d\ldots\ldots$.

Similarly we may replace the greatest and least of the remaining quantities $k, c, d\ldots\ldots$ by two quantities G, l so that the G.M. is unaltered but the A.M. is less than before, and *two* of the new n quantities are now equal to G.

Proceeding in this way we see that when *all* the n quantities have been replaced by G, their A.M. is less than the original A.M.

But the new A.M. $= (G + G + G + \ldots\ldots)/n = G$.

$\therefore$ the original A.M. $> G$, that is $A > G$.

ARITHMETICAL AND GEOMETRICAL PROGRESSION. 187

Cor. From this theorem it follows that if the G.M. of any number of positive quantities is given, their A.M. is least when the quantities are equal. If their A.M. is given, their G.M. is greatest when they are equal. The results of §§ 188, 189 are thus extended to any number of positive quantities.

Ex. Prove that $\left(\dfrac{ma+nb}{m+n}\right)^{m+n} > a^m b^n$ where a, b are positive and m, n are positive integers.

We have to show that

$$\frac{ma+nb}{m+n} > (a^m b^n)^{\frac{1}{m+n}}.$$

But the left-hand side is the A.M. of $m+n$ quantities of which m are equal to a and n are equal to b, and the right-hand side is the G.M. of the same quantities, whence the result follows.

EXERCISES XIV.

43. If n is odd, prove that the product of n terms in A.P. is less than the nth power of the middle term.

44. Prove that if a, b, c be positive and k, l, m any integers
$$\left(\frac{ka+lb+mc}{k+l+m}\right)^{k+l+m} > a^k b^l c^m.$$

45. Prove that $2^n . 1 . 2 . 3 \ldots\ldots n < (n+1)^n$.

46. Between two positive quantities a, b, any number of arithmetic means are inserted and also an equal number of geometric means. Prove that any term in the series of arithmetic means is greater than the corresponding term in the series of geometric means.

47. The A.M. of two quantities is to their G.M. as $5 : 3$. Find the ratio of the quantities.

48. The two quantities between which A is the A.M. and G is the G.M. are $A \pm \sqrt{(A^2 - G^2)}$.

49. Determine m and n in terms of a and b, so that
$$(ma + nb)/(m+n)$$
may be the A.M. between m and n, and the G.M. between a and b.

50. If $\dfrac{a+bx}{a-bx} = \dfrac{b+cx}{b-cx} = \dfrac{c+dx}{c-dx} = \ldots\ldots$, then $a, b, c, d, \ldots\ldots$ are in G.P.

51. Show that if a, b, c be in A.P., then $a^2(b+c)$, $b^2(c+a)$, $c^2(a+b)$ will also be in A.P.

52. If a, b, c...... be a series of quantities in G.P., then the reciprocals of a^2-b^2, b^2-c^2,...... are in G.P.

53. If the pth, q^{th} and r^{th} terms of an A.P. are in A.P., then p, q, r are also in A.P.

54. Prove that the pth, q^{th} and r^{th} terms of a G.P. are in G.P., if p, q, r be in A.P.

55. If the pth, q^{th} and r^{th} terms of an A.P. are in G.P., then the common ratio $= (q-r)/(p-q)$.

56. If the pth, q^{th}, r^{th} and s^{th} terms of an A.P. be in G.P., then $p-q$, $q-r$, $r-s$ will be in G.P.

57. The income-tax is levied on the average of three years' income. Show that, if a man's income increase either in A.P. or G.P., so will his income-tax, the percentage being supposed uniform.

58. If a, b, c be in G.P., and if p, p' be the A.M. and G.M. between a and b, and q, q' be the A.M. and G.M. between b and c, then p, p', q, q' will be proportionals.

59. If a, b, c be in A.P., and if p, p' be the A.M. and G.M. between a and b, and q, q' the A.M. and G.M. between b and c, then

$$p^2 - p'^2 = q^2 - q'^2.$$

60. If a, b, c be in G.P., and if x be the A.M. between a and b, and y the A.M. between b and c, then $a/x + c/y = 2$.

61. If one A.M., a, and two G.M.'s, p and q, be inserted between two given quantities, then $p^3 + q^3 = 2apq$.

62. If one G.M., g, and two A.M.'s, p and q, be inserted between two given quantities, then $g^2 = (2p-q)(2q-p)$.

63. If $s=$ the sum of n terms of a G.P., p their product, and r the sum of the reciprocals, then $p^2 r^n = s^n$.

64. If the nth term of an A.P. be the G.M. between the sum of n terms and twice the common difference, show that the ratio of the first term to the common difference is $1 \pm \sqrt{n}$.

65. A G.P. whose common ratio r is greater than unity, is continued backwards to infinity by writing the terms ar^{-1}, ar^{-2},...... in order in front of its first term. Prove that $a/(1-r)$ is equal and opposite in sign to the sum to infinity of the terms so added.

CHAPTER XV.

HARMONICAL PROGRESSION.

206. DEFINITION. A series of quantities is said to be in **harmonical progression** (H.P.), when their reciprocals are in Arithmetical Progression*.

Thus $\frac{1}{3}, \frac{1}{4}, \frac{1}{5}, \ldots\ldots$ are in H.P. since their reciprocals 3, 4, 5,...... are in A.P.

$\frac{1}{4}, -\frac{1}{2}, -\frac{5}{4}\ldots\ldots$ are in A.P. $\quad 4, -2, -\frac{4}{5}\ldots\ldots$ are in H.P. since their reciprocals

Similarly $a, b, c, d,\ldots$ will be in H.P. when $\frac{1}{a}, \frac{1}{b}, \frac{1}{c}, \frac{1}{d},\ldots$ are in A.P.;

(*i.e.*) when $\quad \dfrac{1}{b} - \dfrac{1}{a} = \dfrac{1}{c} - \dfrac{1}{b} = \dfrac{1}{d} - \dfrac{1}{c} = \ldots\ldots$

207. If a, b, c be three consecutive terms of a harmonical progression, then $a : c = a - b : b - c$.

Since a, b, c are in H.P., $\dfrac{1}{a}, \dfrac{1}{b}$, and $\dfrac{1}{c}$ are in A.P.

$\therefore \quad \dfrac{1}{b} - \dfrac{1}{a} = \dfrac{1}{c} - \dfrac{1}{b}. \qquad \therefore \quad \dfrac{a-b}{ab} = \dfrac{b-c}{bc}.$

$$\therefore \quad \dfrac{ab}{bc} = \dfrac{a-b}{b-c}.$$

That is, $a : c = a - b : b - c$.

---

* The term *Harmonical* is derived from the fact, that musical strings of equal thickness and tension will produce harmony when sounded together, if their lengths be as the reciprocals of the arithmetic series of numbers 1, 2, 3.......

HARMONICAL PROGRESSION.

Harmonical Progression has been defined by some writers, especially in Geometry, by this property; and, accordingly we have the following alternative

DEFINITION. A series of quantities is said to be in **harmonical progression** when, of any three consecutive terms, the first is to the third as the difference between the first and the second is to the difference between the second and the third :—the differences being taken in the same order.

208. The general form of a series in H.P. is
$$\frac{1}{a}, \quad \frac{1}{a+b}, \quad \frac{1}{a+2b}, \quad \frac{1}{a+3b}, \ldots\ldots$$
where the nth term $= \dfrac{1}{a+(n-1)b}$.

Also *a harmonical progression is known when the positions and values of any two of its terms are known; that is, we can find the nth term of a harmonical progression, when the position and values of two other terms are given.*

Ex. 1. Continue the H.P. 2, 1, $\tfrac{2}{3}$, for two terms each way.

The A.P. of the reciprocals of the given harmonical progression is $\tfrac{1}{2}$, 1, $\tfrac{3}{2}$......, of which $\tfrac{1}{2}$ is the common difference.

This series continued for two terms both ways is

$$-\tfrac{1}{2} \mid 0 \mid \tfrac{1}{2} \mid 1\tfrac{3}{2} \mid 2\tfrac{5}{2} \ldots\ldots$$

∴ the H.P. is $\qquad -2 \mid \infty \mid 2 \mid 1\tfrac{2}{3} \mid \tfrac{1}{2} \mid \tfrac{2}{5}.$

Ex. 2. Find the 20th term of an H.P. of which $\tfrac{1}{2}$ is the 4th term, and $\tfrac{1}{5}$ is the 10th term. Find also the nth term.

If x be the required 20th term, $\dfrac{1}{x}$ is the 20th term of the A.P. whose 4th term is 2, and 10th term 5.

If a be the first term and d the common difference of the A.P.,

$a + 3d = 2$; and $a + 9d = 5$; which give $d = \tfrac{1}{2}$ and $a = \tfrac{1}{2}$.

$$\therefore \quad \frac{1}{x} = \tfrac{1}{2} + 19 \times \tfrac{1}{2} = 10, \text{ whence } x = \frac{1}{10}.$$

HARMONICAL PROGRESSION. 191

Thus the 20th term of the given H.P. is $\frac{1}{10}$.

Similarly the nth term of the corresponding A.P. is $\frac{1}{2}+(n-1)\frac{1}{2}=\frac{1}{2}n$.

$\therefore$ the nth term of the given H.P. $=\frac{2}{n}$.

Ex. 3. If L is the lth term, M the mth term, and N the nth term of an H.P., then,
$$MN(m-n) + NL(n-l) + LM(l-m) = 0.$$

If a and d be the first term and the common difference of the corresponding A.P., we have

$$\frac{1}{L} = a + (l-1)d\,; \;\; \frac{1}{M} = a + (m-1)d\,; \text{ and } \frac{1}{N} = a + (n-1)d.$$

$\therefore \dfrac{m-n}{L} = (m-n)\{a + (l-1)d\} = a(m-n) + d(m-n)(l-1)\,;$

$\dfrac{n-l}{M} = (n-l)\{a + (m-1)d\} = a(n-l) + d(n-l)(m-1)\,;$

and $\dfrac{l-m}{N} = (l-m)\{a + (n-1)d\} = a(l-m) + d(l-m)(n-1)\cdot$

Hence, adding, we have $\dfrac{m-n}{L} + \dfrac{n-l}{M} + \dfrac{l-m}{N} = 0\,;$

which gives $MN(m-n) + NL(n-l) + LM(l-m) = 0.$

NOTE 1. This example furnishes us with an interesting relation between the positions and values of any three terms of an H.P.

NOTE 2. This is a simple equation in N from which its value can be easily obtained. To apply the result to a particular case, let us take Ex. 2.

Here $l=4$; $L=\frac{1}{2}$; $m=10$; $M=\frac{1}{5}$; $n=20$; and we are required to find the value of N.

The equation of Ex. 3 becomes

$\frac{1}{5}.N(10-20) + N.\frac{1}{2}(20-4) + \frac{1}{2}.\frac{1}{5}(4-10) = 0.$

Hence $\quad -2N + 8N = \frac{3}{5}$ which gives $N = \frac{1}{10}.$

209. Harmonic Means.

DEFINITIONS. (1) When three quantities are in H.P., the middle one is called the **harmonic mean** of the other two.

The usual abbreviation for this is H.M.

HARMONICAL PROGRESSION.

(2) When any number of quantities are in H.P., the terms intermediate between the two extreme ones are called **harmonic means** of the two extreme terms.

210. The harmonic mean of two quantities is twice their product divided by their sum.

Let a and b be the two quantities, and x their harmonic mean. Then a, x, and b are in H.P.

$$\therefore \frac{1}{a}, \frac{1}{x} \text{ and } \frac{1}{b} \text{ are in A.P.} \quad \therefore \frac{1}{x} - \frac{1}{a} = \frac{1}{b} - \frac{1}{x}.$$

$$\therefore \frac{2}{x} = \frac{1}{a} + \frac{1}{b} = \frac{a+b}{ab}. \quad \therefore x = \frac{2ab}{a+b}.$$

211. To insert a given number of harmonic means between two given quantities.

Let a and b be the two given quantities, and n the number of means to be inserted. We thus have to find $n+2$ terms in H.P., of which a is the first term and b the last term.

Since the reciprocals of these quantities are in A.P., we have to form an A.P. of $n+2$ terms, of which the first term shall be $\frac{1}{a}$ and the last term $\frac{1}{b}$.

Let d be the common difference of this A.P.

Then, $\quad \frac{1}{b} = \frac{1}{a} + (n+1)d$.

$$\therefore d = \frac{1/b - 1/a}{n+1} = \frac{a-b}{(n+1)ab}.$$

Thus the A.P. is

$$\frac{1}{a}, \frac{1}{a} + \frac{a-b}{(n+1)ab}, \frac{1}{a} + \frac{2(a-b)}{(n+1)ab}, \ldots \frac{1}{b};$$

that is, $\frac{1}{a}, \frac{nb+a}{(n+1)ab}, \frac{(n-1)b+2a}{(n+1)ab}, \ldots \frac{1}{b}.$

HARMONICAL PROGRESSION.   193

The reciprocals of these terms are in H.P., and the required harmonic means are thus

$$\frac{(n+1)ab}{nb+a}, \ \frac{(n+1)ab}{(n-1)b+2a}, \ \frac{(n+1)ab}{(n-2)b+3a}, \ldots\ldots$$

Ex. Insert 3 harmonic means between $\frac{1}{2}$ and $\frac{1}{14}$.

We have here to find 5 terms in A.P. of which 2 is the first term and 14 is the 5th term.

If d be the common difference of this A.P., $14 = 2 + 4d$, which gives $d = 3$.

Hence the arithmetical means are 5, 8, and 11; and the required harmonic means are therefore $\frac{1}{5}, \frac{1}{8}, \frac{1}{11}$.

212. The student may be informed that **no concise formula can be found for the sum of a harmonical progression.**

We shall now give the following example to illustrate an artifice which is sometimes useful in solving questions in H.P.

Ex. If a, b, c, d are positive quantities in H.P., prove that $ad > bc$, and that $a + d > c + b$.

Since a, b, c, d are in H.P., they may be represented by the fractions

$$\frac{1}{p-3q}, \ \frac{1}{p-q}, \ \frac{1}{p+q}, \ \frac{1}{p+3q}$$

of which the denominators are in A.P.

(i) $\quad ad = \dfrac{1}{(p-3q)(p+3q)} = \dfrac{1}{p^2 - 9q^2}$; and $bc = \dfrac{1}{p^2 - q^2}$.

Since $p^2 - 9q^2 < p^2 - q^2$, it follows that $ad > bc$.

(ii) $\quad a + d = \dfrac{1}{p-3q} + \dfrac{1}{p+3q} = \dfrac{2p}{p^2 - 9q^2}$;

and $\quad b + c = \dfrac{1}{p-q} + \dfrac{1}{p+q} = \dfrac{2p}{p^2 - q^2}$,

whence clearly $a + d > b + c$.

TUT. ALG. II.   13

EXERCISES XV.

1. Continue the harmonical progression $\frac{1}{5}$, $\frac{2}{11}$, $\frac{1}{6}$ for two more terms each way.

2. Find the 9th term of the H.P., of which $\frac{1}{3}$ is the first term and $\frac{1}{5}$ the second.

3. Find the 3rd term of an H.P. of which the first two terms are
$$\frac{a-b}{a+b} \text{ and } \frac{a^2-b^2}{4ab}.$$

4. Insert 5 harmonic means between $\frac{1}{2}$ and $\frac{1}{3}$.

5. Between what numbers are 12 and 6 the harmonic means?

6. Insert 3 harmonic means between -2 and -3.

7. Find a fourth harmonic to 12, 6, 4.

8. Find two numbers whose difference is 8, and whose harmonic mean is $1\frac{4}{5}$.

9. The sum of three terms of an H.P. is $1\frac{1}{12}$, and the first term is $\frac{1}{2}$; find the series.

10. If the sum of three terms in H.P. be 11, and their product 36, find them.

11. If a, b, c be in H.P., then

   (i) $\dfrac{1}{a}+\dfrac{1}{c}=\dfrac{1}{b-a}+\dfrac{1}{b-c}.$ (ii) $\dfrac{b+a}{b-a}+\dfrac{b+c}{b-c}=2.$

   (iii) $b^2(a-c)^2 = 2\{c^2(b-a)^2 + a^2(c-b)^2\}.$

12. If a, b, c be in H.P., then $a+c-2b : a-c = a-c : a+c$.

13. If a, b, c and d are in H.P., then $ab+bc+cd = 3ad$.

14. Find the harmonic mean between the roots of
$$ax^2+bx+c=0.$$

15. If $b-a$ is a harmonic mean between $c-a$ and $d-a$, then $d-c$ is a harmonic mean between $a-c$ and $b-c$.

16. If $a-b$, $a-c$ and $a-d$ are in H.P., prove that
   (i) $b-c, b-d$ and $b-a$,   (ii) $c-d, c-a$ and $c-b$, and
   (iii) $d-a, d-b$ and $d-c$ are also in H.P.

17. If n harmonic means be inserted between 1 and r, show that
   the first : the last $= n + r : nr + 1$.

18. If the pth term of a harmonic series is q, and the qth term p, show that the pq^{th} term is 1.

19. If the pth term of a harmonic series is qr, and the qth term rp, prove that the rth term is pq.

20. If a_r denotes the rth term of an H.P.,
then
$$\frac{1}{a_1} = \frac{r-1}{a_{r-1}} - \frac{r-2}{a_r}.$$

21. In any H.P., prove that the product of the first two terms is to the product of any two consecutive terms as the difference between the two first is to the difference between the two others.

22. Compare the lengths of the sides of a right-angled triangle, when squares described on them are in H.P.

23. Show that, if p, q, r are in H.P. and have the same sign, the roots of $px^2 + 2qx + r = 0$ will be imaginary.

24. If $a(b-c)x^2 + b(c-a)xy + c(a-b)y^2$ is a perfect square, show that a, b, c are in H.P.

25. Prove that the base of a triangle is cut harmonically by the bisectors of the interior and exterior vertical angles.

26. If A, P, B, Q be four points in a straight line, such that AP, AB, AQ are in H.P., and if O be the middle point of AB, show that $OP \cdot OQ = OA^2$.

27. When a straight line is cut as in Euclid II. 11, prove that the difference of the segments equals half the harmonic mean between them.

213. Three quantities, a, b, c, are in A.P., G.P., or H.P., according as

$$a-b : b-c = a : a, \text{ or } = a : b, \text{ or } = a : c.$$

(i) When $a - b : b - c = a : a$,
$$a - b = b - c;$$
which shows that a, b, c are in A.P.

(ii) When $a - b : b - c = a : b$,
$$ab - b^2 = ab - ac; \text{ or } b^2 = ac.$$
$$\therefore b : a = c : b,$$
which shows that a, b, c are in G.P.

(iii) When $a - b : b - c = a : c$,
then a, b, c are in H.P. [See § 207.]

214. The geometric mean of any two quantities is also the G.M. of their A.M. and H.M.

Let a and b be any two quantities, A their A.M., G their G.M., and H their H.M.

Then $A = \dfrac{a+b}{2}$; $G = \sqrt{ab}$; and $H = \dfrac{2ab}{a+b}$.

$\therefore \quad A \times H = \dfrac{a+b}{2} \times \dfrac{2ab}{a+b} = ab = G^2.$

Hence $A : G = G : H$, which establishes the proposition.

215. The A.M. of two unequal positive quantities, their positive G.M., and their H.M. are in descending order of magnitude.

If a and b be any two positive quantities, A their A.M., G their positive G.M., and H their H.M., then as in § 204 we see that $A > G$.

Again, since $A > G$ and $A : G = G : H$,
it follows that $\qquad G > H$.

Hence $A > G > H$.

Ex. 1. Show that three quantities cannot be in two progressions, unless they are all equal.

For let a, b, c be the three quantities, and suppose a, c unequal. Then since the A.M. of a, $c >$ their G.M. $>$ their H.M.,

therefore b cannot be more than one of the three means of a, c.

216. DEFINITION. The **harmonic mean** (H.M.) of any number of quantities is the reciprocal of the arithmetic mean of their reciprocals.

Hence the harmonic mean of n quantities $a, b, c......$ is a quantity H such that

$\dfrac{1}{H}$ is the A.M. of $\dfrac{1}{a}, \dfrac{1}{b}, \dfrac{1}{c} ...,$

(*i.e.*) such that $\dfrac{n}{H} = \dfrac{1}{a} + \dfrac{1}{b} + \dfrac{1}{c} + ...$

HARMONICAL PROGRESSION.

217. The H.M. of any number of unequal positive quantities is less than their G.M. or A.M.

For let H be the H.M., G the G.M., and A the A.M. of any number of quantities $a, b, c\ldots\ldots$

Then since H is the H.M. of $a, b, c\ldots\ldots$

$$\therefore \frac{1}{H} \text{ is the A.M. of } \frac{1}{a}, \frac{1}{b}, \frac{1}{c}, \ldots\ldots$$

But $\quad\dfrac{1}{G}$ is the G.M. of $\dfrac{1}{a}, \dfrac{1}{b}, \dfrac{1}{c}, \ldots\ldots$

(as may be easily verified), and this is less than their A.M. (§ 205).

$$\therefore \frac{1}{G} < \frac{1}{H} \text{ or } G > H.$$

Also by § 205, $A > G$. Hence $A > G > H$.

CAUTION. If there are *more* than two quantities G is **not** the G.M. between A and H, so that the proof of § 215 is inapplicable.

218. To show that *if half the middle term be subtracted from three quantities in* H.P., *the resulting quantities will be in* G.P.

If a, b, c be in H.P., we have to prove that $a-\frac{1}{2}b, \frac{1}{2}b, c-\frac{1}{2}b$ are in G.P.

Now, $\qquad a-b : b-c = \quad a \ :c.$ $\qquad$ [§ 213.]

$\therefore\ a-b :\ a\ = b-c : c.\qquad$ [*Alternando.*]

$$\therefore \frac{(a-\tfrac{1}{2}b)-\tfrac{1}{2}b}{(a-\tfrac{1}{2}b)+\tfrac{1}{2}b} = \frac{\tfrac{1}{2}b-(c-\tfrac{1}{2}b)}{\tfrac{1}{2}b+(c-\tfrac{1}{2}b)}.$$

$\therefore\ a-\tfrac{1}{2}b : \tfrac{1}{2}b = \tfrac{1}{2}b : c-\tfrac{1}{2}b\qquad$ [§ 78],

from which the required result follows.

NOTE. Conversely it can be easily proved that, if to each of three quantities in G.P. the middle term is added, the resulting quantities will be in H.P.

219. Illustrative Examples. In the solution of examples involving harmonical progression the following property corresponding to those of §§ 193, 197 is often useful :—

If all the terms of an H.P. *be multiplied or divided by the same quantity, the resulting quantities will form an* H.P.

Ex. 1. If a, b, c be in H.P., then (i) bc, ca, ab will be in A.P., and (ii) $a(b+c), b(c+a), c(a+b)$ will also be in A.P.

(i) Since a, b, c are in H.P., $\dfrac{1}{a}, \dfrac{1}{b}, \dfrac{1}{c}$ are in A.P.

$\therefore$ multiplying by abc, we have bc, ca, ab in A.P.

HARMONICAL PROGRESSION.

(ii) Subtracting these three quantities from $ab+bc+ca$, we have $ab + ac$, $bc + ba$, $ca + cb$ in A.P.

∴ $a(b + c)$, $b(c + a)$ and $c(a + b)$ are in A.P.

Ex. 2. If $\dfrac{a}{b+c}$, $\dfrac{b}{c+a}$, $\dfrac{c}{a+b}$ are in A.P., then $b+c, c+a, a+b$ are in H.P.

Here, $1 + \dfrac{a}{b+c}$, $1 + \dfrac{b}{c+a}$, $1 + \dfrac{c}{a+b}$ are in A.P.

∴ $\dfrac{a+b+c}{b+c}$, $\dfrac{a+b+c}{c+a}$, $\dfrac{a+b+c}{a+b}$ are in A.P.

∴ $b+c$, $c+a$, $a+b$ are in H.P.

Ex. 3. If bc, ca, ab are in H.P., then

$$a\left(\dfrac{1}{b}+\dfrac{1}{c}\right), b\left(\dfrac{1}{c}+\dfrac{1}{a}\right), c\left(\dfrac{1}{a}+\dfrac{1}{b}\right) \text{ are in A.P.}$$

Since bc, ca, ab are in H.P.; $\dfrac{1}{bc}, \dfrac{1}{ca}, \dfrac{1}{ab}$ are in A.P.

∴ $\dfrac{ab+bc+ca}{bc}$, $\dfrac{ab+bc+ca}{ca}$, $\dfrac{ab+bc+ca}{ab}$ are in A.P.

Subtracting 1 from these fractions, we have

$$\dfrac{ab+ca}{bc}, \dfrac{ab+bc}{ca}, \dfrac{bc+ca}{ab} \text{ in A.P.}$$

That is, $a\left(\dfrac{1}{b}+\dfrac{1}{c}\right), b\left(\dfrac{1}{c}+\dfrac{1}{a}\right), c\left(\dfrac{1}{a}+\dfrac{1}{b}\right)$ are in A.P.

Ex. 4. If a, b, c be in A.P., and b, c, d in H.P., prove that
$$a : b = c : d.$$

Since b, c, d are in H.P., $b : d = b - c : c - d$.

But since a, b, c are in A.P., $b - c = a - b$.

∴ $a - b : c - d = b : d$; that is, $a - b : b = c - d : d$. [*Alternando.*]

∴ $a : b = c : d$. [*Componendo.*]

Ex. 5. If a is the A.M. between b and c, and b the G.M. between c and a, then c will be the H.M. between a and b.

Since b, a, c are in A.P., $\dfrac{1}{ac}, \dfrac{1}{bc}, \dfrac{1}{ab}$ are in A.P. [Divide by abc.]

And since c, b, a are in G.P., $b^2 = ac$.

HARMONICAL PROGRESSION.

$$\therefore \frac{1}{b^2}, \frac{1}{bc}, \frac{1}{ab} \text{ are in A.P.}$$

$$\therefore \frac{1}{b}, \frac{1}{c}, \frac{1}{a} \text{ are in A.P.}$$

which shows that c is the H.M. of a and b.

Ex. 6. If $a^x = b^y = c^z$, and a, b, c be in G.P., show that x, y, z are in H.P.

Let $a^x = b^y = c^z = k$.
Then $\qquad a = k^{1/x}; \ b = k^{1/y} \ c = k^{1/z}$.
But $ac = b^2$, since a, b, c are in G.P.

$$\therefore \ k^{1/x} \times k^{1/z} = k^{2/y}. \quad \therefore \frac{1}{x} + \frac{1}{z} = \frac{2}{y}.$$

$$\therefore \frac{1}{x} - \frac{1}{y} = \frac{1}{y} - \frac{1}{z},$$

which shows that x, y, z are in H.P.

EXERCISES XV.

28. Simplify $a(b-c)/(a-b)$ when a, b, c are (i) in A.P., (ii) in G.P., and (iii) in H.P.

29. Find the A.M., G.M. and H.M. between 2 and 8, and continue for two more terms the G.P. which these three means form.

30. The difference between two numbers is 18, and 4 times their G.M. is equal to 5 times their H.M. Find the numbers.

31. Sum to infinity the G.P., of which the first term and the common ratio are the A.M. and the H.M. between a and a^{-1}.

32. If the A.M. between two numbers is 1, prove that their H.M. is the square of their G.M.

33. If A be the A.M. and H the H.M. between a and b, show that

$$\frac{a-A}{a-H} \times \frac{b-A}{b-H} = \frac{A}{H}.$$

34. If the G.M. between x and y is to their H.M. as m to n, then $\qquad x : y = m + \sqrt{(m^2-n^2)} : m - \sqrt{(m^2-n^2)}$.

35. If the $(m-n)^{\text{th}}$ and $(m+n)^{\text{th}}$ terms of a G.P. are the A.M. and H.M. of a and b, then the mth term is their G.M.

HARMONICAL PROGRESSION.

36. If a, b, c be in G.P., and if p be the A.M. between a and b, and q the A.M. between b and c, then b will be the H.M. between p and q.

37. If between b and a there be inserted n A.M.'s, and between a and b there be inserted n H.M.'s, prove that the product of the rth A.M. and the rth H.M. is ab.

38. If $b+c$, $c+a$, and $a+b$ be in H.P., then a^2, b^2, c^2 shall be in A.P.

39. If bc, ca, ab are in H.P., then $a(b^{-1}+c^{-1})$, $b(c^{-1}+a^{-1})$, $c(a^{-1}+b^{-1})$ are in A.P.

40. If $a/(b+c)$, $b/(c+a)$, $c/(a+b)$ are in A.P., then $b+c$, $c+a$, and $a+b$ will be in H.P.

41. If a, b, c be in H.P., then $\dfrac{1}{a}+\dfrac{1}{b+c}$, $\dfrac{1}{b}+\dfrac{1}{c+a}$, $\dfrac{1}{c}+\dfrac{1}{a+b}$ will be in H.P.

42. If a, b, c be in A.P., and b, c, a be in H.P., then c, a, b shall be in G.P.

43. If a_1, a_2, a_3 be in A.P., a_2, a_3, a_4 be in G.P., and a_3, a_4, a_5 be in H.P., then a_1, a_3, a_5 are in G.P.

44. If a, b, c be in A.P.; and a, mb and c be in G.P., then a, m^2b, c shall be in H.P.

45. Prove that a, b, c will be in G.P., if $b+a$, $b+b$, $b+c$ be in H.P.

46. If a, b, c be in A.P.; p, q, r in H.P.; and if $\dfrac{p}{r}+\dfrac{r}{p}=\dfrac{a}{c}+\dfrac{c}{a}$, then ap, bq, cr are in G.P.

47. Between every two terms of a G.P., an H.M. is found. Prove that these means form a G.P.

48. If between any two quantities there be inserted $2n-1$ arithmetic, geometric and harmonic means, show that the nth means are in G.P.

49. Of four proportionals the first three are in A.P.; prove that the last three are in H.P.

50. If m, a_1, a_2, n be in A.P.; m, g_1, g_2, n be in G.P.; m, h_1, h_2, n be in H.P.; then $mn+g_1g_2+a_1h_2+a_2h_1$.

51. If a_1, a_2 be two A.M.'s between m and n; g_1, g_2 two G.M.'s, and h_1, h_2 two H.M.'s, then $g_1g_2 : h_1h_2+a_1+a_2 : h_1+h_2$.

52. If m and n be any two numbers, g their G.M., and a_1, h_1, and a_2, h_2 the A.M. and the H.M. between m and g, and g and n respectively, prove that $a_1h_2=g^2=a_2h_1$.

HARMONICAL PROGRESSION.

53. Prove that $a_2a_3 - a_1a_4$ is positive, zero or negative, according as a_1, a_2, a_3, a_4 (all of which are supposed positive) are in A.P., G.P., or H.P.

54. There are three numbers in H.P.; if 1 be subtracted from the first, they form a G.P.; and if 4 be subtracted from the third they form an A.P.; what are the numbers?

[Let a, $a + d$, $a + 2d + 4$ be the numbers, or let $a + 1$, ar, ar^2 be the numbers.]

*55. Show that $a^n + c^n > 2b^n$, whether a, b, c be in A.P., G.P., or H.P., a, b, c, n being taken to be positive integers.

*56. BD and BE, the bisectors of the interior and exterior angles at B of the triangle ABC, meet the base AC at D and E. Prove that the rectangles AE.EC, AB.BC and AD.DC, are in A.P., if the difference of the base angles is a right angle; are in G.P., if one of the base angles is a right angle; and are in H.P., if the vertical angle is a right angle.

57. Prove that $\dfrac{a^n + b^n + c^n + \dots}{a^{n-1} + b^{n-1} + c^{n-1} + \dots}$ will be the A.M. of a, b, $c\dots$ when $n = 1$, and their H.M. when $n = 0$.

58. If A, G, H be the A.M., G.M. and H.M. of three quantities a, b, c prove that
$$(x - a)(x - b)(x - c) = x^3 - 3Ax^2 + 3G^3x/H - G^3,$$
x being any quantity whatever. Hence or otherwise show how to find three quantities whose A.M., G.M. and H.M., have given values.

CHAPTER XVI.

SOME OTHER SIMPLE SERIES.

220. We shall consider in this chapter some simple series which are neither arithmetical nor geometrical, but which may very easily be summed.

221. Arithmetico-Geometrical Series. DEFINITION. A series that is formed by multiplying together, term by term, an arithmetical and geometrical progression, is often called an **arithmetico-geometrical series.**

The general form of such a series is
$$a + (a+d)r + (a+2d)r^2 + \ldots\ldots + \{a+(n-1)d\}r^{n-1} + \ldots$$
The artifice that is used in summing such a series is an extension of that used in summing a geometrical series.

222. To find the sum of n terms of a series whose nth term is the product of the nth terms of an A.P. and a G.P.

Let the nth term be $\{a + (n-1)d\}r^{n-1}$ and let s denote the required sum. Then
$$s = a + (a+d)r + (a+2d)r^2 + \ldots + \{a+(n-1)d\}r^{n-1};$$
$$\therefore rs = \quad ar \quad + (a+d)r^2 + \ldots + \{a+(n-2)d\}r^{n-1}$$
$$+ \{a+(n-1)d\}r^n.$$
$$\therefore (1-r)s = a + dr + dr^2 + \ldots + dr^{n-1} - \{a+(n-1)d\}r^n.$$
$$\therefore (1-r)s = a + dr\frac{1-r^{n-1}}{1-r} - \{a+(n-1)d\}r^n.$$
$$\therefore s = \frac{a - \{a+(n-1)d\}r^n}{1-r} + dr\frac{1-r^{n-1}}{(1-r)^2}.$$

SOME OTHER SIMPLE SERIES.

Cor. The sum may be written in the form

$$\frac{a}{1-r} + \frac{dr}{(1-r)^2} - \frac{dr^n}{(1-r)^2} - \frac{\{a+(n-1)d\}r^n}{1-r},$$

and if r be less than 1, r^n can be made as small as we please by taking n sufficiently large.

Hence the sum to infinity is

$$\frac{a}{1-r} + \frac{dr}{(1-r)^2},$$

it being *assumed* that the terms involving r^n can be made less than any assignable quantity, by making n sufficiently large.

Ex. 1. Find the sum of n terms of the series

$$1 + 2x + 3x^2 + 4x^3 + \ldots\ldots\ldots$$

Let $s = 1+2x+3x^2+4x^3+\ldots\ldots+(n-1)x^{n-2}+nx^{n-1}$.

$\therefore xs = \quad x+2x^2+3x^3+\ldots\ldots+(n-2)x^{n-2}+(n-1)x^{n-1}+nx^n$.

Hence, by subtraction,

$(1-x)s = 1+x+x^2+x^3+\ldots\ldots+x^{n-2} \quad +x^{n-1} \quad -nx^n$.

Here the first n terms of the expression on the right-hand side form a geometrical progression.

Hence $\qquad (1-x)s = \dfrac{1-x^n}{1-x} - nx^n.$

$$\therefore s = \frac{1-x^n}{(1-x)^2} - n\frac{x^n}{1-x}.$$

Ex. 2. Find the sum to infinity of the series

$$1 + \frac{4}{5} + \frac{7}{5^2} + \frac{10}{5^3} + \ldots\ldots$$

In summing an arithmetico-geometrical series to infinity, we generally proceed as follows:

Let $\qquad S = 1 + \dfrac{4}{5} + \dfrac{7}{5^2} + \dfrac{10}{5^3} + \ldots\ldots$ to infinity.

Then, $\quad \dfrac{1}{5}S = \quad \dfrac{1}{5} + \dfrac{4}{5^2} + \dfrac{7}{5^3} + \ldots\ldots \qquad ,,$

By subtraction, $\frac{4}{5} S = 1 + \left(\frac{3}{5} + \frac{3}{5^2} + \frac{3}{5^3} + \ldots \text{to infinity} \right)$.
$$= 1 + \tfrac{3}{5}/(1 - \tfrac{1}{5}) = 1 + \tfrac{3}{4} = 1\tfrac{3}{4}.$$
$$\therefore S = 1\tfrac{3}{4} \times \tfrac{5}{4} = 2\tfrac{3}{16}.$$

223. The artifice of the preceding article can also be employed to sum a series like the following :—

Ex. Sum to infinity
$$1^2 + 2^2 x + 3^2 x^2 + \ldots, x \text{ being less than } 1.$$

Let S denote the sum to infinity.

Then $\quad S = 1 + 4x + 9x^2 + 16x^3 + \ldots$ to infinity.
$\therefore xS = \quad x + 4x^2 + 9x^3 + \ldots \quad$ „
$\therefore (1-x)S = 1 + 3x + 5x^2 + 7x^3 + \ldots \quad$ „ $\quad \ldots\ldots\ldots$(i).

Multiplying by x, we have

$x(1-x)S = \quad x + 3x^2 + 5x^3 + \ldots \quad$ „ $\quad \ldots\ldots\ldots$(ii).

Subtract (ii) from (i).

$(1-x)^2 S = 1 + 2x + 2x^2 + 2x^3 + \ldots$
$$= 1 + \frac{2x}{1-x} = \frac{1+x}{1-x}.$$
$$\therefore S = \frac{1+x}{(1-x)^3}.$$

EXERCISES XVI.

Find the general term and the sum to n terms of the series in Examples 1 to 5. Verify your answer in each case, by putting $n = 1$, and $n = 2$.

1. $1 + 4x + 7x^2 + 10x^3 + \ldots$ 2. $1 + 3.3 + 5.3^2 + 7.3^3 + \ldots$

3. $1 + \dfrac{2}{3} + \dfrac{3}{3^2} + \dfrac{4}{3^3} + \ldots$ 4. $a - (a+1)x + (a+2)x^2 - \ldots$

5. $1 - 3 + 5 - 7 + \ldots$
$\quad [= 1 + 3x + 5x^2 + 7x^3 + \ldots \text{ where } x = -1.]$

SOME OTHER SIMPLE SERIES. 205

Sum the following series to infinity :

6. $\dfrac{a}{r} + \dfrac{a+b}{r^2} + \dfrac{a+2b}{r^3} + \ldots$, where $r > 1$.

7. $\tfrac{1}{2} + \tfrac{3}{4} + \tfrac{5}{8} + \ldots$ 	8. $1 - \tfrac{3}{2} + \tfrac{5}{4} - \tfrac{7}{8} + \ldots$

9. $ar + 2ar^2 + 3ar^3 + \ldots$, where $r < 1$.

10. Sum to infinity the series whose n^{th} term is $(n+3)a^n$, a being less than 1.

11. Find the sum to infinity of the series whose n^{th} term is $\dfrac{n}{2^{n-1}}$.

12. If $x < 1$, sum $1^2 + 5^2 x + 9^2 x^2 + \ldots$ to infinity.

Summation of the powers of the natural numbers.

224. To prove that the sum of the first n integers is $\tfrac{1}{2}n(n+1)$.

This is an A.P. which can be summed in the usual way. We shall here effect the summation by another method, which will be found analogous to that by which we sum the squares, cubes, &c., of the first n natural numbers.

Let $\Sigma(n)$ denote the required sum. Taking the identity,
$$(x+1)^2 - x^2 = 2x + 1,$$
and substituting n, $n-1$, $n-2, \ldots 2, 1$ in order for x, we have

$$\begin{aligned}
(n+1)^2 - n^2 &= 2n &&+ 1 \\
n^2 - (n-1)^2 &= 2(n-1) &&+ 1 \\
(n-1)^2 - (n-2)^2 &= 2(n-2) &&+ 1 \\
&\ldots\ldots\ldots\ldots \\
3^2 - 2^2 &= 2.2 &&+ 1 \\
2^2 - 1^2 &= 2.1 &&+ 1.
\end{aligned}$$

Hence, by addition,
$$(n+1)^2 - 1^2 = 2(1+2+3+\ldots+n) + n = 2\Sigma(n) + n.$$
$$\therefore\ 2\Sigma(n) = (n+1)^2 - 1^2 - n = n^2 + n.$$
$$\therefore\ \Sigma(n) = \tfrac{1}{2}(n^2 + n) = \tfrac{1}{2}n(n+1).$$

SOME OTHER SIMPLE SERIES.

225. To prove that the sum of the squares of the first n integers is $\frac{1}{6}n(n+1)(2n+1)$.

Let $\Sigma(n^2)$ denote the required sum as explained in § 192.

Taking the identity,
$$(x+1)^3 - x^3 = 3x^2 + 3x + 1,$$
and substituting $n, n-1, \ldots 3, 2, 1$ in order for x, we have

$$(n+1)^3 - n^3 = 3n^2 + 3n + 1$$
$$n^3 - (n-1)^3 = 3(n-1)^2 + 3(n-1) + 1$$
$$(n-1)^3 - (n-2)^3 = 3(n-2)^2 + 3(n-2) + 1$$
$$\cdots\cdots\cdots\cdots\cdots\cdots\cdots\cdots\cdots\cdots\cdots\cdots\cdots$$
$$3^3 - 2^3 = 3.2^2 + 3.2 + 1$$
$$2^3 - 1^3 = 3.1^2 + 3.1 + 1.$$

Hence, by addition,
$$(n+1)^3 - 1^3 = 3(1^2 + 2^2 + \ldots + n^2) + 3(1 + 2 + \ldots + n) + n$$
$$= 3\Sigma(n^2) + 3\Sigma(n) + n.$$
$$\therefore 3\Sigma(n^2) = (n+1)^3 - 1 - \tfrac{3}{2}n(n+1) - n$$
$$= \tfrac{1}{2}(n+1)\{2(n+1)^2 - 3n - 2\}$$
$$= \tfrac{1}{2}(n+1)(2n^2 + n) = \tfrac{1}{2}n(n+1)(2n+1).$$

Hence $\quad \Sigma(n^2) = \dfrac{n(n+1)(2n+1)}{6}.$

226. To prove that the sum of the cubes of the first n integers is the square of the sum of the first n integers.

Let $\Sigma(n^3)$ denote the required sum.

Taking the identity,
$$(x+1)^4 - x^4 = 4x^3 + 6x^2 + 4x + 1,$$
and substituting $n, n-1, n-2 \ldots 3, 2, 1$ in order for x, we have

SOME OTHER SIMPLE SERIES. 207

$$(n+1)^4 - n^4 \qquad = 4n^3 \qquad + 6n^2 \qquad + 4n \qquad + 1$$
$$n^4 \qquad - (n-1)^4 = 4(n-1)^3 + 6(n-1)^2 + 4(n-1) + 1$$
$$(n-1)^4 - (n-2)^4 = 4(n-2)^3 + 6(n-2)^2 + 4(n-2) + 1$$
....
$$3^4 \qquad - 2^4 \qquad = 4 \cdot 2^3 \qquad + 6 \cdot 2^2 \qquad + 4 \cdot 2 \qquad + 1$$
$$2^4 \qquad - 1^4 \qquad = 4 \cdot 1^3 \qquad + 6 \cdot 1^2 \qquad + 4 \cdot 1 \qquad + 1$$

Hence, by addition, we have

$$(n+1)^4 - 1^4 = 4\Sigma(n^3) + 6\Sigma(n^2) + 4\Sigma(n) + n$$
$$= 4\Sigma(n^3) + 6 \cdot \tfrac{1}{6}n(n+1)(2n+1) + 4 \cdot \tfrac{1}{2}n(n+1) + n.$$
$$\therefore \; 4\Sigma(n^3) = (n+1)^4 - n(n+1)(2n+1) - 2n(n+1) - (n+1)$$
$$= (n+1)\{n^3 + 3n^2 + 3n + 1 - 2n^2 - n - 2n - 1\}$$
$$= (n+1)(n^3 + n^2) = n^2(n+1)^2.$$
$$\therefore \; \Sigma(n^3) = \frac{n^2(n+1)^2}{4} = \{\tfrac{1}{2}n(n+1)\}^2.$$

NOTE. It can be similarly proved that the sum of the fourth powers of the first n integers is $\tfrac{1}{30}n(n+1)(6n^3+9n^2+n-1)$ which can also be written $\tfrac{1}{6}n(n+1)(2n+1) \times \tfrac{1}{5}(3n^2+3n-1).$

227. Illustrative Examples.

Ex. 1. Find the sum of the square of n numbers in A.P.

Let the numbers be a, $a+d$, $a+2d$,...... $a+(n-1)d$; and let s denote the sum of the squares of these numbers. Then
$$s = a^2 + (a+d)^2 + (a+2d)^2 + \ldots\ldots + \{a+(n-1)d\}^2$$
$$= a^2$$
$$+ a^2 + 2 \cdot ad + d^2$$
$$+ a^2 + 2 \cdot 2ad + (2d)^2$$
$$+ a^2 + 2 \cdot 3ad + (3d)^2$$
....
$$+ a^2 + 2(n-1)ad + \{(n-1)d\}^2$$
$$= na^2 + 2ad\{1+2+3+\ldots+(n-1)\} + d^2\{1^2+2^2+3^2+\ldots\ldots+(n+1)^2\}$$
$$= na^2 + 2ad \cdot \tfrac{1}{2}(n-1)n + d^2 \cdot \tfrac{1}{6}(n-1)n(2n-1)$$
$$= na^2 + n(n-1)ad + \tfrac{1}{6}n(n-1)(2n-1)d^2.$$

208   SOME OTHER SIMPLE SERIES.

Ex. 2. Show that the sum of the cubes of any number of consecutive integers is divisible by the sum of the integers.

Let $a+1,\ a+2,\ a+3,\ldots\ldots b$ be n consecutive integers; and let s and s_3 stand for the sum of these integers and of their cubes respectively.

Then $s_3 = (a+1)^3 + (a+2)^3 + (a+3)^3 + \ldots\ldots + b^3$
$= \{1^3 + 2^3 + 3^3 + \ldots\ldots + a^3 + (a+1)^3 + \ldots\ldots + b^3\}$
$\qquad\qquad\qquad\qquad - \{1^3 + 2^3 + 3^3 + \ldots\ldots + a^3\}$
$= \{\tfrac{1}{2}b(b+1)\}^2 - \{\tfrac{1}{2}a(a+1)\}^2 \quad\ldots\ldots\ldots\ldots\ldots\ldots\ldots\text{(i)}.$

Again, $s = (a+1) + (a+2) + (a+3) + \ldots\ldots + b$
$= \{1 + 2 + 3 + \ldots\ldots + b\} - \{1 + 2 + 3 + \ldots\ldots + a\}$
$= \tfrac{1}{2}b(b+1) - \tfrac{1}{2}a(a+1) \quad\ldots\ldots\ldots\ldots\ldots\ldots\ldots\text{(ii)}.$

Thus $s_3/s = \tfrac{1}{2}b(b+1) + \tfrac{1}{2}a(a+1)$ which is clearly an integer, since the product of two consecutive integers is even.

Hence s is a factor of s_3, as was to be proved.

Ex. 3. Find the sum of 50 terms of $1^2 + 3^2 + 5^2 + \ldots\ldots$

The nth term of this series is the square of the nth term of the A.P., 1, 3, 5, 7,......; and $= (2n-1)^2$. Expanding $(2n-1)^2$, we have $4n^2 - 4n + 1$ for the nth term.

Thus it can easily be verified that
$$\text{the 1st term} = 4 \cdot 1^2 - 4 \times 1 + 1$$
$$\text{the 2nd term} = 4 \cdot 2^2 - 4 \times 2 + 1$$
$$\text{the 3rd term} = 4 \cdot 3^2 - 4 \times 3 + 1$$
$$\ldots\ldots\ldots\ldots\ldots\ldots\ldots\ldots\ldots\ldots\ldots$$
$$\text{the 50th term} = 4 \cdot 50^2 - 4 \times 50 + 1.$$

Hence the sum of 50 terms
$= 4(1^2 + 2^2 + 3^2 + \ldots\ldots + 50^2) - 4(1 + 2 + 3 + \ldots\ldots + 50) + 50$
$= 4 \cdot \tfrac{1}{6} \cdot 50 \cdot 51 \cdot 101 - 4 \cdot \tfrac{1}{2} \cdot 50 \cdot 51 + 50 = 166{,}650.$

Ex. 4. Find the sum of the products of the first n integers taken two and two together.

Taking the identity,
$(a+b+c+d+\ldots\ldots)^2 = a^2+b^2+c^2+d^2+\ldots\ldots+2(ab+ac+bc+\ldots\ldots)$
and putting $a=1,\ b=2,\ c=3\ldots\ldots$, we find that
$(1+2+3+\ldots+n)^2 = 1^2 + 2^2 + 3^2 + \ldots$
$\qquad\qquad\qquad\quad + n^2 + 2(1.2 + 1.3 + \ldots + 2.3 + 2.4 + \ldots).$

SOME OTHER SIMPLE SERIES.

Thus, twice the sum of the products of the first n integers taken two and two together

$= (1 + 2 + 3 + \ldots\ldots + n)^2 - (1^2 + 2^2 + 3^2 + \ldots\ldots + n^2)$
$= \{\tfrac{1}{2}n(n+1)\}^2 - \tfrac{1}{6}n(n+1)(2n+1)$
$= \tfrac{1}{12}n(n+1)\{3n^2 + 3n - 4n - 2\}$
$= \tfrac{1}{12}n(n+1)(n-1)(3n+2)$.

$\therefore$ the required sum $= \tfrac{1}{24}(n-1)n(n+1)(3n+2)$.

Ex. 5. Find the sum of the series

$1.4 + 3.7 + 5.10 + \ldots\ldots$ to n terms.

The nth term t_n is obtained by multiplying the nth terms of the two arithmetic series 1, 3, 5......and 4, 7, 10......

Hence $\quad t_n = \{1 + (n-1)2\} \times \{4 + (n-1)3\}$
$\quad\quad\quad = (2n-1)(3n+1) = 6n^2 - n - 1$.

Here it can easily be verified that

$t_1 = 6.1^2 - 1 - 1$.
$t_2 = 6.2^2 - 2 - 1$.

and so on.

Hence the required sum

$= 6(1^2 + 2^2 + 3^2 + \ldots\ldots + n^2) - (1 + 2 + 3 + \ldots\ldots + n) - n$
$= 6.\tfrac{1}{6}n(n+1)(2n+1) - \tfrac{1}{2}n(n+1) - n$
$= \tfrac{1}{2}n(4n^2 + 5n - 1)$.

NOTE. The result can easily be verified by putting $n = 1$, or $n = 2$.

EXERCISES XVI.

[In solving the following exercises, verify the answer in each case in which the nth term or the sum of n terms is required to be found, by putting $n = 1$, and $n = 2$.]

Sum to n terms :—

13. $1^2 + 3^2 + 5^2 + \ldots\ldots$ \quad\quad 14. $2^2 + 4^2 + 6^2 + \ldots\ldots$

15. Sum to n terms the series whose n^{th} term is $n^2 - 1$.

Find the sum of n terms of the series

16. $1.2 + 2.3 + 3.4 + \ldots\ldots$ \quad 17. $1.3 + 3.5 + 5.7 + \ldots\ldots$

TUT. ALG. II.

SOME OTHER SIMPLE SERIES.

228. Piles of shot and shells.

Three shot or spherical shells of equal size, placed together on the ground, will support a fourth, and four will support a fifth. Hence arise two distinct methods of piling shot, the triangular and the rectangular. The square method is only a particular case of the rectangular method.

229. Triangular Piles. *To find the number of spherical balls in a complete pyramidal pile of which the base is an equilateral triangle.*

The pile is composed of a number of layers or *courses*, the balls in every layer forming an equilateral triangle with one ball more in

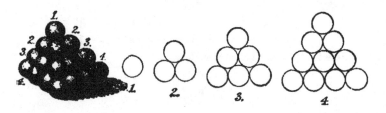

each side than in the layer next above. The pile begins with a single ball at the top.

There are as many interstices in the layer marked (4) as there are balls in the layer marked (3), &c.

Let each side of the base contain n balls.

Then the number of balls at the top $= 1$,

∴ the number of balls in the 2nd layer $= 1 + 2$,
the number of balls in the 3rd layer $= 1 + 2 + 3$,
the number of balls in the 4th layer $= 1 + 2 + 3 + 4$,
....

And the number of balls in the base $= 1 + 2 + 3 + 4 + \ldots + n$
$= \frac{1}{2}n(n+1)$.

Hence the total number of balls in the pile
$= \frac{1}{2}.1(1+1) + \frac{1}{2}.2(2+1) + \frac{1}{2}.3(3+1) + \ldots + \frac{1}{2}.n(n+1)$
$= \frac{1}{2}(1^2 + 2^2 + 3^2 + \ldots + n^2) + \frac{1}{2}(1 + 2 + 3 + \ldots + n)$
$= \frac{1}{12}n(n+1)(2n+1) + \frac{1}{4}n(n+1)$
$= \frac{1}{6}n(n+1)(n+2)$.

SOME OTHER SIMPLE SERIES.

230. Square Piles. *To find the number of spherical balls in a complete pyramidal pile of which the base is a square.*

Let the number of balls in each side of the base be n.

Then the lowest layer consists of n^2 balls, the next layer of $(n-1)^2$ balls, the next of $(n-2)^2$ balls and so on, up to the top layer, which consists of a single ball.

∴ the total number of balls (counting from the top)
$$= 1^2 + 2^2 + \ldots\ldots + n^2 = \tfrac{1}{6}n(n+1)(2n+1).$$

231. Rectangular Piles. *To find the number of spherical balls in a complete pile of which the base is a rectangle.*

The pile consists of a number of layers or *courses*, the balls in every layer forming a rectangle with one ball more in each side than in the layer next above. The pile begins with a *row* of balls at the top.

There are as many interstices in the layer marked (3) as there are balls in the layer marked (2) &c.

Let m and n be the number of balls in the long and the short side of the base.

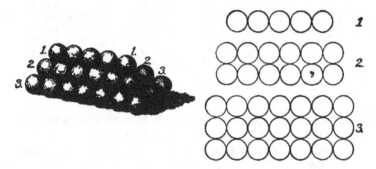

Then the numbers of the balls in the breadths of the layers in order are $n, n-1, n-2, \ldots 3, 2,$ and 1.

And the numbers of balls in the lengths of the layers in order are $m, m-1, m-2, \ldots m-n+3, m-n+2,$ and $m-n+1$.

Hence the total number of balls in the pile (counting from the top)
$$= 1(m-n+1) + 2(m-n+2) + 3(m-n+3) + \ldots\ldots \text{to } n \text{ terms}$$
$$= (m-n)(1+2+\ldots+n) + (1^2+2^2+\ldots+n^2)$$
$$= (m-n).\tfrac{1}{2}n(n+1) + \tfrac{1}{6}n(n+1)(2n+1)$$
$$= \tfrac{1}{6}n(n+1)(3m-n+1).$$

SOME OTHER SIMPLE SERIES.

Ex. 1. Find the number of shot in a complete square pile having 12 courses.

The number of shot in the rth course reckoned downwards $= r^2$, since this course has r shot in each side.

∴ the number of shot required
$$= 1^2 + 2^2 + \ldots + 12^2 = \tfrac{1}{6}.12.13.25 = 650.$$

Ex. 2. Find the number of balls in a complete rectangular pile when there are 15 balls in a single line on the top row, and 12 balls in the breadth of the lowest layer.

The number of balls in the breadth of the lowest layer being 12, the number of layers is also 12.

∴ the number of balls in the pile $= 15.1 + 16.2 + \ldots$ to 12 terms.

Now, the rth term of this series being
$$(15 + r - 1)r = (r + 14)r = r^2 + 14r,$$
the sum of the series
$$= (1^2 + 2^2 + \ldots + 12^2) + 14(1 + 2 + \ldots + 12)$$
$$= \tfrac{1}{6}.12.13.25 + 14.\tfrac{1}{2}.12.13 = 1742,$$
which is the number of balls required.

NOTE.—It may be noted (i) that in a triangular or square pile the number of layers or the number of balls counted on the line from the top to one of the angles at the bottom equals the number of balls counted on any side of the lowest layer; (ii) that, in rectangular piles, the number of layers equals the number of balls in the breadth of the lowest layer; and (iii) that the number of balls in the top row of a rectangular pile is one more than the difference of the numbers of balls in the length and breadth of the base.

232. Incomplete Piles are piles wanting a certain number of courses at the top to complete them.

To find the number of balls in an incomplete pile the base of which is a rectangle.

Let a and b be the number of balls in the sides of the top layer, n the number of the layers.

Then the number of balls in the pile (counting from the top)
$$= ab + (a + 1)(b + 1) + \ldots + (a + \overline{n - 1})(b + \overline{n - 1})$$
$$= ab + \{ab + (a + b) + 1^2\} + \ldots + \{ab + (n - 1)(a + b) + (n - 1)^2\}$$
$$= nab + (a + b)(1 + 2 + \ldots + \overline{n - 1}) + (1^2 + 2^2 + \ldots + \overline{n - 1}^2)$$
$$= nab + (a + b).\tfrac{1}{2}(n - 1)n + \tfrac{1}{6}(n - 1)n(2.\overline{n - 1} + 1)$$
$$= \tfrac{1}{6}n\{6ab + 3(n - 1)(a + b) + (n - 1)(2n - 1)\}.$$

SOME OTHER SIMPLE SERIES.   213

NOTE. By making $a = 1$ in this expression, we find that the number of balls in a complete rectangular pile of n layers having b balls in the top layer $= \frac{1}{6}n(n+1)(3b + 2n - 2)$.

By making $a = b = 1$, we find that the number of balls in a complete square pile of n layers $= \frac{1}{6}n(n+1)(2n+1)$.

Ex. 1. Find the number of balls in an incomplete triangular pile of 15 courses, each side of the base containing 25 balls.

We can obtain a complete triangular pile of 25 courses by placing on the given pile a complete triangular pile of 10 ($= 25 - 15$) courses.

∴ the number of balls required

$$= \tfrac{1}{6}.25.26.27 - \tfrac{1}{6}.10.11.12 = 2705. \qquad [\S\ 229.]$$

EXERCISES XVI.

18. Find the number of balls in a complete triangular pile, each side of the base of which contains 12 balls.

19. Find the number of balls in a complete square pile having 20 balls in each side of the base.

20. The top row of a rectangular pile has 10 balls, and there are 10 courses. Find the number of balls in the pile.

21. Find the number of balls in a finished rectangular pile, the length and breadth of the base layer containing 50 and 24 balls.

22. Find the number of balls in an unfinished triangular pile which has 20 balls in each side of the base, and 7 balls in each side of the top.

23. Find the number of shot in an unfinished square heap of 10 courses, having 10 shot in each side of the top.

24. In an incomplete square heap the top course contains 25 balls and the bottom course 100 balls. Find the number in the heap.

25. In an incomplete triangular heap of 6 courses, the top course contains 210 balls. Find the number of balls in the heap.

26. If the number of balls in a complete triangular pile be 84, find the number of layers.

27. If the middle layer of a complete triangular pile contains 15 balls, find the number of layers.

28. Find the number of balls in an incomplete rectangular pile of 5 courses, the number of balls in the sides of the bottom course being 10 and 12.

29. Find the number of balls in an incomplete rectangular pile of 6 courses, the number of balls in the sides of the top course being 7 and 12.

30. Find the number of balls in an incomplete rectangular pile, the number of balls in the sides of the top course being 7 and 11, and the number of balls in the longer side of the bottom course being 15.

31. If the base of an incomplete rectangular pile of 5 courses contains 352 balls, and the top course 216 balls, how many balls are there in the pile?

32. Find the number of balls in an incomplete triangular pile of 10 layers, when the difference of the number of balls in the bottom and the top course is 135.

33. The number of balls in a complete square pile is to the number of balls in a complete triangular pile of four times the number of courses as 1 is to 24. Find the number of balls in each pile.

34. If n be the number of balls in a complete square pile, show that the number of courses in the pile is the integer next below the cube root of $3n$.

35. If n be the number of balls in a complete triangular pile, show that the number of courses in the pile is the integer next below the cube root of $6n$.

CHAPTER XVII.

LOGARITHMS.

233. Preliminary remarks. Consider the following numbers :

Indices : ......, $-2, -1, 0, 1, 2, 3,$......
Powers : ......, $3^{-2}, 3^{-1}, 3^0, 3^1, 3^2, 3^3,$......
......, $\frac{1}{9}, \frac{1}{3}, 1, 3, 9, 27,$......

The numbers in the first row are connected with those in the third row by means of the number 3 in such a way that any number in the first row is the index of that power of 3 which is the corresponding number in the third row; thus, $\frac{1}{9} = 3^{-2}$, $27 = 3^3$.

'If any other positive number y be taken, we cannot always find a number x so that $3^x = y$ exactly, but we can always find x to such a degree of approximation that 3^x differs from y by as small a quantity as we please. Further, any number lying between a pair of adjacent numbers in the first row will correspond to some number lying between the corresponding pair of adjacent numbers in the third row. Now $3^{-\infty} = 0$ and $3^{+\infty} = +\infty$; hence we can find an endless and continuous series of numbers ranging from $-\infty$ to $+\infty$, each member (x) of which corresponds to some member (y) of another endless and continuous series of numbers ranging from 0 to $+\infty$ and is connected with it by means of the relation $3^x = y$. Any other positive number greater than unity may be taken instead of 3 as the basis of calculation but in all cases the series of indices will range from $-\infty$ to $+\infty$ while the series of powers will range from 0 to $+\infty$.

LOGARITHMS.

Each member of the series of indices is called the **logarithm** of the corresponding member of the series of powers, and all the members of the series of indices constitute a **system of logarithms,** of which the number (in the above case, 3) which is the basis of calculation is called the **base.** Hence we have the following :—

DEFINITIONS :—The **logarithm of a number to a given base** is the *index* of the power to which the base must be raised to produce that number. Thus, if $a^x = y$, then x is called the logarithm of y to the base a, and is usually written $\log_a y$.

Ex. $\log_3 81 = 4$, since $3^4 = 81$.
Similarly $\log_{10} 100 = 2$, since $100 = 10^2$;
$\log_2 8 = 3$, since $8 = 2^3$.

It is obvious from the foregoing remarks that numbers and their logarithms are not proportional. Also the logarithm of $+\infty$ is ∞, and that of 0 is $-\infty$ provided the base is greater than unity.

Since the two relations $a^x = y$ and $x = \log_a y$ are **equivalent,** it follows that *the properties of logarithms can be derived from those of indices.*

Moreover the definition of a logarithm leads at once to the identities
$$\log_a (a^n) = n, \quad a^{\log_a x} = x.$$

234. (i) **The logarithm of unity to any base is zero;** and

(ii) **The logarithm of the base itself is unity.**
For since $a^0 = 1$ and $a^1 = a$, it follows that
$$\log_a 1 = 0 \text{ and } \log_a a = 1.$$

235. The logarithm of a product is equal to the sum of the logarithms of its factors.

Let the product be MN, and let
$$\log_a M = x, \ \log_a N = y;$$
then by definition $\quad M = a^x, N = a^y$;
therefore $\quad MN = a^x \times a^y = a^{x+y}$; [§ 12]
that is, $\log_a MN = x + y = \log_a M + \log_a N$.

COR. In the same way, it may be shown that
$$\log_a MNP\ldots = \log_a M + \log_a N + \log_a P + \ldots$$

236. The logarithm of a quotient is equal to the algebraic difference of the logarithms of the dividend and the divisor.

Let the quotient be $M \div N$,
and let $\log_a M = x$, $\log_a N = y$;
then by definition $M = a^x$, $N = a^y$;
therefore $M \div N = a^x \div a^y = a^{x-y}$;
that is, $\log_a (M \div N) = x - y = \log_a M - \log_a N$.

237. The logarithm of any power of a number is equal to the product of the logarithm of the number and the index of the power.

Let the power be M^n, where n is positive or negative, integral or fractional; and let $\log_a M = x$; then $M = a^x$;
therefore $M^n = (a^x)^n = a^{nx}$; [§ 13]
that is, $\log_a M^n = nx = n \log_a M$.

COR. Hence $\log_a \sqrt[n]{M} = \dfrac{1}{n} \log_a M$.

For $\sqrt[n]{M} = M^{1/n}$.

238. Summary. The foregoing results may be tabulated opposite the corresponding properties of indices as follows:—

Properties of Indices	Properties of Logarithms
$a^0 = 1$	$\log_a 1 = 0$
$a^1 = a$	$\log_a a = 1$
$a^{\frac{1}{2}} = \sqrt{a}$	$\log_a(\sqrt{a}) = \tfrac{1}{2}$
$a^{-1} = 1/a$	$\log_a(1/a) = -1$
$a^\infty = \infty$ (if $a > 1$)	$\log_a \infty = \infty$
$a^{-\infty} = 0$,,	$\log_a 0 = -\infty$
$a^x a^y = a^{x+y}$	$\log_a MN = \log_a M + \log_a N$
$a^x / a^y = a^{x-y}$	$\log_a M/N = \log_a M - \log_a N$
$(a^x)^n = a^{nx}$	$\log_a M^n = n \log_a M$
$\sqrt[n]{a^x} = a^{x/n}$	$\log_a \sqrt[n]{M} = (\log_a M)/n$

Ex. 1. Find $\log_{10} 504$, having given $\log_{10} 2 = \cdot 3010300$,
$\log_{10} 3 = \cdot 4771213$, $\log_{10} 7 = \cdot 8450980$.
Now $504 = 2^3 \cdot 3^2 \cdot 7$.
Hence, by §§ 235, 237, we have
$\log_{10} 504 = 3 \log_{10} 2 + 2 \log_{10} 3 + \log_{10} 7$
$= \cdot 9030900 + \cdot 9542426 + \cdot 8450980$
$= 2 \cdot 7024306$.

Ex. 2. Find $\log_{10} \sqrt[5]{35}$, having given $\log 2$ and $\log 7$ (see Ex. 1).
By § 237, Cor., $\log_{10} \sqrt[5]{35} = \frac{1}{5} \log_{10} 35$.
Now $35 = 10 \times 7 \div 2$.
$\therefore \log_{10} \sqrt[5]{35} = \frac{1}{5} (\log_{10} 10 + \log_{10} 7 - \log_{10} 2)$
$= \frac{1}{5} (1 \cdot 8450980 - \cdot 3010300)$
$= \frac{1}{5} \times 1 \cdot 5440680 = \cdot 3088136$.

239. To find the relation between the logarithms of a number to different bases.

Let N be the number, a and b the bases to which the logarithms of N are calculated.

Let $\log_a N = x$, $\log_b N = y$.
Then $N = a^x$, and $N = b^y$,
$\therefore a^x = b^y$;
$\therefore b^{\frac{y}{x}} = a$, and $a^{\frac{x}{y}} = b$;
$\therefore \frac{y}{x} = \log_b a$, and $\frac{x}{y} = \log_a b$;
$\therefore y = x \log_b a$, and $y = \frac{x}{\log_a b}$,

$\therefore \mathbf{\log_b N = \log_a N \log_b a} = \dfrac{\log_a N}{\log_a b}.$

Hence *the logarithm of any number to the base* b *may be obtained from the logarithm of that number to the base* a *by multiplying it by*

$$\log_b a \text{ or } \frac{1}{\log_a b}.$$

LOGARITHMS.

Cor. Since the multipliers $\log_b a$ and $\dfrac{1}{\log_a b}$ are the same, it follows that

$$\log_b a \times \log_a b = 1.$$

This may also be proved independently thus :—
Let $\log_a b = x$. Then $b = a^x$, $\therefore a = b^{1/x}$.

$$\therefore \log_b a = \frac{1}{x} = \frac{1}{\log_a b}.$$

Ex. Find the value of $\log_2 3 \times \log_3 8$.
By § 239, the required value $= \log_2 8 = \log_2 2^3 = 3 \log_2 2$
$= 3$, since $\log_2 2 = 1$.

240. Natural and Common Logarithms. Logarithms are naturally calculated to a certain base called $e\,(= 2\cdot 7182818...,$ an incommensurable quantity). Such logarithms are used in all theoretical investigations, and are called **Napierian** or **Natural Logarithms.**

[It can be proved by means of a well-known proposition of higher algebra called the Exponential Theorem that if

$$e = 1 + \frac{1}{1} + \frac{1}{1.2} + \frac{1}{1.2.3} + \frac{1}{1.2.3.4} + \ldots ad\ infinitum,$$

then $\quad \log_e (1 + x) = x - \dfrac{x^2}{2} + \dfrac{x^3}{3} - \dfrac{x^4}{4} + \dfrac{x^5}{5} - \ldots ad\ infinitum,$

provided that x lies between 1 and -1. Here by taking sufficient terms of the series, logarithms to base e can be calculated to any degree of approximation.]

In practice however, where numerical calculations are required, Napierian logarithms are inconvenient and, for reasons which we shall shortly give, logarithms calculated to the base 10 (the radix of the ordinary system of numbers) are used instead. These logarithms are called **Briggsian** or **Common Logarithms***.

---

* Logarithms were first discovered by John Napier (1550—1617) Baron of Merchiston. The change of the base from e to 10 is mainly due to Henry Briggs (1556—1630), Savilian Professor at Oxford, although the idea had also occurred to Napier.

Natural logarithms are converted into common logarithms by multiplying them by $\log_{10} e$ or $\dfrac{1}{\log_e 10}$. The value of this multiplier, which is called the **Modulus** of the common system of logarithms, is ·43429448...

In this way tables have been constructed giving the logarithms to the base 10 of all numbers of not more than 5 digits, calculated to seven places of decimals. These are known as **tables of seven figure logarithms.**

The logarithms of commensurable numbers other than powers of 10 are incommensurable. For if $\log_{10} N$ is equal to an exact fraction a/b, then $N = 10^{a/b} = \sqrt[b]{10^a}$, and $N^b = 10^a$. Hence N must be a power of 10, for otherwise no power of N could be equal to a power of 10.

241. Advantages of Common Logarithms. The advantages of the common system of logarithms depend on the fact that its base is the radix of the common scale of notation.

Let M be any number having the same significant figures as N; then

$M = N \times$ some integral power of $10 = N \cdot 10^n$, suppose,

$$\therefore \log_{10} M = \log_{10} N + n \log_{10} 10$$
$$= \log_{10} N + n,$$

that is, the logarithms of N and $N \cdot 10^n$ to the base 10 differ only by an integer n, which may be positive or negative.

Hence if we know the logarithm of any number to base 10 we can at once write down the logarithm of that number multiplied or divided by any power of 10. This is one of the advantages of common logarithms.

Ex. If we are given that $\log_{10} 2 = \cdot 3010300$, we also know that $\log_{10} 2000 = 3 \cdot 3010300$, $\log_{10} 200000 = 5 \cdot 3010300$, and so on.

242. Characteristic and Mantissa. It is found convenient to write a logarithm so that its fractional part may be positive and less than unity, while its integral part is positive or negative, as the case may be.

LOGARITHMS. 221

Ex. 1. Thus
$\log_{10} 24 = 1\cdot3802112$, and $\log_{10} \cdot024 = -1\cdot6197888 = -2 + \cdot3802112$, usually written $\bar{2}\cdot3802112$ for compactness.

The positive fractional part is called the *mantissa*, and the integral part obtained after expressing the mantissa positively is called the *characteristic* of the logarithm.

Ex. 2. If the log be $-2\cdot3010300$, this
$$= -3 + 1 - \cdot3010300 = -3 + \cdot6989700.$$
This is written $\bar{3}\cdot6989700$; $\bar{3}$ or -3 is the *characteristic*, and $\cdot6989700$ the *mantissa*. Hence we have the following definitions :

DEFINITIONS. If the logarithm of a number be positive, the **characteristic** is the integral part, and the **mantissa** the decimal part. If the whole logarithm be negative, arrange it so that the decimal part shall be positive ; and then the integral part (which is negative) is the **characteristic,** and the decimal part is the **mantissa.**

Thus, **mantissæ are always positive.**

From § 241 we have the following property, which is sometimes known as the *rule for the mantissa* :

Two numbers having the same significant figures have the same mantissa to their logarithms and differ only in the characteristics.

A logarithm which is wholly negative may be transformed into one whose mantissa is positive by increasing the negative characteristic by unity and substituting for the negative mantissa its arithmetical complement or defect from unity. This arithmetical complement is most readily found by subtracting the last significant figure of the mantissa from 10 and the other figures before it from 9.

Ex. 3. Having given $\log 2 = \cdot3010300$, $\log 3 = \cdot4771213$ and $\log 7 = \cdot8450980$, find $\log \frac{4}{21}$.

$$\log \tfrac{4}{21} = \log 4 - \log 21 = \log 2^2 - \log 3 \times 7$$
$$= 2\log 2 - \log 3 - \log 7$$
$$= \cdot6020600 - \cdot4771213 - \cdot8450980$$
$$= \cdot6020600 - 1\cdot3222193 = -1 + 1\cdot6020600 - 1\cdot3222193$$
$$= -1 + \cdot2798407 = \bar{1}\cdot2798407.$$

Ex. 4. Find $\log \sqrt[5]{·004\dot{8}}$, having given log 2, log 3 (see Ex. 3), and $\log 11 = 1·0413927$.

Now, $\log \sqrt[5]{·004\dot{8}} = \log (·004\dot{8})^{\frac{1}{5}} = \frac{1}{5}$ of log ·004$\dot{8}$.

Since $·004\dot{8} = \dfrac{48}{9900} = \dfrac{16}{3300} = \dfrac{2^4}{3.11.10^2}$,

therefore, by §§ 235—237,

$\log ·004\dot{8} = 4 \log 2 - \log 3 - \log 11 - 2 \log 10$
$= 1·2041200 - ·4771213 - 1·0413927 - 2$
$= 1·2041200 - 3·5185140 = -3 + 4·2041200 - 3·5185140$
$= -3 + ·6856060 = \bar{3}·6856060$.

∴ required logarithm $= \frac{1}{5}$ of $\bar{3}·6856060$.

Here the negative characteristic is not exactly divisible by 5 and we must make it divisible by 5 by adding -2 to the characteristic and $+2$ to the mantissa.

We therefore write the logarithm thus :

$\log \sqrt[5]{·004\dot{8}} = \frac{1}{5}$ of $(-5 + 2·6856060) = -1 + ·5371212$
$= \bar{1}·5371212$.

243. To determine the characteristic of the logarithm of any number by inspection.

CASE I. Let the number be greater than unity, and let the number of figures in the integral part be n.

Then the number is less than 10^n but is not less than 10^{n-1}.

Hence its logarithm lies between $n-1$ and n; that is, the characteristic is $n-1$.

CASE II. Let the number be less than unity, and let n be the number of zeros before the first significant figure.

Then the number is less than 10^{-n} and not less than 10^{-n-1}.

Hence its logarithm lies between $-(n+1)$ and $-n$, and since the decimal part must be positive, the characteristic must be $-(n+1)$.

LOGARITHMS. 223

We therefore have the following rules for the characteristic :—

RULE 1. *The characteristic of the logarithm of any number greater than unity is one less than the number of digits in the integral part of the number.*

RULE 2. *The characteristic of the logarithm of any number less than unity is negative and numerically one greater than the number of zeros before the first significant figure of the number.*

The mantissa of a logarithm thus depends only on its sequence of digits, and the characteristic only on the position of its first significant figure. In tables of logarithms the mantissa is therefore alone given, and these tables suffice to determine the mantissa for any number of not more than five significant digits, the characteristic being found by inspection.

Ex. 1. Find the characteristic of the logarithm of ·082 to base 3.

Since $3^{-1} = \cdot3333...$; $3^{-2} = \cdot1111...$; and $3^{-3} = \cdot037037$; ·082 lies between 3^{-2} and 3^{-3}, and its logarithm to base 3 is $-3 +$ a positive fraction. Hence the characteristic required is -3.

244. Alternative working rule. As the above rules are a little troublesome to apply in practice the following rule will be found convenient for writing down the characteristic for any number :—

RULE. *If the first figure is in the units' place the characteristic is 0. Add 1 for each place that the first figure is to the left, subtract 1 for each place that the first figure is to the right of the units' place.*

Ex. 1. To find the characteristics of log 7289·2 and log ·000024.

To apply the rule we start at the units' place (the place before the decimal point) and begin mentally counting backwards or forwards as the case may be till we stop at the first significant figure, thus

	(i) 7289·2	(ii) 0·000024,
count thus :	3210	$(-)$ 0·12345 ;

∴ characteristic $= 3$, characteristic $= \bar{5}$.

[In practice the numbers in the second line would not be written down.]

Ex. 2. Given (from the tables) that log 28091 = ·4485672, find the numbers whose logarithms are 3·4485672 and $\bar{2}$·4485672.

[The tables only give the mantissa.]

Starting from the units' place count backwards or forwards as the case may be and put the first significant figure below the last figure counted, thus :

     (i) 3210·   (ii) 0·12

∴ required numbers are 2809·1 and ·028091.

245. Interpolation. Tables of logarithms are used for the following purposes :

(i) To find the logarithm of a given number.

(ii) To find the number whose logarithm is a given number.

If the number whose logarithm is required or is given has not more than 5 significant figures, the information is at once obtainable from the tables, for these contain the *mantissæ* for all numbers from 1 to 99999. When the mantissa is not given in the tables, we must make use of the following principle :

The Theory of Proportional Parts. *If a number be increased by a very small fraction of itself, the increase in the logarithm of the number is approximately proportional to the increase in the number.*

This principle may be conveniently stated in the following practical form :

If N *be any number not less than* 10000, *and* h, k *any numbers not greater than* 1; *then, as far as seven places of decimals, the following relation holds good,*

$$\frac{\log(N+h) - \log N}{\log(N+k) - \log N} = \frac{h}{k}.$$

This rule also applies to mixed numbers less than 10000 if they have five or more figures, because, by multiplying them by some power of 10, they can always be converted into numbers greater than 10000, and the *mantissæ* of their logarithms are unaltered.

LOGARITHMS. 225

The method of applying this principle is known as **interpolation,** and will be most easily understood from the following examples.

Ex. 1. Given log 10686 = 4·0288152 and log 10687 = 4·0288558, find log 10686·32.

The difference in the number of 1 gives a difference in the logarithm of ·0000406.

Hence, by § 245, a difference in the number of ·32 gives a difference in the logarithm of

$$\frac{·32}{1} \times ·0000406 = ·0000129 ;$$

∴ log 10686·32 = 4·0288152 + ·0000129 = 4·0288281.

Ex. 2. Find the number whose logarithm is 2·3456785, having given log 22165 = 4·3456677 and log 22166 = 4·3456873.

At present we need only concern ourselves with the *mantissæ*.

A difference in the mantissæ of ·0000196 is given by a difference of 1 in the number.

Hence a difference in the mantissæ of ·0000108
(*i.e.* ·3456785 − ·3456677)
is given by a difference in the number of $\frac{108}{196}$ = ·551.

Therefore 4·3456785 = log 22165·551,
and 2·3456785 = log 221·65551.

EXERCISES XVII.

The following logarithms are given :—

log 2 = ·3010300, log 11 = 1·0413927,
log 3 = ·4771213, log 13 = 1·1139434.
log 7 = ·8450980,

Other logarithms needed are given in the exercises.

(*Except where otherwise stated, all logarithms in the following exercises are calculated to base* 10.)

1. Find the logarithms of the following numbers :—
   6, 21, 78, 143, 462, 1001.

2. Calculate the logarithms of $\frac{70}{11}, \frac{33}{7}, 47\frac{2}{3}, 20\frac{3}{7}, 166\frac{5}{6}$.

TUT. ALG. II. 15

LOGARITHMS.

3. Find the logarithms of 8, 24, 27, 49, 96, 169, 1331.

4. Find the characteristic of the logarithm of 24 to the base 2, and of 10 to the base $\frac{3}{2}$.

5. Write down the characteristics of
log 723596, log 2134, log ·000312, log ·021.

6. Calculate log ·0025 and log ·00625.

7. Find log 4·5, log 6·75, log 10·125.

8. Find the logarithms of $\frac{1}{24}, \frac{1}{3528}, \frac{4}{65}$.

9. Determine the value of the logarithm of 2401 when the base is the cube root of 7.

10. Calculate the value of $\frac{3}{2} \div 2^{\frac{7}{12}}$, having given
$$\log 1·00113 = ·0004905.$$

11. Calculate to six places of decimals log 65, log ·0065.

12. Find what power of 2 is nearest to 10^8.

13. Show that, if the logarithm of a^2 to the base b^3 be equal to the logarithm of b^3 to the base a^{12}, then each logarithm must be equal to $\frac{2}{3}$.

14. Find the value of the product
$$\log_2 3 . \log_3 4 . \log_4 5 . \log_5 6 . \log_6 7 . \log_7 8.$$

15. Calculate the value of the following logarithms,
$$\log 5250, \log ·004\dot{6}, \log \frac{1}{\sqrt[5]{35}}.$$

16. Find $\log \left(\frac{7}{32}\right)^{\frac{2}{3}}$ to the base 100.

17. Find the number of digits in 2^{36} and 3^{100}.

18. The first and tenth terms of a G.P. are respectively 3 and 65; find the common ratio, having given log 1·407412 = ·1484213.

19. Find log 10632, having given
log 1063 = 3·0265333 and log 1·064 = ·0269416;
and log 324·365, having given
log 32434 = 4·511000 and log 32440 = 4·511081.

LOGARITHMS.

20. Find the number whose logarithm is 3·7680500, having given
log 5·862 = ·7680458 and log 5·8621 = ·7680532.

21. Find the 7th root of 100 to six decimal places, having given
log 193·06 = 2·2856923 and log 19307 = 4·2857148.

22. Given log 3·8862 = ·5895251 and log 3·8863 = ·5895363, calculate approximately log ·03886245.

23. The outside diameter of a spherical shell of iron is 40 centimetres, and the volume of the metal (as estimated from its weight) is 5000 cubic centimetres; find its thickness, having given
π = 3·1416, log 68063 = 4·83291, log 1·8952 = ·27765.

24. Find x to three places of decimals from the equations
(i) $12^x = 96$; (ii) $2^{3x}5^{x-2} = 7^{x-4}$.

25. If a series of numbers are in G.P., their logarithms are in A.P.

26. A certain country determines to reduce its national debt by one per cent. every year. What proportion of the original debt will, on this scheme, be paid off in a hundred years if there are no further borrowings? (log 99 = 1·995635, log 3·660 = ·5635...).

27. If $\log_a m = n$, shew that $\log_{\frac{1}{a}} m = -n$.

28. Find the characteristics of log 300 and log ·02 to base 5.

29. Prove that $\log_b a \cdot \log_c b \cdot \log_a c = 1$.

30. Find the number of digits in e^{1000}, having given
$\log_{10} e$ = ·4342945.

31. Find $\log \sqrt[5]{(·00054 \times 3·6)}$.

32. Find $\log_{10} \sqrt[10]{[2^{10}\sqrt{\{2^{10}\sqrt{(\&c., \text{ad infinitum})\}}}]}$.

33. Find the logarithm of 3·375 to the base 2·25.

34. Find (without the use of logarithm tables) the value of
$$\frac{\log\sqrt{27} + \log 8 - \log\sqrt{1000}}{\log 1·2}.$$

35. Also of $\frac{1}{6}\sqrt{\left\{\frac{3\log 1728}{1 + \frac{1}{2}\log ·36 + \frac{1}{3}\log 8}\right\}}$.

36. Given
log 2 = ·3010300, log 5·76 = ·7604226, and log ·0105 = $\overline{2}$·0211893,
find from these data the logarithms of the digits above 2.

37. Find $\log_{10}$ ·005 and $\log_{10}\left(\frac{64}{35}\right)^{\frac{5}{6}}$.

No.	0	1	2	3	4	5	6	7	8	9	D.	
3431	5354207	4334	4460	4587	4713	4840	4967	5093	5220	5346		
32		5473	5599	5726	5852	5979	6105	6232	6359	6485	6612		
33		6738	6865	6991	7118	7244	7371	7497	7623	7750	7876	126	
34		8003	8129	8256	8382	8509	8635	8762	8888	9015	9141	1	13
35		9267	9394	9520	9647	9773	9900	0026	0152	0279	0405	2	25
36	5360532	0658	0784	0911	1037	1163	1290	1416	1543	1669	3	38
37		1795	1922	2048	2174	2301	2427	2553	2680	2806	2932	4	50
38		3059	3185	3311	3438	3564	3690	3817	3943	4069	4195	5	63
39		4322	4448	4574	4701	4827	4953	5079	5206	5332	5458	6	76
3440		5584	5711	5837	5963	6089	6216	6342	6468	6594	6721	7	88
41		6847	6973	7099	7225	7352	7478	7604	7730	7856	7982	8	101
42		8109	8235	8361	8487	8613	8739	8866	8992	9118	9244	9	113
43		9370	9496	9622	9749	9875	0001	0127	0253	0379	0505		
44	5370631	0758	0884	1010	1136	1262	1338	1514	1640	1766		

246. Use of Logarithmic Tables. An extract from a "seven-figure" table is given above. In using such a table it must carefully be noted that the numbers 0, 1, 2 ......at the head of the columns refer to the fifth figure of the number whose logarithm is required. The tables give only the mantissa of the logarithms, and the decimal point is omitted, as this, together with the necessary characteristic, can be supplied by the reader. Each mantissa is given correct to *seven figures.* It is best to regard the table as giving the logarithms of numbers whose first figure is in the units place; i.e. numbers between 1 and 10.

Hence in reading a table, the decimal point is to be placed between the first and second figures of the number and before the first figure of the logarithm.

The first line of the specimen will then read in full thus :
$$\log 3.4310 = \cdot 5354207$$
$$\log 3.4311 = \cdot 5354334$$
$$\log 3.4312 = \cdot 5354460$$
$$\log 3.4313 = \cdot 5354587 \text{ and so on.}$$

Where the third figure changes in the middle of a line, a rule is placed over the remaining four figures.

LOGARITHMS. 229

Thus in the fifth line of the specimen, we read
$$\log 3{\cdot}4355 = {\cdot}5359900$$
but $\log 3{\cdot}4356 = {\cdot}5360026$ not 5350026,
similarly, $\log 3{\cdot}4357 = {\cdot}5360152$, and so on.

In finding the logarithm of a number of less than five figures we may add zeros for the remaining figures.

Thus $\log 344 = \log 344{\cdot}00 = 2{\cdot}5365584$.

247. Use of a table of proportional parts.
The difference between each logarithm and the next is given in the last column of the tables headed D. This difference represents the increase in the logarithm of any number preceding it when the fifth figure of that number is increased by unity.

Moreover, under each difference is a "table of proportional parts" the use of which is to shorten the work of interpolation, as we shall explain in the examples to be given shortly.

The difference 126 given in the specimen really stands for ·0000126, and the table asserts that ·0000126 is the difference between
$$\log 34310 \text{ and } \log 34311,$$
$$\log 34{\cdot}336 \text{ and } \log 34{\cdot}337,$$
$$\log {\cdot}0034358 \text{ and } \log {\cdot}0034359,$$
and so on.

On referring to such a table, it will be seen that the "difference" remains constant for a considerable series of numbers in succession, thus affording a verification of the Theory of Proportional Parts.

[The actual difference between the tabulated values of log 34310 and log 34311 will be found to be ·0000127 not ·0000126 ; this is due to figures "carrying on" from the eighth place of decimals, as the results are only correct to seven places.]

Ex. 1. To find log 3437·685 and log 3437·687 from the specimen table.

Supplying the necessary characteristic, our work stands thus
$$\log 3437{\cdot}6 \;\;= 3{\cdot}5362553 \quad D = 126$$
$$\text{add pro. part for} \quad 8 = \quad\quad 101$$
$$5 = \quad\quad\; 6\tfrac{3}{8}$$
$$\log 3437{\cdot}685 = 3{\cdot}5362660$$

The proportional part 101 is taken from the table of proportional parts under the difference 126 opposite the number 8. For the part to add for the *seventh* figure 5 we take the number 63 in the same way but cut off its last figure.

In finding log 3437·687 we proceed in the same way, but the proportional part opposite 7 is 88, and as this is nearer to 90 than to 80 we add 9 (*i.e.* ·0000009) instead of 8. Hence we find

log 3437·687 = 3·5362553 + ·0000101 + ·0000009 = 3·5362663.

The rule in such cases is to increase the remaining number by unity if the figure struck off is not less than 5.

Ex. 2. Find from the same specimen table the number whose logarithm is $\bar{1}$·5361247.

Here	$\bar{1}$·5361247	= given logarithm.
From table,	$\bar{1}$·5361163	= log ·34365, Diff. = 126.
By subtraction	84	
	76	= pro. pt. for 6
	8	
	8 (= 7⅔)	= p. p. 6,
∴ required number =	·3436566.	

The result is correct to as many figures as can be found by the use of seven figure logarithms.

248. Advantages of Logarithms. On referring to the results of §§ 235—237 it will be evident that by working with the logarithms of numbers instead of the numbers themselves, the process of

Multiplication is replaced	by	addition (§ 235).
Division	,,	subtraction (§ 236).
Involution	,,	multiplication (§ 237).
Evolution	,,	division (§ 237 Cor.).

In §§ 253, 254 we shall explain how logarithms can be applied to the insertion of geometric means and the solution of certain exponential equations.

The following examples are chosen to illustrate how to deal with logarithms whose characteristics are negative. In some of these examples the logarithms are taken only to 5 places of decimals.

LOGARITHMS.

249. Multiplication by Logarithms.

Ex. 1. Find the continued product of $432 \cdot 17$, $\cdot 67843$, and $\cdot 000503$, correct to 5 significant figures.

Let x be the required product. Then
$$\log x = \log 432 \cdot 17 + \log \cdot 67843 + \log \cdot 000503.$$
Taking the logarithms from the tables, we have to five places

$$\begin{aligned}
\log 432 \cdot 17 &= 2 \cdot 63565 \\
\log \cdot 67843 &= \bar{1} \cdot 83151 \\
\log \cdot 000503 &= \bar{4} \cdot 70157 \\
\hline
\therefore \log x &= \bar{1} \cdot 16873,
\end{aligned}$$

whence the required product $x = \cdot 14748$, approximately.

In the above addition the negative characteristics are to be subtracted from the positive.

[The application of logarithms to multiplication and division forms the principle of the **slide-rule**. The distances of the numbers from the end of the rule are proportional to the logarithms of these numbers.]

250. Division by Logarithms.

Ex. 1. Divide $\cdot 0009197$ by $475 \cdot 25$.

Here $\log (\cdot 0009197 \div 475 \cdot 25) = \log \cdot 0009197 - \log 475 \cdot 25$
$\qquad\qquad\qquad\qquad\qquad\quad = \bar{4} \cdot 96365 - 2 \cdot 67692.$

The subtraction is best written in full thus :

$$\begin{aligned}
&- 4 + \cdot 96365 \\
&- 2 - \cdot 67692 \\
\hline
&- 6 + \cdot 28673 \text{ or } \bar{6} \cdot 28673,
\end{aligned}$$

∴ required quotient $= \cdot 0000019352$ approximately.

Ex. 2. Divide $\cdot 0034567$ by $\cdot 0008912$.

Here $\log (\cdot 0034567 \div \cdot 0008912)$
$\qquad = \log \cdot 0034567 - \log \cdot 0008912$
$\qquad = \bar{3} \cdot 53866 - \bar{4} \cdot 94998$
$\qquad = - 3 + \cdot 53866$
$\qquad\quad\, + 4 - \cdot 94998$
$\qquad\overline{= \quad 0 + \cdot 58868 \text{ or } \cdot 58868,}$

∴ required quotient $= 3 \cdot 8787$.

251. Involution by Logarithms. The use of logarithms enables us to calculate readily the approximate value of a high power of a number when it would be very tedious to do so by actual multiplication.

For if M be the number, we have, by § 237,
$$\log M^n = n \log M,$$
and, knowing $\log M^n$, the tables give the required value of M^n.

Ex. 1. Find the 12th power of ·23628.
$$\log(·23628)^{12} = 12 \log ·23628 = 12 \times \bar{1}·3734270.$$
The working in full would stand thus :
$$\begin{array}{r} -1 + ·3734270 \\ 12 \\ \hline -12 + 4·4811240 = \bar{8}·4811240. \end{array}$$
Now the mantissa of log 302778 = ·4811243.
Hence required 12th power = ·000000030278 nearly.

In practice the working would be abbreviated as in the next example.

Ex. 2. Find the 7th power of ·0826.
$$\begin{array}{r} \log ·0826 = \bar{2}·9169800 \\ 7 \\ \hline \bar{8}·4188600 \end{array}$$
Now the mantissa of log 26234 = ·4188645,
∴ $(·0826)^7 = ·000000026234$, to 5 figures.

252. Evolution by Logarithms. Any root of a number can be extracted approximately by means of a table of logarithms.

For suppose we require to find the mth root of a given number a. We have
$$\log \sqrt[m]{a} = \frac{1}{m} \log a.$$
Hence, all we have to do is to take log a from the tables, and divide by m. The result is the logarithm of the required root, and knowing this, the root itself may be found by a second reference to the tables.

Ex. 1. Find the cube root of ·0003 correct to six places of decimals.

Since $\log 3 = ·4771213$, ∴ $\log ·0003 = \bar{4}·4771213$,
and $\log \sqrt[3]{(·0003)} = \bar{4}·4771213 \div 3$.
$$\begin{array}{r} 3)-6 + 2·4771213 \\ \hline -2 + ·8257071 \text{ or } \bar{2}·8257071 \end{array}$$
whence $\sqrt[3]{(·0003)} = ·066943$ approximately (by the tables).

LOGARITHMS. 233

Ex. 2. Find the value of $\sqrt[10]{20}$, having given

$\log 2 = \cdot 3010300$, $\log 1\cdot 3492 = \cdot 1300763$, and $\log 1\cdot 3493 = \cdot 1301085$.

Let x be the required value.

Then $\log x = \frac{1}{10} \log 20 = \frac{1}{10} \times 1\cdot 3010300 = \cdot 1301030$.

Evidently x lies between $1\cdot 3492$ and $1\cdot 3493$.

Now a diff. in log of $\cdot 0000322$ is given by a diff. in number of $\cdot 0001$.

Hence a diff. in log. of $\cdot 0000267$ ($\cdot 1301030 - \cdot 1300763$) is given by a diff. in number of

$$\frac{267}{322} \times \cdot 0001 = \cdot 000083 \text{ nearly}.$$

Therefore $\log x = \log 1\cdot 349283\ldots$; and $x = 1\cdot 349283$ nearly.

253. Insertion of Geometric Means. The student will have little difficulty in proving that if a series of numbers are in geometrical progression their logarithms are in arithmetical progression. Hence to insert any number of geometric means between two numbers it is only necessary to insert that number of arithmetic means between the logarithms of the numbers; these arithmetic means are the logarithms of the required geometric means.

Ex. Find two mean proportionals between 7 and 100.

Let x, y be the mean proportionals required. Then since

$$\frac{7}{x} = \frac{x}{y} = \frac{y}{100};$$

$\therefore \log 7 - \log x = \log x - \log y = \log y - \log 100$;

$\therefore \log 7$, $\log x$, $\log y$ and $\log 100$ are in arithmetic progression, so that $\log x$ and $\log y$ are the two arithmetic means between $\log 7$ and $\log 100$.

Now $\log 7 = \cdot 8450980$ and $\log 100 = 2$.

The required arithmetic means of the logs are

$$\frac{2\log 7 + \log 100}{3} \text{ and } \frac{\log 7 + 2\log 100}{3},$$

$\therefore \log x = 1\cdot 2300653$ and $\log y = 1\cdot 6150327$.

From the tables we find

$\log 1\cdot 6985 = \cdot 2300656$, diff. for $\cdot 0001 = \cdot 0000255$.
$\log 4\cdot 1212 = \cdot 6150237$, diff. for $\cdot 0001 = \cdot 0000105$.

Hence $x = 16\cdot 98499$, and $y = 41\cdot 21286$.

234   LOGARITHMS.

254. Solution of Exponential Equations.

Ex. 1. Find x from $2^{2x} = 3^{x+1}$, having given log 2 and log 3 (see Ex. 1, § 238).

Taking logarithms of both sides of the given equation, we have
$$2x \log 2 = (x + 1) \log 3,$$
$$\therefore 2x \times \cdot 3010300 = (x + 1) \times \cdot 4771213,$$
$$\therefore \cdot 6020600x = \cdot 4771213x + \cdot 4771213,$$
$$\therefore \cdot 1249387x = \cdot 4771213,$$
$$\therefore x = 3 \cdot 82....$$

Ex. 2. Find what integral power of 2 is nearest to the 12th power of $\frac{3}{2}$.

Let
$$2^x = \left(\frac{3}{2}\right)^{12},$$
then
$$x \log 2 = 12 (\log 3 - \log 2),$$
$$(x + 12) \log 2 = 12 \log 3,$$
$$x + 12 = \frac{12 \log 3}{\log 2} = \frac{12 \times \cdot 4771213}{\cdot 3010300}$$
$$= 19 \cdot 02 ;$$
$$\therefore x = 7 \cdot 02,$$
and the nearest integer is 7. Hence the 7th is the nearest power.

255. Application of formulæ to logarithmic computation. Although logarithms effect a great simplification in all formulæ involving the operations of multiplication and division, involution and evolution, they cannot be conveniently used in formulæ involving the operations of addition and subtraction, for *there is no simple formula for the logarithm of a sum or difference of two numbers in terms of the logarithms of the numbers themselves.* Hence arises the necessity of using *formulæ adapted to logarithmic computation* ; i.e. formulæ not involving the operations of addition and subtraction in the middle of logarithmic work.

Ex. The hypotenuse of a right-angled triangle is 33490 feet long and one of its sides is 366 ft. long; find the other side, having given log 33124 = 4·5201428, log 33488 = 4·5248892, log 33856 = 4·5296356.

LOGARITHMS. 235

Let x be the unknown side ; then $x^2 = 33490^2 - 366^2$
$$= (33490-366)(33490+366)$$
$$= 33124 \times 33856 ;$$
$\therefore 2 \log x = \log 33124 + \log 33856$
$$= 4\cdot5201428 + 4\cdot5296356 = 9\cdot0497784,$$
$\therefore \log x = 4\cdot5248892,$
whence, by the data, $x = 33488$, exactly.

EXERCISES XVII.

The following logarithms are given
$\log 2 = \cdot3010300$, $\log 3 = \cdot4771213$, $\log 7 = \cdot8450980$.

38. Take, from the specimen table on page 228, the logarithms of
    34·359, 343200, ·000344.
39. Take, from the same table, the numbers whose logarithms are
    . $2\cdot5360026$, $\bar{3}\cdot5364322$, $7\cdot5355599$.
40. Determine from the same table the logarithms of
    343126, 344·0075, ·003443787.
41. Determine from the same table the numbers whose logarithms are
    $4\cdot5360000$, $\bar{2}\cdot5355555$, ·5366666.
42. The following entries are given in a table of logarithms

No.	0	1	2	3	4	5	6	7	8	9	D
1482	1708482	8775	9068	9361	9654	9947	0240	0533	0826	1119	293

Construct a table of proportional parts.

Find the logarithms of 1482765 and ·001482489 and the numbers whose logarithms are $10\cdot1709999$ and $\bar{1}\cdot1710218$.

43. Given $\log 2$ and $\log 2000\cdot1 = 3\cdot3010517$, construct a table of proportional parts and find $\log \cdot2000088$.
44. Find the 540th root of ·00007, having given
    $\log 7$ and $\log 9\cdot824394 = \cdot9923057$.
45. Find the fifth root of 24 and the seventeenth power of 18·88812 having given $\log 2$, $\log 3$ and the following extracts from the tables :

No.	0	1	2	3	4	5	6	7	8	9	D
1888	2760020	0250	0480	0710	0940	1170	1400	1630	1860	2090	231
4956	6951313	1401	1488	1576	1663	1751	1839	1926	2014	2102	87

236 LOGARITHMS.

46. Given $\log 1\cdot 3894 = \cdot 1428273$ and $\log 1\cdot 3895 = \cdot 1428586$, find the value of $(13\cdot 89492)^{\frac{1}{8}}$.

47. Find the mean proportional between $33\cdot 549$ and $44\cdot 642$, having given
$\log 3\cdot 3549 = \cdot 5256796$, $\log 4\cdot 4642 = \cdot 6497436$, and $\log 387 = 2\cdot 5877110$.

48. Given $\log 1\frac{1}{5} = \cdot 0791812$ and $\log 2\frac{2}{5} = \cdot 3802112$, find the values of $\sqrt[5]{(3\cdot 6)^3} \times \sqrt[4]{\dfrac{1}{25}} \div \sqrt[3]{8\frac{49}{72}}$, the mantissæ for 46929 and 46930 being 6714413 and 6714506.

49. Find the value of $\sqrt[6]{\{9\sqrt{(3\sqrt{2})}\}}$, given
$\log 2 = \cdot 30103$, $\log 3 = \cdot 47712$, $\log 1626 = 3\cdot 21112$, $\log 1627 = 3\cdot 21139$.

50. Prove that if m and n be respectively the number of square feet in the surface, and of linear feet in an edge, of a regular tetrahedron
$$\log_{10} m = 2\log_{10} n + \cdot 23856.$$
What would be the corresponding equation if m were the number of cubic feet in the volume?

51. A solid cube of lead weighs $126\cdot 44$ lb., 998 ounces of water occupy one cubic foot, and a cubic foot of lead is $11\cdot 352$ times as heavy as a cubic foot of water. Find by logarithms the length of a side of the cube correctly to six places of decimals of a foot. Given logarithms:—

No.	Mantissa	Diff.	No.	Mantissa	Diff.
4	·6020600		11	·0413927	
6	·7781513		19	·2787536	
499	·6981005		129	·1105897	
12644	·1018845	343	56311	·7505932	78

52. If $5^x = 3152$, find by logarithms the value of x, having given $\log 2$, and $\log 312 = 2\cdot 4941546$, $\log 313 = 2\cdot 4955454$.

Verify the result by actual involution.

53. If $7^x = 36^2 \div 10$, what is the value of x?

54. Find the value of x from the equation $18^{8-4x} = (54\sqrt{2})^{3x-2}$ (using base $3\sqrt{2}$).

55. Solve the equation $\left(\dfrac{1}{2}\right)^{x+4} = (25)^{3x+2}$ (given $\log 2$).

56. Find to three places of decimals the value of x from the equation $4^{2x} - 8.4^x + 12 = 0$.

57. Solve the simultaneous equations $18y^x - y^{2x} = 81$, $3^x = y^2$.

CHAPTER XVIII.

INTEREST AND DISCOUNT.

256. An important practical application of logarithms is to be found in the calculation of Compound Interest and the values of annuities. We shall treat in this and the next chapter the principal problems met with in the subject.

257. Interest. DEFINITIONS. In business transactions where money is lent, it is customary for the borrower to pay to the lender a certain sum for the use of it. The sum lent is called the **principal**; the sum paid for the use of it is called the **interest**; and the sum of the principal and the interest is called the **amount**.

The **rate per cent. per annum** is the sum paid as interest for the use of £100 for one year, the interest on any other sum of money being proportional to the principal.

The interest on a loan or investment is payable at regular intervals, generally either once a year or once every six months; sometimes quarterly; the original principal being repaid at the end of a certain time, according to agreement in the case of a loan, or being recoverable by selling the investment.

Ex. Thus, if £100 is borrowed for a certain number of years (say, five), interest being reckoned at 5 per cent. per annum, the borrower pays to the lender £5 at the end of each year for four years, and at the end of the fifth year he pays £105, which is the borrowed principal together with its interest for the fifth year; or, if interest is payable half-yearly, the lender receives £2. 10s. (the half of £5) every six months, the first instalment of interest being paid six months after the time when the money was lent, and the last instalment being paid when the principal is repaid, so that the final payment is £102. 10s.

258. Simple and compound interest. DEFINITIONS. When money is lent or invested for a limited period, it is often found convenient to leave the interest to accumulate until the principal is repaid; in such cases interest may either be *simple* or *compound*.

When interest is **simple,** the unpaid interest merely remains idle till the time of repayment, so that the total amount received by the investor is the same as he would have received if the interest had been paid periodically.

Ex. Thus in the example of the last article the four annual payments of £5 may be allowed to remain unpaid (subject to agreement) until the end of the fifth year, when the lender will receive £125 in all.

If, however, the interest is not paid when due, but allowed to accumulate until the repayment of the principal, it is only just to the lender that he should receive interest on the unpaid interest, which is money the use of which he is losing. The interest therefore when due is added to the borrowed principal and becomes part of it; and when interest again falls due, it is calculated on the combined principal and interest. In such a case the interest is called **compound.**

Thus, if compound instead of simple interest is reckoned in the above example, at the end of the first year the interest £5 is added to the principal; and at the end of the second year, the interest will be calculated on £105 and added to it, forming the principal on which the interest for the third year is calculated, and so on, until the end of the fifth year, when the total amount (in this case, £127. 12s. 6d.) is repaid to the lender.

Since the calculation of Simple Interest is most readily performed by arithmetical rules, we shall merely prove the formulæ referring to it. From these the student will have no difficulty in solving any examples that may be proposed.

INTEREST AND DISCOUNT. 239

259. The ratio of the interest for one year to the principal will be denoted by the letter r, and the *ratio of the amount for one year to the principal* will be denoted by the capital letter R.

With this notation the interest for 1 year
on 1 *pound* is r *pounds* and the amount is R *pounds*,
on 1 *shilling* ,, r *shillings* ,, ,, ,, R *shillings*,
on £100 ,, £100×r ,, ,, ,, £100 × R.

Since amount = principal + interest,

$$\frac{\text{amount}}{\text{principal}} = 1 + \frac{\text{interest}}{\text{principal}}$$

$$\therefore R = 1 + r.$$

If c denotes the rate *per cent.*, then £c is the interest on £100 for one year, so that $c = 100r$. Hence

$$r = \frac{c}{100} \text{ and } R = 1 + \frac{c}{100}.$$

Thus at 5 per cent. $r = \cdot 05 = \frac{1}{20}$, $R = 1\cdot 05 = \frac{21}{20}$.

260. To find the amount of a given sum in any time at simple interest.

Let P pounds be the principal, n the number of years for which the principal is lent, r pounds the interest on £1 for one year, I pounds the interest and M pounds the amount of P pounds in n years.

Since the interest on one pound for one year is r pounds, therefore the interest on £P for one year is Pr pounds and for n years it is nPr pounds. That is

$$I = nPr \quad \text{...................(1).}$$

Also by definition we *always* have

$$M = P + I \quad \text{.................(2).}$$

Hence $M = P + Pnr = P(1 + nr) \quad \text{.........(3).}$

From this equation any one of the quantities P, M, n, r, can be found when the remaining three are given.

240   INTEREST AND DISCOUNT.

If c denote the rate *per cent.*, equations (1), (3) become

$$I = nP\frac{c}{100} \quad \dots\dots\dots\dots\dots(1a).$$

$$M = P\left(1 + \frac{nc}{100}\right)\dots\dots\dots\dots(3a).$$

261. To find the amount of a given sum for any number of years at compound interest payable yearly.

Let P pounds be the principal, n the number of years for which the principal is lent, r pounds the interest on £1 for one year, M pounds the amount at the end of n years.

Since the interest on £1 for one year is r pounds, therefore the interest on £P for one year is Pr pounds, and the amount of £P for one year is $P + Pr$, or $P(1 + r)$ pounds.

Let $1 + r$ be denoted by R, so that R pounds is the amount of £1 for one year.

Then, PR being the principal on which the interest for the second year is calculated, the amount at the end of the second year is $PR \times R$, or PR^2. Similarly the amount at the end of the third year is PR^3; and so on. Proceeding in this way we find that the amount at the end of n years is PR^n.

Hence $\qquad M = PR^n \dots\dots\dots\dots\dots\dots(4).$

Also if I is the interest gained in n years, then by definition,

$$M = P + I \quad \dots\dots\dots\dots\dots(5).$$

Thus $I = M - P$; hence

$$I = P(R^n - 1)\dots\dots\dots\dots\dots (6).$$

If c is the rate *per cent.*, equations (4), (6) become

$$M = P\left(1 + \frac{c}{100}\right)^n \quad \dots\dots\dots\dots(4a).$$

$$I = P\left(1 + \frac{c}{100}\right)^n - P\dots\dots\dots\dots(6a).$$

INTEREST AND DISCOUNT. 241

262. Logarithmic Calculation of Compound Interest. When any three of the quantities M, P, R, n are given, the fourth can be found from the formula obtained in the preceding article. This, however, is a matter of considerable difficulty and involves a large amount of arithmetical work when M, P, R are given or when n is large or fractional. It is customary therefore to use logarithms, and in this way the work is considerably shortened.

Taking logarithms of both sides of the equation $M = PR^n$, we get
$$\log M = \log P + n \log R,$$
an equation from which any one of the four quantities, M, P, R, n, can be readily obtained when the remaining three are given.

[It is better however to start in every case with the equation $M = PR^n$ and to deduce the logarithmic form from it when required.]

Ex. 1. Find the amount and interest of £500 for 10 years at 5 per cent. compound interest, having given
$\log 2 = \cdot 3010300$, $\log 1\cdot 05 = \cdot 0211893$, and $\log 8\cdot 14447 = \cdot 9108630$.

Here $\qquad R = 1 + \dfrac{5}{100} = 1\cdot 05$, $n = 10$, $P = 500$.

By the result of § 261, we have
$$M = 500\,(1\cdot 05)^{10};$$
$\therefore\ \log M = \log 500 + 10 \log 1\cdot 05$
$\qquad\qquad = \log 1000 - \log 2 + 10 \log 1\cdot 05$
$\qquad\qquad = 3 - \cdot 3010300 + \cdot 211893$
$\qquad\qquad = 2\cdot 9108630;$
$\therefore\ M = 814\cdot 447,$

i.e. required amount = £814. 9s., nearly;
also $\qquad I = M - P;$

whence the interest = £314. 9s., nearly.

Ex. 2. In what time will a sum double itself at 4 per cent. compound interest? Given $\log 1\cdot 04 = \cdot 0170330$, $\log 2 = \cdot 3010300$.

Substituting $\quad M = 2P$ and $R = 1\cdot 04$ in $M = PR^n$, we get
$$2P = P\,(1\cdot 04)^n; \qquad \therefore\ 2 = 1\cdot 04^n;$$

TUT. ALG. II.

242   INTEREST AND DISCOUNT.

$$\therefore n \log 1{\cdot}04 = \log 2;$$

$$\therefore n = \frac{{\cdot}30103}{{\cdot}017033} = 17{\cdot}7 \text{ years, nearly.}$$

Ex. 3. Find approximately the rate per cent. at which a sum of money will treble itself in thirteen years, having given
$$\log 3 = {\cdot}4771213 \text{ and } \log 1{\cdot}08822 = {\cdot}0367016.$$

From the equation $M = PR^n$, we have
$$3P = PR^{13}; \quad \therefore 3 = R^{13},$$
$$\therefore 13 \log R = \log 3 = {\cdot}4771213,$$
$$\therefore \quad \log R = {\cdot}0367016 = \log 1{\cdot}08822,$$
$$\therefore \quad R = 1{\cdot}08822;$$

i.e. amount of £1 in 1 year = £1·08822;
$$\therefore \text{interest } \,, \quad ,, \quad ,, \quad = £0{\cdot}08822,$$
$$,, \quad ,, \quad £100 \quad ,, \quad = £8{\cdot}822.$$

Therefore the required rate is 8·822 or between $8\tfrac{3}{4}$ and $8\tfrac{7}{8}$ per cent.

Ex. 4. Find the rate of compound interest in order that a sum of money may double itself in 20 years.

$$\log 2 = {\cdot}3010300$$
$$\log 20705 = 4{\cdot}3160752$$
$$\log 20706 = 4{\cdot}3160962.$$

If P, R have the usual meanings, we have
$$2P = \text{amount after 20 years} = PR^{20} \text{ whence } R^{20} = 2;$$
$$\therefore 20 \log R = \log 2 = {\cdot}3010300;$$
$$\therefore \log R = {\cdot}0150515.$$

This does not lie between the mantissæ of the given logarithms; we therefore add log 2 to it and find
$$\log 2R = \log 2 + \log R = {\cdot}3160815.$$

Now $\log 2{\cdot}0705 = {\cdot}3160752$, diff. for ·0001 = ·0000210, also log $2R$ exceeds log 2·0705 by ·0000063.

Hence if $2R = 2{\cdot}0705 + x$ we have
$$x : {\cdot}0001 = 63 : 210,$$
whence $\qquad\qquad x = {\cdot}00003;$
$$\therefore 2R = 2{\cdot}07053,$$
$$R = 1{\cdot}035265.$$

INTEREST AND DISCOUNT.

Hence the interest on £100 for one year
$$= 100\,(R-1) \text{ pounds} = £3\cdot 5265,$$
and the rate of interest is therefore 3·5265 or rather over $3\frac{1}{2}$ per cent.

Ex. 5. At what rate of compound interest will a given sum be increased elevenfold in 100 years?

$$\log 11 = 1\cdot 0413927,$$
$$\log 1\cdot 1266 = 0\cdot 0517697,$$
$$\log 1\cdot 1267 = 0\cdot 0518083.$$

Here $PR^{100} = 11P$; $\therefore$ $100 \log R = \log 11 = 1\cdot 0413927$;

$\therefore \log R = \cdot 0104139$ to seven places.

Since this does not lie between the given logarithms, we add log 11 and find $\log 11R = 1\cdot 0518066$;

$$\therefore 11R = 11\cdot 266 + x,$$
where $\qquad x : \cdot 001 = 369 : 386$;

$\therefore x = \cdot 000956$ (the last figure cannot be relied on);

$\therefore 11R = 11\cdot 266956$ whence $R = 1\cdot 024269$ nearly;

$\therefore$ rate per cent. $= 100\,(R-1) = 2\cdot 4269$, or rather under $2\frac{1}{2}$.

Note. As a rule, seven-figure logarithms are sufficient for these calculations, but when n is very large, it will be found that more accurate results are given by taking ten-figure logarithms of R.

Ex. 6. Find the amount of £2000 for 100 years at $2\frac{1}{2}$ per cent. compound interest, having given

$$\log 1\cdot 025 = \cdot 0107238564,\ \log 2 = \cdot 3010300,$$
$$\log 2\cdot 3627 = \cdot 3734086,\text{ tabular difference }184.$$

If M pounds is the required amount,
$$M = 2000 \times (1\cdot 025)^{100};$$
$$\therefore \log M = 3 + \log 2 + 100 \log (1\cdot 025)$$
$$= 3\cdot 3010300 + 1\cdot 07238564$$
$$= 4\cdot 3734156.$$

But $\qquad \log 23627 = 4\cdot \underline{3734086}$, diff. for $1 = 184$.
$$ 70$$

Hence if $M = 23627 + x$ we have
$$x : 1 = 70 : 184,$$
whence $\qquad x = 1 \times \dfrac{70}{184} = \dfrac{70}{184}$;

$\therefore$ required amount $= £23627\tfrac{70}{184} = £23627.\ 7s.\ 7d.$ nearly.

244   INTEREST AND DISCOUNT.

Had $\log 1\cdot025$ only been given to 7 places, we should have found $100 \log 1\cdot025 = 1\cdot07239$ *correct to 5 places only*, and M could only have been calculated correct to the nearest pound.

Ex. 7. What would a penny amount to if left at 5 per cent. compound interest for 300 years, and what would 6*s*. 8*d*. amount to in 200 years at the same rate of interest? Given

$$\log 1\cdot05 = \cdot 0211892991, \quad \log 3 = \cdot 4771213,$$
$$\log 2\cdot2739 = \cdot 3567714, \quad \text{diff. for } \cdot 0001 = 190 \times 10^{-7},$$
$$\log 5\cdot7642 = \cdot 7607390, \quad \text{diff. for } \cdot 0001 = 76 \times 10^{-7}.$$

Let M *pence* be the amount of $1d$. in 300 years. Then
$$M = 1 \times (1\cdot05)^{300};$$
$$\therefore \log M = 300 \log 1\cdot05 = 6\cdot3567897\overline{3}.$$

Now $\log 2273900$ (*i.e.* $\log 2\cdot2739 \times 10^6) = 6\cdot3567714$
$$\text{difference obtained} = \overline{\cdot 0000183}$$

Also* diff. for 100 (*i.e.* $\cdot 0001 \times 10^6$) $= \cdot 0000190$.

Hence $M = 2273900 + x$ where
$$x : 100 = 183 : 190 \text{ or } x = 96;$$
$$\therefore M = 2273996.$$

$\therefore$ required amount $= 2273996$ pence $= £9474$. 19*s*. 8*d*.

Since 6*s*. 8*d*. is one third of a pound, it is easiest to find its amount in 200 years in *pounds*. Let this be M. Then

$$M = \frac{1}{3} \cdot (1\cdot05)^{200};$$

$$\therefore \log M = \log \frac{1}{3} + 200 \log 1\cdot05$$
$$= - \cdot 4771213 + 200 \times \cdot 0211892991$$
$$= \overline{1}\cdot 5228787 + 4\cdot 23785982 = 3\cdot 7607385.$$

Now $\quad\quad\quad \log 5764\cdot2 = 3\cdot7607390.$

Hence M is a little *less* than $5\cdot7642 \times 10^3$ or $5764\cdot2$, the difference in logs being $\cdot 0000005$, and we must write $M = 5764\cdot2 - x$.

But the corresponding difference for $\cdot 0001 \times 10^3$, or $\cdot 1$, is $\cdot 0000076$;
$$\therefore x : \cdot 1 = 5 : 76, \text{ whence } x = \cdot 007;$$
$$\therefore M = 5764\cdot2 - \cdot 007 = 5764\cdot193;$$
$$\therefore \text{required amount} = £5764. \ 3s. \ 10\tfrac{1}{4}d.$$

---

* Since $\cdot 0000190$ is the difference between the mantissæ of $\log 2\cdot2739$ and $\log 2\cdot2740$, it is also the difference between those of $\log 2273900$ and $\log 227400$, *i.e.* in this case the difference for 100.

INTEREST AND DISCOUNT. 245

263. Half-yearly and quarterly interest. It is a common practice to pay the interest on an investment more often than once a year. The interest on stocks and shares is usually paid half-yearly, and rents of houses are often paid quarterly. The rate per cent. per annum is then defined as the *sum* of the payments made in one year for the interest of £100. If part of the interest is paid before the end of the year, the investor obtains the use of this portion earlier than he would do if interest were paid yearly. Conversely if no interest is paid till the end of the year, the investor would require a slightly higher rate of interest to compensate him for losing the use of the first instalments of interest in the meanwhile, as the following examples will explain.

Ex. 1. Find what rate of interest payable yearly is equivalent to 5 per cent. payable half yearly.

When £100 is invested at 5 *per cent. payable half-yearly* this means that the investor receives a year's interest, viz. £2. 10s., at the end of six months, and the remaining £2. 10s. at the end of the year. He thus receives the first £2. 10s. *six months earlier* than he would if interest were payable yearly. If however this interest were left unpaid till the end of the year he ought then to be entitled to six moths' interest on *it*, hence the total interest due at the end of 12 months would be

= £2. 10s. + £2. 10s. + interest on £2. 10s. for 6 months
= £2. 10s. + £2 10s. + 1s. 3d. = £5. 1s. 3d. = £5$\frac{1}{16}$.

Thus 5 per cent. payable half-yearly is equivalent to 5$\frac{1}{16}$ per cent. payable yearly.

Ex. 2. A house worth £1000 is let for £65 a year payable quarterly. If the tenant wishes to pay rent once a year only, what must the new rent be in order that the landlord may obtain the same rate of interest as before ? given that

log 1·0162 = ·0069792, tabular difference 427 ;

log 1·0666 = ·0280016, tabular difference 407.

Here under the old agreement the interest on £1000 for one quarter is the quarter's rent £16 5s. or £16·25 ;

∴ interest on £1 for one quarter = £·01625 ;

∴ amount of £1 in one quarter = £1·01625.

Hence at compound interest, the amount of £1000 in 1 quarter is £1000 × 1·01625, in 2 quarters it is £1000 × (1·01625)^2, in 3 quarters it is £1000 × (1·01625)^3, and in 1 year it is £1000 × (1·01625)^4.

In order to obtain the same rate of interest, the rent payable at the end of the year must be equal to the interest on £1000 for the 12 months = (amount) – (principal)

$$= £1000 \times (1\cdot01625)^4 - £1000.$$

The *first term* can be calculated by logarithms.

Calling it M pounds, we have

$$\log M = \log 1000 + 4 \log (1\cdot01625).$$

Now $\log 1\cdot 0162 = \cdot 0069792$, diff. for $\cdot 0001 = \cdot 0000427$;

∴ diff. for $\cdot 00005 = \cdot 0000214$ nearly;

∴ $\log 1\cdot 01625 = \overline{\cdot 0070006}$
$$\underline{\quad 4 \quad}$$
$$\cdot 0280024$$

∴ $\log M = 3\cdot 0280024$ (since $\log 1000 = 3$).

Now $\log 1066\cdot 6 = 3\cdot 0280016$, diff. for $\cdot 1 = \cdot 0000407$.

Hence $M = 1066\cdot 602$ approximately;

∴ required rent = £1066·602 – £1000 = £66·602

$$= £66.\ 12s.\ \text{nearly}.$$

Hence the rent must be raised by £1. 12s.

264. To find the amount of a given sum in a given time at a given rate per cent. if the interest be added to the principal m times a year.

Let P pounds be the principal, c the rate per cent. per annum, n the number of years that the principal remains invested.

Then the interest on £P for 1 year at c per cent.

$$= £Pc \div 100.$$

If this interest is payable m times a year, *this is to be interpreted as meaning* that the interest on £P for one mth of a year is one mth of the interest for a year or $Pc \div 100m$ pounds.

Supposing that the interest at the end of each interval of $1/m$ year is added to the principal, the principal at the

INTEREST AND DISCOUNT. 247

beginning of any one interval will be the amount at the end of the previous interval. Thus

amount of £P in $\dfrac{1}{m}$ year $= P\left(1 + \dfrac{c}{100m}\right)$ pounds,

∴ ,, ,, $\dfrac{2}{m}$,, $= P\left(1 + \dfrac{c}{100m}\right)^2$ pounds,

and so on. But there are mn of these intervals in the time for which the money is lent. Hence the amount of P pounds at the end of n years will be given by

$$M = P\left(1 + \dfrac{c}{100m}\right)^{mn} \quad \ldots\ldots\ldots\ldots(7).$$

NOTE. It is often (indeed usually) the practice not to add the interest to the principal oftener than once a year. In this case, if interest is payable m times a year, the interest on £P for $1/m$ year is $Pc/100m$ pounds, but even if this interest is not drawn when due, no further interest is allowed on *it* till the end of the year when the total amount is

$$= P + m\,\dfrac{Pc}{100m} = P\left(1 + \dfrac{c}{100}\right).$$

This is the principal on which the interest for the second year is calculated and the formula of § 261 now takes the place of (7).

This is practically equivalent to reckoning interest as compound for integral numbers of years and as simple for any period of less than a year. The following example affords a further illustration of this method of computing interest.

Ex. To find the interest on £500 in 2 years 10 months at 4 per cent., supposing the interest added to the principal once a year.

The amount of £500 in 2 years at 4 per cent. compound interest

$$= £500 \times \left(1 + \dfrac{4}{100}\right)^2 = £540\cdot 8 = £540.\ 16s.$$

The interest on this amount for 10 months at 4 per cent. simple interest

$$= £540\cdot 8 \times \dfrac{10}{12}\,\dfrac{4}{100} = £18\cdot 02\dot{6};$$

∴ total amount in 2 years 10 months $= £558\cdot 826$,

and required interest $= £58\cdot 82\dot{6} = £58.\ 16s.\ 6\cdot 4d.$

265. To find the amount of a given sum for a fractional number of years, supposing the interest to be added to the principal once a year.

Let the given period $= n + p/q$ years, n being a whole number and p/q a proper fraction. Let P pounds be the principal, r the ratio of the interest for one year to the principal.

Then the amount of P pounds in n years $= P(1+r)^n$ pounds and the simple interest on this for p/q of a year

$$= P(1+r)^n \times \frac{p}{q} r \text{ pounds.}$$

Hence the required amount in pounds

$$= P(1+r)^n \left(1 + \frac{pr}{q}\right) \quad \ldots\ldots\ldots\ldots\ldots\ldots(8).$$

266. Discount. When an amount due at the end of a certain time is paid before it is due, it is obvious that some reduction should be made, to compensate the debtor for losing the use of his money for that time. An equitable arrangement is obtained by the debtor paying such a sum, as with its interest for the given time, would be equal to the amount of the debt. This sum is called the **present value** of the debt, and the reduction made, that is, the difference of the amount of the debt and its present value, is called the **discount**.

267. To find the present value and discount of a sum of money due at the end of a certain time; *interest being compound.*

It is evident from the definitions of the preceding article that a debt and its present value stand to each other in the same relation as the amount of a loan and the principal. Hence if M be the amount of the debt and V its present value, we get, on substituting V for P in the result of § 261

$$M = VR^n,$$
$$\therefore V = MR^{-n} \quad \ldots\ldots\ldots\ldots\ldots(9).$$

The discount (D) is then given by $D = M - V$; therefore
$$D = M(R^n - 1)/R^n \quad \ldots\ldots\ldots\ldots(10).$$

INTEREST AND DISCOUNT.   249

On adapting the first result for logarithmic calculation we have
$$\log V = \log M - n \log R,$$
whence the present value V may be found, the discount D being subsequently found from the formula $D = M - V$. As this latter involves the operation of subtraction the *logarithm* of D *cannot* be found directly by logarithmic computation.

NOTE. The formulæ for discount are identical with those for interest, and the methods are identical except that usually the principal is the given quantity in interest problems and the amount is the given quantity in discount problems.

Thus "to find the discount on £100 due 5 years hence at 4 per cent." means "to find the interest for 5 years at 4 per cent. on a sum whose *amount* at the end of that time is £100."

But the problem may be regarded from another point of view. For since by § 261, PR^n is the sum which must be paid to liquidate a debt n years *after* it is due, allowing compound interest, it is a natural inference from all our general notions of negative quantities that PR^{-n} should represent the sum required to liquidate the same debt n years *before* it is due, i.e. the present value of P due at the end of n years.

Ex. Find the present value and discount of £500 due 10 years hence at 5 per cent. compound interest, having given
$$\log 1{\cdot}05 = {\cdot}0211893, \ \log 2 = {\cdot}3010300 \text{ and } \log 306{\cdot}9566 = 2{\cdot}4870770.$$

Let P pounds be the present value.

Then the amount of £P in 10 years is £500,

By the result of § 267, we have PR^n = given amount = 500,

where $\qquad R = 1{\cdot}05, \ n = 10;$

$\qquad \therefore \ P = 500 \div (1{\cdot}05)^{10}.$

$\qquad \log P = \log 500 - 10 \log 1{\cdot}05$

$\qquad \qquad = \log 1000 - \log 2 - 10 \log 1{\cdot}05$

$\qquad \qquad = 3 - {\cdot}3010300 - {\cdot}211893$

$\qquad \qquad = 2{\cdot}4870770.$

$\qquad \therefore$ present value = £306·9566 = £306. 19s., nearly,

and since $\qquad D = P - V$

$\qquad$ the discount = £193. 1s., nearly.

250   INTEREST AND DISCOUNT.

268. In the case of simple interest the corresponding results are obtained from the formula of § 260, $M = P(1 + nr)$ whence

$$V = \frac{M}{1 + nr} \quad \ldots\ldots\ldots\ldots\ldots\ldots(11),$$

and $$D = \frac{Mnr}{1 + nr} \quad \ldots\ldots\ldots\ldots\ldots\ldots(12).$$

These formulæ are often used to calculate the discount for intervals of less than a year, n being then a proper fraction; for longer periods, we have seen that interest is generally compound.

In all cases the discount being the interest on V is less than the corresponding interest of M for an equal length of time, as it should be.

269. Banker's Discount. In commerce it is the rule to allow as the discount for any period the *interest of the debt* for that period and not of its present value.

Thus, at simple interest, Mnr would be allowed instead of $Mnr/(1 + nr)$.

In text-books, this is called **Banker's Discount**, and discount as defined above is distinguished by being called **True Discount**.

Another kind of discount is often allowed by tradesmen, namely, a percentage of the debt calculated without regard to time.

Thus booksellers often allow a discount of 2*d*. or 3*d*. in the shilling off the price of a book.

Ex. To find the difference between the true discount and the banker's discount on £100 due six months hence, at 5 per cent. per annum, simple interest.

The banker's discount = interest on £100 for six months at 5 per cent. = £2. 10*s*.

The present value of £100 is evidently such a sum £P as will amount to £100 in 6 months hence;

INTEREST AND DISCOUNT. 251

$$\therefore P + \frac{6}{12} \cdot \frac{5}{100} P = 100 \text{ or } 1\cdot025 P = 100;$$

$$\therefore P = 100 \div 1\cdot 025,$$

or present value $= £97.\ 11s.\ 2\frac{1}{2}d.$,

and true discount $= £100 - £97.\ 11s.\ 2\frac{1}{2}d. = £2.\ 8s.\ 9\frac{1}{2}d.$

Hence the banker's discount exceeds the true discount by $1s.\ 2\frac{1}{2}d.$

270. Reversions. When property is to pass into the possession of any person after the lapse of a certain number of years, that person is said to have a **reversionary interest** in the property. Such a reversionary interest may be sold or purchased.

The sum paid down for it is called the **present value of the reversion**, and the purchaser in return for this payment receives the property when it becomes due. If P pounds is the present value of the reversion of property worth M pounds receivable after n years, this is equivalent to saying that P pounds is the present value of a debt of M pounds due n years hence; so that, as in § 267,

$$PR^n = M \text{ or } P = MR^{-n},$$

R having the usual meaning.

A reversion may depend on various causes, such as the lapse or termination of a lease, the redemption of a loan or investment, or the death of the present owner of the property. If on the last-named cause, the owner's probable expectation of life as given in tables of mortality in *some* cases enables us to form a *fairly* correct, though not perfectly accurate, estimate of the present value of the reversion.

Ex. 1. An estate worth £4000 is leased for 99 years. Find the present value of the reversion to the owner and of the lease to the leaseholder, reckoning interest at $5\frac{1}{2}$ per cent., having given the following extracts from a table of logarithms:

No.	0		5		Diff.
1055	0232525	...		...	
1995	2999429	...	0517	...	218
2000	3010300				

252 INTEREST AND DISCOUNT.

The present value of the reversion is the present value of £4000 due 99 years hence, given by

$$P = 4000 \times (1 \cdot 055)^{-99};$$
$$\therefore \log P = \log 4000 - 99 \log 1 \cdot 055$$
$$= 3 + 2 \log 2 + \log 1 \cdot 055 - 100 \log 1 \cdot 055$$
$$= 3 \cdot 6020600 + \cdot 0232525 - 2 \cdot 32525$$

(the last term being correct to five decimal places only)

$$= 3 \cdot 62531 - 2 \cdot 32525 = 1 \cdot 30006;$$
$$\therefore P = 19 \cdot 955,$$

or present value of reversion = £19·955 = £19. 19s. 1d.

The leaseholder obtains the use of £4000 or its equivalent for 99 years, and then that amount lapses by the reversion of the lease. Hence the present value of the lease is less than £4000 by the present value of the reversion, that is, by £19. 19s. 1d.;

$$\therefore \text{present value of lease} = £3980. \ 0s. \ 11d.$$

The difference £19. 19s. 1d. thus represents the difference between the value of a *freehold* estate of £4000 and one on a 99 years' lease.

Ex. 2. A man aged 65 bequeaths £1000 to his nephew payable at his own death, but the latter being in want of cash sells the reversionary interest in the property. The purchaser reckons that if the man lives 10·55 years (that being the average expectation of life for a male aged 65) he will obtain 6 per cent. on his outlay. Find the price paid for the reversion.

$$\log 1 \cdot 06 = \cdot 0253059, \quad \log 5 \cdot 40783 = \cdot 733023.$$

The price is the present value of £1000 due 10·55 years hence at 6 per cent. compound interest. Let this be P pounds. Then remembering that $R = 1 \cdot 06$, we have

$$P = 1000 \times (1 \cdot 06)^{-10 \cdot 55};$$
$$\therefore \log P = 3 - 10 \cdot 55 \log 1 \cdot 06 = 3 - (10 \cdot 55 \times \cdot 0253059)$$
$$= 2 \cdot 733023, \text{ correct to 6 places only};$$
$$\therefore P = 540 \cdot 783,$$

or present value = £540·783 = £540. 15s. 8d. approximately.

NOTE.—We have obtained this result by using the formula $P = MR^{-n}$ even though n is fractional; this is the simplest plan where logarithms are used. Had we employed the formula (8) of § 265 we should have obtained as answer £540 11s. The very conditions under which interest is paid for the use of capital in commerce would render a discussion of the discrepancy of no practical value.

INTEREST AND DISCOUNT. 253

Ex. 3. A lady aged 50 bequeaths an estate bringing in £500 a year at 4 per cent. interest to her nephew aged 21 for use during his lifetime. Calculate what would be the present value of the young man's life interest in the estate on the supposition that his aunt will live for 20·7 years and that he will live for 38·7 years*, and reckoning interest at 4 per cent. throughout.

We must first find the value of the estate. Since it brings in £500 a year at 4 per cent. its value

$$= £500 \times \frac{100}{4} = £12,500.$$

The value of the life interest is therefore to be regarded as the present value of £12,500 paid in 20·7 years and repaid in 38·7 years.

This is the *difference* of the present values of £12,500 due 20·7 years hence and of the same amount due 38·7 years hence, that is,

present value $= £12500 \,(1·04)^{-20·7} - £12500 \,(1·04)^{-38·7}$
$= £12500 \{(1·04)^{-20·7} - (1·04)^{-38·7}\}$.

The two terms within the brackets must be found separately by logarithmic calculation. Now

$\log (1·04)^{-20·7} = -20·7 \times ·017033 = -·3525831$
$\quad\quad\quad\quad\quad\quad = \bar{1}·6474169,$

whence $\quad (1·04)^{-20·7} = ·444035.$

$\log (1·04)^{-38·7} = -38·7 \times ·017033 = -·6591771$
$\quad\quad\quad\quad\quad\quad = \bar{1}·3408229,$

whence $\quad (1·04)^{-38·7} = ·219191.$

Hence present value $= £12500 \{·444035 - ·219191\}$
$= £12500 \times ·224844$
$= £2810·55$, or £2810. 11s.

---

* These are the average durations of life for a female aged 50 and a male aged 21. The present method of calculating the value of life interests is not however correct. Thus a man may inherit property from someone younger than himself. Then, according to the Tables of mortality the presumption is that the heir will die first, and if he does, his life interest will be *nil*. But the younger man *may* die first, so the older one has certain chances of inheriting the property, the present value of which is calculable by more accurate methods as used by actuaries.

EXERCISES XVIII.

Table of Logarithms.

The *mantissæ* only of the following logarithms are given. The student will readily supply the necessary characteristic and insert the decimal point.

101	0043214	17908	2530471
102	0086002	09	0713
1025	0107239	2	3010300
103	0128372	20705	3160752
104	0170333	06	0962
105	0211893	20789	3178336
106	0253059	90	8545
10325	0263289	24992	3978010
108	0334238	26169	4177871
10824	0343878	70	8037
25	4279	26917	4300267
11000	0413927	18	0428
11266	0517697	28119	4488608
67	8083	40965	6124130
12744	1053058	64082	8067361
45	3398	83	7428
13000	1139434	67683	8304796
13150	1189258	84	4860
51	9588	74725	8734659
13459	1290128	26	4717
60	0451	9	9542425
14888	1728364		
89	8655		

Compound Interest is supposed to be reckoned in the following exercises unless otherwise stated.

1. Find the amount of £100 for 10 years at 6 per cent.

2. Find the interest of £200 for 15 years at 2 per cent.

3. Find the amount of £250 for 12 years at 8 per cent., the interest being payable half-yearly.

4. Find the interest of £1000 for 10 years at 4 per cent., the interest being payable quarterly.

5. The number of births in a certain town every year is 55 per thousand and the number of deaths is 30 per thousand of the population at the beginning of every year. The population in a certain year is 25000, find to the nearest hundred the population 20 years afterwards.

INTEREST AND DISCOUNT.

6. In what time will a sum of money double itself, the rate of interest being 5 per cent.?

7. How many years will it take £100 to accumulate to £1000 at 4 per cent.?

8. In what time would a sum of money, accumulating at 3 per cent. per annum, come to five times its original amount?

9. Find the rate of interest in order that a sum of money may double itself in 20 years.

10. At what rate of interest will a given sum be increased elevenfold in 100 years?

11. Find the present value of a debt of £1000 due 5 years hence, interest being reckoned at the rate of 6 per cent.

12. Find the discount of a debt of £200 due 20 years hence, the rate of interest being 2 per cent.

13. The amount at 8 per cent. simple interest of a certain sum of money for a certain time is £2574, and if the principal were diminished by £975, and the time increased by $12\frac{1}{2}$ years, the amount would be unchanged.
Find the original principal.

14. Work out the corresponding problem to the last example, supposing compound instead of simple interest to be taken.

15. What is the difference between the compound interest on £100 for 2 years according as the interest is paid yearly or half-yearly at 4 per cent.?

(This can be worked as easily by arithmetic as by logarithms.)

16. Define *present worth* and discount, and find the proportion between the interest of £100 for 4 years at $6\frac{1}{4}$ per cent. and the discount of the same sum payable at the end of 4 years at the same rate of interest.

17. Find the difference between the interest and the true discount on £264. 10s. for 3 years at 5 per cent. simple interest.

18. Find by logarithms the amount of £5,500 in 15 years at 5 per cent. compound interest, giving the result in pounds and decimals of a pound.

19. What sum of money at compound interest will amount to £650 at the end of the first year and £676 at the end of the second year?

20. What sum would in 3 years amount to £2811. 18s. at compound interest if the rate were 3 per cent. per annum for the first year, 4 per cent. for the second year and 5 per cent. for the third?

INTEREST AND DISCOUNT.

21. A cottage at the beginning of a year was worth £250, but it was found that by dilapidations at the end of each year it lost 10 per cent. of the value it had at the beginning of that year. After what number of years would the value of the cottage be reduced below £25? (log 3 = ·4771213.)

22. A person borrowed £11,000 for two months at 5 per cent. per annum. At the end of that time the interest was added on and the debt renewed for another two months. This was continually repeated till at the end of 2 years the debt and interest were paid. Find by logarithms what this debt and interest amounted to.

Given logarithms:

No.	Log	No.	Log	Diff.
2	·3010300	13	1·1139434	
7	·8450980	121·51	2·0846120	357
90	1·9542425	217·47	2·3373994	200
·011	$\bar{2}$·0413927			

23. In what time will a sum of money quadruple itself at 4 per cent. compound interest? Given log 2 = ·30103, log 1·3 = ·11394 (no other logs to be assumed).

24. The number of births in a town is 25 in every thousand of the population annually, and the number of deaths 20 in every thousand, in how many years will the population double itself?

Given log 67 = 1·82607.

25. Find the true discount on £142. 1s. 9d. due 18 months hence at 3½ per cent. per annum.

26. Find the discount on £51. 15s. 10d. due 4⅓ years hence at 3 per cent. simple interest.

27. Find the difference between the amounts of £1000 in 100 years at 5 per cent. according to whether simple or compound interest is reckoned.

28. Find correct to the nearest farthing the present value of £10,000 due 8 years hence at 5 per cent. per annum compound interest.

CHAPTER XIX.

ANNUITIES.

271. Annuities. DEFINITIONS. An **annuity** is a series of equal payments usually made at yearly intervals. In some cases it is paid at more frequent intervals. When an annuity is paid for a fixed number of years and then allowed to cease, it is said to be **terminable**; when it is to continue without ever ceasing, it is said to be **perpetual**. If an annuity is not to begin until after the lapse of a certain time, it is called a **deferred** annuity.

If, as in the case of a minor, the annuity is allowed to accumulate, it is said to be **unpaid** or **forborne,** and the yearly sums, together with the interest on each, form the **amount** of the annuity.

In order to secure the payment of an annuity for a certain number of years, a person pays a certain sum which is called the **present value** of the annuity. If the present value of an annuity of A pounds for any number of years is mA pounds, the annuity is said to be worth m **years' purchase.**

Except in certain special cases when a different arrangement is expressly specified, the first payment of an annuity takes place not at the time of purchase but *a year after purchase*, or, if the annuity is payable more than once a year, *at the end of the first interval* reckoned from the time of purchase. When a yearly annuity is **deferred for** m **years,** the first payment takes place $m + 1$ years after the date of purchase.

ANNUITIES.

The application of simple interest to annuities is altogether unpractical and leads to contradictory results. For this reason we shall omit all consideration of it, and in the following articles the interest will be understood to be *compound*.

272. Annuities treated by Geometrical Progression. The present value of an annuity at a given rate of interest may be found by writing down (from § 267) the present values of each of the payments which constitute the annuity and finding their sum by the formula for a geometrical progression. The number of terms in the progression is the number of payments in the annuity.

Ex. 1. Find the present value of an annuity of £10 for six years, reckoning interest at 5 per cent., given

$$\log 1\cdot05 = \cdot 0211893, \quad \log 746215 = 5\cdot8728642.$$

With the usual notation, $R = 1\cdot05$; hence, by § 267,

the present value of £10 due 1 year hence $= £10 \div (1\cdot05)$,
,,   ,,   ,,   ,,   ,, 2 years ,, $= £10 \div (1\cdot05)^2$,
,,   ,,   ,,   ,,   ,, 3 years ,, $= £10 \div (1\cdot05)^3$,

and so on, and there are six payments altogether.

Hence the present value of the annuity in pounds is given by

$$P = 10 \times \left\{ \frac{1}{1\cdot05} + \frac{1}{(1\cdot05)^2} + \ldots \text{ to six terms} \right\}.$$

This is a geometrical progression whose first term is $10 \div 1\cdot05$, and whose common ratio $= (1\cdot05)^{-1}$.

$$\therefore P = \frac{10}{1\cdot05} \cdot \frac{(1\cdot05)^{-6} - 1}{(1\cdot05)^{-1} - 1}$$

$$= 10 \cdot \frac{1 - (1\cdot05)^{-6}}{1\cdot05 - 1} = 10 \cdot \frac{1 - (1\cdot05)^{-6}}{\cdot 05}$$

$$= 200 \{1 - (1\cdot05)^{-6}\} = 200 - 200 (1\cdot05)^{-6}.$$

The second term can be found by logarithmic computation. Now

$$\log (1\cdot05)^{-6} = -6 \log (1\cdot05) = - \cdot 1271358 = \bar{1}\cdot 8728642;$$
$$\therefore (1\cdot05)^{-6} = \cdot 746215,$$
$$\therefore P = 200 - 200 \times (\cdot 746215) = 200 - 149\cdot243$$
$$= 50\cdot 757,$$

i.e. present value $= £50.\ 15s.\ 1\tfrac{3}{4}d.$

ANNUITIES. 259

Ex. 2. A corporation raises £2000 for public works to be repaid with interest at 3 per cent. in equal yearly instalments spread over a period of 10 years. Find what sum must be repaid each year.

$\log 1\cdot 03 = \cdot 0128372$, $\log 7\cdot 44094 = \cdot 8716279$.

Let A pounds be the required sum. Then the whole sum borrowed must be equal to the present value of the ten annual payments, i.e. of £A due 1 year hence, £A due 2 years hence, and so on. That is

$$2000 = \frac{A}{1\cdot 03} + \frac{A}{(1\cdot 03)^2} + \frac{A}{(1\cdot 03)^3} + \text{to ten terms}$$

$$= \frac{A}{1\cdot 03} \cdot \frac{1 - (1\cdot 03)^{-10}}{1 - (1\cdot 03)^{-1}}$$

(since the first term is $A/1\cdot 03$ and the common ratio $(1\cdot 03)^{-1}$),

$$= A \frac{1 - (1\cdot 03)^{-10}}{1\cdot 03 - 1} = A \frac{1 - (1\cdot 03)^{-10}}{\cdot 03} = \frac{100A}{3} \{1 - (1\cdot 03)^{-10}\}.$$

Therefore $\quad A = \dfrac{60}{1 - (1\cdot 03)^{-10}}$.

Now $\log (1\cdot 03)^{-10} = -10 \log 1\cdot 03 = - \cdot 128372 = \overline{1}\cdot 871628$;

$\therefore (1\cdot 03)^{-10} = \cdot 744094$ and $1 - (1\cdot 03)^{-10} = \cdot 255906$;

$\therefore A = \dfrac{60}{\cdot 255906} = 234\cdot 461$ *by long division**.

$\therefore$ sum repayable yearly $= £234\cdot 461 = £234$. $9s$. $2\frac{1}{2}d$. or $2\frac{3}{4}d$.

273. To find the present value of an annuity to continue for any number of years. (*First Method.*)

The present value of an annuity of £A for n years may be found by summing the present values of the n annual payments.

Let the required value be P (pounds, of course, understood).

Now the present value of the first payment is $\dfrac{A}{R}$ (see § 267);

the present value of the second payment is $\dfrac{A}{R^2}$, and so on.

Therefore $\quad P = \dfrac{A}{R} + \dfrac{A}{R^2} + \dfrac{A}{R^3} + \ldots$ to n terms.

---

* We could not save the labour of the long division *except by taking logs a second time*, i.e. by taking from the tables the log of the divisor $\cdot 255906$, because this divisor is not $(1\cdot 03)^{-10}$, but $1 - (1\cdot 03)^{-10}$.

260   ANNUITIES.

Summing the geometrical progression we have
$$P = \frac{A}{R}\frac{1-(1/R)^n}{1-1/R}$$
$$= \frac{A(R^n-1)}{R^n(R-1)} = \frac{A}{r}\frac{R^n-1}{R^n} \text{ as before}$$
$$= \frac{A}{r}\left\{1-\frac{1}{R^n}\right\} = \frac{A}{r}(1-R^{-n}).$$

Cor. The present value of a *perpetual annuity* may be found by putting $n = \infty$ and therefore $R^{-n} = 0$ in the preceding result, giving $P = A/r$ as will be proved more simply in § 279.

274. Amount of an Accumulated Annuity. When the series of payments forming an annuity are invested as soon as they fall due at a given rate of interest, their accumulated amount after any number of years is simply the sum of the amounts of the separate payments, and this can be found by the formula for the sum of a geometrical progression.

Ex. A man puts £10 at the end of every year in the Savings Bank at 2½ per cent. compound interest. How much will his savings amount to in 15 years?

log 1·025 = ·0107239,   log 1·44830 = ·1608585.

At the end of one year he has saved £10.

At the end of two years he has saved another £10 plus the amount of the previous £10 with one year's interest or, in all,
$$£10 + £10 \times (1·025).$$

At the end of three years he has saved another £10 plus the amounts of the two previous sums of £10, one for one year and the other for two, or in all £10 + £10 × (1·025) + £10 × (1·025)².

Proceeding in this way we see that the amount of the savings in 15 years in pounds
$$= 10 + 10 \times 1·025 + 10 \times (1·025)^2 + \ldots \text{ to 15 terms}$$
$$= 10\frac{(1·025)^{15}-1}{1·025-1} = 10\frac{(1·025)^{15}-1}{·025}$$
$$= 400\{(1·025)^{15}-1\}.$$

Now $\log(1·025)^{15} = 15 \log 1·025 = ·1608585$;
$$\therefore (1·025)^{15} = 1·4483;$$
$\therefore$ required amount = £400{1·44830 − 1}
$$= £400 \times ·44830 = £179·320 = £179.\ 6s.\ 5d.$$

ANNUITIES. 261

275. To find the amount of an unpaid annuity after a given number of years. (*First Method.*)

Let M denote the required amount.

The amount due at the end of the first year is A.

At the end of the second year the whole sum due is A together with the amount of the first year's annuity for one year, which is AR. Hence the whole sum due is $A(1+R)$.

Similarly the amount due at the end of the third year is

$$A + A(1+R)R, \text{ or } A(1+R+R^2);$$

and so on.

Finally, the amount due at the end of n years is

$$A(1+R+R^2+\ldots\text{to } n \text{ terms}).$$

Hence $$M = \frac{A(R^n-1)}{R-1} = \frac{A(R^n-1)}{r}.$$

276. Deferred Annuities. The present value of a deferred annuity may be found in like manner as the sum of a geometrical progression the first term of which is the present value of the first payment.

Ex. Find the sum of money which can be repaid with interest at 4 per cent. in six yearly instalments of £10, the first instalment being paid 10 years after the date of borrowing the money.

$$\log 1{\cdot}04 = {\cdot}0170333.$$

The required sum is the present value of the six payments, the first being due 10 years hence. Calling it P pounds, we have

$$P = \frac{10}{(1{\cdot}04)^{10}} + \frac{10}{(1{\cdot}04)^{11}} \ldots\ldots \text{ to 6 terms}.$$

Since the common ratio is $(1{\cdot}04)^{-1}$ the required sum

$$P = \frac{10}{(1{\cdot}04)^{10}} \cdot \frac{1-(1{\cdot}04)^{-6}}{1-(1{\cdot}04)^{-1}} = \frac{10}{(1{\cdot}04)^9} \cdot \frac{1-(1{\cdot}04)^{-6}}{1{\cdot}04-1}$$

$$= \frac{10}{{\cdot}04}\{(1{\cdot}04)^{-9} - (1{\cdot}04)^{-15}\} = 250(1{\cdot}04)^{-9} - 250(1{\cdot}04)^{-15}.$$

Now $\log 250(1{\cdot}04)^{-9} = \log 1000 - 2\log 2 - 9\log 1{\cdot}04$
$\qquad = 3 - {\cdot}6020600 - {\cdot}1532997 = 2{\cdot}2446403,$

$\log 250(1{\cdot}04)^{-15} = 3 - {\cdot}6020600 - {\cdot}2554995 = 2{\cdot}1424405,$

whence $250(1{\cdot}04)^{-9} = 175{\cdot}647$ and $250(1{\cdot}04)^{-15} = 138{\cdot}816;$

$$\therefore P = 175{\cdot}647 - 138{\cdot}816;$$

$\therefore$ amount of loan $= £36{\cdot}831 = £36.$ $16s.$ $7\frac{1}{2}d.$

262   ANNUITIES.

277. To find the present value of a deferred annuity. (*First Method.*)

If an annuity A be deferred for p years and then paid for q years its present value may be obtained directly by summing the series

$$\frac{A}{R^{p+1}} + \frac{A}{R^{p+2}} + \frac{A}{R^{p+3}} + \ldots \text{ to } q \text{ terms.}$$

Hence $P = \dfrac{A}{R^{p+1}} \dfrac{1 - R^{-q}}{1 - R^{-1}}$

$= \dfrac{A}{rR^p}\left\{1 - \dfrac{1}{R^q}\right\} = \dfrac{A}{r}\{R^{-p} - R^{-p-q}\}.$

Cor. The present value of a *perpetual annuity deferred for p years* is found by putting $q = \infty$, and therefore

$$P = \frac{A}{rR^p}.$$

278. Alternative aspect of annuities. A very simple way of obtaining an annuity is to invest a suitable sum of money at a fixed rate of interest payable yearly or at regular intervals, the *interest* constituting the annuity required. If the principal remains perpetually invested, the annuity will be a perpetual one; if it is required to terminate the annuity at any time, this may be done by selling or realising the investment and thus recovering the original principal.

This mode of regarding an annuity enables us to write down the present value of an annuity or the accumulated amount of an unpaid annuity at once without summing a geometrical progression. The results are, of course, the same as would be obtained by the methods of the preceding articles, and it will be convenient to preface the general investigation by numerical examples.

Ex. 1. Reckoning interest at 4 per cent., find the present value of a perpetual annuity of £10.

Since £4 is the annual interest on £100,

∴ £10 ,,   ,,   ,,   ,, ,, £250;

∴ a perpetual annuity of £10 can be obtained by investing £250 at 4 per cent.;

∴ the present value of the perpetual annuity is £250.

ANNUITIES. 263

Ex. 2. If the annuity of Ex. 1 is to terminate after 10 years write down its present value.

The annuity of the last Example can be terminated at any time by selling the investment, when the original principal £250 will be repaid. If the annuity is to run for 10 years we shall thus obtain £250 ten years hence. Instead of receiving the latter sum it will come to the same thing if we deduct its *present value* from the price originally paid for the annuity.

Hence the present value of the annuity alone

= £250 − present value of £250 due 10 years hence
= £250 $\{1 - (1 \cdot 04)^{-10}\}$,

and this may be calculated by logarithms in the usual way, the result being £81·109 or £81. 2s. 2d. nearly.

Ex. 3. If the same annuity is allowed to accumulate at 4 per cent. interest for 10 years, write down its amount at the end of that time (i.e. when the 10th payment falls due).

As before, the annuity represents the yearly interest on £250 at 4 per cent.

Hence the total accumulated amount of the annuity in 10 years is the same as the total accumulated compound interest on £250 in 10 years, supposing this interest left unpaid till the end of the time;

∴ required amount = £250 $\{(1 \cdot 04)^{10} - 1\}$,

and this may be calculated by logarithms in the usual way, the result being £120·06 or £120. 1s. 3d. nearly.

Ex. 4. To find the present value of a perpetual annuity of £20 deferred for 20 years, reckoning interest at 3 per cent.

The sum which must be invested at 3 per cent. to bring in £20 yearly = £20 × $\dfrac{100}{3}$ = £666$\tfrac{2}{3}$.

When the annuity is deferred for 20 years the first payment takes place 21 years from the time of purchase, and the annuity could therefore be obtained by investing £666$\tfrac{2}{3}$ in 20 years' time (since the first payment of interest would be made one year later).

Hence the present value of the annuity is the present value of £666$\tfrac{2}{3}$ due 20 years hence, and therefore

= £666$\tfrac{2}{3}$ × $(1 \cdot 03)^{-20}$
= £369·1175   (by logarithms)
= £369. 2s. 4d.

264 ANNUITIES.

279. To find the present value of a perpetual annuity.

The present value of a perpetual annuity of £A is evidently such a sum as will bring in interest to the amount of £A per annum when invested.

Hence, if r be the ratio of the interest for one year to the principal and P pounds the present value, we have

$$Pr = A,$$

$$\therefore P = \frac{A}{r} = \frac{100A}{c},$$

where c is the rate *per cent.*

Ex. Find the present value of a perpetual annuity of £100, at 5 per cent.

The present value is the principal on which the interest is £100 per annum, or

$$P = \frac{A}{r} = \frac{100}{\cdot 05} = 100 \times \frac{100}{5} = 2000,$$

i.e. present value = £2000.

COR. If a perpetual annuity A is worth m **years' purchase**, mA is the present value of the annuity; hence

$$mA = \frac{A}{r};$$

$$\therefore m = \frac{1}{r} = \frac{100}{c};$$

so that the number of years' purchase of a perpetual annuity is 100 divided by the rate per cent.

It will be noted that the number of years' purchase is the number of years in which the original principal will be repaid (without interest).

A fair test of the credit of a Government or the security of a company is furnished by the number of years' purchase at which its Stocks are quoted. Thus the 2¾ p.c. Consols at 104½ are worth 38 years' purchase; Jamaica 4 p.c. Stock at 120 is worth 30 years' purchase; while Cordova 6 p.c. Stock at 24 is only worth 4 years' purchase.

ANNUITIES. 265

280. To find the present value of an annuity to continue for any period. (*Second Method.*)

Let A pounds be the annuity, n the number of years, r the ratio of the interest, and R that of the amount for one year to the principal at the given rate of interest.

As in the last article A/r pounds is the sum on which the yearly interest is £A.

If this sum remain invested for n years we shall obtain an annuity of £A for n years, and at the end of that time the original principal A/r pounds will be repaid.

Now by § 267 the present value of A/r pounds repaid at the end of n years is

$$\frac{A}{rR^n}.$$

Hence the present value of the annuity alone is

$$P = \frac{A}{r} - \frac{A}{rR^n} = \frac{A(1-R^{-n})}{r}.$$

In numerical calculations of annuities it is best to employ the principle illustrated in the present articles, but not to use the formulæ.

Ex. Find the present value of an annuity of £30, to continue for 40 years, the interest being at the rate of 6 per cent.; having given

$$\log 1\cdot 06 = \cdot 0253059, \quad \log 972219 = 5\cdot 9877640.$$

The sum which must be invested at 6 per cent. to bring in £30 per annum is £30 × 100 ÷ 6 = £500.

The annuity could therefore be obtained by investing £500 and recovering the principal after 40 years.

∴ present value = £500 − present value of £500 due 40 years hence.

Calling this £P, we have

$$P = 500 - \frac{500}{(1\cdot 06)^{40}} = 500\{1 - (1\cdot 06)^{-40}\}.$$

Now $\log (1\cdot 06)^{-40} = -40 \log 1\cdot 06$.
$$= -40 \times \cdot 0253059 = \bar{2}\cdot 9877640$$
$$= \log \cdot 0972219;$$
∴ $P = 500(1 - \cdot 0972219) = 500 \times \cdot 9027781$
$$= 451\cdot 38905$$
∴ present value = £451. 7s. 9½d.

281. To find the amount of an annuity left to accumulate for any number of years *at compound interest.* (*Second Method.*)

An annuity of A pounds may be regarded as the annual interest on A/r pounds, where r is the ratio of the interest for one year to the principal.

Hence the amount of the annuity in n years is the total accumulated interest on A/r pounds in n years supposing this interest to be left unpaid till the end of that time.

Hence if M pounds be the required amount we have

$$M = \frac{A}{r} R^n - \frac{A}{r} = \frac{A}{r}(R^n - 1).$$

Ex. Find the amount of an annuity of £200 left unpaid for 15 years, the rate of interest being $2\tfrac{1}{2}$ per cent.; having given

$$\log 1\cdot025 = \cdot0107239 \text{ and } \log 14483 = 4\cdot1608585.$$

The sum which must be invested at $2\tfrac{1}{2}$ per cent. to bring in £200 annually $= £200 \times 100 \div 2\tfrac{1}{2} = £8000$.

The amount of the annuity in 15 years is therefore the total accumulated interest on £8000, supposing this interest left unpaid till the end of 15 years.

Hence (remembering that $R = 1\cdot025$) the accumulated amount of the unpaid annuity $= £8000\{(1\cdot025)^{15} - 1\}$.

Now $\log (1\cdot025)^{15} = 15 \log 1\cdot025 = 15 \times \cdot0107239$

$$= \cdot1608585 = \log 1\cdot4483.$$

Hence amount $= £8000 (1\cdot4483 - 1) = £(8000 \times \cdot4483)$

$$= £3586.\ 8s.$$

282. To find the present value of an annuity deferred for p years and continuing for q years. (*Second Method.*)

A deferred annuity £A to continue for q years, the first payment taking place at the end of $p+1$ years from the present date, could be obtained by investing a sum of A/r pounds in p years' time, and leaving it invested for q years.

Hence the required value is the difference between the present value of $£A/r$ due p years hence, and the present value of the same sum due $p+q$ years hence, that is

$$P = \frac{A}{rR^p} - \frac{A}{rR^{p+q}} = \frac{A}{r}(R^{-p} - R^{-p-q}).$$

COR. If the deferred annuity is to be **perpetual,** q will be infinite, and R^{-p-q} will be zero since $R>1$. Hence the present value of a deferred perpetual annuity of which the first payment takes place in $p+1$ years' time is given by

$$P = \frac{AR^{-p}}{r}.$$

The last result is otherwise evident from the fact that the annuity could be obtained by investing $£A/r$ in p years' time, and that the present value of that sum is AR^{-p}/r pounds.

283. Leases. A **freehold** estate is one which yields a perpetual annuity, this annuity being the rent of the estate. The value of the freehold is the present value of this perpetual annuity.

It follows from § 279 that the number of years' purchase of a freehold is found by dividing 100 by the rate per cent., and the value of the freehold is the rent multiplied by the number of years' purchase.

A **leasehold** estate is one which is held for a limited number of years, and the rent is in this case an annuity which terminates at the expiration of the lease. When the estate is chargeable with **ground rent,** this must be subtracted from the total rent in order to obtain the annuity which the lease represents. The difference is called the **annual value** of the lease.

Thus if a house whose rent is £100 is held on lease for 80 years with a ground rent of £15, the *annual value* of the lease = £(100−15) = £85 and the lease is equivalent to an annuity of £85 for 80 years.

The difference between the value of a leasehold estate which has n years to run and that of a freehold estate of the same *annual value* is of course equal to the present value of the *reversion* of the lease after n years (see § 270).

ANNUITIES.

Ex. A man purchases the leasehold of a house for £3000, the ground rent being £50 per annum and the lease having 20 years to run. He lets the house for the whole term. What annual rent should he receive to make 5 per cent. on his outlay?

$\log 2 = 0\cdot3010300$, $\log 3 = 0\cdot4771213$, $\log 7 = 0\cdot8450980$,
$\log 11306 = 4\cdot0533090$, $\log 11307 = 4\cdot0533474$.

The rent must be sufficient to bring in an annuity for the 20 years, whose present value is £3000, and to pay the ground rent as well. Hence if £A is the annual value of the annuity, the rent charged must be $(50 + A)$ pounds.

Now £3000 is to be the present value of an annuity of £A for 20 years at 5 per cent. interest.

By the geometrical progression method
$$3000 = \frac{A}{1\cdot05} + \frac{A}{(1\cdot05)^2} + \ldots \text{ to 20 terms};$$

[or by the alternative method £A is the annual interest at 5 per cent. on $20A$ pounds, hence the present value is given by

£$3000 = £20A -$ present value of £$20A$ due 20 years hence.]

Therefore $\quad 3000 = 20A\,(1 - (1\cdot05)^{-20})$;

$$\therefore A = \frac{150}{1 - (1\cdot05)^{-20}}.$$

Now $\quad 1\cdot05 = \dfrac{21}{20}$, $\therefore (1\cdot05)^{-1} = \dfrac{20}{21} = \dfrac{2 \times 10}{3 \times 7}$;

$\therefore \log (1\cdot05)^{-1} = 1 + \log 2 - \log 3 - \log 7$
$\qquad = \bar{1}\cdot9788107$;

$\therefore \log (1\cdot05)^{-20} = 20 \log (1\cdot05)^{-1} = \bar{1}\cdot576214$.

Adding log 3 we find
$\quad \log 3\,(1\cdot05)^{-20} = \cdot0533353$ correct to six places.

This lies between log $1\cdot1306$ and log $1\cdot1307$.

Hence by interpolation we find
$3\,(1\cdot05)^{-20} = 1\cdot130668$; $\therefore (1\cdot05)^{-20} = \cdot376889$;

$$\therefore A = \frac{150}{\cdot623111} = 240\cdot728.$$

And the required rent
$\quad = £(50 + A) = £290.\ 14s.\ 7d.$ nearly.

ANNUITIES. 269

284. Fines. When a lease has run a certain number of years the term of the lease is sometimes extended by the payment of a sum of money called a **fine**. The extension of lease is equivalent to a deferred annuity equal to the rent, commencing at the expiration of the old lease; and the fine is of course the present value of this deferred annuity.

Ex. An estate worth £500 per annum is let on a lease of 99 years, with the option of renewal at the end of 20 years on payment of a fine. Find the fine to be paid if interest be allowed at the rate of 5 per cent.; having given

log $1·05 = ·0211893$, log $2·11858 = ·3260453$ and log $7·9849 = ·9022693$.

At the end of 20 years the unexpired lease has still 79 years to run, but if the lease is renewed, a new lease for 99 years is made out from the date of renewal, that is, the period of the lease is extended 20 years.

Hence the fine to be paid is the present value of an annuity of £500 for 20 years, deferred for 79 years. Let this be P pounds.

Since the first payment of rent for the *additional* term takes place in 80 years' time, the geometrical progression method gives

$$P = 500\{(1·05)^{-80} + (1·05)^{-81} + \ldots \text{ to 20 terms}\}.$$

Otherwise by the alternative method the annual value of the estate represents the yearly interest at 5 per cent. on £500 × 100 ÷ 5 or £10,000, and hence the extension of period is equivalent to the use of £10,000 for a period from 79 to 99 years from the date of renewal.

Hence the fine or amount to be paid for the extension of the lease is the difference between the present values of £10,000 due in 79 years and the same sum due in 99 years.

Both method lead to the equation
$$P = 10,000 \times \{(1·05)^{-79} - (1·05)^{-99}\}.$$

Now log $(1·05)^{-79} = -79$ log $1·05$
$= - 79 \times ·0211893 = \bar{2}·3260453$
$= \log ·0211858$

and log $(1·05)^{-99} = -99 \times ·0211893 = \bar{3}·9022693$
$= \log ·0079849$

$\therefore (1·05)^{-79} = ·0211858$ and $(1·05)^{-99} = ·0079849$
$P = 10,000 (·0211858 - ·0079849)$
$= 10000 \times ·0132009 = 132·009;$

$\therefore$ the fine = £132 approximately.

270   ANNUITIES.

285. Annuities payable in advance. In certain cases the first payment of an annuity is made at once instead of at the end of a year. The following kinds of annuities are of this class:

(i) Premiums on policies of insurance, subscriptions to friendly societies, the members of which receive a fixed sum of money after a certain number of years, or during illness, or relatives receive the money at the member's death, subscriptions to building societies, etc.

(ii) Payments for goods obtained on the hire-purchase system, and paid for in yearly, half-yearly, quarterly, or monthly instalments. Here the first instalment has to be paid before the goods are delivered.

(iii) Subscriptions to clubs, scientific societies, etc.; here a new member's first subscription is always payable immediately after election.

In cases of this kind it is often most convenient to employ the geometrical progression method. Care must be taken to write the first term of the progression correctly, and to avoid confusing the *present value* with the *amount*. It may be noted that the progression for the former involves *negative* powers of R and that for the latter *positive* powers of R.

Ex. A man 25 years of age can insure his life for £1000 by paying an annual premium of £18. Taking interest at 5 per cent., what will be an equitable composition for an annual subscription of £3?

Let the man's expectation of life be n years; (i.e. let the calculation be based on the supposition of his dying in n years' time).

Firstly, the sum £1,000 payable at the end of n years is the *amount* of the n annual premiums of £18 made at the *beginning* of each year for n years.

Now the last payment is made *one year before repayment*, hence its amount at the time of repayment $= 18R$ pounds where $R = 1.05$. Proceeding in this way we obtain the equation

$$1000 = 18R + 18R^2 + 18R^3 + \ldots \text{ to } n \text{ terms}$$
$$= 18R \cdot \frac{R^n - 1}{R - 1} \quad \ldots\ldots\ldots\ldots\ldots\ldots(1).$$

Secondly, if £P be the composition fee payable on joining the society, £P is the *present value* of an annuity of £3, payable at the *beginning* of each year till death, i.e. for n years (presumably). The present value of the first year's subscription (payable at once) is £3, that of the second year's £3 $\div R$, or $3R^{-1}$ pounds;

$$\therefore P = 3 + 3R^{-1} + 3R^{-2} + \ldots \text{ to } n \text{ terms}$$
$$= 3 \cdot \frac{1 - R^{-n}}{1 - R^{-1}} \quad \ldots\ldots\ldots\ldots\ldots\ldots(2).$$

ANNUITIES.

From (1) (remembering that $R = \frac{21}{20}$),

$$R^n - 1 = \frac{1000}{18} \cdot \frac{\frac{1}{20}}{\frac{21}{20}} = \frac{1000}{18 \times 21} = \frac{500}{189};$$

$$\therefore R^n = \frac{689}{189} \text{ and } R^{-n} = \frac{189}{689}.$$

Substituting this value of R^{-n} in (2) we have

$$P = 3 \cdot \frac{1 - \frac{189}{689}}{1 - \frac{20}{21}} = 3 \cdot \frac{\frac{500}{689}}{\frac{1}{21}} = \frac{31500}{689} = 45\tfrac{495}{689};$$

$\therefore$ required composition fee = £$45\tfrac{495}{689}$ = £45. 14s. 4d.

286. Variable Annuities.—In problems where it is required to find the present value of an annuity in which the payments increase in arithmetical progression each year, the present value of each year's payment should be written down and the sum of the series thus obtained will be readily found by the methods of Chap. XVI.

Ex. Find the present worth of a perpetual annuity of £1, payable at the end of the first year, the yearly payments increasing by £1 each year, and the rate of interest being $3\frac{1}{3}$ per cent.

Here
$$P = \frac{1}{R} + \frac{2}{R^2} + \frac{3}{R^3} + \frac{4}{R^4} + \ldots \text{to infinity}.$$

$$\therefore P \times \frac{1}{R} = \frac{1}{R^2} + \frac{2}{R^3} + \frac{3}{R^4} + \ldots \text{to infinity}.$$

Subtracting, we get

$$P\left(1 - \frac{1}{R}\right) = \frac{1}{R} + \frac{1}{R^2} + \frac{1}{R^3} + \ldots \text{ad infin.} = \frac{1/R}{1 - 1/R};$$

$$\therefore P \cdot \frac{R-1}{R} = \frac{1}{R-1}; \qquad \therefore P = \frac{R}{(R-1)^2} = \frac{1+r}{r^2}.$$

Substituting the value of $R (= 1\tfrac{1}{30})$, we get $P = 31 \times 30 = £930$.

NOTE. If the yearly payments were to increase or decrease in *geometrical* progression, each payment bearing a constant ratio to the preceding one, the series of present values would also form a geometrical progression whose sum could be written down by the ordinary formula.

EXERCISES XIX.

Table of Logarithms.

The *mantissae* only of the following logarithms are given. The student will readily supply the necessary characteristic and insert the decimal point.

1025	0107239	29530	4702675
103	0128372	3	4771213
10325	0138901	32433	5109871
1035	0149403	32434	9990
104	0170333	34273	5349500
105	0211893	35628	5517910
1055	0232525	3768895	5762140
114674	0594650	424347	6277212
126935	1035820	436297	6397819
13	1139434	4495184	6527475
1406156	1480336	481017	6821605
16289	2118930	5274702	7221980
2	3010300	64461	8092963
21911	3406622	67556	8296670
21912	6820	702587	8467003
225886	3538887	70892	8505972
2313744	3643210	7106813	8516749
23430	3697723	759918	8807669
25269	4025870	79847	9022593
253	4031205	822702	9152428
26658	4258276	8548	9318668
26659	8439		

Compound Interest is supposed to be reckoned in the following exercises.

1. Write down the number of years' purchase of a freehold estate for the following rates of interest;
  $2\frac{1}{2}$, 3, $3\frac{1}{2}$, 4, $4\frac{1}{2}$, 5, $5\frac{1}{2}$ and 6 per cent.

2. How much should be paid for a perpetual annuity of £70, interest being at the rate of $3\frac{1}{2}$ per cent.?

3. How many years' purchase must be paid for leases to run
  (i) for 29 years, interest being at 3 per cent.,
  (ii) for 37 years, ,, ,, 4 per cent.?
Give each result to three places of decimals.

4. Find the annual payments required to liquidate a debt of £100
  in 30 years with interest at $3\frac{1}{2}$ per cent.,
  in 10 ,, ,, ,, ,, 4 per cent.

ANNUITIES. 273

5. Find the amount to which an annuity of £1 will accumulate
   (i) in 50 years at 5 per cent.,
   (ii) in 33 years at $2\frac{1}{2}$ per cent.

6. Find the present value of an annuity of £300 payable every year for 3 years at 5 per cent. per annum, *simple interest*.

7. Find the amount of an annuity of £100 left unpaid for 20 years, the rate of interest being 4 per cent. per annum.

8. How much should be paid for an annuity of £250 to last for 9 years, interest being reckoned at 4 per cent. ? Express the answer in pounds and decimals of a pound.

9. Find to the nearest shilling the present value of an annuity of £100, payable at the end of each year from the present date and running for 25 years, interest being reckoned at 5 per cent.

10. A person saves £40 a year out of his income and invests it at the end of each year, allowing the whole to accumulate at the rate of 4 per cent. per annum. Find the amount, to the nearest shilling, of his savings at the end of 30 years.

11. A man, aged 30, takes out an insurance policy for £1000, payable at death or his attaining the age of 55, the annual premium being £42. 12s. 6d., payable at the beginning of each year. If he reaches the above age, what is the gain of the insurance company at the time the transaction is ended, interest being reckoned at 4 per cent. ?

12. A man is satisfied with $3\frac{1}{2}$ per cent. interest on his money; how much ought he to be willing to give for a debenture bond of £100 redeemable 10 years hence, and bearing interest at the rate of 4 per cent. per annum, the security being, in his opinion, quite good ?

13. Find the present worth of an annuity of £20 for 5 years at $3\frac{1}{4}$ per cent., to commence at the end of 20 years.

14. Find to the nearest pound how much should be paid now for an annuity of £500, the first instalment of which is paid to the annuitant 5 years hence, and the last instalment 15 years hence, reckoning interest at 5 per cent.

15. Find how much should be paid for the reversion after 10 years of a freehold estate worth annually £4500, reckoning interest at $2\frac{1}{2}$ per cent.

16. What is the present worth of an annuity of £20 payable at the end of the first year, £40 payable at the end of the second year, and so on, increasing £20 each year, to run for 20 years ; interest being at the rate of 5 per cent. ?

17. What is the present worth of a perpetual annuity, £10 payable at the end of the first year, £11 at the end of the second year, and so on, increasing £1 per year ; interest at 4 per cent. ?

TUT. ALG. II.

ANNUITIES.

18. If a perpetual annuity is worth 25 years' purchase, what is the value of a perpetual annuity of £1 at the end of the first year, £2 at the end of the second year, £3 at the end of the third year, and so on?

19. What is the present worth of a perpetual annuity of £10 payable at the end of the first year, £20 at the end of the second, £30 at the end of the third, and so on, increasing £10 each year, interest 5 per cent.?

20. Prove that a debt may be paid off in 20 yearly instalments, each a little over 8 per cent. of the debt, if the interest be 5 per cent. and the first payment be made at the end of the first year.

What difference would be made in the instalments if the payments were half-yearly?

21. Find to the nearest shilling the fine which must be paid at the end of 7 years on the renewal of the lease of a building worth £200 per annum, the term of the lease being 30 years and interest being reckoned at 5 per cent.

22. Find the value of a house of which the rent is paid yearly and is £100 for the first year, but is reduced each year by 5 per cent. of the preceding year's rent, reckoning interest at 3 per cent. per annum.

23. A person sells an estate worth £1200 per annum for 25 years' purchase, and after deducting $1\frac{1}{2}$ per cent. for expenses, invests the remainder in North-Eastern Consols at 155; allowing 2s. 6d. per cent. for brokerage, find the amount of stock he will receive. If he gets 7 per cent. on his investment, what will be the difference of his income supposing the management of his estate to have cost him 10 per cent. of the rental?

24. A man borrows every year £25, upon which he pays interest at the rate of 4 per cent. per annum; in how long a time will the sums that he has paid as interest amount to £91 when added together?

25. By the annual payment of £20, a man finds that he can secure £1000 to his family at death. If money fetches 3 per cent. find the greatest number of years that he can live in order that he may benefit by his investment.

26. The reversion of an estate of £600 per annum is bought for £4000. Within what time must the purchaser enter into possession in order not to lose by his purchase, supposing interest to be 5 per cent.? (Given the logs of 2, 3, and 7.)

27. What value of goods can be paid for in half-yearly instalments of £10, of which the first is paid at the time of purchase and the last is paid three years later, if the security is such that the tradesman requires to make 8 per cent. interest (payable half-yearly) on his capital?

ANNUITIES. 275

28. A person has a capital of £3000 which produces interest at 5 per cent. If he spend £270 in every year, in how many years will he become bankrupt?

29. A man insures his life for a perpetual annuity of £50 to be paid to his heirs, the first payment to take place at his death. Find what would be a fair annual premium for him to pay during his lifetime on the predicted supposition that he will live 29 years and pay 30 premiums, the first premium being paid at the time of taking out the policy and interest being at the rate of 4 per cent.

30. A plot of land is let on a 99 years' lease with a ground-rent of £20, on condition that the leaseholder builds a house of value £200 on it, the lease to date from the completion of the house. What must the annual rent be in order that the leaseholder may make 5 per cent. for his money, if the cost of repairs amounts to $\frac{1}{6}$ of the gross rent?

*31. A Fellow of a certain Society can compound for his annual subscription of £1 either by the payment of £15 at the time the first subscription is due, or by the payment of £10 ten years later. If he is getting 5 per cent. for his investments, find which will be the most advantageous course for him to adopt by comparing the amounts to which his payments on the two plans would accumulate at that rate of interest at the time when the second composition is payable.

*32. An artisan rents a house from a Building Society for £40 a year on the understanding that he is to become the possessor of the house in 20 years time. If the Building Society reckons to pay 5 per cent. on its capital as dividend to the shareholders and the expenses of its management amount to an extra $\frac{1}{2}$ per cent., find what the rent of the house should be if it were to remain the property of the Society.

33. A person wishes to purchase an annuity of £50 to commence in 40 years and continue for 20 years. How much would he have to pay for the annuity if interest be reckoned at $3\frac{1}{2}$ per cent.?

*34. A sum of money borrowed at the beginning of a year was returned by equal instalments, paid at the end of that and of each succeeding year; also interest at the rate of 4 per cent. was paid at the end of each year upon the amount of debt remaining uncancelled at the beginning of that year. The total sum received by the creditor was exactly double of the sum originally borrowed. What was the number of instalments?

35. A man aged 54, in the receipt of a pension of £100 a year net, wishes to commute that for a present payment, interest being reckoned at 5 per cent. How much does he receive if his expectation of life is 17 years?

ANNUITIES.

36. Find the annuity to continue for four years which can be purchased for £750, supposing that a perpetual annuity is worth 25 years' purchase. The first payment is made at the expiration of a year from the purchase of the annuity.

37. The annual subscription to a certain club is £3, but a member can compound for all future subscriptions by a single payment of £60. If he is satisfied with 3 per cent. interest for his money, how long must he live in order that it may be worth his while to compound?

38. A man saves p pounds a year from his income, which he invests annually at $3\frac{1}{2}$ per cent. compound interest. At the end of how many years will he have accumulated a capital the interest of which is equal to his yearly saving?

CHAPTER XX.

PERMUTATIONS.

287. Explanation of "arrangements."

If we are given a number of objects and we have to arrange some or all of them in a line, we may of course place them in various orders according to which object is placed first, which second, and so on, and the question naturally suggests itself, *in how many different orders can the things be placed?*

This is the question which we propose in the present chapter to answer generally. But as it is always desirable to solve problems as simply as possible and without using unnecessary formulæ we shall begin by illustrating the method by a few simple examples.

Ex. In how many different orders can three different coins, say a halfpenny, a penny, and a florin be piled on a table?

First Method. The most natural way of answering the question would be to write down the different arrangements; viz.

Hf.-penny	Hf.-penny	Penny	Penny	Florin	Florin
Penny	Florin	Hf.-penny	Florin	Hf.-penny	Penny
Florin	Penny	Florin	Hf.-penny	Penny	Hf.-penny

There are thus six arrangements possible, and this is the number required.

If we were given a larger number of coins (say half-a-dozen or more) this method would be extremely laborious. It is therefore necessary to adopt a different mode of reasoning.

278   PERMUTATIONS.

Second Method. Suppose the coins laid down, one by one, the lowest of course being laid down first.

Then for the first coin, we have all three coins to choose from, and this may be done in three ways.

When this has been done, there only remain two coins to be chosen from for the second coin, and this can be done in two ways. Thus if the halfpenny were put down first the next coin must be either the penny or the florin. Similarly, if the penny were put down first we should have to take either the halfpenny or the florin; if the florin were put down first we should again have two different ways.

Hence the three ways of choosing the first coin give rise to 3×2 or 6 ways of choosing the first two coins.

When these have been chosen, there is only one coin left and this *must* be put at the top.

Hence no further arrangements of the coins are now possible, so that the total number of ways of piling them is 6.

These different arrangements are called the *permutations* of the three coins taken all together.

288. Fundamental Principle. The second solution depends on the following principle :

If one operation can be performed in m different ways, and a second operation can then be performed in n different ways, there will be mn different ways of performing the two operations in succession.

[The "first operation" of the above example is choosing the bottom coin which can be performed in ($m=$) 3 different ways, and the "second operation" is choosing the next coin which can *then* be performed in ($n=$) different ways, thus giving $m \times n$ or 3×2 ways of performing the first two operations, *i.e.* choosing the two lowest coins; and thus fixing the arrangement of the three.]

The truth of the general principle may easily be inferred.

For, corresponding to *each* of the m ways of performing the first operation, we shall have n ways of performing the second operation. Hence, the number of ways of performing the two operations in succession is mn.

We may generalize this statement as follows :

If one operation can be performed in m ways, and then a second can be performed in n ways, and then a third in p ways, and then a fourth in q ways, and so on, the number of ways of performing all the operations in succession will be $m \times n \times p \times q$.......

This can be shown by repeated application of the preceding result, and may be illustrated by the following example.

PERMUTATIONS. 279

Ex. In how many different orders can a hand of five cards be played out?

In playing the first card there are 5 cards to choose from, and this we can do in 5 ways. *Each* of these 5 alternatives leaves 4 cards from which to play the second card and this can now be done in 4 ways. Hence the total number of ways of playing the first two cards is 5×4 or 20.

Again, corresponding to every one of these 20 ways there are 3 different ways of playing the third card since *each* alternative leaves three cards to choose from. Hence the first 3 cards may be played in 20×3 or 60 ways.

Similarly each of these alternatives leaves him two cards to choose from in playing the fourth card, so that the first four can be played in 60×2 or 120 ways.

And since there is only one card now left this can only be played out in one way.

Hence the total number of ways $= 120 \times 1 = 120$.

NOTE. It is evident that the result (120) also represents the number of ways of arranging the five cards in a line or of laying them in a pile on the table. The two problems are identical; we have only to suppose that as the cards are played out they are placed in a line; each different order of playing the cards gives a different arrangement in the line. These different orders are called the *permutations* of the five cards taken all at a time.

289. We are now in a position to give the following definition:

DEFINITION. The different groups in which a given collection of things can be arranged by varying their order in every possible way are called the **permutations** *or* **arrangements** *of the things taken all together* or *all at a time* *.

Ex. 1. The permutations of two things a, b, taken both at a time are ab and ba †.

Ex. 2. The permutations of three things a, b, c, taken all at a time are abc, acb, bca, bac, cab, cba.

---

* In old treatises on Algebra, the word *permutation* was restricted to the arrangements obtained by taking *all* the given things together, while all other arrangements were designated by the term *variations*. The words *changes* and *alternations* are occasionally found used in one or other of these meanings.

† In this chapter, things are generally represented by the earlier letters of the alphabet.

280   PERMUTATIONS.

290. Permutations not involving all the given things. If, instead of arranging the whole of a given collection of things, we stop short when a certain number of the things have been placed, we shall obtain permutations of the things taken only part at a time.

Ex. 1. In how many ways can 2 scholarships of unequal value be awarded among 16 candidates without giving both to the same candidate?

The first scholarship can be awarded in 16 ways, and when it is given the second can be given in 15 ways, since both scholarships are not to be awarded to one candidate; hence the required number of ways is $16 \times 15 = 240$.

NOTE. If the scholarships were of equal value there would only be 120 ways, because it would not matter whether the first scholarship were awarded to A and the second to B or the second to A and the first to B.

Ex. 2. How many numbers of two different digits can be formed with the figures 1, 2, 3, 4, 5 and 6?

We have six ways of choosing the first digit, since it may be any one of the six given figures. Having chosen the first digit, we have only five ways of choosing the second, since the two digits are to be different. Hence the required number of numbers is $6 \times 5 = 30$.

Ex. 3. A man has 5 schools within his reach. In how many ways can he send 3 of his sons to school, if no two of his sons are to read in the same school?

The first son may be sent to any one of the five schools; and the second to any one of the remaining four. Hence the first two sons may be sent to different schools in 5×4 ways.

The first two sons having been sent to two of the five schools, a school may be selected for the third in three ways. And since three ways may be associated with each of the 5×4 ways previously found, the total number of ways required is $5 \times 4 \times 3 = 60$.

Ex. 4. There are three picture nails on a wall and seven pictures to choose from. In how many different ways can pictures be hung on all the nails?

For the first nail there are 7 pictures to choose from and therefore 7 ways of making the choice. Each of these alternatives leaves 6 pictures to choose from for hanging on the second nail, hence there are 7×6 ways of hanging pictures on the first two nails. Again, each of these alternatives leaves 5 pictures to choose from for the third nail. Therefore the total number of arrangements is $6 \times 7 \times 5$ or 210.

These arrangements are called the *permutations of the 7 pictures taken* 3 *at a time*.

PERMUTATIONS. 281

Ex. 5. There are four soldiers of whom three have to stand in a line. In how many ways can they be arranged?

The first man can be chosen from any of the 4 soldiers. When he has been placed the second can be chosen from any of the remaining 3, and the the third can be chosen from either of the remaining 2. Hence the total number of arrangements possible $= 4 \times 3 \times 2 = 24$.

The following alternative method is helpful in leading up to the next article.

The different arrangements can be formed by first selecting which three soldiers are to stand in the line, and then arranging them in different orders.

Now choosing which three are to stand is the same thing as choosing which of the four men is to be left out, and this can, of course, be done in 4 ways.

When any three men have been selected, the first place can be filled in three ways, the second can then be filled in 2 ways, and the third can then be filled in only one way.

Hence the required number of arrangements $= 4 \times 3 \times 2 = 24$ as before.

These arrangements are called the *permutations of the* 4 *men taken* 3 *at a time.*

291. We are now in a position to give the following extended definition of *permutations* :

DEFINITION. The different groups which can be formed from a given collection of n different things by selecting r of the n things in every possible way and arranging them in every possible order are called the **permutations** *or* **arrangements** *of the* n *things taken* r *at a time.*

Ex. The permutations of three things, a, b, c taken two at a time are bc, cb, ca, ac, ab, ba.

292. Distinction between permutations and combinations. It is often necessary to consider groups of things selected *without reference* to the order in which they are arranged. Such groups are called *combinations* in contradistinction to permutations, and the necessity of this distinction is well brought out in the following examples.

Ex. 1. I have three books; say, a Euclid, an Algebra, and a Trigonometry. In how many ways can I lend two of the books to a friend?

First Method. Whichever two books are lent, one of the three books will be retained. And since this book can be chosen in three

282 PERMUTATIONS.

ways, there are three ways of lending the two books. The three selections are: (i) Euclid and Algebra; (ii) Euclid and Trigonometry; (iii) Algebra and Trigonometry.

Second method. If we had treated the question as in the last article, we should have said that the first book could be chosen in 3 ways, and the second book could then have been chosen in 2 ways making 3 × 2 or 6 ways altogether. These six arrangements are

(i) Euc. and Alg. (ii) Euc. and Trig.
(iii) Alg. and Trig. (iv) Alg. and Euc.
(v) Trig. and Euc. (vi) Trig. and Alg.

But if the books are lent simultaneously, the order in which they are chosen makes no difference, and so the first and fourth arrangement consist of the same selection of two books, so do the second and fifth, and the third and sixth.

Hence in the 6 arrangements each selection occurs *twice* and the number of selections is 6 ÷ 2 or 3.

Ex. 2. There are 40 subscribers to a telephone exchange. In how many ways can two subscribers be put into communication with each other?

Any one of the 40 subscribers can ring up any one of the remaining 39 subscribers, and this can, of course, be done in 40 × 39 ways.

But the same two persons A and B (say) will be put into communication whether A rings up B or B rings up A.

Hence in above the 40 × 39 ways each pair of subscribers occurs twice. If the above distinction be not made the result will be *half* the above number, viz. 20 × 39 or 780.

Ex. 3. How many different hands of three cards can be dealt from a pack of 52 cards?

If the cards are dealt out one by one, the first card may be any one of the 52 cards; the second any one of the remaining 51 and the third any one of the remaining 50, thus giving 52 × 51 × 50 arrangements.

But these arrangements do not all consist of different cards; for the different orders in which the *same* three cards can be dealt out are here considered separately.

Now it is easy to see as in § 287 Ex., that three cards can be arranged in 3 × 2 × 1 different orders. Hence every hand of three cards can be dealt in 6 ways, and therefore occurs 6 times over in the above 52 × 51 × 50 ways.

Hence the required number of different hands

$$= \frac{52 \times 51 \times 50}{6} = 22{,}100.$$

PERMUTATIONS. 283

Otherwise thus. Let N be the required number of hands. Then in each hand the cards can be arranged in $3 \times 2 \times 1$ different ways. Hence $N \times 3 \times 2 \times 1$ is the number of different ways in which three cards can be dealt, when the *order* in which they occur is taken into account. But this is equal to $52 \times 51 \times 50$,

$$i.e.\ N \times 6 = 52 \times 51 \times 50.$$

Hence $N = 52 \times 51 \times 50 \div 6$ as before.

Ex. 4. In how many ways can (i) 5 pennies a florin and a half-crown, (ii) 5 pennies and two florins be placed in a line, so that two silver coins shall not be together.

Let the five pennies be first placed in a line, thus:

1, 1, 1, 1, 1.

Then the florin can be placed on the immediate left of either of the five pennies or on the right of the last penny, *i.e.* in 6 different places.

There are now 6 coins instead of 5 on the table; hence an extra coin could now be placed in 7 different places. But since the half-crown is not to be in the places immediately to the right or left of the florin, two of these places are excluded, and only 5 places are available. Hence there are 6×5 or 30 different arrangements of the coins.

For every arrangement in which the florin is on the left of the half-crown, another arrangement can be obtained in which the florin is on the right of the half-crown by simply interchanging the silver coins. Hence the number of the former arrangements must be equal to that of the latter, and both must be 15 in number. But if the silver coins were *both* florins, no difference would be made in the arrangement by merely interchanging them. Hence there would only be 15 possible arrangements.

293. We are now in a position to give the following definition:

DEFINITION. The different groups, each containing the same number of things, which can be formed by taking r things from a given collection of n different things, without reference to their order of arrangement, are called the **combinations** *or* **selections** *of the* n *things taken* r *at a time.*

Ex. 1. If three things be denoted by a, b, c the combinations of them taken two at a time are bc, ca, ab.

Ex. 2. The combinations of four things a, b, c, d taken three at a time are abc, bcd, cda, dab.

Ex. 3. The combinations of the same four things taken two at a time are ab, ac, ad, bc, bd, cd.

284   PERMUTATIONS

It will be useful to remember that *two groups containing the same set of things but arranged in different orders are* **different permutations formed from the same combination.**

Thus take any four letters a, b, c, d. From the first three of them alone, we can form the six permutations

$$abc,\ acb,\ bac,\ bca,\ cab,\ cba.$$

All these permutations, in no two of which the order of the things a, b, c is the same, arise out of the single combination abc.

The permutations that can be obtained by taking the combinations abd, acd, or bcd can be similarly exhibited.

294. The notations $_nP_r$ and $_nC_r$.

DEFINITION. It will be convenient to use the symbols $_nP_r$ to denote the number of permutations of n different things taken r at a time, and the symbols $_nC_r$ to denote the number of combinations of n different things taken r at a time*.

It is evident that $_nP_r$ will represent the **number of different ways** of selecting and arranging r things from a given collection of n different things, and in like manner $_nC_r$ will represent the number of different ways of merely selecting r of the n things without arranging them.

The number of ways of taking one thing at a time out of n given things is evidently n, and the number of ways of taking all at a time is evidently 1. Hence, with the notation of the present article,

$$_nC_1 = n\ ;\ \text{and}\ _nC_n = 1.$$

Again, since in a selection containing a single thing no rearrangement of order is possible, the number of ways of *arranging* 1 thing taken out of n given things is n.

Hence $_nP_1 = n.$

From the examples of the last few articles, we see by actual counting that

$$_3C_2 = 3,\ _4C_3 = 4,\ _4C_2 = 6,\ _3P_2 = 2,\ _3P_2 = 6,\ _3P_3 = 6,$$

and from § 290, Ex. 5 it is readily seen that $_4P_3 = 4 \times 6 = 24.$

---

* The symbols ^nP_r and $P(n, r)$ are sometimes used instead of $_nP_r$, and the symbols ^nC_r and $C(n, r)$ instead of $_nC_r$.

PERMUTATIONS. 285

295. To prove that the number of permutations of n different things taken r at a time is
$$n(n-1)(n-2)\ldots\ldots(n-r+1).$$

If we suppose that there are r blank spaces, the number of all possible arrangements of the n things taken r at a time is the number of ways of taking the n things one by one, and filling the r blanks in order.

The first blank may be filled up in n different ways, as it may be filled up by any one of the n things.

Having filled up the first blank, we have $n-1$ remaining things to choose from for filling up the second blank space. Hence we have $n-1$ ways of filling up the second blank, for each way of filling up the first blank. Thus there are $n(n-1)$ ways of filling up the first two blanks.

Similarly when the first two blanks have been filled up in any way, there are $n-2$ ways of filling up the third blank, since any one of the remaining $n-2$ things may be taken. Thus there are $n(n-1)(n-2)$ ways of filling the first three blanks.

Proceeding in this way, and noticing that the number of factors at any stage is the same as the number of blanks filled up, we see that the number of ways in which the r blanks can be filled up
$$= n(n-1)(n-2)\ldots\ldots \text{ to } r \text{ factors.}$$
$$= n(n-1)(n-2)\ldots\ldots \{n-(r-1)\}.$$

Hence $\quad {}_nP_r = n(n-1)(n-2)\ldots\ldots(n-r+1),$
as was to be proved.

*296. *Alternative Method.*

Let the n different things be denoted by the n letters $a, b, c, d\ldots\ldots$

Then the permutations of the n things taken r at a time may be divided into the following n different groups:

(1) those in which a stands first;
(2) those in which b stands first;
(3) those in which c stands first; and so on.

Now, each permutation in which a stands first can be obtained by taking a corresponding permutation of the $n-1$ things $b, c, d,\ldots\ldots$

taken $r-1$ at a time, and placing a before it*. Hence the number of permutations in which a stands first is denoted by $_{n-1}P_{r-1}$. Similarly the number of permutations in which b stands first is $_{n-1}P_{r-1}$. The same being true of each of the n groups, we have

$$_nP_r = {_{n-1}P_{r-1}} \times n.$$

Similarly $\quad _{n-1}P_{r-1} = {_{n-2}P_{r-2}} \times (n-1).$

$\quad\quad\quad\quad _{n-2}P_{r-2} = {_{n-3}P_{r-3}} \times (n-2).$

$\quad\quad\quad\quad \dots\dots\dots\dots\dots\dots\dots$

$\quad\quad\quad\quad _{n-r+2}P_2 = {_{n-r+1}P_1} \times (n-r+2).$

And obviously $_{n-r+1}P_1 = n-r+1.$

Multiplying together the vertical columns, and cancelling common factors†, we have

$$_nP_r = n(n-1)(n-2)\dots\dots(n-r+2)(n-r+1),$$

as before.

EXERCISES XX.

1. In how many different ways can I accommodate 2 gentlemen each in a separate room, if I have 10 rooms at my disposal?

2. There are five routes from London to Paris. Show that a man has 20 different ways of going from the one place to the other and of returning by a different route.

3. If three persons enter a room where there are seven vacant seats, in how many ways can they sit down?

4. In how many ways can four persons be arranged in a straight line?

5. How many different arrangements of three letters can be formed out of the letters b, c, d, f, g, h, if no letter is to appear more than once in any arrangement?

6. One family consists of 3 sons and 4 daughters, and another family of 5 sons and 6 daughters. In how many ways can a marriage be arranged between the two families, provided that the youngest son or daughter of either family does not marry the eldest daughter or son of the other?

---

* Thus the permutation abc, of 6 different things a, b, c, d, e, f, taken 3 at a time is obtained by prefixing a to bc, which is a permutation of the 5 things b, c, d, e, f, taken 2 at a time.

† This artifice is analogous to the *Chain Rule* of Arithmetic.

PERMUTATIONS.   287

297. To prove that the number of permutations of n different things taken all together is

$$n(n-1)(n-2) \times \ldots\ldots 3.2.1.$$

Since ${}_nP_r = n(n-1)(n-2)\ldots\ldots$ to r factors
it follows by putting $r = n$ that

$${}_nP_n = n(n-1)(n-2)\ldots\ldots \text{to } n \text{ factors}$$
$$= n(n-1)(n-2)\ldots\ldots 3.2.1$$
$$= \text{the product of the first } n \text{ integers *}.$$

298. Factorial notation †.

It is usual to denote the product of the first n consecutive integers, by the symbol $\lfloor n$ or $n!$ which is read '**factorial n.**'

Thus, $\lfloor 1 = 1;\ \lfloor 2 = 1.2;\ \lfloor 3 = 1.2.3 = 6;$
$\qquad\qquad 4! = 1.2.3.4 = 24;$
$\qquad\qquad 5! = 1.2.3.4.5 = 120.$

From § 297 we have therefore ${}_nP_n = n!$

NOTE. Where continued products are occurring constantly it is sometimes convenient to replace brackets by stops, thus $n \cdot n-1 \cdot n-2$ often means the same as $n(n-1)(n-2)$ and is easier to write.

The following examples will show how factorials can be transformed:

Ex. 1. Prove that $n! = n \cdot n-1 \cdot (n-2)!$

$n! = n \cdot n-1 \cdot n-2 \ldots\ldots 3.2.1$
$\quad = n \cdot n-1 \cdot \{n-2 \cdot n-3 \cdot n-4 \ldots\ldots 3.2.1\}$
$\quad = n \cdot n-1 \cdot (n-2)!$

---

* The last factor 1 is of course not wanted for the multiplication, but it is put in to make up the n factors.

† $n!$ is Kramp's notation. Formerly $\lfloor n$ was much used in English works, but this is now nearly abandoned on account of the difficulty in printing the $\lfloor\ $.

The term *factorial* is due to Arbogaste, a mathematician of Strasburg.

288   PERMUTATIONS.

Ex. 2.   Prove that $(2n)! = n!\, 2^n \{1.3.5\ldots\ldots(2n-1)\}$.

$$(2n)! = 1.2.3.4.5\ldots\ldots 2n$$
$$= \{1.3.5.7\ldots\ldots 2n-1\} \times \{2.4.6.8\ldots\ldots 2n\}$$
$$= \{1.3.5.7\ldots\ldots 2n-1\} \times \{(2.1).(2.2).(2.3)\ldots\ldots(2.n)\}$$
$$= \{1.3.5.7\ldots\ldots 2n-1\}.\,2^n\,(1.2.3.4\ldots\ldots n)$$
$$= n!\, 2^n \{1.3.5.7\ldots\ldots(2n-1)\}.$$

Ex. 3.   Prove that
$n+1.n+2.n+3\ldots\ldots$ to n factors $= 2.6.10.14\ldots\ldots$ to n factors.

$n+1.n+2.n+3\ldots\ldots$ to n factors
$$= n+1.n+2.n+3\ldots\ldots 2n$$
$$= \{1.2.3\ldots\ldots n.n+1.n+2\ldots\ldots 2n\}/(1.2.3\ldots\ldots n)$$
$$= (2n)!/(n!)$$
$$= 2^n \{1.3.5\ldots\ldots 2n-1\} \qquad\qquad [\text{Ex. 2}]$$
$$= 2.6.10\ldots\ldots 4n-2$$
$$= 2.6.10\ldots\ldots \text{to } n \text{ factors.}$$

EXERCISES XX.

Prove the following identities:

7. $\dfrac{15!}{13!} = 210.$ **8.** $\dfrac{1}{5!} + \dfrac{1}{6!} + \dfrac{1}{7!} = \dfrac{50}{7!}.$

9. $(n+5)!/(n+3)! = (n^2 + 9n + 20).$

10. $10! = 2^5.\,5!\,.(1.3.5.7.9).$

11. $100! = 2^{50}.\,50!\,(1.3.5\ldots\ldots 99).$ **12.** $\dfrac{1}{99!} - \dfrac{1}{100!} = \dfrac{1}{100\times 98!}.$

13. $\dfrac{n(n+1)(n+2)\ldots\ldots(2n-2)}{(n-1)!} = \dfrac{(2n-2)!}{(n-1)!^2}.$

299. Factorial expressions for $_nP_n$ and $_nP_r$.

The factorial notation enables us to express the results of §§ 295, 297 in a more concise form.   Thus
$$_nP_n = n(n-1)(n-2)\ldots\ldots 3.2.1,$$
becomes $\qquad\qquad _nP_n = \boldsymbol{n!} \ldots\ldots\ldots\ldots\ldots\ldots\ldots\ldots(\text{i}).$

PERMUTATIONS.

Again $\quad {}_nP_r = n(n-1)(n-2)\ldots\ldots(n-r+1)$.

Multiplying and dividing the right-hand side by $(n-r)!$ we have

$${}_nP_r = n(n-1)(n-2)\ldots\ldots(n-r)\ldots\ldots 3\,.\,2\,.\,1 \div (n-r)!$$

or $\qquad\qquad {}_nP_r = \dfrac{n!}{(n-r)!} \quad\ldots\ldots\ldots\ldots\ldots\text{(ii)}.$

[NOTE. Of the two formulæ for ${}_nP_r$, that involving factorials is not used in numerical calculations.]

COR. To deduce a meaning for 0!

If we put $r = n$ in the above formula for ${}_nP_r$, we find that

$${}_nP_n = \dfrac{n!}{0!}.$$

But ${}_nP_n$ is equal to $n!$

Hence $0! = 1$.

300. *To prove that the number of permutations of* n *different things taken* r *at a time is* $n-r+1$ *times the number of permutations of the* n *things taken* $r-1$ *at a time.*

This easily follows from the formulæ of § 295 or from first principles thus:

If any one of the permutations of n things taken $r-1$ at a time be taken, and one of the $n-(r-1)$ things which it does not contain be placed after it, we shall obtain a permutation of the n things taken r at a time. This can be done in $n-r+1$ ways.

$$\therefore\ {}_nP_r = (n-r+1) \times {}_nP_{r-1}.$$

301. Illustrative Examples.

Ex. 1. In how many ways can the letters of the word *formula* be re-arranged?

Since the seven letters of the word *formula* are all different, the total number of ways in which they may be arranged, taken all together, is ${}_7P_7$.

And $\qquad {}_7P_7 = 7\,.\,6\,.\,5\,.\,4\,.\,3\,.\,2\,.\,1 = 5{,}040.$

Hence the letters may be *re*-arranged in 5,039 ways.

Ex. 2. How many numbers between 1,000 and 10,000 can be formed out of the figures 7, 6, 5, 4, 3, 2, if no figure is to appear more than once in any number?

Since all the required numbers are between 1,000 and 10,000, they evidently consist of four digits each. We therefore have to find all the numbers that can be formed by taking four out of the six given figures.

TUT. ALG. II.

Since every arrangement of 4 figures taken out of these 6 different figures gives us a new number, the required number is

$$_6P_4 = 6 \times 5 \times 4 \times 3 = 360.$$

Ex. 3. If $_nP_2 = 30$, find n.

Since $_nP_2 = n(n-1)$, we have $n(n-1) = 30$.

$$\therefore n^2 - n - 30 = 0, \text{ giving } n = 6 \text{ or } -5.$$

Of these values -5 is inapplicable, since n is necessarily positive.

$$\therefore n = 6.$$

Ex. 4. If $_{3n}P_4 = \dfrac{126}{5} \cdot {}_{2n}P_3$, find n.

$$3n \cdot \overline{3n-1} \cdot \overline{3n-2} \cdot \overline{3n-3} = \dfrac{126}{5} \cdot 2n \cdot \overline{2n-1} \cdot \overline{2n-2}.$$

Here, n and $n-1$, being common factors of both the sides, may be equated to zero. But the values of $n = 0$ or 1 are not solutions of the question, since they make $3n = 0$, or $3n = 3$, and 0 or 3 things *cannot* be arranged 4 at a time.

Removing the factors n and $n-1$, the equation becomes

$$9 \cdot \overline{3n-1} \cdot \overline{3n-2} = \dfrac{126}{5} \cdot 4 \cdot \overline{2n-1}.$$

This quadratic, when simplified, becomes $45n^2 - 157n + 66 = 0$; which gives $n = 3$ or $\frac{22}{45}$. The latter value of n is inadmissible, and there is therefore only one solution, $n = 3$.

[NOTE.—In the solution of questions like the above, note that the common factor on both sides must first be removed.]

Ex. 5. If $_{11}P_r = 990$, find r.

Here $11 \cdot 10 \ldots\ldots (11 - r + 1) = 990$.

$$\therefore 11 \cdot 10 \ldots\ldots \text{to } r \text{ factors} = 11 \cdot 10 \cdot 9,$$

which shows that $r = 3$.

EXERCISES XX.

14. Find the numerical values of $_7P_5$ and $_5P_5$.

15. In how many ways can 6 students sit on a bench?

16. In how many ways may the letters of the words *equation* and *factor* be re-arranged?

17. In how many other ways may the triangle ABC be named?

PERMUTATIONS.

18. In how many ways can the seven prismatic colours be arranged?

19. How many numbers can be formed by taking 4 out of the 5 digits 1, 2, 3, 4, 5?

20. In how many ways can 4 coins selected out of 20 coins be placed in a row?

21. Given that the number of permutations of n things taken 5 at a time is 3 times the number of permutations of $n+1$ things taken 4 at a time find n.

22. How many different signals can be made by hoisting four differently coloured flags, one above the other, when all are hoisted each time?

23. Of eleven books of the same size, a shelf will hold seven; how many different arrangements may be made on the shelf?

24. The number of things is to the number of permutations when taken three together as 1 : 20; what is the number?

25. If $_nP_6 = 10 \cdot {_nP_5}$, find n.

26. If $_{n-1}P_3 : {_{n+1}P_3} = 5 : 12$, find n.

27. If $_{2n}P_3 = 12 \cdot {_nP_2}$, find n.

28. The number of permutations of n things, 4 together, is to the number of permutations of $\frac{1}{2}n$ things, 4 together, as 13 : 2. Find n.

29. If $_{2n+1}P_{n-1} : {_{2n-1}P_n} = 3 : 5$, find n.

30. If the number of permutations of n things taken n at a time is 5,040, find n.

[$_nP_n = n! = 5,040$. To find n from this divide 5,040 by the numbers 1, 2, 3, 4... in succession.]

31. How many signals may be made with 6 different flags which can be hoisted any number at a time?

[$_6P_1 + {_6P_2} + {_6P_3} + {_6P_4} + {_6P_5} + {_6P_6}$].

32. If the number of permutations of 6 things taken r at a time is 360, find r.

33. If $_{m+n}P_2 = 56$; and $_{m-n}P_2 = 12$, find m and n.

34. Prove (i) by general reasoning, and (ii) by the formula for $_nP_r$, that

$$_nP_r = {_{n-1}P_r} + r \cdot {_{n-1}P_{r-1}}.$$

35. How many numbers can be formed by using any number of the digits 1, 2, 3, 4, 5, but using each not more than once in any number?

PERMUTATIONS.

302. Permutations with some of the given things restricted in position.

In § 296 we saw that the number of permutations of n things taken r at a time in which a particular thing stands first is $_{n-1}P_{r-1}$.

Similarly if p particular things have to occupy p particular places the number of permutations is $_{n-p}P_{r-p}$, or, if the p things can be rearranged among themselves, the number is $p!$ times as great.

Ex. 1. How many of the permutations of a, b, c, d, e taken all at a time begin with a?

The number of permutations that begin with a = the number of permutations of b, c, d, e taken all together, since the former permutations can be obtained by prefixing a to the different arrangements of b, c, d, e.

∴ required number $= {}_4P_4 = 4 \times 3 \times 2 \times 1 = 24$.

Ex. 2. Of the permutations of the letters of the word *means* taken all together, how many do not begin with *as*?

The total number of permutations of m, e, a, n, s taken all together $= {}_5P_5 = 120$. And the number of permutations beginning with as = the number of permutations of m, e, n taken all together $= {}_3P_3 = 6$.

∴ required number $= 120 - 6 = 114$.

Ex. 3. How many numbers between 5,000 and 6,000 can be formed by using the figures of the number 12,345,678, each not more than once?

Every number between 5000 and 6000 consists of four figures, and begins with 5.

We are therefore required to find the number of arrangements of the remaining 7 figures 1, 2, 3, 4, 6, 7, 8, in the remaining 3 places of the number.

∴ required number $= {}_7P_3 = 7 \times 6 \times 5 = 210$.

Ex. 4. How many numbers can be formed by using any number of the digits 0, 1, 2, 3, 4, 5, but using each not more than once in each number?

The numbers may have either 1, 2, 3, 4, 5 or 6 digits, but every number beginning with 0 is the same as a number with one less digit formed of the figures 1, 2, 3, 4, 5, and must therefore be excluded. Hence the required number

$= {}_5P_1 + ({}_6P_2 - {}_5P_1) + ({}_6P_3 - {}_5P_2) + ({}_6P_4 - {}_5P_3) + ({}_6P_5 - {}_5P_4) + ({}_6P_6 - {}_5P_5)$
$= 1630$.

PERMUTATIONS. 293

Ex. 5. How many arrangements of the letters of the word *coins* can be made, without changing the place of the vowels in the word?

The consonants alone have to be permuted among themselves, and as the number of consonants is 3, the required number $= {}_3P_3 = 6$.

Ex. 6. How many arrangements of the letters of the word *coins* can be made, if every arrangement is to begin and end with the vowels?

The three intermediate consonants can be arranged among themselves in ${}_3P_3 = 6$ ways.

The number of ways in which the vowels may be arranged is 2, since an arrangement may begin with either vowel, and end with the other.

And as each of the two latter arrangements may be associated with any of the six former,

the required number of arrangements $= 6 \times 2 = 12$.

Ex. 7. How many arrangements of the letters of the word *strange* can be made, if the vowels are to appear only in the odd places?

Since the number of letters in the word *strange* is 7, the total number of places is 7, and the number of odd places is therefore 4. The vowels *a* and *e* are required to occupy two of these four odd places, *e.g.* thus

$$\begin{array}{ccccccc} 1 & 2 & 3 & 4 & 5 & 6 & 7 \\ a & & & & e & & \end{array}$$

The number of arrangements of the two vowels $= {}_4P_2 = 4 \cdot 3 = 12$.

When the vowels have been placed in any one way (*e.g.* the arrangement shown above), there remain 5 places to be filled by the remaining 5 letters *s, t, r, n, g*. And this can be done in

$${}_5P_5 = 120 \text{ ways.}$$

Also each of the 12 arrangements of the vowels may be associated with 120 arrangements of the consonants.

$\therefore$ total number of arrangements $= 120 \times 12 = 1440$.

Ex. 8. In how many ways can 7 different letters be arranged (i) if three of the letters are always all to appear together, (ii) if three of the letters are never all to appear together?

Let the seven letters be *A, B, C, d, e, f, g*, of which let us suppose that (i) *A, B, C* are never to be separated, (ii) *A, B, C* are never all to appear together.

(i) The letters *A, B, C* without being separated can be arranged among themselves in ${}_3P_3$ ways or 6 ways.

Since *A, B, C* are never to be separated, each arrangement may be regarded as one thing, so that we have, as it were, only 5 things (*ABC*), *d, e, f, g*.

Hence the number of permutations for each arrangement of A, B, C
$= {}_5P_5 = 5 \cdot 4 \cdot 3 \cdot 2 \cdot 1 = 120$.
∴ total number of different ways required $= 120 \times 6 = 720$.

(ii) If there were no restriction, the 7 letters might be arranged among themselves in ${}_7P_7$ ways or 5040 ways.

But we have just seen that the number of ways in which the three letters A, B, C all appear together $= 720$.

Hence the number of ways in which the three letters never appear all three together $= 5,040 - 720 = 4,320$.

Ex. 9. In how many ways can a, b, c, d, e be arranged so that no one of them may be removed more than one place to the right from its position in the alphabetical order?

Here d, e alone may appear in the fifth place; c, d, e alone in the fourth place; b, c, d, e in the third place; and a, b, c, d, e in the first place and in the second. Thus there are two ways of filling up the fifth place, and then two ways of filling up the fourth place; and so on, up to the first place which may be filled only in one way. Hence the result is $2 \times 2 \times 2 \times 2 \times 1 = 16$.

Ex 10. Find the sum of all the numbers which can be formed by using all the digits 4, 3, 2, 1 only once.

In 3! cases 4 is in the units' place, in as many cases in the tens' place, in as many cases in the hundreds' place, and in as many cases in the thousands' place. Thus the sum arising from the digit 4 alone is $3! \times (4,000 + 400 + 40 + 4)$ $= 4 \times 3! \times 1,111$. Proceeding similarly with the other three digits, the result required $= 3! \times 1,111 \times (1+2+3+4) = 66,660$.

EXERCISES XX.

36. How many of the permutations of a, b, c, d, e, f taken all at a time begin with a and end with f?

37. How many numbers of 4 digits and how many numbers altogether can be formed by using the figures of 1870 each not more than once?

38. Of the numbers that can be formed by using all the digits 4, 7, 8, 9, 6 only once, how many begin with 96?

39. How many numbers each lying between 2,000 and 3,000 can be formed by using the figures of the number 123,456 not more than once?

40. Of the numbers formed by using all the figures 1, 2, 3, 4, 5, 6 only once, how many are divisible by 5 and how many by 25?

41. Of the numbers formed by using all the figures 1, 2, 3, 5, 6, 7, how many are even?

PERMUTATIONS.

42. Six men are to address a public meeting; in how many ways may they take their turns, so that the shortest of them may not speak first?

43. How many words can be formed of the five letters c, r, e, s, t, so that the vowel may be the central letter?

44. If 8 apples are to be distributed among 8 boys so that the largest apple is always to be given to the biggest boy, in how many ways can the distribution be effected?

45. In how many ways can the letters of the word *using* be arranged (i) if the vowels are to retain their places, (ii) if the consonants are to retain their places?

46. In how many ways can 7 boys stand in a line to receive an electric shock, two only being willing to stand at the extremities of the line?

47. In how many ways can three different copper coins and three different silver coins be arranged in a line so that the silver coins may be in the odd places?

48. Show that the number of ways in which 4 ladies and 4 gentlemen can be arranged, so that no two ladies may be together is $5! \times 4!$

49. In how many ways can 4 Conservatives and 4 Liberals be arranged, so that the Conservatives may always sit together?

50. Find the number of different arrangements that can be made of bars of the seven prismatic colours, so that blue and green shall never come together.

51. Six papers are set in an examination of which two are mathematical. In how many different orders can the papers be arranged, so that the two mathematical papers are not consecutive?

52. Show that the number of ways in which n books can be arranged on a shelf, so that two particular books shall not be together is $(n-2)(n-1)!$

53. A shelf contains 20 books, of which 4 are single volumes and the others form sets of 8, 5, and 3 volumes respectively; show that the number of ways in which the books may be arranged on the shelf, so that the volumes of each set may be together is $7! \, 8! \, 5! \, 3!$

54. What is the sum of the numbers that can be formed with the four figures of 1893, and how many of these numbers are less than 1893?

55. In the arrangements of 1, 2, 3, 4, 5 all together, how many times will each figure occur in each place?

56. If all the permutations of a, b, c, d, e, f, taken all together, were written down in alphabetical order like words arranged in a dictionary, show that *dbcefa* would be the 394th permutation.

$$[394 = 3 \times 5! + 1 \times 4! + 1 \times 3! + 1 \times 2! + 1 \times 1! + 1.]$$

296 PERMUTATIONS.

303. Circular permutations.

We have hitherto only considered what may be called *linear permutations* of n different things, that is arrangements of n different things in a straight line. We shall now consider the *circular permutations* of n things taken all together, that is the number of ways in which n different things can be arranged round a circle.

The following examples show that the number of circular permutations of n different things is $(n-1)!$ or $\frac{1}{2}(n-1)!$, according as clockwise and counter-clockwise order are or are not distinguished.

The meaning of the terms **clockwise** and **counter-clockwise** (derived from the direction of motion of the hands on a clock) is evident from the annexed figures:

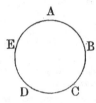   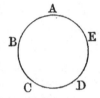

Fig. 1. Fig. 2.

Arrangement of A, B, C, D, E   Arrangement of A, B, C, D, E
in clockwise order.   in counter-clockwise order.

Ex. 1. In how many different ways can 5 persons be seated at a round table?

The number of ways in which 5 persons can be seated in 5 different chairs is 5!

Let one of the ways be that shown in the annexed figure. Then if each person moves the same number of places (one, two or more) to the left or to the right they will occupy different chairs but they will be arranged in the same order. And this may be done in 5 ways, since A may occupy any one of the 5 chairs.

Hence in the 5! arrangements obtained above, the same order occurs 5 times.

∴ required number of ways $= 5!/5 = 4! = 24$.

Fig. 3.

Ex. 2. In how many ways can 5 different beads be strung into a ring?

Reasoning as in Ex. 1, we find that the number of circular arrangements of the 5 beads is 4!

PERMUTATIONS.

But if the beads be arranged in the order ABCDE of Fig. 1, and we turn the ring over on its other side we shall have the arrangement AEDCB of Fig. 2; these two arrangements are therefore to be regarded as identical.

Hence the 4! or 24 circular arrangements give us $\frac{1}{2} \cdot 4!$ or 12 ways of making a ring with the five beads.

Ex. 3. In how many ways can 5 gentlemen and 5 ladies sit down together at a round table, so that no two ladies may be together?

The number of ways in which the five gentlemen may be arranged, clockwise and counter-clockwise order being distinguished, is 4! Let one arrangement be that shown in the annexed figure. Then the ladies have to be put in, one between two gentlemen, so that no two ladies may be together.

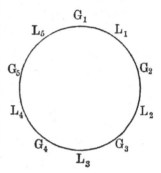

Hence the ladies may be arranged among themselves in 5! ways, for every arrangement of the gentlemen. [The ladies can be arranged in 5! ways and not in 4! ways, since the arrangement in which the lady L_1 comes after G_1 is different from the arrangement in which L_2 comes after G_1.]

Thus the total number of ways required $= 4! \times 5! = 2880$.

This will be the result if clockwise and counter-clockwise order are distinguished. If this distinction is not made, the number of arrangements required is obviously $\frac{1}{2} \times 2880 = 1440$.

EXERCISES XX.

57. In how many ways can 7 persons form a ring?

58. In how many ways can six beads of different colours be strung together on an elastic band so as to form a bracelet?

59. In how many ways can 8 persons be seated at a round table, so that all shall not have the same neighbours in any two arrangements?

60. In how many ways can 5 persons be seated at a round table, so that two of them are never to be separated?

61. In how many ways can 8 different flowers be strung together to form a garland, so that four of them may never be separated?

298   PERMUTATIONS.

304. To prove that the number of permutations of n things r at a time, when each thing may occur any number of times is n^r.

Suppose that we have r blank places. The first blank may be filled up in n ways; and when it is filled up, the second blank can also be filled up in n ways, since we are not prevented from repeating the same thing again. Hence the first two blanks may be filled up in $n \times n$ ways, that is, n^2 ways.

The third blank may also be filled up in n ways, and any one of these ways may be associated with each of the n^2 ways of filling up the first two blanks. Hence the first three blanks may be filled up in $n^2 \times n$ ways, that is, n^3 ways. And so on.

Thus the r blank spaces can be filled up in n^r ways.

COR. The total number of permutations of n different things taken *not more than* r at a time, when any of the n letters may be repeated

$$= n + n^2 + n^3 + \ldots + n^r = n \cdot \frac{n^r - 1}{n - 1}.$$

Ex. 1. If there be p different things to be given to n persons, the whole number of ways in which they may be given supposing each person may have any number of the things will be n^p.

Ex. 2. In the ordinary system of notation, how many numbers are there which consist of 4 digits?

The first digit may be any of the ten except 0, and each of the other three digits may be any of the ten whatever.

Hence altogether a number of 4 digits can be formed in $9 \times 10 \times 10 \times 10$ ways, that is, 9,000 ways.

That is, there are 9,000 different numbers of 4 digits, as is otherwise evident from the fact that these are the numbers 1000 to 9999 inclusive.

Ex. 3. How many numbers of not more than 20 digits are there which can be formed by using the digits 0, 1, 2, 3, 4 any number of times?

The number of permutations of the 5 digits 0, 1, 2, 3, 4 taken 20 at a time is 5^{20}.

PERMUTATIONS. 299

If the first r figures of a permutation are zeros, they may be struck out and the permutation will represent a number with $20-r$ figures.

Hence if the number 0 be included the required number of such numbers is 5^{20}.

If not the required number is $5^{20}-1$.

EXERCISES XX.

62. In how many ways can three different prizes be given to a class of twenty boys, when each boy is eligible for all the prizes?

63. There are 10 candidates for an office and 5 electors. In how many ways can the votes be given?

64. I have five parcels to be sent, and there are three post-offices within my reach. In how many ways can I send them?

65. In how many ways could a five-lettered word be made of the letters A, B, C, D, repetitions being allowed?

66. How many different algebraical quantities can be formed by combining a, b, c, d taken all together, by addition or subtraction, using the signs $+$ and $-$?

67. In the ordinary system of notation, how many numbers are there which consist of 7 figures?

68. The cylinder of a letter-lock contains five rings, each marked with five letters. How many different attempts to open the lock may be made by a person ignorant of the key-word?

69. A signalling apparatus has m arms and each arm is capable of n distinct positions, including the position of rest. Show that the number of signals that can be made is $n^m - 1$.

70. Show that the number of permutations of n things taken r at a time, when repetitions are allowed, but no consecutive repetitions, is $n(n-1)^{r-1}$.

71. How many numbers of 4 digits can be formed with 0, 1, 2, 5, 6, 8 if each of these digits may be repeated? Of these how many are even, and how many are divisible by 5?

72. How many numbers of not more than four digits can be formed with 7, 8, 9?

73. How many numbers of not more than 5 digits can be formed with the digits 0, 1, 2, 3?

74. Find the sum of all the numbers of not more than 3 digits that can be formed with the digits 3, 2, 1.

305. To find the number of permutations of n things taken all together, when the things are not all different.

Let the n things be represented by letters, and suppose that p of them are alike, each being represented by A; q of them are alike, and represented by B; r of them are alike, and represented by C; and so on.

Also let N be the required number of permutations.

If in *any one* of the N permutations, the p letters A were replaced by p new letters (say $A_1, A_2, \ldots A_p$), different from each other, and from all the rest of the n letters, then by changing the arrangement of these p new letters among themselves, and keeping the other letters unaltered in position, we should get, from the single permutation, $p!$ different permutations. Hence if this change were made in *each* of the N permutations, we should get altogether $N \times p!$ permutations.

Similarly, if, in addition, the q letters B were replaced by q new letters $(B_1, B_2, \ldots B_q)$, different from each other and from all the rest of the n letters, we should get $N \times p! \times q!$ permutations from the $N \times p!$ permutations obtained above. By proceeding in this way, we see that if all the letters were changed until no two were alike, we should finally obtain $N \times p! \times q! \times r! \ldots$ permutations.

But the n letters now being all different, the number of permutations thus obtained would be $_nP_n$, that is $n!$

$$\therefore N \times p! \times q! \times r! \ldots = n!$$

$$\therefore N = \frac{n!}{p! \times q! \times r! \ldots}.$$

Ex. 1. (i) Find the number of arrangements that can be made out of the letters of the word *consonant*.

(ii) In how many of these do the two *o*'s come together?

(iii) And how many of these begin with three *n*'s?

(i) In the word *consonant* there are 9 letters of which three are *n*'s, two are *o*'s, and the other letters are all different.

PERMUTATIONS. 301

Hence the required number of arrangements $= \frac{9!}{3! \, 2!} = 30{,}240$.

(ii) Now, consider the two o's as forming one letter, and we have 8 letters of which three are alike, namely the three n's.
Hence the number required $= 8! / 3! = 6{,}720$.

(iii) The number of arrangements beginning with three n's is the number of arrangements of the remaining letters c, o, s, o, a, t taken all together, and $= 6! / 2! = 360$.

Ex. 2. In how many ways can the letters of the word *continue* be arranged,

(i) without changing the order of the vowels;

(ii) without changing the relative position of the vowels and consonants?

(i) Since the four vowels are required to be in a given order, there can be no interchange of positions among them, and they may therefore be treated as *like* letters.

[For if they be replaced by four a's each arrangement of the a's gives rise to the single permutation $o\ i\ u\ e$.]

We have hence to find the number of permutations of 8 letters all together of which the four vowels are alike, and the two consonants n, n are also alike.

Hence the number required $= \frac{8!}{4! \, 2!} = 840$.

(ii) Here the places originally filled by consonants must always be filled by consonants, and those originally filled by vowels must always be filled by vowels.

The consonants can be permuted among themselves in $4!/2!$ ways, and the vowels among themselves in $4!$ ways.

∴ the number required $= \frac{4!}{2!} \times 4! = 288$.

Ex. 3. How many numbers over 200,000 can be formed by arranging the figures of 242302 taken all together?

The total number of numbers formed by using all the six figures of $242302 = 6!/3! = 120$.

Of these numbers, those that are less than 200,000 are the numbers that begin with 0; and the number of such numbers equals the number of arrangements of the remaining figures 2, 4, 2, 3, 2 taken all together and $= 5!/3! = 20$.

Hence the required number $= 120 - 20 = 100$.

Ex. 4. In how many ways can the letters of the word *arrange* be arranged? How many of these arrangements are there in which (i) the two r's come together, (ii) the two r's do not come together, (iii) the two r's and the two a's come together, (iv) the two r's come together but the two a's do not come together, (v) neither the two r's nor the two a's come together?

The number of arrangements of the letters of *arrange* subject to no restrictions $= 7!/(2!\ 2!) = 1260$.

The number of arrangements in which the two r's come together is obtained by treating the two r's as forming a single letter and therefore $= 6!/2! = 360$.

Hence the number of arrangements in which the two r's do not come together $= 1260 - 360 = 900$.

The number of arrangements in which the two r's come together and the two a's also come together $= 5! = 120$.

Hence the number of arrangements in which the two r's come together, but the two a's do not come together $= 360 - 120 = 240$.

The number of arrangements in which neither the two r's nor the two a's come together

$= $(*total* no. in which two a's do not come together)

$-$ (no. in which r's come together and a's do not come together)

$= 900 - 240 = 660$.

EXERCISES XX.

How many arrangements can be made of all the letters of the words:

75. *algebra,* **76.** *insincere,* **77.** *mammalia,* **78.** *proportion?*

79. In how many different ways can the letters in $a^3 b^7 c^4 d$ be written?

80. How many different signals can be made by using 10 flags all together, of which 4 are red, 3 yellow, 2 black and 1 white?

81. Find how many arrangements can be made from the letters of the word *Orion*, supposing (i) that the letters may stand in any order, and (ii) that two consonants may not stand together.

82. How many different permutations can be made of all the letters of the word *engineering*? In how many of these will the three e's stand together, and in how many will they stand first?

83. How many signals can be made by five bugle notes of which two are the same, blown in any order?

PERMUTATIONS. 303

84. How many arrangements can be formed out of the letters of the word *Constantinople*, (i) if the three n's have to come together, (ii) if the relative positions of the vowels and consonants are not changed?

85. There are eight letters of which a certain number are alike, and the rest are different; if 336 words can be formed out of them, how many letters are alike?

86. A word contains 10 letters of which 4 are a's, a certain number are e's, and the rest are unlike. If the whole number of permutations of the letters taken all together is 6,300, find the number of e's.

87. If there be m copies of each of n different books, the number of different orders in which they can be arranged on a shelf is

$$(mn)!/(m!)^n.$$

88. In a shelf there are 8 volumes of which 3 are single volumes, and the others form sets of 2, 3 volumes respectively; find in how many ways the books may be arranged on the shelf, the volumes of each set being in their due order.

89. How many numbers above 5,000 can be formed by arranging the figures of 1566 taken all together?

90. In how many ways can six persons A, B, C, D, E, F address a meeting, so that A may always speak before B, and B always before C? In how many of these ways will A speak immediately before B?

91. I have 3 copies of Homer, 2 copies of Virgil and Dante, and single copies of Shakespere and Milton. In how many ways can I distribute these among 9 friends, if each is to get one book?

92. How many numbers above 20,000 can be formed by arranging the figures of 12013 taken all together?

*93. In a competitive examination, the sixth classes of three provincial schools are arranged in one list. Show that if the number in each class be n, and if the boys of each school preserve the same order relatively to each other as they previously had in a separate examination of that school, there are $(3n)!/(n!)^3$ possible arrangements. Show also that in $(2n+1)!/(n!)^2$ of these, the boys of one particular school will stand together.

94. Show that 504 numbers each ending with two even digits can be formed with the nine digits 1, 2, 3......9, each digit occurring once and only once in each number, and the numbers being subject to the restriction that in reading them from left to right the odd digits shall be met with in ascending order of magnitude.

CHAPTER XXI.

COMBINATIONS.

306. Having defined *combinations* in the last chapter, we shall now investigate the value of $_nC_r$ the number of combinations of n things taken r at a time. We notice that the permutations which can be found by rearranging the *same* collection of things all arise out of a **single** combination.

307. To prove that the number of combinations of n different things taken r at a time is

$$\frac{n(n-1)(n-2)\ldots(n-r+1)}{r!}.$$

Every combination of r different things will produce $r!$ permutations, when the r things are rearranged in every possible way.

$\therefore\ _nC_r$ (the total number of) combinations of n different things r at a time will produce $_nC_r \times r!$ permutations.

But this is the number of permutations of n different things r at a time, and therefore

$$= n(n-1)(n-2)\ldots(n-r+1), \qquad (\S\ 295)$$

i.e. $\qquad _nC_r \times r! = n(n-1)(n-2)\ldots(n-r+1).$

$$\therefore\ _nC_r = \frac{n(n-1)(n-2)\ldots(n-r+1)}{r!}$$

$$= \frac{n(n-1)(n-2)\ldots(n-r+1)}{1\,.\,2\,.\,3\ldots r}.$$

COMBINATIONS. 305

308. *Alternative Method (without assuming the formula for the number of permutations).*

Let the n different things be denoted by the letters A, B, C, D......

Every combination containing the letter A can be obtained by combining A with each of the combinations of the remaining $n-1$ letters B, C, D... taken $r-1$ at a time.

Hence the number of combinations in which a particular letter, say A, appears is $_{n-1}C_{r-1}$. Hence if all the combinations of the n letters taken r at a time be written out, every letter appears in them $_{n-1}C_{r-1}$ times, and the total number of letters in them is therefore $n \times {}_{n-1}C_{r-1}$. But the total number of letters in them also equals $r \times {}_nC_r$, since each combination contains r letters.

Hence $\qquad r \times {}_nC_r = n \times {}_{n-1}C_{r-1}.$

Since this relation is true for all values of n and r, provided r is not greater than n, we have, in succession,

$$_nC_r = \frac{n}{r} \times {}_{n-1}C_{r-1}.$$

$$_{n-1}C_{r-1} = \frac{n-1}{r-1} \times {}_{n-2}C_{r-2}.$$

$$_{n-2}C_{r-2} = \frac{n-2}{r-2} \times {}_{n-3}C_{r-3}.$$

$$................................$$

$$_{n-r+2}C_2 = \frac{n-r+2}{2} \times {}_{n-r+1}C_1.$$

And obviously $\quad _{n-r+1}C_1 = \dfrac{n-r+1}{1}.$ [§ 294]

Now, multiplying together the vertical columns, and cancelling all the common factors, we have

$$_nC_r = \frac{n(n-1)(n-2)......(n-r+1)}{1 \cdot 2 \cdot 3......r}.$$

309. The product of r consecutive integers is divisible by $r!$

For if n be the greatest of the integers, we have to prove that $n(n-1)(n-2)......(n-r+1)$ is divisible by $r!$ This follows from the fact that the number of combinations of n things taken r at a time is necessarily an integer.

TUT. ALG. II.

310. Factorial expression for $_nC_r$.

If we multiply the numerator and denominator of the expression for $_nC_r$ by $(n-r)!$, we get

$$_nC_r = \frac{n(n-1)\ldots\ldots(n-r+1) \times (n-r)(n-r-1)\ldots\ldots 3.2.1}{r!(n-r)!}$$

or
$$_nC_r = \frac{n!}{r!(n-r)!}.$$

311. The number of combinations of n things taken r at a time is equal to the number when taken $n-r$ at a time.

This property is most readily deduced from the result of the preceding article.

For
$$_nC_r = \frac{n!}{r!(n-r)!}.$$

And
$$_nC_{n-r} = \frac{n!}{(n-r)!\{n-(n-r)\}!} = \frac{n!}{(n-r)!\,r!},$$

$$\therefore \; _nC_r = {_nC_{n-r}}.$$

This result may also be obtained by the following reasoning:

For every collection of r things which is selected out of the n given things, it is evident that one collection of $n-r$ things is left behind. Also, any change in the former collection will produce a corresponding change in the latter.

Therefore the number of different ways of taking r things out of n things must be just the same as the number of different ways of taking $n-r$ things as was to be proved.

Cor. 1. The above reasoning shows that $_nC_r$ also represents the number of ways of dividing n things into two groups of r and $n-r$ things respectively. Two such groups are sometimes called **complementary combinations**.

Cor. 2. If the number of combinations of n things taken r at a time is equal to the number of combinations of n things taken s at a time, then either $r = s$; or $n = r + s$.

COMBINATIONS. 307

Ex. 1. A person has 10 friends. On how many nights might he invite a different party of 4?
The number required $= {}_{10}C_4 = (10 \cdot 9 \cdot 8 \cdot 7)/(1 \cdot 2 \cdot 3 \cdot 4) = 210$.

Ex. 2. If ${}_{20}C_{r+11} = {}_{20}C_{3r+1}$, find r.
Here, either $r+11 = 3r+1$; or $(r+11)+(3r+1) = 20$.
$\therefore r = 5$ or 2.

Ex. 3. If ${}_nC_4 = {}_nC_5$, find n.
n obviously $= 4+5$ or 9.

Ex. 4. Find the number of combinations of 20 things taken 17 at a time.
$${}_{20}C_{17} = {}_{20}C_3 = 20 \cdot 19 \cdot 18/(1 \cdot 2 \cdot 3) = 1{,}140.$$

[The student should note that ${}_{20}C_3$ has been taken instead of ${}_{20}C_{17}$ to simplify the work of arithmetical calculation, since the numerator and denominator of the expression for ${}_{20}C_{17}$ contain 17 factors, while those of ${}_{20}C_3$ contain only 3 factors.]

Ex. 5. If ${}_nC_{n-4} = 15$, find n.
$${}_nC_{n-4} = {}_nC_4 = n(n-1)(n-2)(n-3)/(1 \cdot 2 \cdot 3 \cdot 4) = 15.$$
$\therefore (n^2 - 3n)(n^2 - 3n + 2) = 360$.

Solve the equation by putting $y = n^2 - 3n$, and complete the solution. $n = 6$ is the only admissible value.

Ex. 6. If ${}_nP_r = 720$; and ${}_nC_r = 120$, find r.
$${}_nP_r = r! \times {}_nC_r. \quad \therefore 720 = r! \times 120.$$
$\therefore r! = 6$, giving $r = 3$.

Ex. 7. In how many ways can 8 different things be divided into two parcels, of which the first is to hold 3 and the second 5?
The number required $= {}_8C_3 = (8 \cdot 7 \cdot 6)/(1 \cdot 2 \cdot 3) = 56$.

Ex. 8. In how many ways can 4 different things be divided into two parcels of which each is to hold an equal number?

Let the things be denoted by the letters a, b, c, d.

Then the number of pairs of complementary combinations $= {}_4C_2 = 6$.

These are ab, cd; ac, bd; ad, bc; bc, ad; bd, ac; cd, ab. But, *if the order of selecting the parcels be not distinguished*, the first and last pair give the same two parcels, similarly the second and fifth, and also the third and fourth.

Hence the number of ways required $= 6/2 = 3$.

COMBINATIONS.

Ex. 9. A train in going from A to B stops at 6 intermediate stations. How many sets of four different single third class tickets available on the train can be selected?

The number of stations being 8 in all, the total number of single third class tickets available for selection is the number of selections of two out of the 8 stations, and therefore $= {}_8C_2 = 8 . 7/(1 . 2) = 28$.
Hence the required result
$$= {}_{28}C_4 = 28 . 27 . 26 . 25/(1 . 2 . 3 . 4) = 20{,}475.$$

Ex. 10. If n straight lines terminate in a point and no two of them are in the same line, how many different Euclidean angles do they make?

There will be as many different Euclidean angles (or angles under two right angles) as there are selections of two out of the n given straight lines.

Hence the required number $= {}_nC_2 = \tfrac{1}{2} n (n-1)$.

[*If the lines were produced through the point* we should have four times as many angles, i.e. $2n (n-1)$.]

Ex. 11. How many diagonals has a decagon?

The number of lines obtained by joining the ten corners two by two = the number of selections of two out of ten corners
$$= {}_{10}C_2 = (10 . 9)/(1 . 2) = 45.$$
But of these lines 10 are the sides of the decagon.
Hence the number of diagonals $= 45 - 10 = 35$.

Ex. 12. There are n points in a plane, of which no three are in a straight line, except p, which are all in one straight line. Find the number (i) of straight lines formed by joining the points; (ii) of triangles formed by them.

(i) If no three points were in the same straight line, the number of lines would be ${}_nC_2 = n(n-1)/1 . 2$. But p of the points being in one straight line, we have only this one straight line instead of the $p(p-1)/(1 . 2)$ straight lines that would otherwise have been formed by joining the p points in pairs.

Hence the required number of lines $= \tfrac{1}{2} n(n-1) - \tfrac{1}{2} p(p-1) + 1$.

(ii) If no three points were in the same straight line we should get
$$\frac{n(n-1)(n-2)}{1 . 2 . 3}$$
triangles formed by taking the n points in threes. But p of the points being in one straight line, we lose the $\dfrac{p(p-1)(p-2)}{1 . 2 . 3}$ triangles which they would otherwise have formed.

Hence the required number $= \tfrac{1}{6} n(n-1)(n-2) - \tfrac{1}{6} p(p-1)(p-2)$.

EXERCISES XXI.

1. Find $_6C_3$, $_{17}C_{15}$ and $_{41}C_{40}$.

2. A father with eight children takes them three at a time to the Zoological Gardens as often as he can, without taking the same three children together more than once. How often can he go?

3. Out of 20 men, in how many ways can 2 men meet each other?

4. In a school where a different set of five monitors is appointed daily, there are 11 boys eligible for the office. How many weeks will elapse before the same five boys will be in office together again?

5. How many triangles can be formed with 10 straight lines in a plane of which no two are parallel, and no three meet in a common point?

6. From a company of 20 soldiers, four are placed on guard every two hours. For what length of time can different sets be selected?

7. If $_nC_{20}=\,_nC_{35}$, find n.   8. If $_{20}C_{r+4}=\,_{20}C_{2r-4}$, find r.

9. If $_{22}C_{r+3}=\,_{22}C_{3r-5}$, find r.   10. If $_nC_3=\,_nC_{13}$, find $_nC_{15}$.

11. Find n (i) if $_nC_2=66$, (ii) if $_nC_3=220$, (iii) if $_nC_4=210$.

12. Show that the number of combinations of n different things r at a time equals the number of permutations of n things taken all together, of which r are alike and $n-r$ are alike.

13. How many segments are formed if a finite straight line be cut at five points, at six points, and at n points?

14. If $_nC_3=\frac{5}{18}\cdot\,_nC_5$, find n.

15. If $_{n/2}C_4=3\frac{3}{4}\cdot\,_{n/3}C_3$, find n.

16. If $_nP_r=840$, and $_nC_r=35$, find n and r.

17. If $_nP_3=6\cdot\,_nC_4$, find n.

18. There are 15 candidates for three vacancies; in how many ways can a person vote for 3 of the candidates, and for 2 of them?

19. There are 10 candidates for three vacancies, and an elector can vote for any number of candidates not greater than the number of vacancies. In how many ways is it possible to vote?

20. Show that a polygon of m sides has $\frac{1}{2}m(m-3)$ diagonals.

21. Find the number of different triangles which can be formed by joining the angular points of a polygon of m sides.

22. In how many ways can 10 marbles be divided between 2 boys, so that one of them may get 2 and the other 8?

23. In how many ways can 10 marbles be divided into 2 parcels each containing 5?

24. In how many ways can 10 marbles be divided between 2 boys, so that each may have an equal number?

*25. Show that
$$_{4n}C_{2n} : {}_{2n}C_n = \{1 . 3 . 5 \ldots (4n-1)\} : \{1 . 3 . 5 \ldots (2n-1)\}^2.$$

*26. If n points in a plane be joined in all possible ways by indefinite straight lines, and if no two of the straight lines be coincident or parallel, and no three pass through the same point (with the exception of the n original points), then the number of points of intersection, exclusive of the n points, will be $\frac{1}{8}n(n-1)(n-2)(n-3)$.

312. To prove that $_nC_r + {}_nC_{r-1} = {}_{n+1}C_r$.

$$_nC_r + {}_nC_{r-1}$$
$$= \frac{n(n-1)\ldots\ldots(n-r+1)}{1 . 2 . 3 \ldots\ldots r} + \frac{n(n-1)\ldots\ldots(n-r+2)}{1 . 2 . 3 \ldots\ldots(r-1)}$$
$$= \frac{n(n-1)(n-2)\ldots\ldots(n-r+2)}{1 . 2 . 3 \ldots\ldots(r-1)} \times \left\{\frac{n-r+1}{r} + 1\right\}$$
$$= \frac{n(n-1)(n-2)\ldots\ldots(n-r+2)}{1 . 2 . 3 \ldots\ldots(r-1)} \times \frac{n+1}{r}$$
$$= \frac{(n+1)n(n-1)\ldots\ldots(n-r+2)}{1 . 2 . 3 \ldots\ldots r} = {}_{n+1}C_r.$$

The above result may also be proved as follows:

The $_{n+1}C_r$ combinations of $n+1$ things taken r together may be divided into two groups by taking

(i) those combinations which contain a particular thing,

(ii) those which do not contain that thing.

The number in the first group = the number of ways in which $r-1$ things can be selected out of the remaining n things = $_nC_{r-1}$.

And the number in the second group = the number of ways in which r things can be selected out of the remaining n things = $_nC_r$.

Hence $\quad\quad {}_{n+1}C_r = {}_nC_r + {}_nC_{r-1}.$

COMBINATIONS. 311

313. Relation between $_nC_r$ and $_nC_{r-1}$.

By the formula for $_nC_r$,

$$_nC_{r-1} = \frac{n(n-1)(n-2)\ldots\ldots(n-r+2)}{1.2.3\ldots\ldots(r-1)};$$

and $\quad _nC_r = \dfrac{n(n-1)(n-2)\ldots\ldots(n-r+2)(n-r+1)}{1.2.3\ldots\ldots(r-1)r}.$

$\therefore \quad _nC_r = \dfrac{n-r+1}{r} \times {}_nC_{r-1}.$

This formula enables us to find the ratio of any of the quantities $_nC_1, {}_nC_2, {}_nC_3, \ldots\ldots {}_nC_n$ to that which stands next before it.

Ex. 1. If $n = 20$, and $r = 5$; then

$$_{20}C_5 = \frac{20-5+1}{5} \cdot {}_{20}C_4 = \frac{16}{5} {}_{20}C_4.$$

Ex. 2. If $_nC_r : {}_nC_{r-1} = 9 : 2$ and $_nC_{4r-1} = {}_nC_{r+1}$, find n and r.
The second equation gives $n = (4r-1)+(r+1) = 5r$.
The first equation gives $(n-r+1)/r = \frac{9}{2}$.
Hence $r = 2$; and $n = 10$.

314. To find the value of r that makes $_nC_r$ the greatest.

By § 313,

$$_nC_r = \frac{n-r+1}{r} \times {}_nC_{r-1}.$$

We can thus obtain $_nC_r$ by multiplying $_nC_{r-1}$ by $\dfrac{n-r+1}{r}$.

Hence $\quad\quad _nC_r >, =, \text{ or } < {}_nC_{r-1},$

according as $\quad \dfrac{n-r+1}{r} >, =, \text{ or } < 1\;;$

that is, as $\quad n - r + 1 >, =, \text{ or } < r\,;$

that is, as $\quad n + 1 >, =, \text{ or } < 2r\,;$

that is, as $\quad r <, =, \text{ or } > \tfrac{1}{2}(n+1).$

Thus $\quad {}_nC_r > {}_nC_{r-1}$ if $r < \tfrac{1}{2}(n+1)$;

$\quad\quad\quad\quad {}_nC_r = {}_nC_{r-1}$ if $r = \tfrac{1}{2}(n+1)$;

and $\quad\quad {}_nC_r < {}_nC_{r-1}$ if $r > \tfrac{1}{2}(n+1)$.

(i) Let n be even. Then $\tfrac{1}{2}(n+1)$ is a fraction; and the greatest value r can have less than $\tfrac{1}{2}(n+1)$ is $\tfrac{1}{2}n$.

Hence ${}_nC_r$ *is greatest when* $r = \tfrac{1}{2}n$.

(ii) Let n be odd. Then $\tfrac{1}{2}(n+1)$ is an integer and putting $r = \tfrac{1}{2}(n+1)$ we see that ${}_nC_{\tfrac{1}{2}(n+1)} = {}_nC_{\tfrac{1}{2}(n-1)}$.

Hence ${}_nC_r$ *is greatest when* $r = \tfrac{1}{2}(n+1)$, *or* $\tfrac{1}{2}(n-1)$.

Thus if $n=24$, ${}_nC_r$ is greatest when $r = \tfrac{1}{2}n = 12$; and if $n=25$, ${}_nC_r$ is greatest when $r = \tfrac{1}{2}(n-1)$ or $\tfrac{1}{2}(n+1) = 12$ or 13, the values being the same in these two cases.

NOTE. It is evident that ${}_nP_r$ is greatest when $r = n$, or $r = n-1$. (For ${}_nP_n = {}_nP_{n-1}$ evidently.)

EXERCISES XXI.

27. Show that the number of ways of dividing 17 things into two sets of 11 and 6 things is greater than the number of ways of dividing them into two sets of 13 and 4 things.

28. Find the greatest number of combinations that can be made (i) out of 8 things; (ii) out of 7 things.

29. A person wishes to make up as many different parties as he can out of 20 persons, each party consisting of the same number; how many should he invite at a time? To how many of these parties will the same person be invited?

30. Six yachts are to be so arranged in a regatta that there may be the greatest possible number of different matches; how many must sail in each match, and how many matches will there be?

31. Show that the number of arrangements which can be formed from $2n$ things which are all either a's or b's is greatest when the number of a's is equal to the number of b's.

32. If ${}_nP_r : {}_nP_{r-1} = 8 : 1$; and ${}_nC_r : {}_nC_{r-1} = 8 : 3$; find n and r.

33. If ${}_nC_r : {}_nC_{r+1} : {}_nC_{r+2} = 1 : 2 : 3$, find n and r.

COMBINATIONS. 313

Combinations under Restrictions.

315. The number of combinations of n things r at a time in which p particular things always occur is $_{n-p}C_{r-p}$.

If the p things be first set aside, $n-p$ things are left.

Hence, if $r-p$ things be selected in every possible way from these, and added to the p things, we get all the required selections.

Hence the required number $= {_{n-p}C_{r-p}}$.

316. The number of combinations of n things taken r at a time in which p particular things never occur is $_{n-p}C_r$.

Since p of the things are not to be included in any selection, the r things have to be selected from $n-p$ things.

Hence the number required $= {_{n-p}C_r}$.

Ex. 1. In how many ways can four fruits be selected out of ten fruits, so as always to include the largest fruit?

If we first select the largest fruit, 9 fruits are left from which to select the remaining 3 fruits. And this can be done in $_9C_3$ ways or 84 ways, which is therefore the required number.

Ex. 2. In how many ways can four fruits be selected out of ten fruits, so as always to exclude the smallest fruit?

We have here to select 4 out of 9 fruits, since the smallest fruit is not to be included in any selection.

Hence the required number of ways
$$= {_9C_4} = (9 . 8 . 7 . 6)/(1 . 2 . 3 . 4) = 126.$$

Ex. 3. A boat's crew consists of 8 men, of whom 2 can only row on one side and 2 only on the other. In how many ways can a selection be made so that 4 men may row each side?

We have to divide the 4 men who can row on either side into two parties of two and two to complete the two sides.

Hence the number required $= {_4C_2} = (4 . 3)/(1 . 2) = 6.$

Ex. 4. Out of 50 persons, 5 are teachers. In how many ways can 5 persons be selected so as to include at least 1 teacher?

The total number of ways in which 5 persons can be selected out of 50 is $_{50}C_5$.

But the number of ways in which 5 persons can be selected so as to include no teacher is $_{45}C_5$.

∴ the number of ways in which 5 persons can be selected so as to include at least 1 teacher $= {_{50}C_5} - {_{45}C_5}$

$$= \frac{50.49.48.47.46}{5.4.3.2.1} - \frac{45.44.43.42.41}{5.4.3.2.1}$$

$= 2,118,760 - 1,221,759$

$= 897,001.$

EXERCISES XXI.

34. Of ten balls of different colors one is white. If they be formed into as many sets of four as possible, in how many of these will the white one be found?

35. In how many ways can an eleven be selected from 25 boys,
(i) 6 of them being always excluded,
(ii) 5 of them being always included,
(iii) 6 of them being always excluded, and 5 always included?

36. How many groups of 4 men can be selected from 12 men, so as always to include a particular man?

37. Show that in $_{2n}C_n$ the number of combinations in which a particular thing occurs is equal to the number in which it does not occur.

38. Show that in $_{3n}C_n$ the number of combinations in which a particular thing occurs is half the number of combinations in which it does not occur.

39. In an assembly of 100 persons, 15 are Europeans. In how many ways can a committee of 5 persons be formed so as to include at least 1 European?

40. A boat's crew consists of 8 men, of whom 3 can row only on one side and 2 only on the other. Find the number of ways in which the crew can be selected.

41. Twenty persons are to travel by an omnibus which can carry 12 inside and 8 outside. If 4 of these will not travel inside, and 5 will not travel outside, in how many different ways can the parties travel?

317. Permutations and Combinations formed from different sets. The student should verify the two following important theorems:—

(i) *If there be several sets containing* n_1, n_2, n_3...... *things respectively, the number of combinations formed by taking one out of each set* $= n_1.n_2.n_3...... =$ *the product of the numbers in the several sets.* [See § 288.]

(ii) *If m different things of one kind, and n different things of another kind are given, the number of permutations that can be formed, containing r of the first and s of the second is* $_mC_r \times {}_nC_s \times {}_{s+r}P_{s+r}$.

NOTE. In proving (ii) the formula for permutations must not be used till the selections required by the question have been made.

Ex. 1. In how many ways can a committee of 3 teachers and 2 students be formed from 5 teachers and 15 students?

The teachers can be selected in $_5C_3$ ways $= 10$ ways; and the students can be selected in $_{15}C_2$ ways $= 105$ ways.

Hence the required number of ways $= 10 \times 105 = 1,050$.

Ex. 2. If an alphabet consists of 5 vowels and 15 consonants, how many words of 3 different vowels and 2 different consonants can be made?

Three vowels and 2 consonants can be selected in 1050 ways [Ex. 1]; and when they have been selected in any one way, the five letters can be arranged among themselves in $5! = 120$ ways.

Hence the required number of words $= 1,050 \times 120 = 126,000$.

Ex. 3. How many words of 5 different letters can be formed out of 10 consonants and 5 vowels, if *a* is always to be one of the vowels and the word is to contain at least 2 consonants?

There are altogether three cases. The word may contain (i) two consonants, *a*, and two other vowels, (ii) three consonants, *a*, and one other vowel, or (iii) four consonants, and *a*.

Hence the number of words required

$$= {}_{10}C_2 \times {}_4C_2 \times 5! + {}_{10}C_3 \times {}_4C_1 \times 5! + {}_{10}C_4 \times 5! = 115,200.$$

Ex. 4. In how many of the permutations of *n* different things taken *r* together, do *p* particular things occur, *p* being $< r$?

r things can be selected out of *n* things so as to include *p* particular things in $_{n-p}C_{r-p}$ ways. And the *r* things can afterwards be arranged in $r!$ ways.

Hence the number required $= {}_{n-p}C_{r-p} \times r! = \dfrac{(n-p)! \, r!}{(r-p)! \, (n-r)!}$.

316   COMBINATIONS.

Ex. 5. Find the number of ways in which $2p$ persons may be seated at two round tables, p persons being seated at each.

The number of ways of selecting p persons out of $2p$ persons

$$={}_{2p}C_p=\frac{2p\,!}{p\,!\,p\,!}.$$

Also the number of ways in which p persons can sit at each round table $=(p-1)!$

Hence the total number of ways $=\dfrac{(2p)!}{p!\,p!}\times(p-1)!\times(p-1)!=\dfrac{(2p)!}{p^2}$.

Ex. 6. A crew of an eight-oar has to be chosen from 11 men of whom 5 can row on the stroke side only, 2 on the bow side only, and the remaining 4 on either side. How many different selections can be made? And in how many ways can the crew be arranged?

Let A, B, C, D, E be the men that can row only on the stroke side; M and N, the men that can row only on the bow side; and P, Q, R and S, the men that can row on both sides.

We have to select 4 men for the stroke side and 4 for the bow side. Here there are three cases.

(*a*) For the stroke side 4 men may be selected from A, B, C, D, E. For the bow side 4 men may now be selected from the remaining 6 men M, N, P, Q, R, S.

(*b*) For the stroke side 3 men may be selected from A, B, C, D, E and one from P, Q, R, S. For the bow side 4 men may now be selected from M, N and from the 3 remaining men of P, Q, R, S.

(*c*) For the stroke side 2 men may be selected from A, B, C, D, E and 2 from P, Q, R, S. For the bow 4 men may now be selected from M, N and from the 2 remaining men of P, Q, R, S.

[No other case is possible since four men cannot be obtained for the bow side under any other arrangement.]

Thus the number of ways in which the crew can be selected in the first case $={}_5C_4\times{}_6C_4=75$.

in the second case $={}_5C_3\times{}_4C_1\times{}_5C_4=200$,

and in the third case $={}_5C_2\times{}_4C_2\times{}_4C_4=60$.

Hence total number of ways of selection $=75+200+60=335$.

Now the four men on either side may be arranged in $4!$ ways.

Hence total number of arrangements of crew

$=335\times4!\times4!=192{,}960$.

COMBINATIONS. 317

EXERCISES XXI.

42. Out of ten men of whom 4 are ratepayers and the rest labourers, in how many ways can a Parish Council of five be formed, consisting of 2 ratepayers and 3 labourers?

43. In how many ways can 2 black balls and 2 white balls be drawn out of a bag containing 5 of each color?

44. Of ten balls of different colors, two are white. If they be formed into as many sets of four as possible, in how many of these will a white ball be found?

45. How many words of three letters in which the first and third letters are different consonants, and the second is a vowel can be formed with an alphabet of 20 consonants and 5 vowels?

46. From 21 consonants and 5 vowels, how many words of 5 letters can be formed, each containing 3 consonants and 2 vowels, all the letters in each word formed being different?

47. How many 3 letter words are there which can be formed of a, b, c, d, e, f, each word containing one vowel *at least?*

48. Out of 5 masters and 3 boys, a committee of 6 is to be chosen. In how many ways can this be done, (i) when there are 4 masters in the committee; (ii) when there is a majority of masters?

49. There are 3 capitals, 6 consonants and 4 vowels; in how many ways can a word be made beginning with one of the capitals, and containing 3 consonants and 2 vowels?

50. There are $n-1$ sets containing $2a, 3a,\ldots\ldots na$ things respectively; show that the number of combinations which can be formed by taking a out of the first, $2a$ out of the second, and so on is

$$(na)!/(a!)^n.$$

51. In how many ways can a word of 6 letters be formed out of 18 consonants and 6 vowels, if a, b, c are to be three of the letters, and the word is to contain at least 3 consonants? How many of these words begin with a? and in how many of these do b and c come together?

52. Of the permutations of 10 things taken 7 at a time, in how many do 3 particular things occur?

53. How many numbers can be formed from the figures of 123456 taken 4 at a time, if every number is to contain the figures 3 and 4?

54. How many numbers divisible by 25 can be formed,
  (i) from the figures of 123456 taken four at a time?
  (ii) from the figures of 123450 taken five at a time?

318   COMBINATIONS.

55. In the permutations of n things taken r at a time, show that the number of permutations in which p particular things occur is

$$_{n-p}P_{r-p} \times {_rP_p}.$$

56. In how many ways can 50 gentlemen and 20 ladies be arranged so that no two ladies shall be together?

57. Five ladies and three gentlemen are going to play at croquet; in how many ways can they divide themselves into sides of four each, so that the gentlemen may not all be on one side?

58. Find the number of ways in which 10 different men can be drawn up in two ranks of 5, supposing 3 particular men must be in the front and 3 other men in the rear.

59. Out of $2n$ men who have to sit down half on each side of a long table, p particular men desire to sit on one side, and q on the other; find the number of ways in which this may be done.

60. How many different sets of ladies' and gentlemen's doubles may be arranged at a lawn-tennis party of 3 ladies and 3 gentlemen?

61. A has 3 books and B has 5 books; in how many ways can they exchange their books, each keeping the number he had at first?

62. A cricket team consisting of eleven players is to be selected from two sets consisting of 6 and 8 players respectively. In how many ways can the selection be made, on the supposition that the set of 6 shall contribute not fewer than 4 players?

318. To prove that the total number of combinations of n different things taken any number at a time is $2^n - 1$.

Each thing may either be selected or left; and may therefore be disposed of in 2 ways.

Since either way of disposing of any one thing may be associated with either way of disposing of each of the other $(n-1)$ things, the total number of ways of disposing of all the n things is $2 \times 2 \times 2 \ldots$ to n factors $= 2^n$.

But this includes the case in which all the things are left out, which is inadmissible.

∴ required number of combinations $= 2^n - 1$.

Cor. We thus see that

$$_nC_1 + {_nC_2} + {_nC_3} + \ldots + {_nC_n} = 2^n - 1.$$

COMBINATIONS. 319

Ex. 1. In how many ways can a person purchase one or more of 3 books?

He has two ways of dealing with the first book, since he may either purchase it or not purchase it; similarly, he has two ways of dealing with the second book and two ways of dealing with the third.

Hence the number of ways in which he can deal with the books is $2 \times 2 \times 2 = 8$.

But this includes the case in which no book is purchased.

Hence the number required $= 7$.

Ex. 2. How many different weighings can be formed with a pound, a half-pound, a four-ounce, and a two-ounce weight? Also what will be their total weight?

Since in weighing a body each weight can be dealt with in two ways, the number of different weighings (the case in which all the weights are rejected being excluded) $= 2^4 - 1 = 15$.

Each of the above four weights occurs in 2^3 or 8 different weighings, since it may be taken with any of the $2^3 - 1$ different combinations of the other three weights, or may be taken by itself.

$\therefore$ the total weight $= 2^3$ (1 lb. + 8 oz. + 4 oz. + 2 oz.) = 15 lbs.

Ex. 3. How many different products can be formed with five letters a, b, c, d, e?

The number of ways of selecting one or more of these five letters $= 2^5 - 1$.

But 5 of these selections are inadmissible, since a single letter is not a product.

Hence the number of different products required $= 2^5 - 5 - 1 = 26$.

[Thus the number of ways in which *two or more* things can be selected out of n different things is $2^n - n - 1$.]

Ex. 4. At an election, there was one candidate more than the number to be elected, and each person could dispose of his votes in 14 ways; required the number of members to be elected.

Let n be the number of candidates.

Then $_nC_1 + {}_nC_2 + \ldots + {}_nC_{n-1} = 14$.

$\therefore 2^n - 1 - {}_nC_n = 14$. $\therefore 2^n = 16$. $\therefore n = 4$.

$\therefore$ the number of members $= 3$.

Ex. 5. Show that the number of combinations of 8 things of which 4 are alike taken 4 together $= 2^4$.

Let the 8 things be a, b, c, d, e, e, e, e. Then the number that we require $= {}_4C_0 + {}_4C_1 + {}_4C_2 + {}_4C_3 + {}_4C_4$, since to every selection of p things from a, b, c, d, we can add $4 - p$ of the e's so as to make the total number 4.

EXERCISES XXI.

63. In how many ways can I invite one or more of 4 friends to dinner?

64. A house has 5 windows in front; what is the total number of signals that can be given by having one or more of the windows open?

65. How many different condiments can be formed out of salt, pepper, mustard, oil, vinegar and ketchup, taken one, two, three, or more at a time?

66. How many different tints of color can be formed by mixing the seven prismatic colors always in the same proportion and in every possible way?

67. How many different conjunctions are possible between two or more of 5 planets?

68. How many products can be formed with 8 prime numbers?

69. In an examination a minimum is to be secured in each of 9 subjects for a pass. In how many ways can a student fail?

70. In how many ways can a sum of money be drawn from a bag containing a sovereign, a half-sovereign, a crown, a shilling and a penny?

Find the total value of the different sums of money that could thus be drawn.

71. Show that 127 different weighings can be made by means of the seven weights, 1, 2, 4, 8, 16, 32 and 64 lbs., taken any number at a time.

72. At an election the number of candidates was one more than twice the number to be elected, and each elector by voting for one or two or three, or as many persons as were to be elected can dispose of his vote in 15 ways; required the number of members to be elected.

73. Show that the number of combinations of $2n$ things of which n are alike taken n together $= 2^n$.

74. There are $3p$ things of which $2p$ are alike, and the rest, unlike; find the number of combinations of them taken $2p$ at a time.

75. Prove that every integral number of pounds from 1 to 121 can be weighed with the five weights 1 lb., 3 lbs., 9 lbs., 27 lbs., and 81 lbs., when each weight may be put into either scale.

COMBINATIONS. 321

319. To find the total number of combinations of n things taken any number at a time when the things are not all different.

Let there be p like things of one sort, q like things of a second sort, r like things of a third sort, and so on. We shall now show that the total number of selections that can be made by taking any number of the things is

$$(p+1)(q+1)(r+1)\ldots\ldots -1.$$

Of the p like things we may take either 0, 1, 2, 3,...... or p, and reject all the rest, hence they can be disposed of in $p+1$ ways.

Similarly, the q things can be disposed of in $q+1$ ways; the r things in $r+1$ ways; and so on.

Hence the number of ways in which all the things may be disposed of is $(p+1)(q+1)(r+1)\ldots\ldots$

But this includes the case where *all* the things are rejected, which is inadmissible.

Hence the total number of admissible ways is

$$(p+1)(q+1)(r+1)\ldots\ldots -1.$$

Ex. 1. If, of $p+q+r$ things, p be alike, q be alike, and the rest be all unlike, the total number of combinations is $(p+1)(q+1)2^r - 1$.

The p like things may be disposed of in $p+1$ ways, the q like things in $q+1$ ways, and the r unlike things in 2^r ways. Hence all the $p+q+r$ things may be disposed of in $(p+1)(q+1)2^r$ ways. But this includes the case where all the things are rejected. Hence the number of admissible ways $= (p+1)(q+1)2^r - 1$.

Ex. 2. How many factors has 720?

$$720 = 16 \times 9 \times 5 = 2^4 \times 3^2 \times 5.$$

The factor 2 may be treated in 5 ways, as we may take 0, 1, 2, 3, or 4 of the four 2's; similarly, the factor 3 may be treated in 3 ways, and the factor 5 in 2 ways.

Hence the required number of factors $= 5 \times 3 \times 2 = 30$.

In this result 1 and 720 are included as factors, hence, if these two are excluded, the number of factors is 28.

[NOTE. Hence the number of ways in which two integers can be found whose product is 720, is $\frac{1}{2}$ of $30 = 15$.]

TUT. ALG. II.

EXERCISES XXI.

76. In how many ways can a selection be made out of 5 pears and 6 apples, so as to take at least one of each kind?

77. In how many ways can a selection be made from 5 **rubies**, 4 diamonds, and 3 emeralds, if at least one of each kind is to be taken?

78. Prove that 10 different things can be divided into two parcels in $\frac{1}{2}(2^{10}-2)$ different ways.

79. In how many ways can I select one volume or more from 24 copies of *Euclid*, and 25 copies of *Algebra*?

80. Having five sovereigns and five shillings in my pocket, I am asked for a subscription; between how many different amounts have I choice?

81. In how many ways can two booksellers divide between them 10 copies of one book, 19 of another, and 24 of a third?

82. In a book-stall, where 4 copies of one book, 5 **copies of** another book, and **single copies** of 5 other different books **are exposed** for sale, how many choices have I in purchasing books?

83. If there be m sorts of things, and n things of each sort, the number of ways in which a selection can be made from them is

$$(n+1)^m - 1.$$

84. How many factors has 80, and how many has 72?

85. In how many ways can two integers be multiplied to give 98 as their product?

86. Show that 144 has 15 factors, and is obtained by multiplying two integers in 8 ways. Explain why, in this case, the latter number is not $\frac{1}{2}$ of the former number.

87. If a, b, c, d, e, f be algebraic quantities, show that the number of different factors of $a^p b^q c^r def$ (including 1 and the given number) is $2^3(p+1)(q+1)(r+1)$.

88. Show that, if there are n algebraic quantities $a, b, c, d, \ldots$, the number of pairs of factors of $a^p b^q c^r de \ldots$ subject to the condition that the factors in any pair have no common factor, and that the factor 1 is excluded is $2^{n-1} - 1$.

89. Find the number of factors of 360.

90. Also of 1470.

COMBINATIONS.

320. To find the number of ways of dividing a given collection of things into two or more groups containing given numbers of the things.

From § 311, Cor. 1, it follows that the number of ways of dividing $m+n$ things into two sets of m and n things respectively is $_{m+n}C_m$ or $(m+n)!/(m!\,n!)$.

We shall now prove that *the number of ways in which* $m+n+p$ *things can be divided into three sets of* m, n, p *things respectively is*

$$\frac{(m+n+p)!}{m!\,n!\,p!}.$$

For the $m+n+p$ things can be divided into two groups of m things and $n+p$ things in

$$\frac{m+n+p!}{m!\,n+p!} \text{ ways.}$$

And the $n+p$ things can be divided into two groups of n things and p things respectively in

$$\frac{(n+p)!}{n!\,p!} \text{ ways.}$$

Hence the $m+n+p$ things can be divided into three sets of m, n, p things respectively in

$$\frac{(m+n+p)!}{m!\,(n+p)!} \times \frac{(n+p)!}{n!\,p!} \text{ ways} = \frac{(m+n+p)!}{m!\,n!\,p!} \text{ ways.}$$

The result and its proof may easily be extended for any number of sets.

NOTE. When the numbers in the several sets are *equal* $(m=n=p)$ *and no distinction is made between the different sets*, the 3! different orders in which the three sets can be arranged are to be regarded as equivalent. Hence $3m$ things can be divided into three *equal* sets in

$$\frac{(3m)!}{3!\,(m!)^3} \text{ ways.}$$

In like manner and with the same premises, $m+2n$ things can be divided into sets of m, n, and n things in

$$\frac{(m+2n)!}{2!\,m!\,(n!)^2} \text{ ways,}$$

and so on.

Ex. In how many ways can a product of 6 factors be written as the product of three pairs of factors?

The number required $= \dfrac{6!}{2!\,2!\,2!\,3!} = 15$.

EXERCISES XXI.

91. In how many ways can 52 cards be divided among **four players**, so that each may have 13?

92. In how many ways can 52 cards be divided into 4 parcels of 13 each?

93. In how many ways can a product of 120 factors be decomposed into 6 products of 20 factors each?

94. In how many ways can 8 things be divided into groups of 4 pairs?

95. How many different parcels of 4 can be made out of 8 coins, and in how many of these will a particular coin appear?

96. Find the number of ways in which mn different things can be distributed among m persons so that each person may have n of them.

321. Combinations with repetitions.

We shall now find the number of combinations of n things taken r at a time when each of the things may be repeated any number of times (up to r times). The corresponding result for permutations has been obtained in § 304.

The notation $_nH_r$ in which H is the initial letter of *Homogeneous* is used to denote the number of combinations of n things taken r at a time when the things may be repeated, for the reason that this number is the number of powers and products of r dimensions that can be formed from n letters a, b, c, d...and their powers, or the number of terms in a *homogeneous* expression of the rth degree in the n variables a, b, c, d....

To prove that the number of combinations of n *things taken* r *at a time when repetitions are allowed, is the same as the number of combinations of* n + r − 1 *things taken* r *at a time without repetitions.*

We have to prove that $_nH_r = _{n+r-1}C_r$,

where $_nH_r$ denotes the number of combinations of n things taken r at a time with repetitions.

Let the n things be denoted by A, B, C, D......

Write down all the combinations required and add to each of them the n letters A, B, C, D......

While we do not alter the number of combinations by doing this, we secure that each letter appears *at least once* in every combination, of which the number of letters is now $r + n$.

Hence the number of combinations of n things taken r at a time, repetitions being allowed, is the same as the number of combinations of n things taken $r+n$ at a time, repetitions being allowed, but with the restriction that each thing appears *at least once* in every combination.

To find the number of these ways, take $r+n$ units in a line which have of course $n+r-1$ intervals between them, and divide them into n groups by distributing $n-1$ partition marks between them in any manner, *e.g.* thus:

1, 1, 1, | 1, 1, | 1, | 1, 1, 1, 1, 1, | 1, | 1, 1......

Then if the units in the first group are replaced by A's, those in the second by B's and so on we obtain a selection of $n+r$ letters in which each letter occurs at least once, viz.:

A,A,A, | B,B, | C, | D,D,D,D,D, | E, | F,F......

Since different positions of the partition marks give different combinations, the required number of combinations is the number of ways of selecting $n-1$ intervals for the partition marks.

Therefore the number required $={}_{n+r-1}C_{n-1}={}_{n+r-1}C_r$.

Hence ${}_nH_r = {}_{n+r-1}C_r$ .... (1)

$$= \frac{(n+r-1)(n+r-2)......(n+2)(n+1)n}{1.2.3......r}$$

$$= \frac{n(n+1)(n+2)......(n+r-1)}{1.2.3......r} \quad .... (2).$$

Ex. 1. Find the number of combinations of the five letters A, B, C, D, E taken 4 at a time, when the letters may be repeated.

The number required $= {}_5H_4 = \dfrac{5.6.7.8}{1.2.3.4} = 70$.

Ex. 2. Find the number of homogeneous products and powers of 6 dimensions that can be formed out of the 5 letters a, b, c, d, e.

The homogeneous products required can be found by taking the possible combinations of the 5 letters taken 6 at a time, with repetitions, and multiplying the 6 letters in each combination.

Hence the number required $= {}_5H_6$

$$= {}_{5+6-1}C_6 = {}_{10}C_6 = {}_{10}C_4 = \frac{10.9.8.7}{1.2.3.4} = 210.$$

Ex. 3. To find the number of terms in $(a+b+c)^7$.

The number of terms required = the number of homogeneous products and powers of 3 letters taken 7 at a time

$$= {}_3H_7 = {}_{3+7-1}C_7 = {}_9C_7 = {}_9C_2 = 36.$$

Ex. 4. Out of a large number of pence, shillings and sovereigns, in how many ways can a person select five coins?

The number required = the number of combinations of 3 things taken 5 at a time with repetitions $= {}_3H_5 = {}_{3+5-1}C_5 = {}_7C_5 = {}_7C_2 = 21$.

326   COMBINATIONS.

Ex. 5. In how many ways can an examiner assign a total of 50 marks to 6 questions, if he has to give to every question an integral number of marks greater than 4?

The number of ways required is the same as the number of ways of distributing $50-4 \cdot 6$ or 26 marks among the 6 questions, without giving less than *one* mark to any question. Now take 26 units in a line.

The number of ways in which $6-1$ or 5 partition marks can be distributed among the 25 intervals between the 26 units is the number required, and $=_{25}C_5 = 53,130$.

Ex. 6. Show that $_nH_r = {}_{n-1}H_r + {}_nH_{r-1}$.

The number of combinations of n things taken r at a time with repetitions = the sum of the number of combinations in which a particular thing, say a, occurs at least once $(= {}_nH_{r-1})$, and of the number of those in which the particular thing does not occur $(= {}_{n-1}H_r)$.

$$\therefore {}_nH_r = {}_nH_{r-1} + {}_{n-1}H_r.$$

EXERCISES XXI.

97. Find the number of combinations of the letters a, b, c, d taken 3 at a time, repetitions being allowed.

98. Find the number of combinations five at a time which can be formed from five of each of the letters A, B, C, D, E.

99. Find the number of homogeneous products and powers that can be formed by taking 6 different letters four at a time.

100. Find the number of homogeneous powers and products of the 7th degree that can be formed from 4 letters, and also from 2 letters.

101. Find the number of terms in $(a+b)^{15}$, and in $(a+b+c)^3$.

102. Find the number of terms in $(a+b+c+d+e)^7$.

103. In how many ways can a train of three carriages be made up, when there are first, second, and third-class and composite carriages to choose from? In how many of these will there be at least one third-class compartment? [The order of arrangement of the carriages is not taken into account.]

104. Find the number of ways in which I can select 10 coins out of a large number of sovereigns, crowns, shillings, pence and halfpence. In how many of these will each coin be represented?

105. In how many ways can 7 like fruits be divided among four persons, (i) subject to the condition that each is to get at least one fruit, (ii) if one or more may be left without a fruit?

106. How many triangles can be formed, having every side either 10 ft., 11 ft., or 12 ft. long?

107. Show that the number of ways in which m things can be put into $n+1$ pigeon-holes, there being no restriction as to the number in each hole, is $(m+n)!/m!\,n!$

322. Permutations and combinations of things that are not all different.

The general formulæ in these cases being rather intricate, we shall solve the following examples to illustrate the manner of proceeding in any particular case.

Ex. 1. In how many ways can a permutation of six letters be formed out of the letters of the word *Atlantean*?

There are 9 letters of 5 sorts a, a, a; t, t; n, n; l; e.

The selections may be classified, and the permutations obtained as follows:

	Combinations.	Permutations.
(1) Three letters alike, two alike and one different. [a, a, a with either 2 t's and $n, l,$ or e, or with two n's and $t, l,$ or e.]	$1 \times 2 \times 3 = 6.$	$6 \times \dfrac{6!}{3!\,2!} = 360.$
(2) Three letters alike, the other 3 different. [a, a, a with 3 letters selected out of the 4 letters t, n, l, e.]	$1 \times {}_4C_3 = 4.$	$4 \times \dfrac{6!}{3!} = 480.$
(3) Two letters alike, two others alike, and two more alike. [2 a's with 2 t's and 2 n's.]	1	$1 \times \dfrac{6!}{2!\,2!\,2!} = 90.$
(4) Two letters alike, two others alike, and the other two different. [2 a's, 2 t's and nl, le or ne; 2 a's, 2 n's and tl, te or le; or 2 t's, 2 n's and al, ae or le.]	${}_3C_2 \times {}_3C_2 = 9.$	$9 \times \dfrac{6!}{2!\,2!} = 1{,}620.$
(5) Two letters alike and the other four different. [2 a's with t, n, l, e; 2 n's with a, t, l, e; or 2 t's with a, n, l, e.]	3	$3 \times \dfrac{6!}{2!} = 1{,}080.$
Total	23	3,630

Hence the number of permutations required $= 3{,}630.$

Ex. 2. In how many ways can 4 letters be selected out of the letters of the word *therefore*?

There are altogether nine letters of six sorts e, e, e; r, r; t; h; f; o. The selections may be classified as follows:

(i) Three alike, one different.

(ii) Two alike, two others alike.

(iii) Two alike, the other two different.

(iv) All four different.

328   COMBINATIONS.

The selection (i) can be made in $_5C_1$ ways, since any one of the five different letters r, t, h, f, o may be taken with three e's. The selection (ii) can be made in 1 way, since the only way in which two like letters may be combined with the two other like letters is to take the two r's with two of the three e's. The selection (iii) can be made in $2 \times {_5C_2}$ ways, since either two e's may be combined with 2 letters selected out of the five different letters r, t, h, f, o; or the two r's may be taken with two letters selected from the five different letters e, t, h, f, o. The selection (iv) can be made in $_6C_4$ ways, since any four of the 6 different letters may be taken.

Hence the number required
$$= {_5C_1} + 1 + 2 \times {_5C_2} + {_6C_4} = 5 + 1 + 20 + 15 = 41.$$

EXERCISES XXI.

108. Find the number of combinations of the letters of *notation* taken 4 at a time, and also 5 at a time.

109. How many words can be formed with the letters composing the word *examination* taken 3 at a time?

110. Find the number of combinations in the letters of the word *alliteration* taken four at a time.

111. In how many ways can a selection of 3 things be made out of 9 things of which 4 are of one sort, 2 are of another sort, and the remaining 3 are different?

112. How many words each consisting of two consonants and two vowels can be made out of the letters of the word *devastation*? In how many of these will the two t's be together?

113. How many different numbers can be made out of 1 unit, 2 twos, 3 threes, 4 fours, 5 fives taken 5 at a time?

114. In how many ways can 3 coins be drawn from a bag containing 4 sovereigns, 2 crowns, a shilling, and a sixpenny piece?

115. Two points are taken on each side of a triangle, and the six points thus given are joined in every possible way. How many triangles are there *in general* in the resulting figure, when all the lines in it are produced indefinitely?

CHAPTER XXII.

MATHEMATICAL INDUCTION.

323. Mathematical Induction is an important method of establishing the truth of certain theorems in algebra and other branches of mathematics, by extending the proof step by step to all the successive cases which present themselves.

The method will best be understood from the illustrative examples given below, and it will be seen that the proof may in every case be divided into the following steps.

(i) The proposed theorem is verified by trial in the simplest possible case or cases (which we will call the first case).

(ii) It is then proved *perfectly generally* that if the theorem hold good *in any one case* (which we will call the nth case), it must also hold good in the *next* following case (the $n+1$th case), *whatever be the assumed value of n.*

(iii) By combining (i) and (ii) it is then inferred that

since the theorem holds in the 1st case, ∴ it holds in the 2nd case;
since the theorem holds in the 2nd case, ∴ it holds in the 3rd case,

and so on, and the reasoning being perfectly general, the theorem is extended to every possible case, and its universal truth established.

We shall illustrate the method by some examples.

Ex. 1. Prove, by induction, that the sum of the first r odd integers is r^2.

(i) The result is true when $r=1$, since $1=1^2$.

(ii) Let us assume it to be true when r has a certain value n, or, in other words, that the sum of the first n odd integers is n^2 for some *one* value of n.

The sum of the first $n+1$ odd integers can be obtained by adding the $n+1$th odd number to the assumed sum of the first n odd integers, and it therefore $= n^2 + (2n+1) = (n+1)^2$.

Now, this result is of the same form as the result we assumed to be true for n terms, $n+1$ having taken the place of n.

Hence if the result be true for a certain number of terms, it is true for the next higher number.

(iii) Since by (i) it is true for a single term,
$\therefore$ by (ii) it is true for two terms,

and since it is true for two terms,
$\therefore$ by (ii) it is true for three terms,

and so on. Thus the theorem is true universally.

CAUTION. In the second step, *we must not write* $n+1$ *for* n *in the assumed result* and say that "therefore the sum of the first $n+1$ odd integers is $(n+1)^2$," for the property is only assumed to hold good *for one particular value of* n, and *not* when this value is increased by unity; *that is what we have to prove*. Thus we have verified that the theorem is true when $n=1$. If we were to write $n+1$ for n in this result *without further investigation* we should merely be able to assert the truth of the corresponding property (that the sum of the first $n+1$ odd integers is $(n+1)^2$) *for the case* when $n+1=1$, *whereas what we require is to deduce it for the case when* n$=1$ or n$+1=2$.

NOTE. The process is called *induction*, as it establishes a general truth from the examination of particular cases. But the result obtained by this method must not, on this account, be considered more or less uncertain, but must be accepted as absolutely correct. It is impossible that there can be any exception to the theorem, for, if the proposition be true in any single instance, it is true in the next higher case. As the result is proved by reasoning step by step, the process is sometimes known as *successive induction*.

Ex. 2. Prove, by induction, that
$$1.2.3.4 + 2.3.4.5 + \ldots\ldots + r(r+1)(r+2)(r+3)$$
$$= \tfrac{1}{5} r(r+1)(r+2)(r+3)(r+4).$$

(i) The result is true for one term,
since $\qquad 1.2.3.4 = \tfrac{1}{5}.1.2.3.4.5.$

(ii) Let the result to be proved be assumed true for n terms.

Then the sum of $n+1$ terms
$= \tfrac{1}{5} n(n+1)(n+2)(n+3)(n+4) + (n+1)(n+2)(n+3)(n+4)$
$= (n+1)(n+2)(n+3)(n+4)(\tfrac{1}{5}n+1)$
$= \tfrac{1}{5}(n+1)(n+2)(n+3)(n+4)(n+5)$

which is of the *same form* as the assumed sum of n terms, $n+1$ having taken the place of n.

Hence if the result be true for n terms, it is true for $n+1$ terms.

MATHEMATICAL INDUCTION.

(iii) Applying (ii) to the case proved in (i) we see that the result is true for two terms, and therefore for three terms, and so on.

The result is thus true universally.

Ex. 3. Prove by induction that the number of permutations of n things taken all together is $n!$

The result is true if $n=1$ since one thing can be arranged in 1 way.

Let it be assumed that $_mP_m = m!$

Let $a, b, c, \ldots\ldots k$ be m letters, and let l be another letter.

Then from any single permutation of the m letters, such as $abc\ldots\ldots k$ it is possible to derive $m+1$ permutations of the $m+1$ letters $a, b, c, \ldots\ldots k, l$, by placing the letter l before a, between a and b, between b and c, and so on, or finally after k.

$\therefore \;\; _{m+1}P_{m+1} = (m+1)\,_mP_m = (m+1)\,m!$ (by assumption) $= (m+1)!$

Putting $m = 1, 2, 3, \ldots\ldots$ we see that

since $\quad _1P_1 = 1, \quad \therefore \;\; _2P_2 = 2!$
since $\quad _2P_2 = 2! \quad \therefore \;\; _3P_3 = 3!$
since $\quad _3P_3 = 3! \quad \therefore \;\; _4P_4 = 4!$

and since the method is general, it follows that $_mP_m = m!$ universally.

Ex. 4. Prove by induction that
$$1^2 + 2^2 + 3^2 + \ldots\ldots + n^2 = \tfrac{1}{6} n(n+1)(2n+1).$$

Since $\qquad 1^2 = \tfrac{1}{6} \cdot 1 (1+1)(2 \cdot 1 + 1),$
the proposed result is true if $n=1$.

Let it be assumed that when n has some value m
$$1^2 + 2^2 + 3^2 + \ldots\ldots + m^2 = \tfrac{1}{6} m(m+1)(2m+1).$$

Adding $(m+1)^2$ to both sides we have
$$1^2 + 2^2 + 3^2 + \ldots\ldots + m^2 + (m+1)^2 = \tfrac{1}{6} m(m+1)(2m+1) + (m+1)^2$$
$$= \tfrac{1}{6}(m+1)(2m^2 + m + 6m + 6) = \tfrac{1}{6}(m+1)(2m^2 + 7m + 6)$$
$$= \tfrac{1}{6}(m+1)(m+2)(2m+3) = \tfrac{1}{6}(m+1)\{(m+1)+1\}\{2(m+1)+1\}.$$

That is the sum of $m+1$ terms has the same form as that assumed for m terms (with $m+1$ substituted for m of course).

But the sum of 1 term has this form.

$\therefore$ the sum of 2 terms has this form,

and so on, thus the proposed result holds good in every case.

Similarly it may be proved by induction that
$$1^3 + 2^3 + 3^3 + \ldots\ldots + n^3 = \{\tfrac{1}{2} n(n+1)\}^2.$$

Ex. 5. Prove, by induction, that $x^n - y^n$ is divisible by $x - y$ for positive integral values of n.

Taking the identity $x^{n+1} - y^{n+1} = x^n(x-y) + y(x^n - y^n)$, the right-hand expression consists of two terms of which the first contains $x-y$ as a factor, and the second contains $x^n - y^n$.

Hence $x^{n+1} - y^{n+1}$ is divisible by $x-y$, if $x^n - y^n$ be divisible by $x-y$.

But $x-y$ is clearly divisible by $x-y$. We therefore infer that $x^2 - y^2$ is divisible by $x-y$, which shows that the same thing is true of $x^3 - y^3$; and so on, which establishes the theorem.

*324. It happens occasionally that any one case can be proved to be true, only by showing that the preceding two cases are true.

Ex. 1. Show, by induction, that $(3+\sqrt{5})^n + (3-\sqrt{5})^n$ is divisible by 2^n, n being a positive integer.

Let us assume that the theorem is true for two consecutive integral values of n; that is,

let $\qquad (3+\sqrt{5})^{m-2} + (3-\sqrt{5})^{m-2} = M \times 2^{m-2}$
and $\qquad (3+\sqrt{5})^{m-1} + (3-\sqrt{5})^{m-1} = N \times 2^{m-1}$

where M and N are integers.

Then, if $p = 3 + \sqrt{5}$, and $q = 3 - \sqrt{5}$,

$$(3+\sqrt{5})^m + (3-\sqrt{5})^m = p^m + q^m$$
$$= (p^{m-1} + q^{m-1})(p+q) - pq(p^{m-2} + q^{m-2})$$
$$= (p^{m-1} + q^{m-1})(3+\sqrt{5} + 3 - \sqrt{5}) - (3+\sqrt{5})(3-\sqrt{5})(p^{m-2} + q^{m-2})$$
$$= 6N \times 2^{m-1} - 4M \times 2^{m-2} = 3N \times 2^m - M \times 2^m = 2^m(3N - M),$$

which shows that $(3+\sqrt{5})^m + (3-\sqrt{5})^m$ is divisible by 2^m.

Thus if the theorem be true for two consecutive integral values of n, it is true for the next higher value also.

But the theorem is true when $n=1$, and $n=2$,

since $\qquad (3+\sqrt{5})^1 + (3-\sqrt{5})^1 = 6 = 2^1 \times 3$
and $\qquad (3+\sqrt{5})^2 + (3-\sqrt{5})^2 = 9+5+6\sqrt{5}+9+5-6\sqrt{5} = 2^2 \times 7$.

It is therefore true when $n=3$. And as the theorem is true when $n=2$, and $n=3$, it is true when $n=4$; and so on.

Thus the theorem is true for all positive integral values of n.

EXERCISES XXII.

Prove, by mathematical induction, the following theorems:

1. $(-1)^n = -1$ or $+1$ according as n is odd or even.
2. $1 + 2 + 3 + \ldots + n = \tfrac{1}{2} n (n+1)$.
3. $a + (a+b) + (a+2b) + \ldots$ to n terms $= na + \tfrac{1}{2} n (n-1) b$.
4. $\tfrac{1}{2} + \tfrac{1}{4} + \tfrac{1}{8} + \ldots$ to n terms $= 1 - (\tfrac{1}{2})^n$.
5. $a + ar + ar^2 + \ldots$ to n terms $= a(r^n - 1)/(r-1)$.
6. $1 . 2 + 2 . 3 + 3 . 4 + \ldots + n(n+1) = \tfrac{1}{3} n (n+1)(n+2)$.
7. $1^3 + 2^3 + 3^3 + \ldots + n^3 = (1 + 2 + 3 + \ldots n)^2$.
8. $1^4 + 2^4 + 3^4 + \ldots + n^4 = \tfrac{1}{30} n (n+1)(6n^3 + 9n^2 + n - 1)$.
9. $\dfrac{1}{1 . 2} + \dfrac{1}{2 . 3} + \dfrac{1}{3 . 4} + \ldots$ to n terms $= \dfrac{n}{n+1}$.
10. $\dfrac{1}{1 . 2 . 3} + \dfrac{1}{2 . 3 . 4} + \dfrac{1}{3 . 4 . 5} + \ldots$ to n terms $= \dfrac{n(n+3)}{4(n+1)(n+2)}$.
11. $1 . 2 . 3 . 4 . 5 + 2 . 3 . 4 . 5 . 6 + 3 . 4 . 5 . 6 . 7 + \ldots$ to n terms
$= \tfrac{1}{6} n (n+1)(n+2)(n+3)(n+4)(n+5)$.
12. $x^n + y^n$ is divisible by $x + y$ for positive odd values of n.
13. $x^n - y^n$ is divisible by $x + y$ for positive even values of n.
14. The product of three consecutive integers is divisible by 3!.
15. The product of four consecutive integers is divisible by 4!.
*16. The product of r consecutive integers is divisible by $r!$.
17. The difference between an integer and its cube is divisible by 6.
18. The cubes of the natural numbers beginning with unity leave, when divided by 6, the remainders 1, 2, 3, 4, 5, 0 recurring in order.

CHAPTER XXIII.

THE BINOMIAL THEOREM.*

POSITIVE INTEGRAL INDEX.

325. The Binomial Theorem is an important formula, first announced by Sir Isaac Newton, by means of which any power or root of a binomial expression can be readily expressed as a series.

The Binomial Theorem for a positive integral index, enables us to raise a binomial expression to any power without the trouble of actual multiplication.

We commence by finding an expression for the product of n different binomial factors.

326. To investigate the product of any number of binomial factors of the form $x + a_r$.

By actual multiplication, we have

$(x + a_1)(x + a_2) = x^2 + x(a_1 + a_2) + a_1 a_2$;

$(x + a_1)(x + a_2)(x + a_3) = x^3 + x^2(a_1 + a_2 + a_3)$
$\qquad\qquad + x(a_1 a_2 + a_2 a_3 + a_3 a_1) + a_1 a_2 a_3$;

and $(x + a_1)(x + a_2)(x + a_3)(x + a_4) = x^4 + x^3(a_1 + a_2 + a_3 + a_4)$
$\qquad + x^2(a_1 a_2 + a_1 a_3 + a_1 a_4 + a_2 a_3 + a_2 a_4 + a_3 a_4)$
$\qquad + x(a_1 a_2 a_3 + a_1 a_2 a_4 + a_1 a_3 a_4 + a_2 a_3 a_4) + a_1 a_2 a_3 a_4.$

These results suggest that if there are n factors, the product

$(x + a_1)(x + a_2)(x + a_3)\ldots\ldots(x + a_n)$
$= x^n + s_1 \cdot x^{n-1} + s_2 \cdot x^{n-2} + s_3 \cdot x^{n-3} + \ldots\ldots + s_r x^{n-r} + \ldots\ldots + s_n;$

---

* Fractional and negative indices are dealt with in Chap. XXVIII. after the discussion of *Limiting Values and Theory of Infinite Series*.

where $s_1 =$ the sum of the terms $a_1, a_2, a_3 \ldots \ldots a_n$;
$s_2 =$ the sum of their products, two at a time;
$s_3 =$ the sum of their products, three at a time;
....
$s_r =$ the sum of their products, r at a time;
....
$s_n =$ the product of all the a's.

To prove this we notice that every term of the product
$$(x + a_1)(x + a_2)(x + a_3)\ldots\ldots(x + a_n)$$
may be obtained by multiplying together n letters one taken from each of the n factors, and the product itself is the sum of all such combinations of these letters.

The term involving x^n can only be obtained by taking the x term from each of the factors, and this can only be done in one way.

∴ the coefficient of x^n is unity.

The terms involving x^{n-1} can be obtained by taking one of the a terms from any one factor and associating it with the x term in each of the remaining factors.

∴ the coefficient of $x^{n-1} =$ the sum of the letters a.

The terms involving x^{n-2} can be obtained by taking two of the a terms from any two factors and associating them with the x terms in each of the remaining factors.

∴ the coefficient of $x^{n-2} =$ sum of products of a's taken 2 together.

Generally the terms involving x^{n-r} can be obtained by taking the a terms from any r factors and associating them with the x terms in each of the remaining factors.

∴ the coefficient of $x^{n-r} =$ sum of products of a's taken r together.

Finally the term independent of x can only be obtained by taking all the a terms.

∴ the constant term $=$ product of all the a's.

327. To deduce the Binomial Theorem for a positive integral index.

In the above result
$$(x+a_1)(x+a_2)\ldots(x+a_n) = x^n + s_1 x^{n-1} + s_2 x^{n-2} + \ldots + s_r x^{n-r} + \ldots + s_n$$
we notice that since the terms in s_r are the combinations of n letters a taken r at a time, the number of terms in $s_r = {}_nC_r$, whatever be the value of r.

Now let $a_1 = a_2 = \ldots = a_n$ and put each of these $=a$.

Then evidently $s_r = {}_nC_r a^r$.

Also the left-hand side becomes $(x+a)(x+a)\ldots$ to n factors, that is $(x+a)^n$. Therefore
$$(x+a)^n = x^n + {}_nC_1 a x^{n-1} + {}_nC_2 a^2 x^{n-2} + \ldots + {}_nC_r a^r x^{n-r} + \ldots + {}_nC_n a^n.$$

This result is called the **Binomial Theorem.**

If now we substitute the values of ${}_nC_1, {}_nC_2 \ldots {}_nC_r \ldots$ we obtain the form in which the Binomial Theorem is *usually* written :—
$$(x+a)^n = x^n + nax^{n-1} + \frac{n \cdot n-1}{1 \cdot 2} a^2 x^{n-2} + \ldots$$
$$+ \frac{n \cdot n-1 \ldots n-r+1}{r!} a^r x^{n-r} + \ldots + a^n.$$

We have deduced this theorem *indirectly* from the corresponding formula for the product of n *different* binomial factors.

The following proof by Mathematical Induction is the simplest *direct* proof of the Binomial Theorem.

328. The Binomial Theorem. To prove that when n is a positive integer
$$(x+a)^n = x^n + nx^{n-1}a + \frac{n(n-1)}{1 \cdot 2} x^{n-2} a^2 + \ldots$$

the $r+$ 1th term being
$$\frac{n(n-1)(n-2)\ldots(n-r+1)}{r!} x^{n-r} a^r.$$

Since $\qquad (x+a)^2 = x^2 + 2xa + a^2$,
the theorem is easily seen to be true when $n=2$.

Assume therefore that it is true when n has some particular value m, that is, that
$$(x+a)^m = x^m + {}_mC_1 x^{m-1} a + {}_mC_2 x^{m-2} a^2 + \ldots$$
$$+ {}_mC_{r-1} x^{m-r+1} a^{r-1} + {}_mC_r x^{m-r} a^r + \ldots + a^m,$$

THE BINOMIAL THEOREM.

where $_mC_r$ is the number of combinations of m things taken r at a time, so that $_mC_1 = m$,

and generally $_mC_r = \dfrac{m \cdot m-1 \cdot m-2 \ldots\ldots m-r+1}{r!}$.

Multiply both sides by another factor $x + a$, thus
$$x(x+a)^m = x^{m+1} + {}_mC_1 x^m a + {}_mC_2 x^{m-1} a^2 + \ldots\ldots$$
$$+ {}_mC_r x^{m-r+1} a^r + \ldots\ldots,$$
$$a(x+a)^m = x^m a + {}_mC_1 x^{m-1} a^2 + \ldots\ldots$$
$$+ {}_mC_{r-1} x^{m-r+1} a^r + \ldots\ldots + a^{m+1}.$$
$$\therefore (x+a)^{m+1} = x^{m+1} + ({}_mC_1 + 1) x^m a + ({}_mC_2 + {}_mC_1) x^{m-1} a^2 + \ldots\ldots$$
$$+ ({}_mC_r + {}_mC_{r-1}) x^{m+1-r} a^r + \ldots\ldots + a^{m+1}.$$

In the new series for $(x+a)^{m+1}$ we find* that

the second term $= ({}_mC_1 + 1) x^m a = (m+1) x^{m+1-1} a$,

the third term $= ({}_mC_2 + {}_mC_1) x^{m-1} a^2$
$$= {}_{m+1}C_2 x^{m+1-2} a^2,$$

the $r+1$th term $= ({}_mC_r + {}_mC_{r-1}) x^{m+1-r} a^r = {}_{m+1}C_r x^{m+1-r} a^r$

since (by § 312) $_mC_r + {}_mC_{r-1} = {}_{m+1}C_r$.

Therefore the successive terms of the series for $(x+a)^{m+1}$ are of the same form as was assumed for the terms of $(x+a)^m$, but with $m+1$ written for m.

But we have seen that the assumed form holds for $(x+a)^2$. Therefore it holds for $(x+a)^3$. Therefore it holds for $(x+a)^4$, and so on for all positive integral values of the index.

The following alternative proof is practically an abbreviated form of the proofs of §§ 326, 327.

---

* It is thus obvious that the coefficient of the $r+1$th term of $(1+x)^{n+1}$ equals the sum of the coefficients of the rth and $r+1$th terms of $(1+x)^n$.

329. Alternative proof of the Binomial Theorem.

Every term in the continued product of n factors, each equal to $a+b$, is obtained by multiplying together n letters, one taken from each factor; hence all the terms of the continued product can be obtained by taking, in every possible way, n letters one from each factor, and multiplying them all together.

Now, we can take the letter a from every one of the factors, without taking the letter b at all. This can be done in only one way. Hence a^n is a term of the product.

The letter b can be taken from one factor, and the letter a from the remaining $n-1$ factors; and this can be done in ${}_nC_1$ ways. Hence ${}_nC_1 \cdot a^{n-1}b$ is a term of the product.

Again, the letter b may be taken from 2 factors and the letter a from the remaining $n-2$ factors; and this can be done in ${}_nC_2$ ways. Hence ${}_nC_2 \cdot a^{n-2}b^2$ is a term of the product.

And, generally, b can be taken from r factors (r being a positive integer not greater than n), and a from the remaining $n-r$ factors; and the number of ways in which this can be done is ${}_nC_r$. Hence ${}_nC_r \cdot a^{n-r}b^r$ is a term of the product.

And, finally, b can be taken from every one of the n factors, a being not taken at all. Hence b^n is a term of the product.

Thus $(a+b)^n = a^n + {}_nC_1 a^{n-1}b + {}_nC_2 a^{n-2}b^2 + \ldots\ldots$
$ + {}_nC_r a^{n-r}b^r + \ldots\ldots + b^n.$

$= a^n + na^{n-1}b + \dfrac{n(n-1)}{1 \cdot 2} a^{n-2}b^2 + \ldots\ldots$
$ + \dfrac{n!}{r!\,(n-r)!} a^{n-r}b^r + \ldots\ldots + nab^{n-1} + b^n.$

330. Standard forms of the Binomial Theorem.

DEFINITION. When a quantity is expressed as a series it is said to be **expanded.**

Thus in the identity

$$(a+b)^n = a^n + na^{n-1}b + \dfrac{n(n-1)}{1 \cdot 2} a^{n-2}b^2 + \ldots\ldots$$
$$+ \dfrac{n!}{r!\,(n-r)!} a^{n-r}b^r + \ldots\ldots + nab^{n-1} + b^n \ldots\ldots(\mathbf{1}),$$

$(a+b)^n$ is said to be expanded, and the series on the right-hand side is called the **expansion** of $(a+b)^n$ in ascending powers of b and descending powers of a.

THE BINOMIAL THEOREM.

The $r+1$th term is known as the **general term** (see § 192). Any term whatever may be derived from it by giving the proper value to r; for example, the second term, by making $r=1$, the third term by making $r=2$, the tenth term by making $r=9$, and so on. In deriving the first and last term, we must note that $0!=1$. [See § 299, Cor.]

The general term may generally be written in the forms

$$\frac{n(n-1)(n-2)\ldots\ldots(n-r+1)}{1\cdot 2\cdot 3\ldots\ldots r}a^{n-r}b^r,$$

$$_nC_r\, a^{n-r}b^r \quad \text{or} \quad \frac{n!}{r!\,(n-r)!}a^{n-r}b^r,$$

of which the first is most generally used.

The following points should be noted:

(i) In the general term $_nC_r a^{n-r}b^r$, the index of b is the same as the suffix of C, and the sum of the indices is n.

(ii) In $\dfrac{n(n-1)(n-2)\ldots\ldots(n-r+1)}{1\cdot 2\cdot 3\ldots\ldots r}a^{n-r}b^r$, the index of b is the last factor of the denominator, and the sum of the indices of a and b is n.

(iii) If, in (**1**), we make $a=1$ and $b=x$, we have

$$(1+x)^n = 1 + nx + \frac{n(n-1)}{1\cdot 2}x^2 + \frac{n(n-1)(n-2)}{1\cdot 2\cdot 3}x^3 + \ldots\ldots + x^n$$

which is often taken as the standard form; and, conversely, the expansion of $(a+b)^n$ can be derived from it as follows:

$$(a+b)^n = \left\{a\left(1+\frac{b}{a}\right)\right\}^n = a^n\left(1+\frac{b}{a}\right)^n$$

$$= a^n\left\{1 + n\frac{b}{a} + \frac{n(n-1)}{1\cdot 2}\frac{b^2}{a^2} + \ldots\ldots + \frac{b^n}{a^n}\right\}$$

$$= a^n + na^{n-1}b + \frac{n(n-1)}{1\cdot 2}a^{n-2}b^2 + \ldots\ldots + b^n.$$

(iv) The expansion of $(1-x)^n$ may be obtained from that of $(1+x)^n$, by putting $-x$ for x; thus

$$(1-x)^n = \{1+(-x)\}^n = 1 + n(-x) + \frac{n(n-1)}{1\cdot 2}(-x)^2$$
$$+ \frac{n(n-1)(n-2)}{1\cdot 2\cdot 3}(-x)^3 + \ldots\ldots + (-x)^n$$

THE BINOMIAL THEOREM.

$$\therefore (1-x)^n = 1 - nx + \frac{n(n-1)}{1.2} x^2 - \frac{n(n-1)(n-2)}{1.2.3} x^3 + \ldots + (-x)^n$$

in which the odd terms are positive and the even negative and the last term is $+x^n$ or $-x^n$ according as n is even or odd.

(v) The binomial series for a positive integral index is finite, the number of terms in the expansion of $(1+x)^n$ being $n+1$.

This property is evident from the proofs of §§ 327—329 or from the fact that one factor of the numerator of the general term

$$\frac{n(n-1)(n-2)\ldots(n-r+1)}{1.2.3\ldots r} x^r$$

vanishes if $r+1 > n+1$.

If, for example, $r+1 = n+2$; then $n-r+1$, (a factor of the numerator) $= 0$, which shows that there is no $n+2$th term.

331. In the expansion of $(x+a)^n$, the coefficients of terms equidistant from the beginning and the end are equal. For

$$(x+a)^n = x^n + {}_nC_1 x^{n-1}a + {}_nC_2 x^{n-2}a^2 + \ldots + {}_nC_r x^{n-r}a^r + \ldots$$
$$+ {}_nC_{n-r} x^r a^{n-r} + \ldots + {}_nC_{n-2} x^2 a^{n-2} + {}_nC_{n-1} x a^{n-1} + a^n.$$

The coefficients of terms equidistant respectively from the beginning and end when arranged in pairs are

$${}_nC_1 \text{ and } {}_nC_{n-1}, \; {}_nC_2 \text{ and } {}_nC_{n-2} \ldots {}_nC_r \text{ and } {}_nC_{n-r} \ldots$$

And since ${}_nC_r = {}_nC_{n-r}$, the result follows.

NOTE 1. This result might have been anticipated, since the expansion of $(a+x)^n$ equals that of $(x+a)^n$.

NOTE 2. Thus the coefficients in order are the same when read backwards as when read forwards.

The theorem enables us to write down, *without calculation*, the coefficients of the terms after the middle of the expansion of $(a+x)^n$ where n is a positive integer, since after that stage the coefficients are merely repeated in reverse order.

Since $n+1$, the total number of terms is even or odd according as n is odd or even, it follows, that if n be even, there will be one middle term; but, if n be odd, there will be two middle terms with equal coefficients, since the first of them is as far from the beginning as the second from the end.

332. Illustrative Examples.

Ex. 1. Expand $(2a+3x)^4$.

Putting $2a$ for a, $3x$ for b, and 4 for n in the expansion of $(a+b)^n$ we have

$$(2a+3x)^4 = (2a)^4 + 4.(2a)^3.3x + \frac{4.3}{1.2}(2a)^2.(3x)^2 + \frac{4.3.2}{1.2.3}(2a).(3x)^3 + (3x)^4$$
$$= 16a^4 + 96a^3x + 216a^2x^2 + 216ax^3 + 81x^4.$$

Ex. 2. Expand $(1-2x)^8$.

$(1-2x)^8 = 1 + {}_8C_1(-2x) + {}_8C_2(-2x)^2 + {}_8C_3(-2x)^3 + \ldots\ldots$ to 9 terms.

Since ${}_nC_r = {}_nC_{n-r}$, ${}_8C_5 = {}_8C_3$; ${}_8C_6 = {}_8C_2$; ${}_8C_7 = {}_8C_1$; and also ${}_8C_8 = 1$.

Hence $(1-2x)^8 = 1 + 8(-2x) + 28(-2x)^2 + 56(-2x)^3 + 70(-2x)^4$
$\qquad\qquad + 56(-2x)^5 + 28(-2x)^6 + 8(-2x)^7 + (-2x)^8$
$= 1 - 16x + 112x^2 - 448x^3 + 1120x^4 - 1792x^5 + 1792x^6 - 1024x^7 + 256x^8.$

Ex. 3. Expand $\left(c - \dfrac{1}{2c}\right)^4$.

$\left(c - \dfrac{1}{2c}\right)^4 = \left\{c\left(1 - \dfrac{1}{2c^2}\right)\right\}^4 = c^4\left(1 - \dfrac{1}{2c^2}\right)^4$

$= c^4\left\{1 + {}_4C_1\left(-\dfrac{1}{2c^2}\right) + {}_4C_2\left(-\dfrac{1}{2c^2}\right)^2 + {}_4C_1\left(-\dfrac{1}{2c^2}\right)^3 + \left(-\dfrac{1}{2c^2}\right)^4\right\}$

$= c^4 - 2c^2 + \dfrac{3}{2} - \dfrac{1}{2c^2} + \dfrac{1}{16c^4}.$

Ex. 4. Find the value of $(1+\sqrt{5})^5 + (1-\sqrt{5})^5$.

The terms in the two expansions are numerically equal, but the second, fourth, and sixth terms are opposite in sign.

Hence $(1+\sqrt{5})^5 + (1-\sqrt{5})^5 = 2\{1 + {}_5C_2(\sqrt{5})^2 + {}_5C_4(\sqrt{5})^4\}$
$\qquad\qquad = 2 + 100 + 250 = 352.$

Ex. 5. Find the 21st term in the expansion of $(a^2 - a)^{23}$.

The 21st term can be inferred from the general term which is the $(r+1)$th, by making $r = 20$.

The $(r+1)$th term $= {}_{23}C_r(a^2)^{23-r}(-a)^r.$

Thus the 21st term $= {}_{23}C_{20}(a^2)^3(-a)^{20} = {}_{23}C_3 a^6 . a^{20} = 1771a^{26}.$

Ex. 6. Find the coefficient of x^9 in the **expansion** of $(1-2x)^{12}$.
The term involving x^r in the expansion of $(1+x)^n$ is $_nC_r x^r$.
∴ the term involving x^9 in the expansion of $(1-2x)^{12}$ is
$$_{12}C_9(-2x)^9 = {}_{12}C_3(-2x)^9.$$
Hence the coefficient of $x^9 = \dfrac{12 \cdot 11 \cdot 10}{1 \cdot 2 \cdot 3}(-2)^9 = -112{,}640.$

Ex. 7. Find the coefficient of x^{10} in the **expansion** of $\left(x - \dfrac{2}{x}\right)^{20}$.
Suppose that x^{10} occurs in the $(r+1)$th term.
The $(r+1)$th term $= {}_{20}C_r x^{20-r}\left(-\dfrac{2}{x}\right)^r = {}_{20}C_r(-2)^r x^{20-2r}.$
And since this term contains x^{10} by supposition
$10 = 20 - 2r$ which makes $r = 5.$
Hence the required coefficient $= {}_{20}C_5(-2)^5 = -496{,}128.$
We may also proceed as follows:—
$$(x - 2/x)^{20} = x^{20}(1 - 2/x^2)^{20}.$$
Since the terms in order of $(x-2/x)^{20}$ are those of $(1-2/x^2)^{20}$ in order multiplied by x^{20}, the coefficient of x^{10} in the expansion of $(x-2/x)^{20}$ will be that of $x^{10} \div x^{20}$ or $1/(x^2)^5$ in the expansion of
$$(1 - 2/x^2)^{20}.$$
Hence the required coefficient $= {}_{20}C_5(-2)^5 = -496{,}128$, as before.

Ex. 8. Find the term independent of x in the expansion of
$$\left(3x - \dfrac{1}{3x}\right)^{2r}.$$
Let the term be the $(p+1)$th.
The $(p+1)$th term $= {}_{2r}C_p \cdot (3x)^{2r-p}\left(-\dfrac{1}{3x}\right)^p$
$= (-1)^p {}_{2r}C_p \cdot 3^{2r-2p} \cdot x^{2r-2p}.$
Since the term is independent of x, the exponent is 0.
Hence $2r - 2p = 0$; ∴ $p = r.$
Hence the required term
$= (-1)^r {}_{2r}C_r \cdot 3^0 \cdot x^0 = (-1)^r {}_{2r}C_r = (-1)^r \dfrac{(2r)!}{(r!)^2}.$

EXERCISES XXIII.

1. Find the coefficient of x^3 in $(x-1)(x-2)(x+7)(x-6)$.

2. Find the coefficients of x^2 and of x in
$$(x-3)(x-5)(x+7)(x-2).$$

Expand the following binomials:—

3. $(2x+3y)^5$.
4. $(a+3x)^6$.
5. $\left(\dfrac{x}{2}+2y\right)^6$.
6. $(5+4x^2)^4$.
7. $\left(x+\dfrac{1}{x}\right)^7$.
8. $\left(3xy-\dfrac{2x}{y}\right)^5$.
9. $(xy^{\frac{1}{2}}+\tfrac{2}{3}x^{\frac{1}{2}}y)^6$.
10. $(1-\tfrac{1}{2}x)^{10}$.

11. Write out the first five terms and the last two terms of the expansion of $(\tfrac{1}{2}a-\tfrac{1}{3}b)^{12}$.

12. Employ the Binomial Theorem to find 98^3, 999^4.

Write down and simplify:—

13. The 5th term of $(\tfrac{1}{2}x-2y)^7$.
14. The 6th term of $(x^3+3xy)^9$.
15. The 6th term of $(a^2-b^3)^{12}$.
16. The 12th term of $(a^2+ax)^{15}$.
17. The 7th term of $(1-\tfrac{1}{2}x)^{10}$.
18. The 10th term of $\left(\dfrac{a}{b}-\dfrac{2b}{a^2}\right)^{13}$.
19. The 25th term of $\left(x^{\frac{1}{2}}-\dfrac{y^{\frac{3}{2}}}{x^{\frac{1}{2}}}\right)^{26}$.
20. The general term of $(\tfrac{1}{2}+\tfrac{2}{3}x)^{n+3}$.
21. The middle term of $\left(x+\dfrac{1}{x}\right)^8$.
22. The middle term of $\left(\dfrac{y\sqrt{x}}{3}-\dfrac{3}{x\sqrt{y}}\right)^{16}$.
23. The two middle terms of $(a+b)^{11}$.
24. The two middle terms of $(\tfrac{1}{2}x-y)^9$.

25. Show that the middle term of $\left(x-\dfrac{1}{x}\right)^{2n}$ is
$$(-2)^n \cdot \dfrac{1 \cdot 3 \cdot 5 \ldots (2n-1)}{n!}.$$

Find the expansions of:—

26. $(1+\sqrt{x})^4+(1-\sqrt{x})^4$. **27.** $(a+2\sqrt{b})^5+(a-2\sqrt{b})^5$.

28. $(2\sqrt{a}+3)^6+(2\sqrt{a}-3)^6$.

29. Find the coefficient of x^8 in $(1+x^2)^{10}$.

30. Find the coefficient of x^{13} in $(ax-x^2)^{10}$.

31. Show that the coefficients of x^m and x^n in $(1+x)^{m+n}$ are equal.

32. Write down the coefficient of x^{18} in the expansion of
$$(a^4-bx^3)^{10}.$$

33. Show that in the series for $(a^2-x/a^3)^{10}$ there is no term containing a^{12}.

34. Find the coefficient of x^r in $(x+x^{-1})^{2n}$.

35. Find the term independent of x in $(\sqrt{x}-2/x^2)^{10}$.

36. Find the term independent of x in $(x^2-x^{-4})^{6r}$.

37. Show that there will be no term containing x^{2r} in the expansion of $(x+x^{-2})^{n-3}$, unless $\tfrac{1}{3}(n-2r)$ is a positive integer.

38. Show that, if n be even, the coefficient of the middle term of $(1+x)^n$ is
$$\frac{1 \cdot 3 \cdot 5 \ldots\ldots (n-1)}{1 \cdot 2 \cdot 3 \ldots\ldots \tfrac{1}{2}n} 2^{\tfrac{n}{2}};$$

and that, if n be odd, the coefficient of each of the two middle terms is
$$\frac{1 \cdot 3 \cdot 5 \ldots\ldots n}{1 \cdot 2 \cdot 3 \ldots\ldots \tfrac{1}{2}(n+1)} 2^{\tfrac{n-1}{2}}.$$

39. Show that the coefficient of the $(r+1)$th term in the expansion of $(1+x)^{n+1}$ equals the sum of the coefficients of the rth and $(r+1)$th terms in that of $(1+x)^n$.

333. To find the greatest coefficient in the expansion of $(1+x)^n$.

The coefficient of the $(r+1)$th term is $_nC_r$.

(i) If n be even, the greatest value of $_nC_r$ is when $r=\tfrac{1}{2}n$. [§ 314.] Hence the coefficient of the middle term, i.e. the $(\tfrac{1}{2}n+1)$th is greater than any other.

(ii) If n be odd, the greatest value of $_nC_r$ is when $r=\tfrac{1}{2}(n-1)$, or when $r=\tfrac{1}{2}(n+1)$, the values being the same in the two cases. Hence the coefficients of the two middle terms, i.e. the $\tfrac{1}{2}(n+1)$th and $\tfrac{1}{2}(n+3)$th term are equal and each is greater than any other coefficient.

THE BINOMIAL THEOREM.   345

334. To find the greatest term in the expansion of $(1+x)^n$, n being a positive integer.

The rth term of the expansion of $(1+x)^n$ is
$$\frac{n(n-1)\ldots\ldots(n-r+2)}{1.2.3\ldots\ldots(r-1)} x^{r-1}$$
and the $(r+1)$th term is
$$\frac{n(n-1)\ldots\ldots(n-r+2)(n-r+1)}{1.2.3\ldots\ldots(r-1)r} \cdot x^r.$$
Hence the $(r+1)$th term is obtained by multiplying the rth term by
$$\frac{n-r+1}{r} \cdot x \text{ that is } \left(\frac{n+1}{r}-1\right)x.$$

∴ denoting the rth term and the $(r+1)$th term by t_r and t_{r+1}, and supposing x positive, we see that
$$t_{r+1} >, =, \text{ or } < t_r,$$
according as $\left(\dfrac{n+1}{r}-1\right)x >, =, \text{ or } < 1,$

that is, according as $\dfrac{n+1}{r} - 1 >, =, \text{ or } < \dfrac{1}{x};$

that is, according as $\dfrac{n+1}{r} >, =, \text{ or } < 1+\dfrac{1}{x};$

that is, according as $n+1 >, =, \text{ or } < r\left(1+\dfrac{1}{x}\right);$

that is, according as $r <, =, \text{ or } > \dfrac{(n+1)x}{x+1}.$

CASE I. If $(n+1)x/(x+1)$ be an integer, denote it by p.

Then for all values of r from 1 to $p-1$, the multiplying factor is greater than unity, and the terms increase.

When $r = p$, the multiplying factor $= 1$, and therefore the $(p+1)$th term equals the pth.

And for all values of r from $p+1$ to n, the multiplying factor is less than unity, and the terms decrease.

Hence, the pth term is equal to the $(p+1)$th, and these are the greatest terms.

THE BINOMIAL THEOREM.

CASE II. If $(n+1)x/(x+1)$ be not an integer, denote its integral part by q.

Then for all values of r from 1 to q, the multiplying factor is greater than unity, and the terms increase.

And for all values of r from $q+1$ to n, the multiplying factor is less than unity, and the terms decrease.

Hence, the $(q+1)$th is the greatest term.

In particular if $(n+1)x/(x+1)$ is a proper fraction or $x<1/n$ the first term is the greatest, and if the same quantity $>n$ or $x>n$ the last term is the greatest.

NOTE 1. To find the greatest term in the expansion of $(x+a)^n$, we have to express $(x+a)^n$ in the form of $x^n(1+a/x)^n$, and find the greatest term in the expansion of $(1+a/x)^n$, since the terms of the latter expansion, when multiplied by the same quantity x^n, give the terms of the former expansion.

NOTE 2. In finding the numerically greatest term in the expansion of $(1-x)^n$, or $(x-a)^n$, we may ignore the negative sign, and proceed as in finding the greatest term in the expansion of $(1+x)^n$ or $(x+a)^n$, since the magnitude of any term is not changed by a change of sign.

The student is recommended to work out examples independently of the above formulæ.

Ex 1. Find the greatest term in $(5-2x)^{17}$, if $x=\frac{1}{2}$.

Since $(5-2x)^{17}=5^{17}(1-\frac{2}{5}x)^{17}$, we require the greatest term in the expansion of $(1-\frac{2}{5}x)^{17}$.

Denoting the $(r+1)$th and the rth term of the expansion of $(1-\frac{2}{5}x)^{17}$ by t_{r+1} and t_r, and disregarding signs, we have *numerically*

$$t_{r+1}=t_r \times \frac{17-r+1}{r} \cdot \frac{2x}{5} = t_r \cdot \frac{18-r}{5r} \cdot 2x = t_r \cdot \frac{18-r}{5r}, \text{ (since } 2x=1\text{).}$$

Hence $t_{r+1}>, =, \text{ or } <t_r$,

according as $\frac{18-r}{5r}>, =, \text{ or } <1;$

that is, according as $18-r>, =, \text{ or } <5r;$

that is, according as $18>, =, \text{ or } <6r;$

that is, according as $r<, =, \text{ or } >3.$

Hence the 3rd and the 4th term are numerically equal, and they are greater than any other term.

THE BINOMIAL THEOREM.

Ex. 2. Find the condition that the greatest term in the expansion of $(1+x)^n$ may have the greatest coefficient.

(i) Let n be even. Then the greatest coefficient is obtained when $_nC_r$ is greatest; that is, when $r = \frac{1}{2}n$.

Hence the greatest term must be $_nC_{\frac{1}{2}n} \cdot x^{\frac{1}{2}n}$,

$$\therefore \quad _nC_{\frac{1}{2}n} \cdot x^{\frac{1}{2}n} > {}_nC_{\frac{1}{2}n-1} \cdot x^{\frac{1}{2}n-1}, \text{ and } _nC_{\frac{1}{2}n} \cdot x^{\frac{1}{2}n} > {}_nC_{\frac{1}{2}n+1} \cdot x^{\frac{1}{2}n+1};$$

$$\therefore \quad \frac{n - \frac{1}{2}n + 1}{\frac{1}{2}n} \cdot x > 1, \quad \text{and} \quad \frac{n - (\frac{1}{2}n + 1) + 1}{\frac{1}{2}n + 1} \cdot x < 1;$$

$$\therefore \quad x > \frac{\frac{1}{2}n}{\frac{1}{2}n + 1}, \text{ and } \frac{\frac{1}{2}n}{\frac{1}{2}n + 1} \cdot x < 1;$$

$$\therefore \quad x > \frac{n}{n+2}, \text{ and } x < \frac{n+2}{n}.$$

Hence x lies between $\dfrac{n}{n+2}$ and $\dfrac{n+2}{n}$.

(ii) Let n be odd. Then the greatest coefficients are obtained when $r = \frac{1}{2}(n-1)$ and $r = \frac{1}{2}(n+1)$.

$\therefore$ the greatest term is either $_nC_{\frac{1}{2}(n-1)} \cdot x^{\frac{1}{2}(n-1)}$ or $_nC_{\frac{1}{2}(n+1)} \cdot x^{\frac{1}{2}(n+1)}$,

$$\therefore \quad \frac{n - \frac{1}{2}(n-1) + 1}{\frac{1}{2}(n-1)} \cdot x > 1; \text{ and } \frac{n - \{\frac{1}{2}(n+1) + 1\} + 1}{\frac{1}{2}(n+1) + 1} \cdot x < 1.$$

Hence x lies between $\dfrac{n-1}{n+3}$ and $\dfrac{n+3}{n-1}$.

Ex. 3. If three consecutive coefficients of $(1+x)^n$ are 6, 15, and 20; find n.

Let the terms be the rth, $(r+1)$th, $(r+2)$th.

Then $_nC_{r-1} = 6$; $_nC_r = 15$; $_nC_{r+1} = 20$.

$$\therefore \quad \frac{n+1}{r} - 1 = \frac{15}{6} = \frac{5}{2}, \text{ and } \frac{n+1}{r+1} - 1 = \frac{20}{15} = \frac{4}{3}. \qquad \text{[See § 313.]}$$

$$\therefore \quad \frac{n+1}{r} = \frac{7}{2}; \text{ and } \frac{n+1}{r+1} = \frac{7}{3} \quad\ldots\ldots\ldots\ldots\ldots\text{(i).}$$

By division, $(r+1)/r = \frac{3}{2}$; whence $r = 2$.

Hence, from (i), $n = 6$.

EXERCISES XXIII.

40. If three consecutive coefficients of $(1+x)^n$ are 35, 21, 7; find n.

41. If three consecutive coefficients of $(1-x)^n$ are -20, 190, and -1140; find n.

42. Of what expansion are 14, 84, 280 and 560 four consecutive terms?

43. Find the binomial expansion of which 4, 7, 7, and $4\frac{3}{8}$, are four consecutive terms?

44. If the 2nd and the 3rd term in $(a+b)^n$ are in the same ratio as the 3rd and 4th in $(a+b)^{n+3}$; find n.

45. If two consecutive coefficients of an expanded binomial are equal, show that the next preceding and the next succeeding coefficients are also equal.

46. If the first three terms of $(a+b)^n$ are in A.P., find n, it being given that $a=2b$.

47. If the rth, $(r+1)$th and $(r+2)$th coefficients of $(1+x)^n$ are in A.P., show that $n^2 - n(4r+1) + 4r^2 - 2 = 0$.

48. Apply the result in the preceding question to show that $n=7$ or 14, if the 5th, 6th and 7th coefficients of $(1+x)^n$ are in A.P.

49. If the 9th, 10th and 11th coefficients of $(1+x)^n$ are in A.P., find n.

50. If a, b, c, d be any consecutive coefficients of an expanded binomial, shew that $(bc+ad)(b-c) = 2(ac^2 - b^2d)$.

51. Find the term with the greatest coefficient in the expansion of
(i) $(1+x)^{75}$; (ii) $(1-x)^{33}$; (iii) $(a+b)^{22}$.

Find the greatest term in the expansion of

52. $(1+x)^7$, when $x=\frac{1}{2}$.   **53.** $(1-2x)^8$, when $x=1$.

54. $(1-\frac{2}{3}x)^8$, when $x=1\frac{1}{2}$.   **55.** $(\frac{1}{2}-\frac{1}{3}x)^4$, when $x=1$.

56. $(2+\frac{2}{7}x)^{11}$, when $x=14$.   **57.** $(a-3x)^9$, when $x=1\frac{1}{2}$, and $a=2$.

58. $(3-2)^n$, when $n=6$.

59. Find the limits between which x must lie if the greatest term of $(1+x)^{72}$ contains the greatest coefficient.

60. Find x if the greatest terms of $(1+x)^{51}$ are the terms containing the greatest coefficients and are equal.

THE BINOMIAL THEOREM.

335. Some other properties of binomial coefficients.

In the remaining articles of this chapter we shall write the Binomial Theorem in the form :—
$$(1 + x)^n = C_0 + C_1 x + C_2 x^2 + \ldots\ldots + C_r x^r + \ldots\ldots + C_n x^n,$$
where C_0, C_1, C_2,......C_n stand respectively for ${}_nC_0$, ${}_nC_1$. ${}_nC_2$......, ${}_nC_n$.

336. The sum of the coefficients of all the terms in the expansion of $(1 + x)^n$ is 2^n.

In the expansion
$$(1 + x)^n = C_0 + C_1 x + C_2 x^2 + \ldots\ldots + C_r x^r + \ldots\ldots + C_n x^n$$
(which is true for all values of x), put $x = 1$; then
$$2^n = C_0 + C_1 + C_2 + \ldots\ldots + C_r + \ldots\ldots + C_n.$$

NOTE. The result of § 318 follows readily from this theorem.

337. The sum of the coefficients of the odd terms = the sum of the coefficients of the even terms = 2^{n-1}.

In the expansion
$$(1 + x)^n = C_0 + C_1 x + C_2 x^2 + \ldots\ldots + C_r x^r + \ldots\ldots + C_n x^n$$
(which is true for all values of x), put $x = -1$.

$\therefore \ 0 = C_0 - C_1 + C_2 - C_3 + \ldots\ldots$

$ = (C_0 + C_2 + C_4 + \ldots\ldots) - (C_1 + C_3 + C_5 + \ldots\ldots)$

$ = $ (sum of odd coefficients) $-$ (sum of even coefficients),

which proves the first result.

Again, since the sum of all the coefficients $= 2^n$

the sum of the odd coefficients,

$=$ the sum of the even coefficients

$= \tfrac{1}{2} \cdot 2^n = 2^{n-1}.$

338. Expansion of Multinomials.

An expression consisting of three or more terms may be raised to any power by the repeated use of the Binomial Theorem.

Ex. 1. Expand $(2+x+x^2)^3$.

$$(2+x+x^2)^3 = (\overline{2+x}+x^2)^3$$
$$= (2+x)^3 + 3(2+x)^2(x^2) + 3(2+x)(x^2)^2 + (x^2)^3$$
$$= (2^3 + 3 \cdot 2^2 \cdot x + 3 \cdot 2 \cdot x^2 + x^3) + 3x^2(4+4x+x^2)$$
$$\quad + 3x^4(2+x) + x^6$$
$$= 8 + 12x + 18x^2 + 13x^3 + 9x^4 + 3x^5 + x^6.$$

Ex. 2. Find the coefficient of x^2 in the expansion of $(1+x+x^2)^5$.

$$(1+x+x^2)^5 = (\overline{1+x}+x^2)^5 = (1+x)^5 + 5(1+x)^4 x^2 + \ldots\ldots$$

It is evident that all the powers of x which appear after the second term are higher than x^2.

Hence the coefficient of x^2 in the expansion of $(1+x+x^2)^5$
$$= \text{that of } x^2 \text{ in } (1+x)^5 + 5(1+x)^4 x^2$$
$$= 10 + 5 = 15.$$

339. Miscellaneous Examples.

Ex. 1. Show that when n is any positive integer except unity, $3^{3n} - 26n - 1$ is divisible by 676.

$$3^{3n} = 27^n = (1+26)^n$$
$$= 1 + 26n + {}_nC_2 \cdot 26^2 + {}_nC_3 \cdot 26^3 + \ldots\ldots + 26^n;$$
$$\therefore \; 3^{3n} - 26n - 1 = {}_nC_2 \cdot 26^2 + {}_nC_3 \cdot 26^3 + \ldots\ldots + 26^n$$
$$= 26^2 \{{}_nC_2 + {}_nC_3 \cdot 26 + \ldots\ldots + 26^{n-2}\}$$
$$= 26^2 \times \text{an integer};$$

which proves the result.

Ex. 2. To prove that, if n be any positive integer, the integral part of $(3+\sqrt{7})^n$ is an odd number.

[The meaning of the example may be easily seen by taking a few simple cases; thus, $3+\sqrt{7}$ lies in value between 5 and 6, and its integral part is therefore the odd number 5. Similarly $(3+\sqrt{7})^2$ lies in value between 31 and 32, and its integral part is therefore the odd number 31; and so on.]

THE BINOMIAL THEOREM.

Let the integral part of $(3+\sqrt{7})^n$ be I, and the remaining part F.

$$\therefore I + F = 3^n + {}_nC_1 \cdot 3^{n-1}\sqrt{7} + {}_nC_2 \cdot 3^{n-2} \cdot 7 + \ldots \ldots \ldots \ldots (i).$$

Now $3 - \sqrt{7}$ is evidently a positive quantity less than 1 and therefore so is $(3-\sqrt{7})^n$. Denote this by G.

$$\therefore G = (3-\sqrt{7})^n = 3^n - {}_nC_1 \cdot 3^{n-1}\sqrt{7} + {}_nC_2 \cdot 3^{n-2} \cdot 7 - \ldots \ldots \ldots (ii).$$

Now adding (i) and (ii) we find that the irrational terms on the right-hand side disappear, and we have

$$I + F + G = 2\{3^n + {}_nC_2 \cdot 3^{n-2} \cdot 7 + {}_nC_4 \cdot 3^{n-4} \cdot 7^2 + \ldots\} = \text{an even integer}.$$

Therefore $F + G = $ an even integer $- I = $ an integer.

Since F and G are each of them <1, $F+G < 2$, and as $F+G$ is an integer, it follows that $F+G=1$.

Again, $I+F+G$ being an even integer, we have $I+1=$ an even integer; that is, I is an odd integer, as was proved.

NOTE. A similar result holds for $(a+\sqrt{b})^n$, if a is the integer next greater than $\sqrt{b}$ so that $a - \sqrt{b}$ is a proper fraction.

Ex. 3. If I be the integral part and F the remaining part of
$$(5\sqrt{2}+7)^{2n+1},$$
show that $F(I+F)=1$, n being a positive integer.

$$(5\sqrt{2}+7)^{2n+1} = (5\sqrt{2})^{2n+1} + {}_{2n+1}C_1 \cdot (5\sqrt{2})^{2n} \cdot 7$$
$$+ {}_{2n+1}C_2 \cdot (5\sqrt{2})^{2n-1} \cdot 7^2 + {}_{2n+1}C_3 \cdot (5\sqrt{2})^{2n-2} \cdot 7^3 + \ldots\ldots$$

And $(5\sqrt{2}-7)^{2n+1} = (5\sqrt{2})^{2n+1} - {}_{2n+1}C_1 \cdot (5\sqrt{2})^{2n} \cdot 7$
$$+ {}_{2n+1}C_2 \cdot (5\sqrt{2})^{2n-1} \cdot 7^2 - {}_{2n+1}C_3 (5\sqrt{2})^{2n-2} \cdot 7^3 + \ldots\ldots$$

By subtraction, we have

$$(5\sqrt{2}+7)^{2n+1} - (5\sqrt{2}-7)^{2n+1} = 2\{{}_{2n+1}C_1 \cdot 50^n \cdot 7 + {}_{2n+1}C_3 \cdot 50^{n-1} \cdot 7^3 + \ldots\}$$
$$= \text{an even integer}.$$

Now $(5\sqrt{2}-7)^{2n+1}$ is evidently a positive quantity less than 1; denote it by G.

$\therefore I + F - G = $ an even integer.

$\therefore F - G = $ an even integer $- I$.

And as F and G are both <1, their difference cannot be an integer, and must therefore $=0$.

$$\therefore F = G.$$
$$\therefore F(I+F) = G(I+F) = (5\sqrt{2}+7)^{2n+1} \times (5\sqrt{2}-7)^{2n+1}$$
$$= (50-49)^{2n+1} = 1.$$

352 THE BINOMIAL THEOREM.

Ex. 4. Sum the series $1 + 2\,_nC_1 + 3\,_nC_2 + \ldots\ldots + (n+1)\,_nC_n$.

The sum $= 1 + 2n + 3\dfrac{n(n-1)}{1\,.\,2} + 4\dfrac{n(n-1)(n-2)}{1\,.\,2\,.\,3} + \ldots\ldots$

$\quad = \left\{ 1 + n + \dfrac{n(n-1)}{1\,.\,2} + \dfrac{n(n-1)(n-2)}{1\,.\,2\,.\,3} + \ldots \text{ to } n+1 \text{ terms} \right\}$

$\quad\quad + \left\{ n + 2\,.\,\dfrac{n(n-1)}{1\,.\,2} + 3\,\dfrac{n(n-1)(n-2)}{1\,.\,2\,.\,3} + \ldots \text{ to } n \text{ terms} \right\}$

$\quad = (1+1)^n + n\left\{ 1 + (n-1) + \dfrac{(n-1)(n-2)}{1\,.\,2} + \ldots \text{ to } n \text{ terms} \right\}$

$\quad = (1+1)^n + n(1+1)^{n-1} = 2^n + n\,.\,2^{n-1} = 2^{n-1}(2+n)$.

Ex. 5. If the coefficients in the expansion of $(1+x)^n$, where n is a positive integer, are C_0, C_1, C_2, ... show that

$$C_0 + \dfrac{C_1}{2} + \dfrac{C_2}{3} + \dfrac{C_3}{4} + \ldots\ldots = \dfrac{2^{n+1}-1}{n+1}.$$

$C_0 = 1$; $C_1 = n$; $C_2 = \dfrac{n(n-1)}{1\,.\,2}$; $C_3 = \dfrac{n(n-1)(n-2)}{1\,.\,2\,.\,3}$; ......

$\therefore C_0 + \dfrac{C_1}{2} + \dfrac{C_2}{3} + \dfrac{C_3}{4} + \ldots\ldots$

$\quad = 1 + \dfrac{n}{2!} + \dfrac{n(n-1)}{3!} + \dfrac{n(n-1)(n-2)}{4!} + \ldots\ldots$

$\quad = \dfrac{1}{n+1}\left\{ (n+1) + \dfrac{(n+1)n}{2!} + \dfrac{(n+1)n(n-1)}{3!} \right.$

$\quad\quad\quad\quad\quad\quad\quad\quad \left. + \dfrac{(n+1)n(n-1)(n-2)}{4!} + \ldots\ldots \right\}$

$\quad = \dfrac{1}{n+1}\left\{ 1 + (n+1) + \dfrac{(n+1)n}{2!} + \ldots\ldots - 1 \right\}$

$\quad = \dfrac{1}{n+1}\{(1+1)^{n+1} - 1\} = \dfrac{2^{n+1}-1}{n+1}.$

Ex. 6. Prove that the sum of the squares of the coefficients of the expansion of $(1+x)^n$ is $(2n)!/(n!)^2$.

$(1+x)^n = C_0 + C_1 x + C_2 x^2 + \ldots\ldots + C_r x^r + \ldots\ldots + C_n x^n$;

and $(1+x)^n = C_n + C_{n-1} x + C_{n-2} x^2 + \ldots\ldots + C_{n-r} x^r + \ldots\ldots + C_0 x^n$

since $C_n = C_0$; $C_1 = C_{n-1}$; $C_2 = C_{n-2}$; ...... $C_r = C_{n-r}$; ......

THE BINOMIAL THEOREM.

Now the coefficient of x^n in the product of the two series
$$= C_0^2 + C_1^2 + C_2^2 + C_3^2 + \ldots\ldots + C_r^2 + \ldots\ldots + C_n^2.$$
Hence $C_0^2 + C_1^2 + C_2^2 + \ldots\ldots + C_n^2$
= the coefficient of x^n in the product of $(1+x)^n$ and of $(1+x)^n$
= the coefficient of x^n in the expansion of $(1+x)^{2n}$
$= (2n)!/(n!)^2$.

Ex. 7.   To prove that
$$_{m+n}C_r = {_mC_r} + {_mC_{r-1}} \cdot {_nC_1} + {_mC_{r-2}} \cdot {_nC_2} + \ldots\ldots + {_nC_r}.$$
$$(1+x)^m = 1 + {_mC_1}x + {_mC_2}x^2 + {_mC_3}x^3 + \ldots\ldots + {_mC_m}x^m;$$
and $\quad(1+x)^n = 1 + {_nC_1}x + {_nC_2}x^2 + {_nC_3}x^3 + \ldots\ldots + {_nC_n}x^n.$

Here the coefficient of x^r in the product of the two series = the coefficient of x^r in the expansion of $(1+x)^{m+n}$.

$\therefore\ {_mC_r} + {_nC_1} \cdot {_mC_{r-1}} + {_nC_2} \cdot {_mC_{r-2}} + \ldots\ldots + {_nC_r} = {_{m+n}C_r}.$

This result is known as *Vandermonde's Theorem* and can easily be deduced from the Theory of Combinations.

Ex. 8.   To find the result arising from multiplying the successive terms of the expansion of $(1+x)^n$ by the corresponding terms of an A.P., whose first term is a, last term l and the number of terms $n+1$.

Let d be the common difference so that
$$l = a + nd;\ \therefore\ d = (l-a)/n.$$
Then the result $= a + n(a+d)x + \dfrac{n(n-1)}{1 \cdot 2}(a+2d)x^2 + \ldots\ldots$

$\qquad= a\left\{1 + nx + \dfrac{n(n-1)}{1 \cdot 2}x^2 + \ldots\ldots\right\}$

$\qquad+ d\left\{nx + 2\dfrac{n(n-1)}{1 \cdot 2}x^2 + 3 \cdot \dfrac{n(n-1)(n-2)}{1 \cdot 2 \cdot 3}x^3 + \ldots\ldots\right\}$

$\qquad= a(1+x)^n + dnx\left\{1 + (n-1)x + \dfrac{(n-1)(n-2)}{1 \cdot 2}x^2 + \ldots\right\}$

$\qquad= a(1+x)^n + (l-a)x(1+x)^{n-1}.$

$\qquad= (1+x)^{n-1}(a + ax + lx - ax) = (1+x)^{n-1}(a+lx).$

EXERCISES XXIII.

61.   Find the sum of the coefficients of $(1+x)^6$.

62.   If n be a positive integer, prove that the middle coefficient of $(1+x)^{2n}$ equals the sum of the squares of the coefficients of $(1+x)^n$.

THE BINOMIAL THEOREM.

If C_0, C_1, C_2......C_n are the coefficients in the expansion of $(1+x)^n$, prove the following relations:

63. $C_0{}^2 - C_1{}^2 + C_2{}^2 - \ldots + (-1)^n C_n{}^2 = 0$ if n is odd, or
$$= (-1)^{\frac{n}{2}} n!/(\tfrac{1}{2}n)!^2 \text{ if } n \text{ is even.}$$

64. $C_1 + 2C_2 + 3C_3 + \ldots + nC_n = n\,2^{n-1}$.

65. $C_1 - 2C_2 + 3C_3 - \ldots + n(-1)^{n-1}C_n = 0$.

66. $C_0 C_r + C_1 C_{r+1} + C_2 C_{r+2} + \ldots + C_{n-r} C_n$
$$= \frac{2n(2n-1)\ldots(n-r+1)}{(n+r)!}.$$

67. $C_0 C_1 + C_1 C_2 + C_2 C_3 + \ldots + C_{n-1} C_n$
$$= \frac{2 . 1 . 2 . 3 \ldots (2n-1)}{(n+1)\{1 . 2 . 3 \ldots (n-1)\}^2}.$$

68. $C_2 + 2C_3 + 3C_4 + \ldots + (n-1)C_n = 1 + (n-2)2^{n-1}$.

69. Show that $49^n + 16n - 1$ is divisible by 64.

70. Prove that if n be any positive integer, the integral part of $(2+\sqrt{3})^n$ is an odd number.

71. If I be the integral part and F the fractional part of
$$(3\sqrt{3}+5)^{2n+1},$$
then shall $\qquad F(1+F) = 2^{2n+1}$.

72. Show that any two consecutive coefficients of the expansion of $(1+x)^n$ are to each other as their distances from the beginning and end respectively.

73. Show that the sum of the products two at a time of the coefficients in the expansion of $(1+x)^n$ is $\tfrac{1}{2}\{2^{2n} - (2n!)/(n!)^2\}$.
[Begin as in Ex. 4, § 227, and apply Ex. 6, § 339.]

74. If n be a prime number, prove that every term in the expansion of $(1+x)^n$, except the first and last, is divisible by n.

75. If $(1+x+x^2)^{3n} = C_0 + C_1 x + C_2 x^2 + C_3 x^3 + \ldots$,
prove that $C_0 - C_1 + C_2 - C_3 + C_4 - \ldots = 1$.

76. If, in the expansion of $(a+x)^n$, s_1 be the sum of the odd terms, and s_2 the sum of the even terms, show that
$$s_1{}^2 - s_2{}^2 = (a^2-x^2)^n, \text{ and } 4s_1 s_2 = (a+x)^{2n} - (a-x)^{2n}.$$

CHAPTER XXIV.

IMAGINARY AND COMPLEX QUANTITIES.

340. Necessity for the introduction of imaginary quantities.

In Elementary Algebra, starting with the conception of positive integral numbers and the four fundamental rules of Arithmetic, viz. Addition, Subtraction, Multiplication and Division, we are first of all led to introduce fractional numbers, so as to be able to apply the operation of Division to any two numbers; we are led also to introduce negative quantities in order that the operation of Subtraction may be applied to any two positive quantities.

Then in like manner we introduce irrational numbers in order that the operation of evolution can be applied to any number whatever.

Now the idea of a negative quantity once being introduced we are just as before led to apply to it all the rules of arithmetic including Evolution, and in order that we may be able to give a meaning to the square root, for example, of a negative quantity we introduce **imaginary quantities**; for the square root of a negative quantity cannot be of the same nature as any of the quantities already introduced which have their squares positive. But further, since

$$\sqrt{ab} = \sqrt{a}\,\sqrt{b},$$

we need only introduce a quantity to represent $\sqrt{-1}$ in order to have an expression for the square root of any negative quantity.

In fact $\quad \sqrt{-a} = \sqrt{a \times (-1)} = \sqrt{a} \times \sqrt{-1}.$

356 IMAGINARY AND COMPLEX QUANTITIES.

341. Solution of quadratic equations.

To show how the square roots of negative quantities do in fact enter into analysis we may take the solution of quadratic equations as follows.

The roots of $ax^2 + bx + c = 0$ are
$$(-b \pm \sqrt{b^2 - 4ac})/2a,$$
and the factors of $ax^2 + bx + c$ are
$$a\left(x - \frac{-b + \sqrt{b^2 - 4ac}}{2a}\right)\left(x - \frac{-b - \sqrt{b^2 - 4ac}}{2a}\right). \quad (\S\ 157)$$

If, in these examples, $b^2 - 4ac$ be positive and an exact square, the roots and factors are easily seen to be ordinary positive or negative numbers. If $b^2 - 4ac$ be positive, but not a square number, the roots and factors will be positive and negative quantities as before, but quantities of which the values cannot be determined exactly, though, as has been explained in § 17, approximate values can be obtained to any degree of accuracy. If, however, $b^2 - 4ac$ be negative, we cannot think of any quantity which when multiplied by itself can produce this, since every conceivable quantity gives a positive result on being raised to an even power. It is on this account that the square root of a negative quantity, like $\sqrt{-5}$, is called an **imaginary** or **impossible** quantity.

In assigning now a meaning to an expression like $\sqrt{-5}$, we have to take care that the meaning that we give is not inconsistent with the fundamental laws proved for other quantities. What we shall therefore understand by an expression like $\sqrt{-5}$ is that, when multiplied by itself by the ordinary rules of Algebra which we shall apply to it *agreeably to the general principles of symbolical algebra*, it produces a negative quantity, which, in this case, is -5.

Quantities which are not imaginary are known as **real**.

DEFINITION. By $\sqrt{-1}$ is meant an expression which, when multiplied by itself, produces -1.

We shall assume that imaginary quantities obey the fundamental laws of algebra quite like any real quantity.

IMAGINARY AND COMPLEX QUANTITIES. 357

342. Properties of purely imaginary quantities.

Purely imaginary quantities like $\sqrt{-a}$ can always be expressed by means of the factor $\sqrt{-1}$ and are best expressed by it. Thus

$$\sqrt{-a} = \sqrt{(a \times -1)} = \sqrt{a}\sqrt{-1}.$$

It thus appears that the single expression $\sqrt{-1}$, which is generally indicated by the symbol ι, is sufficient to denote the imaginary element in all imaginary quantities.

(i) **To prove that $O\iota = O$.**

$$0\iota = (m-m)\iota = m\iota - m\iota = 0.$$

(ii) **If $a\iota = O$, then $a = O$.**

Squaring both sides, we have $-a^2 = 0$, which gives $a = 0$.

(iii) **A real quantity cannot be equal to an imaginary quantity, unless both are zero.**

If possible, suppose $a\iota = b$ where a and b are real quantities not equal to zero.

Then $a\iota . a\iota = b . b$; that is, $-a^2 = b^2$.

Since the square of every real quantity is positive, we thus obtain that a negative quantity equals a positive quantity, which is absurd.

Hence $a\iota$ is not equal to b, which establishes the proposition.

(iv) **Positive integral powers of ι admit only of four different values, of which two are real, ± 1, and the other two imaginary, $\pm \iota$.**

[Thus $\iota^2 = -1$; $\iota^3 = \iota^2 \times \iota = -1 \times \iota = -\iota$; $\iota^4 = (\iota^2)^2 = (-1)^2 = +1$,
$\iota^5 = \iota \times \iota^4 = \iota$; $\iota^6 = \iota^2 \times \iota^4 = -1$; $\iota^7 = \iota^3 . \iota^4 = -\iota$; $\iota^8 = \iota^4\iota^4 = 1.$]

Every integer is exactly divisible by 4, or leaves a remainder 1, 2, or 3 on division by 4. Hence every positive integer is of one of the forms $4m$, $4m+1$, $4m+2$ or $4m+3$ where m is zero or a whole number.

Now, $\quad \iota = \iota$; $\iota^2 = -1$; $\iota^3 = \iota^2 \times \iota = -1 \times \iota = -\iota$;

and $\quad\quad \iota^4 = \iota^2 \times \iota^2 = -1 \times -1 = +1.$

Hence $\quad\quad \iota^{4m} = (\iota^4)^m = (1)^m = +1$;

$\iota^{4m+1} = \iota^{4m} . \iota = (+1) . \iota = \iota$;

$\iota^{4m+2} = \iota^{4m} . \iota^2 = (+1) . (-1) = -1$;

$\iota^{4m+3} = \iota^{4m} . \iota^3 = (+1) . (-\iota) = -\iota.$

Thus the values of positive integral powers of ι recur in a cycle of 4.

[NOTE. The product of $\sqrt{-a}$ and $\sqrt{-b}$ is

$$\iota\sqrt{a} \times \iota\sqrt{b} = \iota^2\sqrt{ab} = -\sqrt{ab}.]$$

358   IMAGINARY AND COMPLEX QUANTITIES.

343. Complex Quantities.—Definition.

An expression of the form $a + b\iota$ where a and b are real* is called a **complex quantity**.

A complex quantity is thus the sum or difference of a real quantity and a purely imaginary quantity.

344. If $a + b\iota = 0$, then $a = 0$ and $b = 0$.

Since $a + b\iota = 0$, $a = -b\iota$. Hence a and b must both be equal to zero, as otherwise a real quantity would be equal to an imaginary quantity. [§ 342, (iii).]

345. If $a + b\iota = a' + b'\iota$, then $a = a'$; and $b = b'$.

By transposition, we have $(a - a') + (b - b')\iota = 0$.

∴ $a - a' = 0$; and $b - b' = 0$. [§ 344.]

Hence $a = a'$; and $b = b'$.

This is called **equating real and imaginary parts.**

NOTE. Hence for two complex quantities to be equal, it is necessary and sufficient that the real parts and the purely imaginary parts are separately equal.

346. *The sum (or difference) of two or more complex quantities is another complex quantity.*

$(a \pm b\iota) + (c \pm d\iota) + (e \pm f\iota)\ldots = (a + c + e + \ldots) \pm (b + d + f + \ldots)\iota.$

NOTE. If $a + c + e + \ldots = 0$, the sum becomes purely imaginary; and if $b + d + f + \ldots = 0$, the sum becomes real.

347. *The product of two or more complex quantities is another complex quantity.*

$(a + b\iota)(c + d\iota) = ac + ad\iota + bc\iota + bd\iota^2 = (ac - bd) + (ad + bc)\iota,$

which is a complex quantity. Denoting this by $p + q\iota$, we have

$$(a + b\iota)(c + d\iota)(e + f\iota) = (p + q\iota)(e + f\iota),$$

which is a complex quantity, being the product of two complex quantities. The proposition may be similarly proved for more factors.

---

* In the remaining portion of this chapter, in every complex quantity of the form $a + b\iota$, a and b will be taken to be real.

IMAGINARY AND COMPLEX QUANTITIES. 359

348. *The quotient of two complex quantities is another complex quantity.*

$$\frac{a+b\iota}{c+d\iota} = \frac{(a+b\iota)(c-d\iota)}{(c+d\iota)(c-d\iota)} = \frac{ac - ad\iota + bc\iota - bd\iota^2}{c^2 - d^2\iota^2}$$

$$= \frac{ac+bd - (ad-bc)\iota}{c^2+d^2} = \frac{ac+bd}{c^2+d^2} - \frac{ad-bc}{c^2+d^2}\iota$$

which proves the proposition.

NOTE. When $ac+bd=0$, the quotient becomes purely imaginary; and when $ad-bc=0$, the quotient becomes real.

349. Conjugate complex quantities.

DEFINITION. Two complex quantities are said to be **conjugate,** when they differ only in the sign of the imaginary part. Thus $a+b\iota$ and $a-b\iota$ are conjugate.

350. The sum and the product of two conjugate complex quantities are both real.

For $\qquad (a+b\iota) + (a-b\iota) = 2a,$
and $\qquad (a+b\iota)(a-b\iota) = a^2 - b^2\iota^2 = a^2 + b^2.$

351. *If the sum and the product of two complex quantities are both real, the quantities are conjugate.*

Let $\qquad (a+b\iota) + (c+d\iota) = e$ ....(i),
and $\qquad (a+b\iota)(c+d\iota) = f$ ....(ii),
where e and f are real.

Then $\qquad (a+c) + (b+d)\iota = e + 0\iota$......(iii). [From (i).]
And $\quad (ac-bd) + (ad+bc)\iota = f + 0\iota$...(iv). [From (ii).]
Hence $b+d=0$, which gives $b = -d$....
And $\qquad ad+bc = 0 \qquad$ [From (iii), and § 345.]
$\qquad\qquad\qquad\qquad\qquad$ [From (iv), and § 345.]

Substituting in the last equation $-d$ for b, we have $ad - cd = 0$, which makes $d = 0$; or $a - c = 0$.

If $d=0$, then $b=0$, and the given quantities are real.

If $a-c=0$, $a=c$. And as $b=-d$, $c+d\iota=a-b\iota$, which shows that $a+b\iota$ and $c+d\iota$ are conjugate.

Ex. 1. Resolve 5 into a pair of imaginary factors.
$$5 = 4+1 = 4-(-1) = (2+\sqrt{-1})(2-\sqrt{-1}) = (2+\iota)(2-\iota).$$

Ex. 2. Factorise completely $a^4+a^2x^2+x^4$ using surd and imaginary coefficients.

$$a^4+a^2x^2+x^4 = (a^4+2a^2x^2+x^4) - a^2x^2$$
$$= (a^2+x^2)^2 - (ax)^2 = (a^2+x^2+ax)(a^2+x^2-ax)$$
$$= (a^2+x^2+2ax-ax)(a^2+x^2-2ax+ax)$$
$$= \{(a+x)^2 - (\sqrt{(ax)})^2\}\{(a-x)^2 - (\sqrt{(ax)}\cdot\iota)^2\}$$
$$= \{a+x+\sqrt{(ax)}\}\{a+x-\sqrt{(ax)}\}\{a-x+\sqrt{(ax)}\cdot\iota\}\{a-x-\sqrt{(ax)}\cdot\iota\}.$$

Ex. 3. Remove* ι from the denominator of $\dfrac{2}{4+3\iota}$.

$$\frac{2}{4+3\iota} = \frac{2(4-3\iota)}{(4+3\iota)(4-3\iota)} = \frac{8-6\iota}{(4)^2-(3\iota)^2} = \frac{8-6\iota}{16+9} = \frac{1}{25}(8-6\iota).$$

Ex. 4. If $a^2+b^2=0$ and a and b are real, show that $a=b=0$.

Here if a^2 be positive, then b^2 must be negative so that a^2+b^2 may vanish and b must therefore be imaginary; and if a^2 be negative, a itself will become imaginary. Hence $a^2=0$ and therefore also $b^2=0$. Therefore $a=0$ and $b=0$.

It may be similarly shown that, if $a^2+b^2+c^2=0$ and a, b, c are real, then $a=b=c=0$, and generally if the sum of the squares of any number of real quantities is zero, each of the quantities is zero.

Ex. 5. If $x^2+y^2+z^2 = xy+yz+zx$; and x, y and z are real, show that $x=y=z$.

$$2x^2+2y^2+2z^2-2yz-2zx-2xy = 0.$$
$$\therefore (y-z)^2 + (z-x)^2 + (x-y)^2 = 0.$$

Hence $y-z=0$, $z-x=0$, $x-y=0$, which gives $x=y=z$.

Ex. 6. *Show that unity has three cube roots and that two of them are complex quantities.*

If we solve the equation $x^3=1$, the roots of this equation will be the cube roots required.

Now, $\qquad x^3-1=0, \quad \therefore (x-1)(x^2+x+1) = 0.$

---

* Prof. Chrystal suggests that the name 'realise' might be given to this process. Compare the name with 'rationalise.'

IMAGINARY AND COMPLEX QUANTITIES.

$\therefore$ either $\quad x-1=0$; which gives $x=1$;
or $x^2+x+1=0$ which gives $x=\tfrac{1}{2}\{-1\pm\sqrt{(1-4)}\}=\tfrac{1}{2}(-1\pm\sqrt{-3})$.
We thus find that unity has three cube roots, which are
$$1,\ \tfrac{1}{2}(-1+\sqrt{3}.\iota),\ \tfrac{1}{2}(-1-\sqrt{3}.\iota),$$
of which the last two are complex quantities.

NOTE 1. The two complex cube roots of unity are generally denoted by ω_1 and ω_2. It will be found that, if we square either of these, we obtain the other, and the cube roots of unity may, therefore, be represented by $1, \omega$ and ω^2, where ω is either of the imaginary cube roots of unity.

NOTE 2. Since the sum of the three cube roots of unity
$$=1+\tfrac{1}{2}(-1+\iota\sqrt{3})+\tfrac{1}{2}(-1-\iota\sqrt{3})=0,$$
we get $\quad 1+\omega+\omega^2=0.$

NOTE 3. Since $\omega\,.\,\omega^2=\omega^3=1$, it is clear that the product of the two imaginary cube roots of unity is unity.

Ex. 7. Find the value of $(1+\omega)^3+(1-\omega-\omega^2)^3$, where ω is either of the imaginary cube roots of unity.

The expression $=(1+\omega+\omega^2-\omega^2)^3+\{2-(1+\omega+\omega^2)\}^3$
$\qquad=(0-\omega^2)^3+(2-0)^3 \qquad$ [See Note 2 above.]
$\qquad=(-\omega^2)^3+2^3=-\omega^6+8=-1+8=7.$

Ex. 8. Show that the successive positive integral powers of ω recur in a cycle of 3.

Every integer is exactly divisible by 3 or leaves a remainder 1 or 2 on division by 3. Hence every positive integer is of the form $3r$, $3r+1$ or $3r+2$, where r is zero or a whole number.

Now $\qquad\qquad \omega^{3r}=(\omega^3)^r=1,$
$\qquad\qquad\qquad \omega^{3r+1}=\omega^{3r}\,.\,\omega=\omega\,;$
and $\qquad\qquad \omega^{3r+2}=\omega^{3r}\,.\,\omega^2=\omega^2.$

Thus the values recur in a cycle and are $1, \omega, \omega^2$.

Ex. 9. Resolve $a^2+b^2+c^2-bc-ca-ab$ into factors.

Since $\omega+\omega^2=\omega^4+\omega^2=-1$, and $\omega^3=1,$
the expression $=a^2+\omega^3 b^2+\omega^3 c^2+(\omega^4+\omega^2)\,bc+(\omega+\omega^2)\,ca+(\omega+\omega^2)\,ab$
$\qquad =a\,(a+\omega^2 b+\omega c)+\omega b\,(a+\omega^2 b+\omega c)+\omega^2 c\,(a+\omega^2 b+\omega c)$
$\qquad =(a+\omega b+\omega^2 c)\,(a+\omega^2 b+\omega c).$

NOTE. Hence
$$a^3+b^3+c^3-3abc=(a+b+c)\,(a+\omega b+\omega^2 c)\,(a+\omega^2 b+\omega c).$$

352. Square roots of complex quantities.—
To find the square root of $a + b\iota$.

Let $\sqrt{(a + b\iota)} = x + y\iota$.

By squaring, $a + b\iota = x^2 - y^2 + 2xy\iota$.

Equating possible and impossible parts, we have
$$x^2 - y^2 = a \quad \text{...........................(i)},$$
and
$$2xy = b \quad \text{...........................(ii)}.$$

By squaring and adding (i) and (ii), we have
$$(x^2 + y^2)^2 = a^2 + b^2,$$
$$\therefore x^2 + y^2 = +\sqrt{(a^2 + b^2)}* \quad \text{............(iii)}.$$
$$\therefore x^2 = \tfrac{1}{2}\{\sqrt{(a^2 + b^2)} + a\},$$
and $\qquad y^2 = \tfrac{1}{2}\{\sqrt{(a^2 + b^2)} - a\},$

[From (i) and (iii).]

$$\therefore x = \pm\sqrt{[\tfrac{1}{2}\{\sqrt{(a^2 + b^2)} + a\}]};$$
and $\qquad y = \pm\sqrt{[\tfrac{1}{2}\{\sqrt{(a^2 + b^2)} - a\}]} \quad \text{.........(iv)}.$

Since $2xy = b$, if b be positive xy is positive, and therefore x and y must be taken to be of the same sign; and if b be negative, xy is negative, and x and y must therefore be taken to be of opposite signs. Hence

$$\sqrt{(a + b\iota)} = +\sqrt{[\tfrac{1}{2}\{\sqrt{(a^2 + b^2)} + a\}]} + \iota\sqrt{[\tfrac{1}{2}\{\sqrt{(a^2 + b^2)} - a\}]}$$
or $\qquad = -\sqrt{[\tfrac{1}{2}\{\sqrt{(a^2 + b^2)} + a\}]} - \iota\sqrt{[\tfrac{1}{2}\{\sqrt{(a^2 + b^2)} - a\}]},$

if b be positive;

$$= +\sqrt{[\tfrac{1}{2}\{\sqrt{(a^2 + b^2)} + a\}]} - \iota\sqrt{[\tfrac{1}{2}\{\sqrt{(a^2 + b^2)} - a\}]}$$
or $\qquad = -\sqrt{[\tfrac{1}{2}\{\sqrt{(a^2 + b^2)} + a\}]} + \iota\sqrt{[\tfrac{1}{2}\{\sqrt{(a^2 + b^2)} - a\}]},$

if b be negative.

Note. The values of x and y are real since $a^2 + b^2$ is greater than a^2, and $\sqrt{(a^2 + b^2)}$ is therefore greater than a. Hence this result shows that *the square root of a complex quantity is a complex quantity*.

---

* Since x and y are real, $x^2 + y^2$ is positive, and the negative sign should not therefore be prefixed to $\sqrt{(a^2 + b^2)}$.

IMAGINARY AND COMPLEX QUANTITIES.

Ex. 1. Find the square root of $3+4\iota$.

Let $$\sqrt{(3+4\iota)} = x+y\iota.$$
$$\therefore 3+4\iota = x^2 - y^2 + 2xy\iota.$$
$$\therefore x^2 - y^2 = 3 \dots \dots \dots \dots \dots \dots \dots \dots \dots \dots \dots \text{(i)},$$
and $$2xy = 4 \dots \dots \dots \dots \dots \dots \dots \dots \dots \text{(ii)}.$$

Squaring and adding (i) and (ii), we have
$$(x^2+y^2)^2 = 25. \quad \therefore x^2+y^2 = 5.$$
But $$x^2 - y^2 = 3.$$
Hence $$x^2 = 4 \text{ and } y^2 = 1. \quad \therefore x = \pm 2;\ y = \pm 1.$$

Since $2xy = 4$, we must have x and y of the same sign, and we therefore take $x = 2,\ y = 1$; or $x = -2,\ y = -1$.

Hence the square root required equals $2+\iota$ or $-2-\iota$.

Ex. 2. Extract the square root of ι.

Let $$\sqrt{\iota} = x+y\iota. \quad \therefore \iota = x^2 - y^2 + 2xy\iota.$$
$$\therefore x^2 - y^2 = 0;\ \text{and}\ 2xy = 1. \quad \therefore x^2 + y^2 = +1.$$
Hence $$x^2 = \tfrac{1}{2},\ \text{and}\ y^2 = \tfrac{1}{2}.$$
$$\therefore x = \pm\sqrt{\tfrac{1}{2}} = \pm\tfrac{1}{2}\sqrt{2};\ \text{and}\ y = \pm\tfrac{1}{2}\sqrt{2}.$$

And since $2xy$ is positive, x and y must be of the same sign.

Hence the required square root equals
$$+\tfrac{1}{2}\sqrt{2} + \tfrac{1}{2}\sqrt{2}\iota,\ \text{or}\ -\tfrac{1}{2}\sqrt{2} - \tfrac{1}{2}\sqrt{2}\iota = \pm\tfrac{1}{2}\sqrt{2}(1+\iota).$$

353. The modulus of a complex quantity.

DEFINITION. The **modulus of** $a+b\iota$ is the positive value of the square root of $a^2 + b^2$. It is generally written **mod $(a+b\iota)$**.

Thus $$\text{mod}\,(12+5\iota) = +\sqrt{(12^2+5^2)} = 13;$$
$$\text{mod}\,(5) = \text{mod}\,(5+0\iota) = +\sqrt{(5^2+0^2)} = 5;$$
and $$\text{mod}\,(-5) = +\sqrt{\{(-5)^2\}} = 5.$$

Hence *the modulus of a real quantity is its absolute numerical value.*

Since $\text{mod}\,(a-b\iota) = +\sqrt{\{a^2+(-b)^2\}} = +\sqrt{(a^2+b^2)}$
$$= \text{mod}\,(a+b\iota) = +\sqrt{\{(a+b\iota)(a-b\iota)\}},$$

it follows that *a complex quantity and its conjugate have the same modulus, which is the positive value of the square root of their product.*

354. If a complex quantity vanishes, its modulus vanishes; and conversely a complex quantity vanishes if its modulus vanishes.

(i) If $a + b\iota = 0$, then $a = 0$, and $b = 0$. [§ 344.]
$$\therefore \sqrt{(a^2 + b^2)} = 0,$$
which establishes the first part of the proposition.

(ii) If $\sqrt{(a^2 + b^2)} = 0$, then $a^2 + b^2 = 0$.
$$\therefore a = 0 \text{ and } b = 0. \quad [\S 351, \text{Ex. 4.}]$$
$\therefore a + b\iota = 0$, which proves the second part.

355. The modulus of the product of two complex quantities is the product of their moduli.

Let $a + b\iota$ and $c + d\iota$ be two complex quantities.

The product of these quantities
$$= (a + b\iota)(c + d\iota) = (ac - bd) + (bc + ad)\iota.$$

$\therefore$ The modulus of the product of these quantities
$$= \sqrt{\{(ac - bd)^2 + (bc + ad)^2\}}$$
$$= \sqrt{(a^2c^2 + b^2d^2 - 2abcd + b^2c^2 + a^2d^2 + 2abcd)}$$
$$= \sqrt{\{c^2(a^2 + b^2) + d^2(a^2 + b^2)\}} = \sqrt{\{(a^2 + b^2)(c^2 + d^2)\}}$$
$$= \sqrt{(a^2 + b^2)} \times \sqrt{(c^2 + d^2)}$$
$$= \text{the product of the moduli of } a + b\iota \text{ and } c + d\iota.$$

It can hence be easily proved that the modulus of the product of three (or more) complex quantities is the product of their moduli.

356. The modulus of the quotient of two complex quantities is the quotient of their moduli.

This follows from § 355. It can be directly proved thus:

$$\text{Mod}\left(\frac{a + b\iota}{c + d\iota}\right) = \text{mod}\,\frac{(a + b\iota)(c - d\iota)}{(c + d\iota)(c - d\iota)}$$
$$= \text{mod}\,\frac{\{(ac + bd) + (bc - ad)\iota\}}{c^2 + d^2} = \frac{\sqrt{\{(ac + bd)^2 + (bc - ad)^2\}}}{\sqrt{(c^2 + d^2)^2}}$$

IMAGINARY AND COMPLEX QUANTITIES.   365

$$= \frac{\sqrt{(a^2c^2 + b^2d^2 + b^2c^2 + a^2d^2)}}{c^2 + d^2} = \frac{\sqrt{(a^2 + b^2)}\sqrt{(c^2 + d^2)}}{c^2 + d^2}$$

$$= \frac{\sqrt{(a^2 + b^2)}}{\sqrt{(c^2 + d^2)}} = \frac{\text{mod}(a + b\iota)}{\text{mod}(c + d\iota)}.$$

Ex. Find the modulus of $\frac{(2+\iota)(3+\iota)}{1+\iota}$.

The required modulus $= \text{mod}(2+\iota) \times \text{mod}(3+\iota) \div \text{mod}(1+\iota)$
$$= \sqrt{(2^2+1^2)}\sqrt{(3^2+1^2)} \div \sqrt{(1^2+1^2)}$$
$$= \sqrt{5}\sqrt{10} \div \sqrt{2} = \sqrt{5}\sqrt{5} = 5.$$

357. Geometrical interpretation of $\sqrt{-1}$.

Let AOA' and BOB' be two straight lines at right angles to each other, AOA' being drawn from right to left. Take OM_1 in OA equal to a unit of length, and take OM_2 and OM_3 in OA' and OB respectively equal to OM_1 in magnitude. Then OM_1 being equal to $+1$, $OM_2 = -1$. To

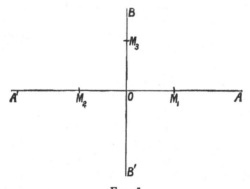

Fig. 1.

multiply OM_1 by 2 is to take twice the line without reversing its direction. To multiply OM_1 by -3 is to take thrice the line and also to reverse its direction. To multiply OM_1 by $(-3)^2$ is to take three times OM_1 and reverse its direction and then repeat the process which, in effect, will be to take OM_1 nine times without changing the direction.

IMAGINARY AND COMPLEX QUANTITIES.

Let us now try to make out whether any meaning can be assigned to the operation of multiplying OM_1 by $\sqrt{-1}$. The meaning of multiplying a magnitude like OM_1 by $\sqrt{-1}$ is suggested by the fact that two successive multiplications by $\sqrt{-1}$ produce the same result as a single multiplication by -1. Let us suppose that OM_1 revolves from right to left, that is, in the direction opposite to that in which the hands of a clock move. If, according to the usual convention, this rotation which we shall call counter-clockwise be taken to be positive, clockwise rotation, that is, revolving from left to right, must be considered to be negative. If now OM_1 revolves counter-clockwise through *one* right angle, it occupies the position OM_3; if from this position it revolves in the same direction through one right angle more, it becomes OM_2, which means that its direction becomes reversed, its magnitude remaining unaltered; that is, it becomes multiplied by -1.

(1) Hence the meaning of multiplying any length OR by i is to rotate this length counter-clockwise through *one* right angle, since the same amount of further rotation reverses its direction without changing its magnitude. Similarly to multiply OR by $-i$ is to rotate it clockwise through one right angle.

Since the effect of multiplying by $-i$ OR, which we shall suppose contains a units of length, is the same as that of multiplying it by i, and *reversing* its direction, we have

$$a \times -i = -ai.$$

(2) The result of dividing OR by i is obtained by rotating OR clockwise through one right angle, since, when this quotient is rotated counter-clockwise through one right angle—that is, multiplied by i—the original direction is got back. Similarly to divide OR by $-i$ is to rotate it counter-clockwise through one right angle, that is, to multiply it by i. Hence we have

$$a/i = a \times -i; \text{ and } a/-i = a \times i.$$

(3) Since the effect of rotating a units of length counter-clockwise through one right angle is the same as that of rotating the unit of length counter-clockwise

through one right angle, and then multiplying it by a, we have
$$ai = ia.$$

(4) If we rotate a units of length counter-clockwise through one right angle, rotate b units of length counter-clockwise through one right angle and add the two results, the effect is the same as that of rotating $a+b$ units of length counter-clockwise through one right angle. Hence we have
$$ai + bi = (a+b)i.$$
Similarly $\quad ai - bi = (a-b)i.$

(5) To multiply ai by bi, we do to ai what is done to the unit to get bi; that is, we multiply ai by b and then rotate the result obtained counter-clockwise through one right angle, the final result being ab units rotated through two right angles. The same result will be obtained if we multiply bi by ai. Hence we have
$$ai \times bi = bi \times ai = -ab = abii.$$
It can be similarly shown that
$$\frac{a}{i} \times \frac{b}{i} = \frac{b}{i} \times \frac{a}{i} = -ab = \frac{ab}{ii}.$$

(6) To divide ai by bi, we divide ai by b and then rotate the result obtained clockwise through one right angle, the final result being a/b units in the original direction. The same result will be obtained if we divide a/i by b/i. Hence
$$\frac{ai}{bi} = \frac{a}{b} = \frac{a/i}{b/i}.$$

The following results relating to purely imaginary quantities which we proved in § 342 can also be immediately deduced from the figure:—

(i) $\quad 0i = 0.$ $\qquad$ (ii) If $ai = 0$, then $a = 0$.

(iii) A real quantity cannot equal an imaginary quantity, since they are represented by lines in different directions.

(iv) Positive integral powers of i admit only of four different values, ± 1 and $\pm i$.

358. Graphic representation of Complex Numbers.

If, in the adjoining figure, OM represents a units of length and MP b units, we can get from O to P by first moving forwards a units from O to M and then, turning counter-clockwise through one right angle, move units b from M to P.

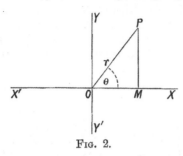

Fig. 2.

If now the angle MOP contains θ units of angle, and OP r units of length,

$$OM = r \cos \theta,$$
$$MP = r \sin \theta,$$

so that $\qquad a + bi = r (\cos \theta + i \sin \theta),$

where $r(\cos \theta + i \sin \theta)$ may be taken as showing that, to get from O to P, the unit of length in the original direction OX has to be rotated through an angle θ and altered in length in the ratio $r : 1$.

Since $r^2 = a^2 + b^2$, the quantity r which is taken to be positive is the **modulus** of $a + bi$, and the angle θ is called the **argument** of the complex number.

The argument of $a + bi$ may be denoted by

$$\arg. (a + bi).$$

If the modulus vanishes, that is, P coincides with O, then both a and b vanish, and hence the complex quantity vanishes. The converse of this proposition is also evidently true.

If two complex numbers $a + bi$ and $a' + b'i$ are equal,—that is, are represented by the same radius vector OP, **vector** being a directed straight line involving the ideas of length and direction—then $a = a'$ and $b = b'$, since the horizontal components of the vector must be equal to each other as also the vertical components.

IMAGINARY AND COMPLEX QUANTITIES.   369

359. Addition and subtraction of complex numbers.

Let the radius vector OP represent $a + bi$, and the vector OP', $a' + b'i$. From P draw PP'' parallel and equal to OP' and join OP''. Draw PM, $P'M'$ and $P''M''$ perpendicular to $X'X$ and PR parallel to $X'X$. Then
since $OM' = PR = MM''$,
and $P'M' = P''R$,
the sum of the horizontal components of OP and OP' equals the horizontal component of OP'', and the sum of the vertical

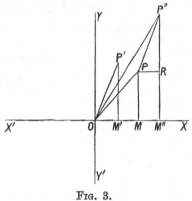

Fig. 3.

components of OP and OP' equals the vertical component of OP''. Hence OP'' represents the sum of the two complex numbers, so that

$$OP'' = \text{mod } \{(a + a') + (b + b') i\};$$
and $$XOP'' = \text{arg. } \{(a + a') + (b + b') i\}.$$

COR. It follows from this that the sum of two conjugate complex numbers is real.

Since OP'' is less than $OP + PP''$, we deduce that the modulus of the sum of two complex quantities is less than the sum of their moduli.

The addition of any number of complex numbers and the subtraction of complex numbers can be similarly represented. and it can be easily shown that

(i) *the sum or difference of any number of complex numbers is a complex number,*

and (ii) *the modulus of the sum of any number of complex numbers is less than the sum of their moduli.*

TUT. ALG. II.   24

360. Multiplication and division of complex numbers.

Let it be required to find the product of $a + bi$ and $a' + b'i$. Representing these complex numbers by

$$r_1 (\cos \theta_1 + i \sin \theta_1) \text{ and } r_2 (\cos \theta_2 + i \sin \theta_2),$$

we find that, if we rotate the unit of length through an angle θ_1, and alter it in length in the ratio of $r_1 : 1$, and then rotate the result thus obtained through an angle θ_2 and alter it in length in the ratio $r_2 : 1$, the effect will be the same as that of rotating the unit of length through an angle $\theta_1 + \theta_2$ and altering it in length in the ratio $r_1 r_2 : 1$. Hence
$$r_1 (\cos \theta_1 + i \sin \theta_1) r_2 (\cos \theta_2 + i \sin \theta_2)$$
$$= r_1 r_2 \{\cos (\theta_1 + \theta_2) + i \sin (\theta_1 + \theta_2)\}.$$

The result can also be obtained as follows:

The product
$$= r_1 r_2 \{\cos \theta_1 \cos \theta_2 - \sin \theta_1 \sin \theta_2 + i (\sin \theta_1 \cos \theta_2 + \cos \theta_1 \sin \theta_2)\}$$
$$= r_1 r_2 \{\cos (\theta_1 + \theta_2) + i \sin (\theta_1 + \theta_2)\}.$$

Similarly
$$r_1 (\cos \theta_1 + i \sin \theta_1) \, r_2 (\cos \theta_2 + i \sin \theta_2) \, r_3 (\cos \theta_3 + i \sin \theta_3)$$
$$= r_1 r_2 r_3 \{\cos (\theta_1 + \theta_2 + \theta_3) + i \sin (\theta_1 + \theta_2 + \theta_3)\}$$
and
$$r_1 (\cos \theta_1 + i \sin \theta_1) \, r_2 (\cos \theta_2 + i \sin \theta_2) \ldots r_n (\cos \theta_n + i \sin \theta_n)$$
$$= r_1 r_2 \ldots r_n \{\cos (\theta_1 + \theta_2 + \ldots + \theta_n) + i \sin (\theta_1 + \theta_2 + \ldots + \theta_n)\}*.$$

Thus the product of any number of complex numbers is a complex number whose modulus is the product of their moduli, and whose argument is the sum of their arguments.

The result
$$r_1 (\cos \theta_1 + i \sin \theta_1) \, r_2 (\cos \theta_2 + i \sin \theta_2)$$
$$= r_1 r_2 \{\cos (\theta_1 + \theta_2) + i \sin (\theta_1 + \theta_2)\}$$

---

* If, in this equality, we put $r_1 = r_2 = \ldots\ldots = r_n$, and $\theta_1 = \theta_2 = \ldots\ldots = \theta_n = \theta$, we get that $(\cos \theta + i \sin \theta)^n = \cos n\theta + i \sin n\theta$, a result which is known as *Demoivre's Theorem*.

IMAGINARY AND COMPLEX QUANTITIES. 371

shows that the quotient of two complex numbers is a complex number whose modulus is the quotient of their moduli, and whose argument is the difference of their arguments.

361. No New Quantities need be introduced to enable us to perform the ordinary operations on complex numbers.

It is important to notice that the application of the operations of arithmetic to complex quantities does not necessitate the introduction of any further quantities of a new kind.

As regards Addition, Subtraction, Multiplication and Division this has been already proved. (§§ 359, 360.)

We shall now show that the nth root of a complex number is a complex number.

Suppose the number is $a + bi$, then we have to solve the equation
$$x^n = a + bi,$$
x being an integer.

If we put $b/a = \tan \theta$ and $a^2 + b^2 = r^2$ this becomes
$$x^n = r(\cos \theta + \iota \sin \theta),$$
and *one* value of x satisfying this equation is given by
$$x = r^{\frac{1}{n}}(\cos \theta/n + \iota \sin \theta/n),$$
for by Demoivre's Theorem
$$(\cos \theta/n + \iota \sin \theta/n)^n = \cos \theta + \iota \sin \theta,$$
and this is a complex number of the usual type.

The student will now see the importance of this introduction, the set of numbers we consider has now a completeness it entirely lacked before.

We may add here that it can be proved that all the roots of an equation whose coefficients are complex numbers are themselves complex numbers, but the proof is much too difficult for the present work.

362. The nth roots of unity.

We have seen that one nth root of a complex quantity is also a complex quantity, but inasmuch as the equation
$$x^n = a + bi$$
must have n roots we infer that every quantity has n nth roots.

We proceed to consider more particularly the nth roots of unity, i.e. the roots of the equation
$$x^n = 1.$$
Since $\left(\cos\dfrac{2r\pi}{n} + \iota \sin\dfrac{2r\pi}{n}\right)^n = \cos 2r\pi + \iota \sin 2r\pi = 1$ (§ 350)
when r is any integer, we infer that
$$\cos\frac{2r\pi}{n} + \iota \sin\frac{2r\pi}{n}$$
is an nth root of unity for all integral values of n.

Moreover *all the nth roots are obtained by giving r the values*
$$0, 1, 2 \ldots n-1.$$

For (i) this gives us n expressions, each satisfying the equation
$$x^n = 1,$$
and (ii) they are all different, for if
$$\cos\frac{2p\pi}{n} + \iota \sin\frac{2p\pi}{n} = \cos\frac{2q\pi}{n} + \iota \sin\frac{2q\pi}{n},$$
where p and q are each less than n, then equating real and imaginary parts we have
$$\cos\frac{2p\pi}{n} = \cos\frac{2q\pi}{n}$$
$$\sin\frac{2p\pi}{n} = \sin\frac{2q\pi}{n}.$$

$\therefore$ the angles $\dfrac{2p\pi}{n}$, $\dfrac{2q\pi}{n}$ are coterminal, and since each lies between 0 and 2π they are equal, $\therefore p = q$. See Tut. Trig. § 80.

IMAGINARY AND COMPLEX QUANTITIES. 373

Hence the nth roots of unity are the n quantities

$$1,\ \cos\frac{2\pi}{n}+\iota\sin\frac{2\pi}{n},\ \cos\frac{4\pi}{n}+\iota\sin\frac{4\pi}{n}\ \ldots\ \cos\frac{(n-1)2\pi}{n}+\iota\sin\frac{(n-1)2\pi}{n}.$$

Thus, for example, the cube roots are

$$1,\ \cos\frac{2\pi}{3}+\iota\sin\frac{2\pi}{3},\ \cos\frac{4\pi}{3}+\iota\sin\frac{4\pi}{3},$$

i.e. $1,\ \dfrac{-1+\iota\sqrt{3}}{2},\ \dfrac{-1-\iota\sqrt{3}}{2}.$ [§ 351.]

The fourth roots are

$$1,\ \cos\frac{2\pi}{4}+\iota\sin\frac{2\pi}{4},\ \cos\frac{4\pi}{4}+\iota\sin\frac{4\pi}{4},\ \cos\frac{6\pi}{4}+\iota\sin\frac{6\pi}{4}$$

i.e. $1,\quad \iota,\quad -1,\quad -\iota.$

363. Geometrical Representation of the nth roots of unity.

Take a circle $ABA'B'$ of unit radius having its centre at the origin, then a point P on this circle represents (§ 8) the quantity

$$\cos\theta + \iota\sin\theta$$

where θ is the $\angle AOP$ measured as usual, and the point A represents unity.

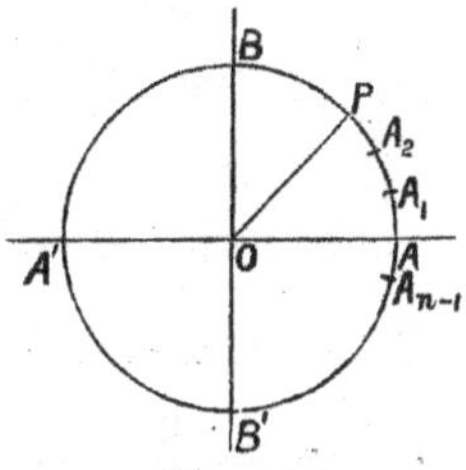

Now divide the circumference into n equal parts, starting from A, and let $A_1 A_2 \ldots A_{n-1}$ be the remaining points of section.

Then

$$\angle AOA_1 = \frac{2\pi}{n},\ \ AOA_2 = \frac{4\pi}{n}\ \text{etc.}$$

FIG. 4.

∴ the points $A, A_1, A_2 \ldots A_{n-1}$ represent the nth roots of unity.

Thus, for example, the points $ABA'B'$ represent the fourth roots of unity.

EXERCISES XXIV.

1. Multiply $2+3\iota$ by $2+\iota$.
2. Multiply $3\iota^{\frac{2}{3}}+3\iota^{\frac{1}{3}}$ by $2\iota^{\frac{2}{3}}+2\iota^{\frac{1}{3}}$.
3. Divide $\dfrac{c+d\iota}{c-d\iota}$ by $\dfrac{c-d\iota}{c+d\iota}$.
4. Find the value of $(\iota-1)^3$.
5. Write down the square, cube and fourth powers of $a+b\iota$, and show that they are of the same form.
6. Simplify $(3+\iota)^4+(3-\iota)^4$.
7. Find the fourth power of $-\sqrt{\{-3\sqrt{(-4)}\}}$.
8. Show that $\{\tfrac{1}{2}(-1+\iota\sqrt{3})\}^2+\{\tfrac{1}{2}(-1-\iota\sqrt{3})\}^2=-1$.
9. Resolve 17 into a pair of factors.
10. Simplify $(x-1)(x+\tfrac{1}{2}-\tfrac{1}{2}\iota\sqrt{3})(x+\tfrac{1}{2}+\tfrac{1}{2}\iota\sqrt{3})$.
11. Resolve x^4+x^2+1 into four factors of the first degree.
12. Resolve x^4-y^4 into four factors of the first degree.
13. Show that $\dfrac{1}{(1-\iota)^2}-\dfrac{1}{(1+\iota)^2}=\iota$.
14. Prove that
    (i) if $2(a^2+b^2)=(a+b)^2$, then $a=b$.
    (ii) if $3(a^2+b^2+c^2)=(a+b+c)^2$, then $a=b=c$.
    (iii) if $4(a^2+b^2+c^2+d^2)=(a+b+c+d)^2$, then $a=b=c=d$.
15. (i) Show that, if $a=\tfrac{1}{2}(-1+\iota\sqrt{3})$,
    $$a^{-1}=\tfrac{1}{2}(-1-\iota\sqrt{3}).$$
    (ii) Show that $\dfrac{1+2\iota+3\iota^2}{1-2\iota+3\iota^2}=-\iota$.

Realise the denominators of the following four fractions:

16. $\dfrac{4+\sqrt{-2}}{4-\sqrt{-2}}$.   17. $\dfrac{26}{5+3\sqrt{-3}}$.   18. $\dfrac{1}{(1+2\iota)^2}$.

19. $\dfrac{a+b\iota}{a-b\iota}-\dfrac{a-b\iota}{a+b\iota}$.

20. Express $\dfrac{(1+\sqrt{-1})^2}{1-\sqrt{-1}}$ in the form of $a+b\iota$.

IMAGINARY AND COMPLEX QUANTITIES.   375

21. Show that, if a and b are real, $a^2 + b^2$ is not less than $2ab$.

Find the square roots of

22. $-1 + 2\sqrt{-2}$.
23. $-1 + 4\sqrt{-5}$.
24. 2ι.
25. -2ι.
26. $-\iota$.
27. $2a + 2\sqrt{(a^2 + b^2)}$.
28. Prove that $\sqrt{(4 + 3\iota\sqrt{20})} + \sqrt{(4 - 3\iota\sqrt{20})} = 6$.
29. Prove that $\sqrt{\iota} + \sqrt{-\iota} = \sqrt{2}$ and that $\sqrt{\iota} - \sqrt{-\iota} = \iota\sqrt{2}$.
30. Show that $\dfrac{1}{\sqrt{(3 + 4\iota)}} + \dfrac{1}{\sqrt{(3 - 4\iota)}} = \dfrac{4}{5}$.
31. Show that $x^2 + x + 1 = (x - \omega)(x - \omega^2)$.
32. Prove that $x^2 + xy + y^2 = (x - y\omega)(x - y\omega^2)$.
33. Show that (i) $(1 - \omega)(1 - \omega^2) = 3$;
    (ii) $(1 - 2\omega + \omega^2)(1 + \omega - 5\omega^2) = 18$.
34. Prove that (i) $\dfrac{a + \omega b + \omega^2 c}{b + \omega c + \omega^2 a} = \omega$;
    (ii) $(1 - \omega^2 + \omega^4)^2 + (1 + \omega^2 + \omega^4)^2 = 4\omega$.
35. Find the moduli of the following expressions:

(i) -7;   (ii) $1 + \iota$;   (iii) $(1 + \iota)^2$;   (iv) $(40 + 9\iota)(24 + 7\iota)$;

(v) $\dfrac{3 + 4\iota}{12 + 5\iota}$;   (vi) $\dfrac{(3 + \iota)(7 - \iota)}{2 - \iota}$;   (vii) $\dfrac{3 - 4\iota}{3 + 4\iota} + \dfrac{12 - 5\iota}{12 + 5\iota}$.

36. Prove that unless n is divisible by 100, the number of days in the year n A.D. is $365 + \frac{1}{4}\{1 + (-1)^n\}\{1 + \iota^n\}$.

37. Find and represent geometrically the sixth roots of unity.

38. Find the cube roots of -1 and represent them geometrically; also the fourth roots of -1.

39. Find the fifth roots of unity by using the method of Art. 13.

40. Show that the fifth roots of unity other than unity are the roots of the reciprocal equation $x^4 + x^3 + x^2 + x + 1 = 0$, and solve this equation.

41. Find all the eighth roots of unity.

CHAPTER XXV.

RATIONAL FUNCTIONS.

364. Functions.

DEFINITION. A quantity which varies with x and is constant when x is constant is called a **function** of x, and is usually denoted by some such notation as $f(x)$, $\phi(x)$, &c.

If the function involves only finite integral powers of x it is said to be a **rational function,** and if in addition the denominator is unity it is said to be a **rational integral function** of the variable x.

The general type of such a function is manifestly

$$p_0 x^n + p_1 x^{n-1} + p_2 x^{n-2} \ldots\ldots p_{n-1} x + p_n,$$

where n, of course, is a positive integer, and the function is said to be *linear, quadratic, cubic*, &c., according as n is equal to *one, two, three*, &c. In general it is said to be of the **nth degree.**

We may write any equation of the nth degree in x in the form

$$f(x) = 0,$$

and then a value of x which makes $f(x)$ vanish is called a **root** of this equation.

RATIONAL FUNCTIONS. 377

365. The Remainder Theorem. *If a rational integral function $f(x)$ of x be divided by $x-a$ then the remainder is $f(a)$.*

Let Q be the quotient and R the remainder after the ordinary process of division has been carried out, so that R does not involve x, then we have the identity

$$f(x) = Q(x-a) + R,$$

which is true for all values of x. On putting $x = a$ the right-hand side reduces to R, which does not depend on the value of x, and we therefore obtain

$$f(a) = R.$$

From this very important result two far-reaching corollaries may be inferred.

I. *If $f(x)$ vanishes when $x = a$ then $f(x)$ is divisible exactly by $x - a$.*

For in this case $f(a) = 0$ by hypothesis, $\therefore R = 0$. Thus if a is a root of $f(x) = 0$, then $f(x)$ is divisible *by $x - a$.*

II. Conversely, *if $f(x)$ is divisible by $x - a$ then a is a root of $f(x) = 0$.*

In this case $R = 0$ by hypothesis, and $\therefore f(a) = 0$, or a is a root of the equation.

366. *Ex.* 1. $x^n - a^n$ is divisible by $x - a$ if n be a positive integer, for on putting $x = a$ the expression vanishes.

Ex. 2. If n be even then $x^n - a^n$ is divisible by $x + a$, for in this case $(-a)^n - a^n = 0$.

Ex. 3. If n be odd then $x^n + a^n$ is divisible by $x + a$.

Ex. 4. $x^3 - 6x^2 + 11x - 6 = (x-1)(x-2)(x-3)$, for it vanishes when $x = 1$, 2, or 3 and it is of the third degree, so it has no further factor.

Ex. 5. To find the remainder when $f(x)$ is divided by $(x-a)(x-b)$. The remainder is of the form $Lx + M$ where L and M do not involve x, and this may clearly be written in the form $p(x-a) + q(x-b)$ where p and q do not involve x, $\therefore$ we have identically

$$f(x) = Q(x-a)(x-b) + p(x-a) + q(x-b),$$

putting $x = a$ we get $f(a) = q(a-b)$,

and putting $x = b$ we get $f(b) = p(b-a)$.

the remainder is $\{f(a)(x-b) - f(b)(x-a)\}/(a-b)$.

If $f(a) = 0$ and $f(b) = 0$ the remainder is of course zero, agreeing with what we could at once infer from the remainder theorem.

367. Definition. An expression in any number of letters is said to be **symmetric** when it is unaltered by interchanging any two of the letters.

Thus $(a+b+c)$, $(a^2+b^2+c^2)$, $(bc+ca+ab)$ are all symmetric in a, b, and c.

Ex. Find the condition that $\lambda a + \mu b + \nu c$ should be symmetric in a, b, and c.

Since it must be unaltered on interchanging b and c we find
$$\lambda a + \mu b + \nu c = \lambda a + \mu c + \nu b.$$

Hence $\mu(b-c) = \nu(b-c)$, and this has to hold for all values of a, b, and c. $\therefore \mu = \nu$. Similarly $\lambda = \mu$.

$\therefore \lambda = \mu = \nu$. $\therefore \lambda(a+b+c)$ is the only symmetric expression of the first degree. Similarly $a^2+b^2+c^2$ and $bc+ca+ab$ are symmetric functions of the second degree. The general expression for a symmetric function of the second degree is
$$\lambda(a^2+b^2+c^2) + \mu(bc+ca+ab)$$
where λ and μ are independent of a, b, and c.

The sum, difference, or product of two symmetric expressions is clearly another symmetric expression.

368. Definition. *An expression is said to be* **skew symmetric** *when only its sign is changed by interchanging any two letters involved in it.*

Thus $b-c$ is skew symmetric in b and c;
$$a^2(b-c) + b^2(c-a) + c^2(a-b)$$
is skew symmetric in a, b and c.

The product or quotient of two skew symmetric expressions is symmetrical. For if P and Q be the expressions, they become $-P$ and $-Q$ on interchanging two letters;

$$\therefore \text{ since } (-P)(-Q) = PQ \text{ and } \frac{-P}{-Q} = \frac{P}{Q},$$

their product and quotient are unaltered by interchanging two letters. Hence they are symmetrical.

369. Factors of expressions in three or more letters.

Ex. 1. To prove that
$$a^2(b-c)+b^2(c-a)+c^2(a-b) = -(b-c)(c-a)(a-b).$$

The expression on the left vanishes when $a=b$, when $b=c$, and when $c=a$. ∴ $(b-c)$, $(c-a)$, $a-b$ are all factors. Hence since the expression is only of the third degree we have
$$a^2(b-c)+b^2(c-a)+c^2(a-b) = A(b-c)(c-a)(a-b)$$
where A is a number.

Since this holds for all values of a, b and c put $a=1$, $b=0$, $c=-1$, and we obtain
$$2 = A(1)(-2)(1).$$
$$\therefore A = -1.$$
$$\therefore a^2(b-c)+b^2(c-a)+c^2(a-b) = -(b-c)(c-a)(a-b).$$

We could also have found A by equating coefficients, say of b^2c.

Ex. 2. Find all the factors of
$$a^4(b-c)+b^4(c-a)+c^4(a-b).$$

This expression vanishes when $b=c$,
$$\therefore b-c \text{ is a factor.}$$

Similarly $c-a$, $a-b$ are factors.

Hence $a^4(b-c)+b^4(c-a)+c^4(a-b) = (b-c)(c-a)(a-b)X$.

Since the expression on the left is skew symmetric of the fifth degree and $(b-c)(c-a)(a-b)$ skew symmetric of the third, X must be symmetric and of the second.
$$\therefore X = A(a^2+b^2+c^2)+B(bc+ca+ab)$$
where A and B are numbers.

To find A and B we proceed as in (i), only now we need two equations.

Putting $\quad a=1, \ b=0, \ c=-1$
we get $\quad 2 = -2\{A.2-B\}$. ∴ $2A-B = -1$,
and putting $\quad a=2, \ b=1, \ c=0$
$$14 = (-2)\{A.5+B.2\}, \quad \therefore 5A+2B = -7.$$

Solving these two equations for A and B we get $A = -1$, $B = -1$.
∴ $a^4(b-c)+b^4(c-a)+c^4(a-b)$
$$= -(b-c)(c-a)(a-b)\{a^2+b^2+c^2+bc+ca+ab\}.$$

Ex. 3. Find the factors of
$$a^4(b^2-c^2)+b^4(c^2-a^2)+c^4(a^2-b^2).$$
Using the result of Ex. 1, we have
$$a^4(b^2-c^2)+b^4(c^2-a^2)+c^4(a^2-b^2) = -(b^2-c^2)(c^2-a^2)(a^2-b^2)$$
$$= -(b-c)(c-a)(a-b)(b+c)(c+a)(a+b).$$

These factors could also have been obtained by noticing that the expressions vanish not only when $b=c$ but when $b=-c$.

Ex. 4. Shew that $(a+b+c)^3 - a^3 - b^3 - c^3 = 3(b+c)(c+a)(a+b)$.

The expression on the left vanishes when $b=-c$. $\therefore$ $b+c$ is a factor; similarly $c+a$, $a+b$ are factors, and since it is only of the third degree $(a+b+c)^3 - a^3 - b^3 - c^3 = A(b+c)(c+a)(a+b)$ where A is a number.

Putting $a=2$, $b=1$, $c=0$ we find $18 = A.3.2$. $\therefore$ $A=3$. Or we can find A by equating coefficients of b^2c.

Ex. 5. Establish the identity
$$(a+b)^7 - a^7 - b^7 = 7ab(a+b)(a^2+ab+b^2)^2.$$

The expression on the left vanishes when $a=0$, when $b=0$ and when $a+b=0$, $\therefore$ a, b and $a+b$ are factors.

The remaining factor must be symmetrical and of the fourth degree, i.e. of the form
$$A(a^4+b^4)+B(a^3b+ab^3)+Ca^2b^2.$$
$$\therefore (a+b)^7 - a^7 - b^7 = ab(a+b)\{A(a^4+b^4)+B(a^3b+ab^3)+Ca^2b^2\}$$
$$= (a^2b+ab^2)\{A(a^4+b^4)+B(a^3b+ab^3)+Ca^2b^2\}.$$

Equating coefficients of a^6b, a^5b^2, a^4b^3 we find
$$7=A$$
$$\frac{7.6}{1.2}=A+B$$
and $$\frac{7.6.5}{1.2.3}=B+C,$$
i.e. $A=7$, $A+B=21$, $B+C=35$.
$$\therefore A=7, \quad B=14, \quad C=21,$$
and hence
$$(a+b)^7 - a^7 - b^7 = 7ab(a+b)\{a^4+b^4+2(a^3b+ab^3)+3a^2b^2\}$$
$$= 7ab(a+b)\{(a^2+b^2)^2+2ab(a^2+b^2)+a^2b^2\}$$
$$= 7ab(a+b)(a^2+ab+b^2)^2$$
as required.

RATIONAL FUNCTIONS. 381

370. If a rational integral function of x of the nth degree vanish for n different values of x, it can be resolved into n different factors of the first degree.

Let the function $p_0 x^n + p_1 x^{n-1} + p_2 x^{n-2} + \ldots + p_n$ vanish for n different values of x, viz. a, β, γ, ...

Since the function vanishes when a is substituted for x, $x - a$ is a factor of the function by the remainder theorem.

Dividing the function by $x - a$, we have $p_0 x^{n-1}$ for the first term of the quotient, which is of the $(n-1)$th degree. Hence $p_0 x^n + p_1 x^{n-1} + \ldots + p_n = (x - a)(p_0 x^{n-1} + \ldots)$.

Again, since the function vanishes when $x = \beta$,

$$(x - a)(p_0 x^{n-1} + \ldots)$$

vanishes when $x = \beta$.

But $x - a$ cannot vanish when $x = \beta$, which shows that $x - \beta$ is a factor of $p_0 x^{n-1} + p_1 x^{n-2} + \ldots$. If the division be actually performed, $p_0 x^{n-2}$ would be the first term of the quotient. Hence

$$p_0 x^n + p_1 x^{n-1} + \ldots = (x - a)(x - \beta)(p_0 x^{n-2} + \ldots).$$

Similarly, as the function vanishes when $x = \gamma$, δ, ..., we have

$p_0 x^n + p_1 x^{n-1} + \ldots$
$\quad = (x - a)(x - \beta)(x - \gamma)(x - \delta) \ldots (p_0 x^{n-r} + \ldots),$

where r is the number of the factors of the first degree.

When $r = n$, the number of the factors of the first degree will be n, and $p_0 x^{n-r}$ will become reduced to p_0.

Hence $p_0 x^n + p_1 x^{n-1} + \ldots = p_0 (x - a)(x - \beta) \ldots$

to n factors of the first degree.

It is thus evident that if

$$p_0 x^n + p_1 x^{n-1} + p_2 x^{n-2} + \ldots + p_{n-1} x + p_n x = 0$$

has n different roots, the function on the left side of the equation can be resolved into n different linear factors.

371. Every equation of the nth degree has n roots.

We *assume* in the following proof that every equation of the form $f(x) = 0$ has at least *one* root real, imaginary*, or complex. Let a be a root of $f(x) = 0$.

Then $f(x)$ is divisible by $x - a$. [§ 365.]

If $f_1(x)$ be the quotient, $f(x) = (x-a)f_1(x)$, where $f_1(x)$ is a rational integral function of the $(n-1)$th degree.

Let β be a root of $f_1(x) = 0$.

Then $f_1(x)$ is divisible by $x - \beta$.

If $f_2(x)$ be the quotient, $f_1(x) = (x - \beta)f_2(x)$, where $f_2(x)$ is a rational integral function of the $(n-2)$th degree.

Hence $f(x) = (x - a)(x - \beta)f_2(x)$.

Proceeding thus, we get ultimately

$$f(x) = p_0(x-a)(x-\beta) \ldots (x-\lambda),$$

which is the product of n simple factors with the coefficient p_0 which does not contain x.

$\therefore f(x)$ will vanish when x has any of the n values $a, \beta, \ldots \lambda$. Hence $f(x) = 0$ has n roots.

372. A function of x of the nth degree cannot vanish for more than n values of x, unless the coefficient of every power of x is zero.

Let the function $p_0 x^n + p_1 x^{n-1} + \ldots + p_n$ vanish for the n values $a, \beta, \gamma, \ldots$. Then

$$p_0 x^n + p_1 x^{n-1} + \ldots = p_0(x-a)(x-\beta)(x-\gamma) \ldots$$

Let now r be another value, the $(n+1)$th value, for which the function vanishes. Then

$$0 = p_0(r-a)(r-\beta)(r-\gamma) \ldots$$

---

* The proof of this, the fundamental theorem of algebra, is beyond the scope of the present work.

RATIONAL FUNCTIONS. 383

But evidently none of the factors $r-\alpha$, $r-\beta$, $r-\gamma$, ... can be zero, since, by supposition, the values α, β, γ, ... are all different from r. Hence $p_0 = 0$.

The function is thus reduced to $p_1 x^{n-1} + ...$, which is of the $(n-1)$th degree. This can vanish only for $n-1$ values of x, unless p_1 is also zero. It can be similarly shown that the other coefficients vanish, which establishes the proposition.

From this proposition we deduce the

COROLLARIES. **I. The maximum number of linear factors in a rational integral expression in x of the nth degree is n.**

II. An equation of the nth degree cannot have more than n roots.

373. The Principle of Undetermined Coefficients.

If two functions of x of the nth degree are equal for more than n values of x, the coefficients of like powers of x in the two functions are equal, and the functions are equal for all values of x.

If $ax^n + bx^{n-1} + cx^{n-2} + ... = a_1 x^n + b_1 x^{n-1} + c_1 x^{n-2} + ...$
for more than n values of x, then
$$(a-a_1)x^n + (b-b_1)x^{n-1} + (c-c_1)x^{n-2} + ... = 0$$
for more than n values of x.

Hence, by § 372, $a - a_1 = 0$; $b - b_1 = 0$; $c - c_1 = 0$; ... ;
$$a = a_1;\ b = b_1;\ c = c_1;\ ....$$
Hence the functions are equal for all values of x.

From this we deduce the **Principle of Undetermined Coefficients** for functions of a finite number of terms.

If two functions of a finite number of terms are equal for all values of any letter involved in them, then the coefficients of like powers of this letter in the two functions are equal.

Ex. If $ax^4 + bx^3 + dx + e = a_1 x^4 + c_1 x^2 + d_1 x + e_1$ for five values of x, then $a = a_1$, $b = 0$, $0 = c_1$, $d = d_1$, and $e = e_1$.

374. Relations between the roots of the coefficients in an equation.

If $a_1, a_2, a_3 \ldots a_n$ be the roots of the equation
$$p_0 x^n + p_1 x^{n-1} + p_2 x^{n-2} \ldots + p_n = 0 \ldots \ldots \ldots (A),$$
we have seen that the above expression is identically equal to
$$p_0 (x - a_1)(x - a_2) \ldots (x - a_n).$$

Further, by the theorem (§ 326) for the product of any number of binomial factors this last is equal to
$$p_0 \{x^n - s_1 x^{n-1} + s_2 x^{n-2} \ldots (-1)^r s_r x^{n-r} \ldots (-1)^n s_n\} \ldots \ldots (B),$$
where s_r is the sum of the products of the a's taken r at a time.

Thus, equating coefficients of like powers of x in the expressions A and B, we get

$$p_0 = p_0,$$

$$-p_0 s_1 = p_1, \quad \text{or} \quad s_1 = -\frac{p_1}{p_0},$$

$$p_0 s_2 = p_2, \qquad s_2 = \frac{p_2}{p_0},$$

$$\vdots$$

$$(-1)^r p_0 s_r = p_r, \qquad s_r = (-1)^r \frac{p_r}{p_0},$$

$$\vdots$$

$$(-1)^n p_0 s_n = p_n, \qquad s_n = (-1)^n \frac{p_n}{p_0}.$$

375. We shall now apply the preceding to some examples.

Ex. 1. To find the sum of the cubes of the roots of the equation
$$3x^3 - 6x^2 + 36x - 10 = 0.$$

Let α, β, γ be the roots, then
$$\alpha + \beta + \gamma = 2, \quad \beta\gamma + \gamma\alpha + \alpha\beta = 12, \quad \alpha\beta\gamma = 3\tfrac{1}{3}.$$
Hence $\alpha^3 + \beta^3 + \gamma^3 = (\alpha + \beta + \gamma)^3 - 3(\beta\gamma + \gamma\alpha + \alpha\beta)(\alpha + \beta + \gamma) + 3\alpha\beta\gamma$
$$= 2^3 - 3 \cdot 12 \cdot 2 + 3 \cdot 3\tfrac{1}{3} = -54.$$

RATIONAL FUNCTIONS.

Ex. 2. Find the sum of the reciprocals of the roots of
$$3x^4 - 5x^3 + 6x^2 - 3x - 4 = 0.$$

Here if a, β, γ, δ be the roots
$$\frac{1}{a} + \frac{1}{\beta} + \frac{1}{\gamma} + \frac{1}{\delta} = \frac{a\beta\gamma + a\beta\delta + a\gamma\delta + \beta\gamma\delta}{a\beta\gamma\delta}$$
$$= \frac{3/3}{-4/3} = -\frac{3}{4}.$$

Ex. 3. If a, β, γ be the roots of the equation
$$ax^3 + bx^2 + cx + d = 0,$$
explain how to calculate $\sigma_n = a^n + \beta^n + \gamma^n$.

Here we note that
$$aa^3 + ba^2 + ca + d = 0.$$
$$\therefore aa^n + ba^{n-1} + ca^{n-2} + da^{n-3} = 0.$$

Consequently adding this to the corresponding equations in β and γ we find
$$a\sigma_n + b\sigma_{n-1} + c\sigma_{n-2} + a\sigma_{n-3} = 0$$
if only $n > 3$.

If $n = 3$ we find $\quad a\sigma_3 + b\sigma_2 + c\sigma + 3d = 0.$

Now $\quad \sigma_2 = (a + \beta + \gamma)^2 - 2(\beta\gamma + \gamma a + a\beta)$
$$= \frac{b^2 - 2ac}{a^2} \text{ and } \sigma_1 = -\frac{b}{a},$$

and the last equation enables us to calculate σ_3 when σ_1 and σ_2 are known.

Then from the equation
$$a\sigma_4 + b\sigma_3 + c\sigma_2 + d\sigma_1 = c$$
we can find σ_4 from $\sigma_1, \sigma_2, \sigma_3$ and so on.

By this process we can calculate the σ's as far as we like.

For example $\quad a\sigma_3 = -b\sigma_2 - c\sigma_1 - 3d$
$$= -\frac{b(b^2 - 2ac)}{a^2} + \frac{bc}{a} - 3d$$
$$= \frac{-b^3 + 3abc - 3a^2d}{a^2}.$$
$$\therefore \sigma_3 = \frac{-b^3 + 3abc - 3a^2d}{a^3}.$$

Ex. 4. In the same equation to find how to calculate
$$f(a)+f(\gamma)$$
when f is a rational integral function.

Suppose when $f(x)$ is divided by ax^3+bx^2+cx+d that Q is the quotient and $R \equiv P_1x^2+P_2x+P_3$ the remainder, then
$$f(x) = Q(ax^3+bx^2+cx+a) + P_1x^2+P_2x+P_3.$$
$$\therefore f(a) = P_1a^2 + P_2a + P_3,$$
and $\quad f(a)+f(\beta)+f(\gamma) = P_1(a^2+\beta^2+\gamma^2) + P_2(a+\beta+\gamma) + 3P_3,$

the value of which can be written down at once.

376. Equation whose roots are functions of the roots of a given equation.

We frequently require to find the equation whose roots are certain functions of the roots of a given equation.

Thus, for example, we might want to find the equation whose roots are a^2, β^2, γ^2 where $a\beta\gamma$ are the roots of the equation
$$ax^3+bx^2+cx+d=0.$$

The equation required is
$$(y-a^2)(y-\beta^2)(y-\gamma^2) = 0$$
or $\quad y^3 - y^2(a^2+\beta^2+\gamma^2) + y(\beta^2\gamma^2+\gamma^2a^2+a^2\beta^2) - a^2\beta^2\gamma^2 = 0,$

and the coefficients being symmetric functions of a, β, γ can be calculated in terms of a, b, c, d.

In many cases we may proceed more shortly, as for example to find the equation whose roots are the roots of a given equation with the signs changed. Here if y be a root of the new equation and x one of the old we have
$$y = -x. \quad \therefore x = -y,$$
but $\quad ax^3+bx^2+cx+d=0.$
$$\therefore ay^3 - by^2 + cy - d = 0,$$

and in general it is quite clear that we derive the new equation from the old one by merely changing the sign of x.

377. Examples.

Ex. 1. To find the equation whose roots are the reciprocals of a given equation.

Let $f(x) = 0$ be the equation, then if y be a root of the required equation

RATIONAL FUNCTIONS. 387

$$y = \frac{1}{x}.$$

$$\therefore x = \frac{1}{y}.$$

$\therefore f\left(\dfrac{1}{y}\right) = 0$ is the equation we want;

e.g. taking the quadratic

$$ax^2 + bx + c = 0$$

we derive

$$a\frac{1}{y^2} + b\frac{1}{y} + c = 0$$

or

$$cy^2 + by + a = 0.$$

Ex. 2. To find the equation whose roots are those of a given equation multiplied by any given quantity.

Let k be the given quantity, then $y = kx$.

$$\therefore x = \frac{y}{k},$$

and the equation is

$$f\left(\frac{y}{k}\right) = 0.$$

Ex. 3. *To find the equation whose roots are those of a given equation, each increased by the same quantity.*

Let $f(x) = 0$ be the given equation and h the given quantity; then when x is a root of the given equation $x + h = y$ is a root of the equation sought.

$$\therefore x = y - h \text{ and } f(x) = 0. \quad \therefore f(y - h) = 0,$$

and this is the equation required.

Thus the equation whose roots are those of

$$x^2 + 2x + 3 = 0,$$

each increased by unity is found by writing $y - 1$ for x in the equation, i.e. it is

$$(y - 1)^2 + 2(y - 1) + 3 = 0$$

or

$$y^2 + 2 = 0.$$

The readers will notice that in the numerical example we obtained an equation in which the second term was wanting. It is easy to see that *every equation can be transformed so as to get rid of the second term.*

For let $a_0x^n + a_1x^{n-1} + a_2x^{n-2} \ldots a_n = 0$
be the given equation. The equation whose roots are the roots of this increased by h is

$$a_0(y-h)^n + a_1(y-h)^{n-1} + a_2(y-h)^{n-2} \ldots + a_n = 0$$

or $\quad a_0(y^n - ny^{n-1}h + \ldots) + a_1(y-h)^{n-1} \ldots + a_n = 0$,

and the coefficient of y^{n-1} is

$$-na_0h + a_1. \quad \therefore \text{ if } h = a_1/na_0,$$

the new equation has no second term, so we effect the transformation required by increasing each of the roots by a_1/na_0.

Ex. 4. Required to transform $x^3 + 6x^2 + 2x + 1 = 0$ into one which shall lack the term in x^2.

Increasing the roots by h we get the new equation

$$(x-h)^3 + 6(x-h)^2 + 2(x-h) + 1 = 0$$

or $\quad x^3 - x^2(3h-6) + x(3h^2 - 12h + 2) - h^3 + 6h^2 - 2h + 1 = 0$.

Hence $3h - 6 = 0$. $\therefore h = 2$, and the required equation is

$$x^3 + x(-10) + 21 = 0 \text{ or } x^3 - 10x + 21 = 0.$$

378. In any equation with *real* coefficients, complex roots occur in pairs.

Let $f(x) = 0$ be an equation with real coefficients; and let $a + bi$ be an imaginary root. We shall shew that $a - bi$ is also a root.

The *real* factor of $f(x)$ corresponding to these two roots is

$$(x - a - bi)(x - a + bi) = (x-a)^2 - (bi)^2 = (x-a)^2 + b^2.$$

Let $f(x)$ be divided by $(x-a)^2 + b^2$; and let Q be the quotient and $Rx + R'$ be the remainder.

Then $\quad f(x) = \{(x-a)^2 + b^2\} Q + Rx + R$.

If, in this identity, we substitute $a + bi$ for x, then $f(x) = 0$, since $a + bi$ is a root of $f(x) = 0$; and also

$$(x-a)^2 + b^2 = 0,$$

since one of its factors $x - a - bi$ is zero.

$$\therefore R(a + bi) + R' = 0.$$

RATIONAL FUNCTIONS.

Equating the real and imaginary parts of this equality to zero [§ 345], we have

$$Ra + R' = 0; \text{ and } Rb = 0.$$

And since b is not zero, $R = 0$; and $\therefore R' = 0$.

$$\therefore Rx + R' = 0 \text{ in the identity}$$

$$f(x) = \{(x-a)^2 + b^2\} Q + Rx + R'.$$

Hence $f(x)$ is divisible by $(x-a)^2 + b^2$; that is, by

$$(x - a - bi)(x - a + bi).$$

Thus $f(x)$ is divisible by $x - a + bi$, and therefore $a - bi$ is a root of $f(x) = 0$.

NOTE. It can be similarly proved that, when $a + \sqrt{b}$ is a root of $f(x) = 0$ where $f(x)$ has rational coefficients, $a - \sqrt{b}$ is also a root.

Ex. 1. Solve $x^3 - 4x^2 + 6x - 4 = 0$, having given that $1 + i$ is a root.

Since $1 + i$ is a root, $1 - i$ is also a root.

Hence $x^3 - 4x^2 + 6x - 4$ is divisible both by $x - 1 - i$ and $x - 1 + i$.

And $(x - 1 - i)(x - 1 + i) = (x - 1)^2 - (i)^2 = x^2 - 2x + 2.$

Dividing out, we have $x - 2 = 0$.

Hence the remaining root is 2.

Ex. 2. Form an equation with rational coefficients, of which one of the roots is $\sqrt{5} - \sqrt{3}$.

Let $\qquad x = \sqrt{5} - \sqrt{3}.$

Then $\qquad x - \sqrt{5} = -\sqrt{3}.$

Squaring, $\qquad x^2 - 2\sqrt{5}x + 5 = 3.$

$\qquad \therefore x^2 - 2\sqrt{5}x + 2 = 0.$

[This equation has $\sqrt{5} + \sqrt{3}$ also for a root.]

$\qquad \therefore x^2 + 2 = 2\sqrt{5}x.$

Squaring again, $\qquad x^4 + 4 + 4x^2 = 20x^2.$

$\qquad \therefore x^4 - 16x^2 + 4 = 0.$

EXERCISES XXV.

1. Show that $x^{31} + 7x^{30} - 8$ is divisible by $x - 1$.
2. Show that $x^4 + 4x^3 + 6x^2 - 8$ is divisible by $x + 2$.
3. Prove that $x^n - 1$ is divisible by $x - 1$, and obtain the quotient.
4. Find the remainder when
$$x^4 - 3x^3 + 4x^2 - 6x + 7 \text{ is divided by } x - 1.$$

Establish the identities

5. $ab(a - b) + bc(b - c) + ca(c - a) = -(b - c)(c - a)(a - b)$.
6. $a^3(b - c) + b^3(c - a) + c^3(a - b) = -(b - c)(c - a)(a - b)(a + b + c)$.
7. $(a + b + c)^3 - a^3 - b^3 - c^3 = 3(b + c)(c + a)(a + b)$.
8. $(a + b + c)^3 - (b + c)^3 - (c + a)^3 - (a + b)^3 + a^3 + b^3 + c^3 = 6abc$.
9. $a(b^3 - c^3) + b(c^3 - a^3) + c(a^3 - b^3) = (b - c)(c - a)(a - b)(a + b + c)$.
10. $(a + b)^5 - a^5 - b^5 = 5ab(a + b)(a^2 + ab + b^2)$.
11. $(a + b + c)^5 - a^5 - b^5 - c^5 = 5(b + c)(c + a)(a + b)$
$\qquad (a^2 + b^2 + c^2 + bc + ca + ab)$.
12. $(bc + ca + ab)^3 - b^3c^3 - c^3a^3 - a^3b^3 = 3abc(b + c)(c + a)(a + b)$.
13. $(b - c)^3 + (c - a)^3 + (a - b)^3 = 3(b - c)(c - a)(a - b)$.
14. $(b - c)^5 + (c - a)^5 + (a - b)^5 = 5(b - c)(c - a)(a - b)$
$\qquad (a^2 + b^2 + c^2 - bc - ca - ab)$.

Apply the Principle of Undetermined Coefficients to prove the following three identities:

15. $(a + b - c)^2 = a^2 + b^2 + c^2 + 2ab - 2bc - 2ca$.
16. $(a + b + c)^3 = \Sigma(a^3) + 3\Sigma a^2 b + 6\Sigma abc$.
17. $\sqrt[3]{(m^6 - 12m^5 + 54m^4 - 112m^3 + 108m^2 - 48m + 8)} = m^2 - 4m + 2$.
18. If $ax^3 + bx^2 + cx + d$ is a perfect cube, show that $b^2 = 3ac$ and $c^2 = 3bd$.
19. If $m^6 + 6m^5 - 40m^3 + pm - q$ is a perfect cube, show that $p = 96$ and $q = 64$.
20. If $x^6 - 6x^5 + 15x^4 + nx^3 + px^2 + qx + r$ is a perfect cube, find p, q, r, n.
21. Express $2x^3 + x^2 + x + 1$ in ascending powers of $x - 3$.
22. Frame an equation of the fourth degree with real coefficients, two of whose roots are $-1 + \sqrt{-7}$ and $-2 + \sqrt{-3}$.
23. Frame an equation of the third degree with real coefficients, one of whose roots $-\frac{1}{2}(1 + \sqrt{-23})$, and another root, -3.

RATIONAL FUNCTIONS.

24. Frame an equation of the fourth degree with rational coefficients, one of whose roots is $\tfrac{1}{2}(-1+\sqrt{105})$, and two other roots are 0 and -1.

25. Frame an equation with rational coefficients, one of whose roots is $\sqrt{5}-\sqrt{2}$.

26. Solve $x^4-5x^3+4x^2+8x-8=0$, having given that one of the roots is $1+\sqrt{5}$.

27. Solve $(x-4)^2+2(x-4)=2/x-1$, having given that $2+\sqrt{3}$ is one of the roots.

28. Solve $x^4-2x^3+x^2-4x+4=0$, having given that one of the roots is $\tfrac{1}{2}(-1+\sqrt{-7})$.

29. Solve $x^4+12x^3+78x^2+252x+272=0$, having given that $-3+5i$ is one of its roots.

30. Solve $x^3-x^2-22x+40=0$, if one of the roots is double another root.

31. Solve $x^3-15x^2+71x-105=0$, if the roots are in arithmetical progression.

32. Solve $x^4-8x^3+11x^2+32x-60=0$, if the sum of two of the roots is zero.

33. Find the sum of the squares of the reciprocals of the roots of $x^3+10x^2-3x+2=0$.

34. If a, b, c are the roots of $x^3+qx^2+r=0$, find the values of

(i) $\dfrac{1}{a}+\dfrac{1}{b}+\dfrac{1}{c}$. (ii) $\dfrac{1}{bc}+\dfrac{1}{ca}+\dfrac{1}{ab}$.

(iii) $\dfrac{1}{a^2}+\dfrac{1}{b^2}+\dfrac{1}{c^2}$. (iv) $\dfrac{1}{a^3}+\dfrac{1}{b^3}+\dfrac{1}{c^3}$. (v) $a^3+b^3+c^3$.

(vi) $(a+b)(b+c)(c+a)$. (vii) $\dfrac{1}{a+b}+\dfrac{1}{b+c}+\dfrac{1}{c+a}$.

35. If a, b, c, d are the roots of $x^4+px^3+qx^2+r=0$, find the values of (i) $a^2+b^2+c^2+d^2$. (ii) $a^{-2}+b^{-2}+c^{-2}+d^{-2}$. (iii) $a^4+b^4+c^4+d^4$.

36. Find the equation whose roots are less by 2 than those of $x^4-10x^3+4x^2-5x+24=0$.

37. Transform $x^3+6x^2+9x-12=0$ into another equation without the second term.

38. Transform $x^3-x^2-34x-56$ into an equation whose roots are thrice the roots of the given one.

39. Transform $x^4+x^3+2x^2+3x-7=0$ into an equation lacking the second term.

40. Transform $2x^4+8x^3+x^2+x+1=0$ into one with no term in x^3.

CHAPTER XXVI.

GRAPHIC REPRESENTATION OF FUNCTIONS.
CONTINUITY.

379. Change in the value of a function.

In the last chapter we have been concerned with the values of x which make a rational integral function of x vanish; we now proceed to shortly consider the values of the function for other values of the variable.

Take, for example, a cubic function having three real roots a, β, γ, then we may write it in the form

$$a(x-a)(x-\beta)(x-\gamma), \qquad (\S\ 370)$$

and we shall suppose that $a < \beta < \gamma$. If a be positive then

when $x < a$, $f(x)$ is negative,
 ,, $x = a$, $f(x) = 0$,
 ,, $x > a$ and $< \beta$, $f(x)$ is positive,
 ,, $x = \beta$, $f(x) = 0$,
 ,, $x > \beta$ and $< \gamma$, $f(x)$ is negative,
 ,, $x = \gamma$, $f(x) = 0$,
 ,, $x > \gamma$, $f(x)$ is positive;

and thus we derive a general idea as to the value of the function for different values of x. The changes in this value may be best represented to the eye by what is called *graphical representation*—a method we shall now explain.

380. Graphic Representation.

Take a line XOX' of unlimited length and O as origin, just as in Elementary Analytical Geometry, with which we suppose the student to be acquainted.

Measure off on OX a length $ON = x$, with the usual convention as regards sign.

Through N draw a line NP perpendicular to ON, so that PN is equal to the value of the function for this particular value of x. Of course if $f(x)$ be positive we measure PN upwards, and if $f(x)$ be negative we measure PN downwards.

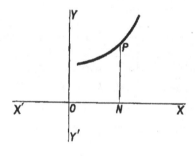

Fig. 5.

Thus by taking any number of points N we get an equal number of points P, and the variation in the distance of P from the axis represents the variation in the value of the function.

All the points P as we know from analytical geometry lie on a curve whose equation is

$$y = f(x),$$

and this curve, when we are able to trace it, gives us all we require.

381. Important features to be observed in the construction of the graph.

In general the equation
$$f(x) = 0$$
gives the points in which the curve cuts the axis of x, and the reader will perceive from the diagram that the function has opposite signs for two values of x, one of which is a little less than, and the other a little greater than, the same root of the equation
$$f(x) = 0.$$
Of course to obtain the curve the most obvious and satisfactory way is to find a very large number of points on it, e.g. to find the values of the function for values of x differing by 1, say, and then having obtained a large number of such points on the curve we can trace it with ease and accuracy. But this method, although excellent theoretically, is naturally very tedious in practice, and we can often form a very good idea as to the form of the curve by paying attention to the following points:—

(I) *The values of x which make y vanish.*

(II) *The values of x which make y infinite.*

(III) *The values of x which make y take its greatest and least values.*

(IV) *The enquiry as to whether y increases or decreases with x at different parts of the curve.*

And of course if the function have numerical coefficients we find some points on the curve.

For the functions we have to deal with (I) and (II) are fairly simple problems, and as regards (III) and (IV) we have not the methods at hand in this treatise to deal with them satisfactorily.

Of course if the curve happens to be one with whose geometrical properties we are familiar it is comparatively easy to trace it.

We shall now give some general examples of these methods.

GRAPHIC REPRESENTATION OF FUNCTIONS. CONTINUITY. 395

382. Graph of $y = ax^2 + 2bx + c$, *where a, b and c are numbers*

Here $\qquad y = a\left(x + \dfrac{b}{a}\right)^2 + c - \dfrac{b^2}{a}$ ....(A).

(i) If $b^2 > ac$ the curve cuts the axis in two real points given by the equation

$$ax^2 + 2bx + c = 0,$$

and from (A) we perceive that it is a parabola having the point $-\dfrac{b}{a}, -\dfrac{b^2 - ac}{a}$ for vertex, and the line $x = -\dfrac{b}{a}$ for axis. (Fig. 6.)

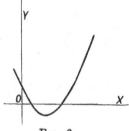

Fig. 6.

(ii) If $b^2 < ac$ the curve is the same in appearance but it now lies all on one side of the axis. (Fig. 7.)

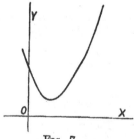

Fig. 7.

(iii) If $b^2 = ac$ the curve meets the axis in two coincident points, i.e. it touches this line. (Fig. 8.)

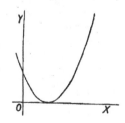

Fig. 8.

If we make $x = 0$ we get $y = c$, and if x be infinite $y = \infty$ and has the same sign as a.

Supposing a and c to be positive, the figures for the three cases are as shown.

And since we have taken the axis of the parabola to the right of YOY', b must be supposed negative.

The student will have no difficulty in modifying this for the cases in which

(i) a and c are positive and b positive,

(ii) a is positive, c is negative, and b is positive, and so on; in (ii) since $b^2 + ac$ will be positive always, the curve always meets the axis in real points.

In all cases of this kind there is a limiting value for y, for since the equation $ax^2 + 2bx + c = y$ must have real roots we must have

$$b^2 > a(c - y), \text{ i.e. } {}^*y > c - \frac{b^2}{a}.$$

This is evident geometrically, since $y = c - b^2/a$ gives the vertex of the parabola.

Let us consider a numerical example of the case discussed.

$y = x^2 - 4x + 5$. (Fig. 9.)

Here $y = (x - 2)^2 + 1$.

∴ the least value of y is 1.

The axis of the parabola is the line $x = 2$.

To enable us to trace the curve we calculate y for values of x differing by $\frac{1}{2}$ as in the table:

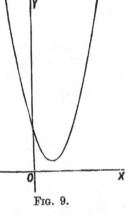

Fig. 9.

x	0	$\frac{1}{2}$	$-\frac{1}{2}$	1	-1	$1\frac{1}{2}$	$-1\frac{1}{2}$	2	-2	$2\frac{1}{2}$	3	$3\frac{1}{2}$
y	5	$3\frac{1}{4}$	$7\frac{1}{4}$	2	10	$1\frac{1}{4}$	$13\frac{1}{4}$	1	17	$1\frac{1}{4}$	2	$3\frac{1}{4}$

---

* a is supposed positive.

383. Graph of $y = \dfrac{a+bx}{A+Bx}$. (Fig. 10.)

Here when y is given there is only one value for x since x only appears to the first degree.

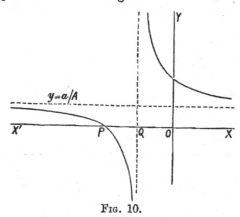

Fig. 10.

$y = 0$ when $x = -a/b$,
$y = \infty$,, $x = -A/B$;
also   when $x = 0$, $y = a/A$,
and   ,, $x = \infty$, $y = b/B$.

Supposing P and Q to represent
$$x = -a/b \text{ and } x = -A/B,$$
then the student will easily be able to see that, when a, b, A, B are all positive,

to the left of P y is positive,
between P and Q y is negative,
and   to the right of Q y is positive.

Consequently the march of the function is somewhat as in the diagram.

The curve is a rectangular hyperbola having its asymptotes parallel to OX and OY. (Smith's *Conics*, § 174.)

398 GRAPHIC REPRESENTATION OF FUNCTIONS. CONTINUITY.

384. We shall now consider some numerical examples.

Ex. 1. Take $y = \dfrac{x}{2-x}$. (Fig. 11.)

Here
$y = 0$ when $x = 0$;
$y = \infty$ when $x = 2$.

When
$x = \infty$, $y = \dfrac{1}{\dfrac{2}{x} - 1} = -1$.

, When x is
between $-\infty$ and $\quad 0\quad y$ is negative
.... $\quad 0\ ...\quad 2\quad y$ is positive,
.... $\quad 2\ ...\ +\infty\quad y$ is negative.

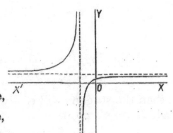

Fig. 11.

To get the curve somewhat accurately in the neighbourhood of the origin we find the value of y for values of x differing by $\frac{1}{2}$ thus:

$x =$	0	$\frac{1}{2}$	$-\frac{1}{2}$	1	-1	$1\frac{1}{2}$	$-1\frac{1}{2}$	2	-2	$2\frac{1}{2}$	$-2\frac{1}{2}$	3	-3	$3\frac{1}{2}$	$-3\frac{1}{2}$
$y =$	0	$\frac{1}{3}$	$-\frac{1}{5}$	1	$-\frac{1}{3}$	3	$-\frac{3}{7}$	∞	$-\frac{1}{2}$	-5	$-\frac{5}{9}$	-3	$-\frac{3}{5}$	$-\frac{7}{3}$	$-\frac{7}{11}$

Ex. 2. $\qquad y = \dfrac{x+1}{2x+3}$. (Fig. 12.)

Here $y = 0$ when $x = -1$;
$y = \infty$ when $x = -\frac{3}{2}$.
$x = 0$ gives $y = \frac{1}{3}$.

and $\quad x = \infty$ gives $y = \frac{1}{2}$.

Between $\ -\infty\ $ and $\ -\frac{3}{2}\ y$ is positive,
.... $\quad -\frac{3}{2}\ ...\ -1\ y$ is negative,
.... $\quad -1\ ...\ +\infty\ y$ is positive.

Fig. 12.

To get the curve accurately near the origin we tabulate the function as before:

$x =$	0	$\frac{1}{2}$	$-\frac{1}{2}$	1	-1	$1\frac{1}{2}$	$-1\frac{1}{2}$	2	-2	$2\frac{1}{2}$	$-2\frac{1}{2}$	3	-3	$3\frac{1}{2}$	$-3\frac{1}{2}$
$y =$	$\frac{1}{3}$	$\frac{3}{8}$	$\frac{1}{4}$	$\frac{2}{5}$	0	$\frac{5}{12}$	∞	$\frac{3}{7}$	1	$\frac{7}{16}$	$\frac{3}{4}$	$\frac{4}{9}$	$\frac{2}{3}$	$\frac{9}{20}$	$\frac{5}{8}$

GRAPHIC REPRESENTATION OF FUNCTIONS. CONTINUITY. 399

385. Graph of $y = \dfrac{ax^2 + 2bx + c}{Ax^2 + 2Bx + C}$.

This example is important as being a good illustration of the present methods and is interesting in connection with the theory of Quadratic equations.

Various cases arise according as the equations

$$ax^2 + 2bx + c = 0 \text{ and } Ax^2 + 2Bx + C = 0$$

have real or imaginary roots.

If $ax^2 + 2bx + c = 0$ has imaginary roots the curve never cuts the axis.

The equation $Ax^2 + 2Bx + C = 0$ gives the value of x for which y becomes infinite. Hence if this equation has imaginary roots y is always finite.

Further, when y is given x is the root of a quadratic equation, viz.

$$(a - Ay)x^2 + 2(b - By)x + c - Cy = 0,$$

i.e. for a given value of y there are two values of x if this equation has real roots. If it has imaginary roots for $y = \kappa$ say, then the line $y = \kappa$ does not cross the curve at all. When $x = 0$, $y = \dfrac{c}{C}$, and when x is very large, we have

$$y = \frac{a + 2b/x + c/x^2}{A + 2B/x + C/x^2} = \frac{a}{A},$$

ultimately when x becomes infinite; this means that as x becomes numerically larger and larger (either positive or negative) the curve constantly approaches the line

$$y = a/A,$$

and at a very great distance becomes very close to it. The curve meets $y = a/A$ in one finite point, for the equation

$$\frac{a}{A} = \frac{ax^2 + 2bx + c}{Ax^2 + 2Bx + C}$$

is in reality only of the first degree.

400 GRAPHIC REPRESENTATION OF FUNCTIONS. CONTINUITY.

386. Cases in which both numerator and denominator have not real roots.

(*a* and *A* are supposed positive.)

(i) WHEN THE EQUATIONS

$$ax^2 + 2bx + c = 0 \text{ AND } Ax^2 + 2Bx + C = 0$$

BOTH HAVE IMAGINARY ROOTS. (Fig. 13.)

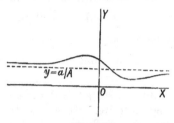

FIG. 13.

In this case y never vanishes nor becomes infinite; further, both numerator and denominator are always positive (§ 183); $\therefore$ y is always positive. When x is very large the curve approaches $y = a/A$ and crosses this line in one point, consequently it must be of the form shown in the figure.

(ii) WHEN $ax^2 + 2bx + c = 0$ HAS REAL ROOTS AND $Ax^2 + 2Bx + C = 0$ IMAGINARY ROOTS. (Fig. 14.)

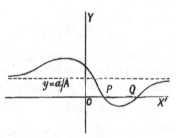

FIG. 14.

Suppose P and Q to be points such that OP, OQ represent the roots of $ax^2 + 2bx + c = 0$; then to the left of P the numerator is positive, between P and Q it is negative, to the right of Q it is positive, and the denominator is always positive. Thus y vanishes at P and Q, between P and Q it is negative and elsewhere positive. Hence since y is always finite the curve must be something like the one shown. (Fig. 14.)

NOTE. In cases (i) and (ii) we can easily settle at which end of the line $y = a/A$ the curve is above $y = a/A$, and which end below.

Thus if it meets OY above $y = a/A$ and $y = a/A$ to the right of OY, then it is as shown.

If it meets OY above $y = a/A$ and $y = a/A$ to the left of OY, then it is below $y = a/A$ at the negative end and above it on the positive end, and so on.

(iii) WHEN $ax^2 + 2bx + c = 0$ HAS IMAGINARY ROOTS AND $Ax^2 + 2Bx + C = 0$ HAS REAL ROOTS. (Fig. 15.)

Suppose R and S represent the roots of the denominator, then it is easily seen that to the left of R y is positive, at R

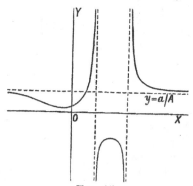

FIG. 15.

y becomes infinite; between R and S it is negative, at S it is infinite, and to the right of S positive. As it approaches the line $y = a/A$ the figure must be as shown (Fig. 15). Of the two branches one to the left of R and the other to the right of S only one meets $y = a/A$, and we can easily settle which it is by finding whether the curve meets the line in a point to the left of R or the right of S.

387. Cases in which both numerator and denominator have real roots.

It only remains to discuss the case in which both numerator and denominator have real roots.

In this case we have

$$y = \frac{a\,(x-\alpha)(x-\beta)}{A\,(x-\gamma)(x-\delta)},\qquad (\S\ 360)$$

where α, β are the roots of the numerator and γ, δ those of the denominator. Let P and Q represent the roots of the numerator and R, S those of the denominator, then we have four distinct cases arising according to the relative positions of these four points.

(iv) *Suppose P and Q are both between R and S*, then to the left of R y is positive, between R and P it is negative, between P and Q it is positive, between Q and S negative, and to the right of S it is always positive. The curve is as shown (Fig. 16).

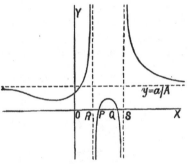

Fig. 16.

We remark that the branch to the left of R cannot come as low down as the branch between R and S because if it did we should have more than two values of x for the same value of y.

(v) *Suppose one of P and Q is between R and S and the other outside*, then the reader will easily perceive that the curve is somewhat as shown (Fig. 17).

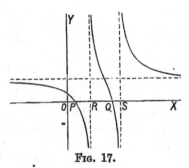

Fig. 17.

GRAPHIC REPRESENTATION OF FUNCTIONS. CONTINUITY. 403

(vi) *Suppose P and Q are both outside R, S and both on the same side of R* (Fig. 18).

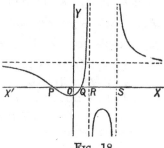

Fig. 18.

(vii) *P and Q both outside R, S and on opposite sides* (Fig. 19).

This exhausts all cases in which P, Q, R, S have general positions, i.e. when no two of them coincide.

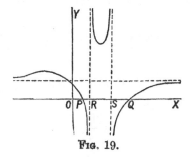

Fig. 19.

388. Maximum and minimum values of
$$\frac{ax^2 + 2bx + c}{Ax^2 + 2Bx + C}.$$

If $y = \dfrac{ax^2 + 2bx + c}{Ax^2 + 2Bx + C}$, then x is given by the quadratic equation

$$x^2(a - Ay) + 2x(b - By) + c - Cy = 0,$$

and if y be a possible value of the above expression this quadratic must have real roots,

i.e. $\qquad (b - By)^2 - (a - Ay)(c - Cy) \geqq 0,$

or $\qquad y^2(B^2 - AC) - y(2Bb - Ac - Ca) + (b^2 - ac) \geqq 0.$

Let y_1 and y_2 be the roots of
$$y^2(B^2 - AC) - y(2Bb - Ac - Ca) + b^2 - ac = 0, \text{ or } J = 0.$$
If these roots are real, then we must have
$$(B^2 - AC)(y - y_1)(y - y_2) \geqq 0.$$
Thus if $B^2 - AC > 0$, i.e. *if the denominator have real roots* (§ 159), y MUST NOT *lie between* y_1 *and* y_2.

If $B^2 - AC < 0$, i.e. *if the denominator have imaginary roots*, y MUST *lie between* y_1 *and* y_2.

389. General Determination of the side from which the graph approaches the line $y = a/A$.

We have $\quad y = \dfrac{ax^2 + 2bx + c}{Ax^2 + 2Bx + C} = \dfrac{a + \dfrac{2b}{x} + \dfrac{c}{x^2}}{A + \dfrac{2B}{x} + \dfrac{C}{x^2}}.$

Consequently when x is so large that $\dfrac{1}{x^2}$ may be **neglected** we have

$$y = \dfrac{a + \dfrac{2b}{x}}{A + \dfrac{2B}{x}}, \text{ and } \left(A + \dfrac{2B}{x}\right)\left(A - \dfrac{2B}{x}\right) = A^2;$$

$$\therefore y = \dfrac{1}{A^2}\left(a + \dfrac{2b}{x}\right)\left(A - \dfrac{2B}{x}\right)$$

$$= \dfrac{1}{A^2}\left\{aA + \dfrac{2}{x}(Ab - Ba)\right\}$$

$$= \dfrac{a}{A} + \dfrac{2}{A^2 x}(Ab - Ba),$$

neglecting $\dfrac{1}{x^2}$ and other powers of $\dfrac{1}{x}$ throughout*.

Thus when x becomes infinite $y = \dfrac{a}{A}$; but when x is very large the value of $y - \dfrac{a}{A}$ is $\dfrac{2}{A^2 x}(Ab - Ba)$ approximately.

---

* As to this see § 151.

GRAPHIC REPRESENTATION OF FUNCTIONS. CONTINUITY. 405

If $Ab - Ba$ be positive,

y is a little less than $\dfrac{a}{A}$ when x is very large and negative,
and

a little greater than $\dfrac{a}{A}$ when x is very large and positive;

thus in this case the graph approaches the line $y = a/A$ from the lower side on the left and from the upper side on the right.

If $Ab - Ba$ be negative,

y is a little greater than a/A when x is very large and negative,

and a little less than a/A when x is very large and positive; consequently in this case the graph approaches the line $y = a/A$ from the upper side on the left, and from the lower side on the right.

Hence if the roots of the equation J be real there are always limits between which y cannot lie or between which it must lie, and these limits are clearly maximum and minimum values of y.

Conversely, when there are such limits for y J must have real roots. Now, on examining the diagrams in cases (i)—(vii), we see that y takes all values in one case only, viz. case (v); $\therefore$ in this case only has J imaginary roots. Hence:—"The equation J has imaginary roots only when $ax^2 + 2bx + c = 0$ and $Ax^2 + 2Bx + C = 0$ have both real roots, and the roots of one equation have one root of the other between them."

When y takes the value y_1 or y_2 it is clear from the equation J that the corresponding values of x become equal, and therefore the curve has two tangents parallel to the axis of x, viz. $y = y_1$ and $y = y_2$.

If the roots of J are imaginary then the relation

$$y^2(B^2 - AC) - y(2Bb - Ac - Ca) + (b^2 - ac) > 0$$

is satisfied for all values of y, for in case (v) $B^2 - AC > 0$

(§ 388). Hence all values of y are possible in this case, and in this case only. (Compare Fig. 17.)

Ex. Consider $y = \dfrac{2x^2 - 22x + 61}{7x^2 - 78x + 219}$.

It is easy to see that both numerator and denominator have imaginary roots, and therefore y never vanishes and never becomes infinite so long as we only consider real values of the variable.

The quadratic giving x in terms of y is
$$x^2(2-7y) - 2x(11-39y) + 61 - 219y = 0,$$
and if this has real roots we must have
$$(11-39y)^2 - (2-7y)(61-219y) \geq 0,$$
i.e. $-(12y^2 - 7y + 1) \geq 0$;
$\therefore 12y^2 - 7y + 1 \leq 0$. $\therefore (4y-1)(3y-1) \leq 0$,
and y must always lie between $\tfrac{1}{3}$ and $\tfrac{1}{4}$.

Of course if y has either of these values the two values of x become equal. The student will easily find that $y = \tfrac{1}{3}$ gives $x = 6$ and $y = \tfrac{1}{4}$ gives $x = 5$.

The graph then lies entirely between the lines $y = \tfrac{1}{3}$ and $y = \tfrac{1}{4}$.

390. We shall now give some numerical examples, of which the first is discussed at greater length as a guide to the reader.

Ex. 1. $\quad y = \dfrac{(x-1)(x-3)}{(x-2)(x-4)}$. (Fig. 20.)

This belongs to case (v) in the general discussion, consequently there are no limiting values for y. This we verify as follows:

The values of x corresponding to a given value of y are the roots of
$$x^2(1-y) - x(4-6y) + 3 - 8y = 0.$$
$\therefore$ in order that x may be real we must have
$$(4-6y)^2 > 4(1-y)(3-8y),$$

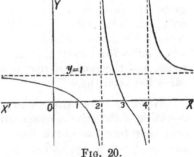

Fig. 20.

or $\quad (2-3y)^2 > (1-y)(3-8y)$, i.e. $4 - 12y + 9y^2 > 3 - 11y + 8y^2$,
i.e. $y^2 - y + 1 > 0$.

As the roots of $y^2 - y + 1 = 0$ are imaginary and the coefficient of y^2 is positive, the expression is always positive. The equation

$y^2 - y + 1 = 0$ is of course equivalent to the equation J in the general discussion.

When x is infinitely great we have $y=1$, hence the graph approaches the line $y=1$ in the manner shown.

It cuts $x=0$ in the point $y=\frac{2}{3}$.

When $x=1$, $y=0$, between $x=1$ and $x=2$ the numerator of y is negative and the denominator positive, thus y is negative. When $x=2$ y becomes infinite. Between $x=2$ and $x=3$ the numerator is negative, and so also is the denominator. $\therefore$ y is positive. y again vanishes when $x=3$, and between $x=3$ and $x=4$ is negative. It becomes infinite when $x=4$, and when $x>4$ it is positive, and gradually approaches $y=1$ from the upper side. Hence the curve is as shown. It meets the line $y=1$ in the finite point $x=2\frac{1}{2}$.

Ex. 2. $\qquad y = \dfrac{(x-2)(x-4)}{(x-1)(x-3)}.$

This again belongs to case (v), and there are no limiting values for y, as we leave the reader to verify. The discussion is similar to (i) and the graph is as shown (Fig. 21)

$\left(\text{when } x \text{ is large } y = 1 - \dfrac{2}{x}\right).$

Ex. 3. $\quad y = \dfrac{(x-1)(x-2)}{(x-3)(x-4)}.$

(Fig. 22.)

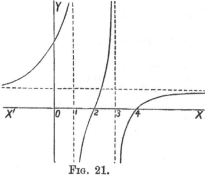

Fig. 21.

This belongs to case (vi), and there are limiting values of y which the reader will easily find to be $-7 \pm 4\sqrt{3}$ or $-·07$ and $-13·9$; when x is very large $y = 1 + \dfrac{4}{x}$. From the figure it follows that one value of x corresponding to a limiting value of y lies between 1 and 2 and the other between 3 and 4. The one between 1 and 2 corresponds to $-·07$, otherwise we should have values of y giving rise to

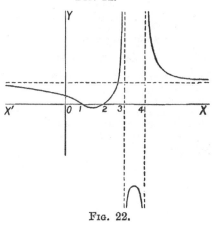

Fig. 22.

four values of x. The discussion is similar to (i), and the graph is as shown. The part between $x=3$ and $x=4$ does not come within a distance 13·9 of the axis, and it is drawn somewhat out of scale so as to appear in the figure.

Ex. 4. $\qquad y = \dfrac{(x-3)(x-4)}{(x-1)(x-2)}$. (Fig. 23.)

This belongs to the same case as (3), but the figure is modified. In fact in this curve the ordinate corresponding to any value of x is the reciprocal of the ordinate in (3).

The limiting values of y will be found to be the reciprocals of those in (3), i.e. $-13\cdot 9$ and $-\cdot 07$.

When x is large $y = 1 - \dfrac{4}{x}$.

The part between $x=1$ and $x=2$ is out of scale so as to be in the diagram.

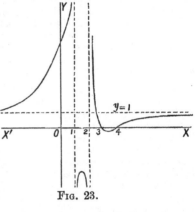

Fig. 23.

Ex. 5. $\qquad y = \dfrac{(x-1)(x-4)}{(x-2)(x-3)}$. (Fig. 24.)

This belongs to case (vii). The limiting values of y are 1 and 9. If we attempt to find an approximation for y when x is large we find the coefficient of $\dfrac{1}{x}$ zero, so the method fails.

In fact the curve does not meet $y=1$ in any finite point, and approaches it from the lower side at both ends. $y=1$ is one of the limiting values for y and $y=9$ is the other. The graph is as shown.

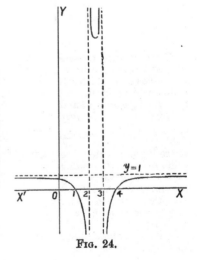

Fig. 24.

GRAPHIC REPRESENTATION OF FUNCTIONS. CONTINUITY. 409

Ex. 6. $y = \dfrac{(x-2)(x-3)}{(\quad - \quad -4)}$. (Fig. 25.)

This we leave to the reader.

Ex. 7. $y = \dfrac{x^2 - x + 1}{x^2 + x + 1}$. (Fig. 26.)

Here both numerator and denominator have imaginary roots, and the example belongs to case (i).

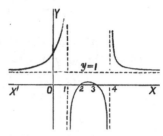

FIG. 25.

The quadratic for x is
$$x^2(1-y) - x(1+y) + 1 - y = 0.$$
∴ as x is real we must have
$$(1+y)^2 \not< 4(1-y)^2,$$
i.e. $3y^2 - 10y + 3 \not> 0$.
∴ y lies between $\tfrac{1}{3}$ and 3.
When x is infinite $y = 1$.

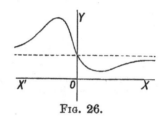

FIG. 26.

Other values of y are given in the table:

$x=$	0	1	-1	2	-2	3	-3	4	-4
$y=$	1	$\tfrac{1}{3}$	3	$\tfrac{3}{7}$	$\tfrac{7}{3}$	$\tfrac{7}{13}$	$\tfrac{13}{7}$	$\tfrac{13}{21}$	$\tfrac{21}{13}$

Ex. 8. $y = \dfrac{2x^2 - 3x + 1}{x^2 + 1}$. (Fig. 27.)

Here the numerator has real roots and the denominator imaginary roots.

Therefore y never becomes infinite, and this belongs to case (ii).

The quadratic for x when y is given is
$$x^2(2-y) - 3x^2 + 1 - y = 0.$$

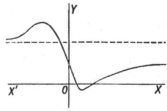

FIG. 27.

∴ 9 ≮ 4 (1 − y) (2 − y),

and y lies between the limits

$$\frac{+3-\sqrt{10}}{2} \text{ and } \frac{+3+\sqrt{10}}{2}.$$

$x = \infty$ gives $y = 2$, and $y = 0$ when $x = 1$ or $\frac{1}{2}$.

Other values of y are given below.

x	0	+1	−1	2	−2	3	−3	4	−4
y	1	0	3	$\frac{3}{5}$	3	1	$\frac{14}{5}$	$\frac{2\frac{1}{7}}{}$	$\frac{45}{17}$

Ex. 9. $y = \dfrac{x^2 + 4x + 5}{2x^2 - 5x + 2}$.   (Fig. 28.)

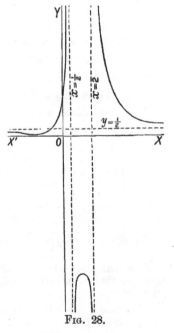

Here the numerator has imaginary and the denominator real roots.

Therefore y never vanishes, but does become infinite, viz. when $x = \frac{1}{2}$ or 2.

Since

$$x^2(1 - 2y) + x(4 + 5y) + 5 - 2y = 0$$

must have real roots, we find that

$$(4 + 5y)^2 ≮ 4(1 - 2y)(5 - 2y).$$

Hence the limiting values for y

are $\dfrac{-44 - 2\sqrt{493}}{9}$

and $\dfrac{-44 + 2\sqrt{493}}{9}$.

$x = \infty$ gives $y = \frac{1}{2}$, and other values of y are given below.

Fig. 28.

x	0	1	−1	2	−2	3	−3	4	−4
y	$\frac{5}{2}$	−10	$\frac{2}{9}$	∞	$\frac{1}{20}$	$\frac{26}{5}$	$\frac{2}{35}$	$\frac{37}{14}$	$\frac{5}{54}$

391. Graph of $y = ax^3 + bx^2 + cx + d$.

The values of x which make y vanish are the roots of the equation
$$ax^3 + bx^2 + cx + d = 0,$$
and since imaginary roots occur in pairs either all three roots are real, say a, β, γ, or only one is real.

We shall discuss these two cases separately.

(I) If $\qquad y = a(x-a)(x-\beta)(x-\gamma)$
and $\qquad a < \beta < \gamma,$
then we have already discussed the sign of y for all values of x, so we are only concerned with the magnitude which can be calculated when $a\beta\gamma$ and a are known.

The values of x which make y infinite are $-\infty$ and $+\infty$, and supposing a positive the former makes $y = -\infty$, and the latter $y = +\infty$.

Let P, Q, R represent the vanishing points of the function, then to the left of P, since y is negative, the

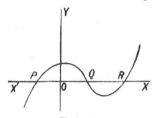

Fig. 29.

curve is below the line XOX', between P and Q it is above this line, between Q and R it is again below, and to the right of R it is always above the line.

Hence it is something like the curve shown in the figure (Fig. 29).

(II) If only one root be real the expression may be written
$$y = a(x-a)(x^2 + bx + y),$$
where the roots of the equation $x^2 + bx + y = 0$ are imaginary.

If P represent the point a then, since
$$x^2 + bx + y$$

412 GRAPHIC REPRESENTATION OF FUNCTIONS. CONTINUITY.

is always positive, when $x < a$ y is negative and when $x > a$ y is positive.

The curve is either of the same form as the foregoing, but only cuts the axis in one real point (Fig. 30), or else there is no twist in it (Fig. 31). This point can only be settled in our state of knowledge by considering numerical examples.

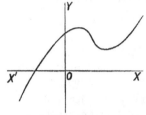

FIG. 30.

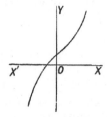

FIG. 31.

A numerical example of each case will now be given.

Ex. 1. $\qquad y = x^3 - 3x.$ (Fig. 32.)

Here $\qquad y = 0$

when $\qquad x = 0, \ x = \sqrt{3}, \ x = -\sqrt{3}.$

Further $\quad x = \ \ 1, \ -1, \ \ \ 2, \ -2$

give $\qquad y = -2, \ +2, \ +2, \ -2$

and noticing that y increases very rapidly indeed with x we have a diagram like the one shown.

Ex. 2. $\quad y = x^3 - 4x^2 + 5x.$ (Fig. 33.)

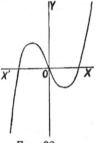

FIG. 32.

The only real value of x giving $y = 0$ is $x = 0$.

The values 0, 1, 2, 3 of x give $y = 0, 2, 1, 6$, and hence it is clear that the curve has a twist in it.

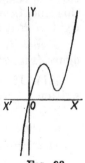

FIG. 33.

GRAPHIC REPRESENTATION OF FUNCTIONS. CONTINUITY. 413

392. Graph of $y = \sqrt{f(x)}$.

Suppose we know the shape of the graph for $f(x)$, then it is not difficult to deduce some general ideas as to that for $\sqrt{f(x)}$.

Viz. for a value of x which makes $f(x)$ positive gives $\sqrt{f(x)}$ two equal and opposite values; while a value which makes $f(x)$ infinite makes $\sqrt{f(x)}$ infinite also. If $y = f(x)$ approach the line $y = a$ for infinite values of x, then $y = \sqrt{f(x)}$ approaches both the lines $y = \pm \sqrt{a}$.

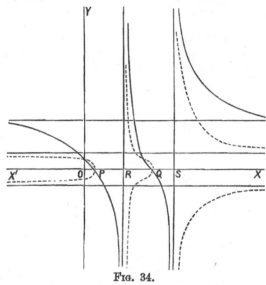

Fig. 34.

Thus consider

$$y = \sqrt{\frac{a(x-\alpha)(x-\beta)}{A(x-\gamma)(x-\delta)}}$$ (Fig. 34), (see § 386, case (ii)),

where $\delta > \beta > \gamma > \alpha$.

Taking the figure from § 387 the dotted line represents our graph and the continuous one (Fig. 34) the graph of

$$y = \frac{a(x-\alpha)(x-\beta)}{A(x-\gamma)(x-\delta)}.$$

To the left of P
$$\sqrt{\frac{a(x-\alpha)(x-\beta)}{A(x-\gamma)(x-\delta)}}$$
is real and lies between $+\sqrt{a/A}$ and $-\sqrt{a/A}$. Between P and R it is imaginary. Between R and Q it is real and has two infinite branches both approaching the line $x=\gamma$. Between Q and S it is imaginary, and to the right of S real again. It is quite clear that such a curve is always symmetrical with respect to the axis and meets $y=f(x)$ on the line $y=1$, since $y=\sqrt{y}$ at these points.

Ex. 1. The graph of $y=\sqrt{x}$ is a parabola, viz. $y^2=x$.

For each value of x there are two equal and opposite values of y; $y=0$ when $x=0$ and $y=\infty$ when $x=\infty$. For negative values of x, y is imaginary. (Fig. 35.)

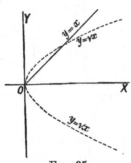

Fig. 35.

Ex. 2. The graph of $y=\sqrt{ax+b}$ is a parabola for all values of a and b for $y^2=ax+b$.

Ex. 3. The graph of $y=\sqrt{a^2-x^2}$ is a circle whose centre is at the origin, and whose radius is a.

Ex. 4. The graph of $y=\sqrt{a+bx-x^2}$ is a circle having its centre on the axis of x.

The student should construct the graphs of each case of
$$y=\sqrt{\frac{ax^2+2bx+c}{Ax^2+2Bx+C}}$$
in the same way as we have treated case (v).

GRAPHIC REPRESENTATION OF FUNCTIONS. CONTINUITY. 415

393. Example of Discontinuity.

By considering the graph of a function $f(x)$ it is clear that in general $f(x)$ cannot change sign without vanishing (i.e. if $f(x)$ is positive when $x = a$ and negative when $x = b$), then in general $f(x)$ vanishes for some value of x between a and b), for the curve to get from one side of the line OX to the other must generally cross OX.

But exceptional cases certainly arise, thus $\dfrac{1}{x-1}$ is negative when $x = 0$ and positive when $x = 2$, but so far from vanishing in the interval it becomes infinite. When the change of sign occurs the function $\dfrac{1}{x-1}$ suddenly changes from a very large negative value to a very large positive value, although x may only alter by very little. Thus when $x = \cdot 99$, $\quad \dfrac{1}{x-1} = -100$;

and when $x = 1\cdot 01$, $\quad \dfrac{1}{x-1} = 100$.

We say that $\dfrac{1}{x-1}$ is *discontinuous* when $x = 1$, and roughly speaking in such a case we mean that there is a very large change in the value of the function corresponding to a small change in the variable. Thus in the example when x changes by $\cdot 02$,

$$\dfrac{1}{x-1} \text{ changes by } 200.$$

We shall enter into this more fully in the next article, but meanwhile we remark that the consideration of the graph cannot afford a rigorous proof of a property of the corresponding function, for as we have never laid down a definition of the word *curve* we cannot without further examination identify its properties with those of a function. It will be evident in fact that any property of the function which is intuitively evident on examining the graph must have been tacitly assumed in drawing the graph.

394. Continuity. DEFINITION. A function of x, $f(x)$ is said to be **continuous** for the value a when corresponding to every finite quantity ϵ, however small, we can choose a quantity h, such that for values of the variable which differ from a by less than h the value of the function differs from A by less than ϵ where $f(a) = A$.

Thus $\qquad f(x) \sim f(a) < \epsilon,$
whenever $\qquad x \sim a < h,$
where the sign $\sim$, as usual, means difference between, expressed positively.

The above is the accurate definition. It may be expressed shortly (and less precisely) by saying that to any small change in the variable corresponds a small change in the function, or the function changes gradually with the variable.

DEFINITION. A function is said to be **continuous between a and b** when it is continuous for all values of x lying between a and b.

Ex. x^2 is a continuous function of x for all values of x and when $x = 2$, $x^2 = 4$; suppose we want to find within what limits x^2 differs from 4 by less than $\cdot 1$.

We have $\qquad \sqrt{4 \cdot 1} = 2 \cdot 0248 \ldots\ldots$ and $\sqrt{3 \cdot 9} = 1 \cdot 9748 \ldots\ldots$

Thus $\qquad \sqrt{4 \cdot 1} - 2 > \cdot 0248,$

$\qquad\qquad 2 - \sqrt{3 \cdot 9} > \cdot 0251.$

Hence when $\qquad x - 2 < \cdot 0248 \quad x^2 - 4 < \cdot 1,$
and when $\qquad 2 - x < \cdot 0251 \quad 4 - x^2 < \cdot 1.$

Therefore certainly when x differs from 2 by less than $\cdot 0248$, x^2 differs from 4 by less than $\cdot 1$,

or $\qquad x^2 \sim 4 < \cdot 1$ whenever $x \sim 2 < \cdot 0248.$

395. If a finite number of functions are continuous for a given value of x so also is their sum.

Let the functions $f_1(x)$, $f_2(x)$, $f_3(x) \ldots f_n(x)$ be continuous for the value a, then taking ϵ/n to be the change in each function, we have

GRAPHIC REPRESENTATION OF FUNCTIONS. CONTINUITY. 417

$f_1(x) \sim f_1(a) < \epsilon/n$, whenever $x \sim a < h_1$,
$f_2(x) \sim f_2(a) < \epsilon/n$, „ $x \sim a < h_2$,
....
$f_n(x) \sim f_n(a) < \epsilon/n$, „ $x \sim a < h_n$,

where the h's are not necessarily the same, but different for different functions. If, however, h be the smallest of them then all the inequalities on the left hold when

$$x \sim a < h.$$

Hence, adding

$$\{f_1(x) + f_2(x) \ldots f_n(x)\} \sim \{f_1(a) + f_2(a) \ldots f_n(a)\} < \epsilon,$$

whenever $x \sim a < h$.

Hence $f_1(x) + f_2(x) \ldots + f_n(x)$ is continuous when $x = a$.

396. If a finite number of functions be continuous when $x = a$ so also is their product *if the functions are finite.*

Consider first two functions $f_1(x)$ and $f_2(x)$ then, by hypothesis, however small ϵ_1 and ϵ_2 may be, we have

$f_1(x) \sim f_1(a) < \epsilon_1$, whenever $x \sim a < h_1$,
and $f_2(x) \sim f_2(a) < \epsilon_2$, „ $x \sim a < h_2$,

so that if h be the smaller of the two quantities h_1 and h_2 both relations hold whenever $x \sim a < h$.

Hence $f_1(x)f_2(x) \sim f_1(a)f_2(a) < \epsilon_1 f_2(a) + \epsilon_2 f_1(a) + \epsilon_1 \epsilon_2$
if only $x \sim a < h$.

But if $f_1(a)$ and $f_2(a)$ be both finite the right-hand side may be made less than any finite quantity ϵ, however small, by sufficiently diminishing ϵ_1 and ϵ_2; consequently

$$f_1(x)f_2(x) \sim f_1(a)f_2(a) < \epsilon \text{ if only } x \sim a < h;$$

∴ $f_1(x)f_2(x)$ is a continuous function of x.

If we have three functions f_1, f_2, f_3, then $f_1 f_2 f_3$ is continuous; for by the above $f_1 f_2$ is, and then applying the above again we see that $f_1 f_2 f_3$ is continuous. So on, for any finite number of functions. Q. E. D.

397. Continuity of any rational integral function.

First, if m be a positive integer and a a constant, then ax^m is continuous.

For x is continuous being the variable itself, and therefore so also is

$$x^m = x \times x \times x \ldots \text{ for } m \text{ factors};$$

also a does not alter with x, and therefore ax^m is continuous.

Now any rational integral function of x is the sum of a finite number of terms like ax^m, each of which is continuous, therefore the whole function is continuous. (§ 395.)

398. If a **continuous** function of x be positive when x has the value a, then it is positive for **all values of x lying within a certain range on both sides of a.**

Let $f(x)$ be the function, and suppose $f(a) = A$, so that A is positive; then by the definition of continuity we can choose h, so that $\quad f(x) \sim f(a) < A$,
whenever $\quad\quad\quad\quad x \sim a < h$.

Now any quantity which differs from a positive quantity by less than that quantity is obviously positive, and therefore $f(x)$ is positive whenever x differs from a by less than h, i.e. when x lies between $a - h$ and $a + h$.

Similarly, *if $f(x)$ be negative when $x = b$ it is negative for all values of x lying within a certain range on both sides of b.*

This perhaps may be made clearer by a reference to the graphical method. Measure off x along the line Ox (Fig. 36) and let $Oa = a$, $Oa_1 = a - h$ and $Oa_2 = a + h$.

$y = f(x)$ is positive at a by hypothesis, and we have shown that it is positive for all points lying between a_1 and a_2.

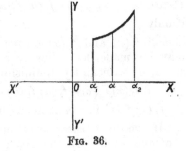

Fig. 36.

GRAPHIC REPRESENTATION OF FUNCTIONS. CONTINUITY. 419

399. If a continuous function of x be positive when $x = a$ and negative when $x = b$, then it vanishes for some value of x between a and b.

Let $y = f(x)$ be the function of x, and suppose for convenience that $b > a$.

Let $Oa = a$ and $O\beta = b$ (Fig. 37), then at a y is positive, and therefore by the last article we can find a point γ to the right of a such that y is positive for all points between a and γ. (§ 398.)

Again, γ must be between a and β because y is negative at β. We shall show that at γ y is zero.

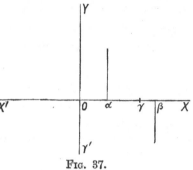

FIG. 37.

For if y were positive at γ we could choose a point δ to the *right* of γ (§ 398), such that y is positive for all points between γ and δ, and therefore the range in which y is positive would extend beyond γ.

Secondly, if y were negative we could choose a point ϵ to the *left* of γ (§ 398), such that for all points in $\gamma\epsilon$ y is negative, and therefore the range in which y is positive would not extend so far as γ.

Hence, at γ y is zero, and if $O\gamma = c$ we have $f(c) = 0$. Exactly the same method applies if $b < a$.

Ex. 1. The function $(x-1)(x-3)$ is positive when $x = 0$ and negative when $x = 2$, and it vanishes for $x = 1$ between 0 and 2.

Ex. 2. The function $\dfrac{1}{x-1}$ is negative when $x = 0$ and positive when $x = 2$, but it does not vanish for any intermediate value of x. It is in fact not continuous for $x = 1$ since it here changes sign by suddenly passing from $-\infty$ to $+\infty$.

400. A continuous function cannot overleap any value intermediate between two values acquired by it.

i.e. if $f(a) = A$ and $f(b) = B$, then we shall show that if C lie between A and B then there is a value of c between a and b, such that $f(c) = C$.

Suppose for convenience that $B > A$, then $B > C > A$.

Consider the continuous function
$$f_1(x) = f(x) - C,$$
when $x = a$ $f_1(x) = A - C$, and is negative;
when $x = b$ $f_1(x) = B - C$, and is positive.

Thus as $f_1(x)$ is negative when $x = a$ and positive when $x = b$, it vanishes for some intermediate value of x, say c, and we have
$$f_1(c) = f(c) - C = 0.$$
$$\therefore f(c) = C.$$

Exactly the same method applies if $B < A$.

EXERCISES XXVI.

1. Show that the expression $y = ax + b$ is represented by a straight line and discuss the following cases:

(i) $x + 1$, (ii) $3x + 4$, (iii) $-5x + 6$, (iv) $-x - 2$.

Represent graphically the following functions, find the real values of x for which they vanish, their limiting values and the values of x, if any, for which they are positive:

2. $x^2 + 4x + 2$.   3. $3x^2 + 4x - 5$.   4. $x^2 - a^2$.
5. $x^2 - 6x + 9$.   6. $x^2 + 2x + 7$.   7. $5x^2 + 12x + 13$.
8. $3x^2$.   9. $-x^2 + 4x - 3$.   10. $-2x^2 - 3x - 4$.
11. $-x^2 - 6x + 5$.   12. $x^2 + 10x + 26$.

Represent graphically the following, find whether they become zero or infinite for finite values of x, find their limiting values and state to which of the seven cases of
$$y = (ax^2 + 2bx + c)/(Ax^2 + 2Bx + C)$$
each example belongs.

GRAPHIC REPRESENTATION OF FUNCTIONS. CONTINUITY.

13. $\dfrac{(x+1)(x+2)}{(x+3)(x+4)}$.

14. $\dfrac{(x+3)(x+4)}{(x+1)(x+2)}$.

15. $\dfrac{(x+1)(x+3)}{(x+2)(x+4)}$.

16. $\dfrac{(x+2)(x+4)}{(x+1)(x+3)}$.

17. $\dfrac{(x+1)(x+4)}{(x+2)(x+3)}$.

18. $\dfrac{(x+2)(x+3)}{(x+1)(x+4)}$.

19. $\dfrac{(2x-3)(4x-5)}{(x-2)(3x-5)}$.

20. $\dfrac{x(x-2)}{(x-1)(x-3)}$.

21. $\dfrac{3x^2-10x+9}{7x^2-22x+19}$.

22. $\dfrac{x^2-1}{5x^2+8x+5}$.

23. $\dfrac{5x^2-12x+7}{x^2-2}$.

*24. $\dfrac{10x^2+14x+5}{7x^2+10x+3}$.

25. $\dfrac{2x^2+2x+5}{x^2+1}$.

26. $\dfrac{x^2+1}{x^2+12x+6}$.

27. $\dfrac{x^2+x+1}{x^2-x+1}$.

28. In connection with the graphic representation of
$$y=(ax^2+2bx+c)/(px+q),$$
show that (i) the graph is always a hyperbola,

(ii) there are always two real quantities between which y cannot lie unless the numerator has real roots and that of the denominator lies between them.

29. Discuss the graphic representation of
$$y=(px+q)/(ax^2+2bx+c),$$
and show that there are always real limiting values for y unless the denominator has real roots and the root of the numerator lies between them.

Construct the graphs of the following. Are there limits between which y cannot lie, or between which it must lie?

30. $\dfrac{x}{2-x}$.

31. $\dfrac{x-2}{2x-3}$.

32. $\dfrac{3x-4}{4x-5}$.

33. $\dfrac{1}{x-4}$.

34. $\dfrac{2x+3}{x^2+2x+3}$.

35. $\dfrac{5x^2+4x+1}{4x+1}$.

36. $\dfrac{3x-2}{2x^2-2x+5}$.

37. $\dfrac{2x^2-2x+5}{2x-1}$.

38. $\dfrac{x^2+3x+1}{x}$.

39. $\dfrac{x^2+3x+1}{x^2}$.

40. $\dfrac{x^2}{x^2+5x+2}$.

CHAPTER XXVII.

LIMITING VALUES AND THE ELEMENTARY THEORY OF INFINITE SERIES.

401. If in any rational integral function of a variable x (§ 370) an arbitrary finite value be given to x, it will always be found that the function itself acquires a determinate finite value. This, however, is not always the case if the function is *fractional, i.e.* has a denominator involving x. It is necessary to discuss such exceptional cases.

Ex. Find the value of
$$y \equiv \frac{x^2 - 3x + 2}{x^2 - 1} \text{ (i) when } x = 3, \text{ (ii) when } x = 1.$$

(i) The value is $\quad \dfrac{9 - 9 + 2}{9 - 1} = \dfrac{1}{4}$.

(ii) Substituting $x = 1$, the expression is
$$\frac{1 - 3 + 2}{1 - 1} = \frac{0}{0}.$$

Now the elementary operations of Algebra include only addition, subtraction, multiplication, and division *by any quantity other than zero or infinity*. Hence the expression $\frac{0}{0}$ is meaningless or technically speaking *indeterminate*. The above function therefore has no *value* when $x = 1$ and will require closer consideration.

Substitute in order the values
$$x = \quad \cdot 9, \quad \cdot 99, \quad \cdot 999, \quad 1\cdot 001, \quad 1\cdot 01, \quad 1\cdot 1,$$
then $\quad y = -\cdot 52, \; -\cdot 506\ldots, \; -\cdot 5006\ldots, \; -\cdot 499, \; -\cdot 492, \; -\cdot 42.$

These values indicate that as x acquires values approaching 1, y acquires values approaching $-\frac{1}{2}$. Suppose it can be shown that y can be made as nearly equal to $-\frac{1}{2}$ as we please by taking x sufficiently near the value 1: then $-\frac{1}{2}$ would be said to be **the limit of y when $x = 1$.**

LIMITING VALUES AND THEORY OF INFINITE SERIES.

Before proceeding to the formal definition of a limit, some terms of frequent occurrence must be explained.

(1) A quantity u is said to *tend to zero*, to *become vanishingly small*, or to *decrease indefinitely*, if it can be shown to diminish to a value numerically smaller than any arbitrarily selected finite quantity ϵ, however small ϵ may be.

(2) The positive numerical or *absolute* value of any quantity u is denoted by $|u|$. Thus the numerical value of -1, written $|-1|$ is 1. This notation will be found very useful in the sequel.

Observe that an inequality of the form $|x-a| < \delta$ states that x lies between $a-\delta$ and $a+\delta$.

The formal definition of a limit may now be given as follows.

DEFINITION.—A function $f(x)$ approaches a **limit** l when x approaches a, if $|f(x)-l| < \epsilon$ for **all** values of x such that $|x-a| < h$, where ϵ is any finite quantity however small, and h is some quantity depending on ϵ.

The notation employed is
$$\operatorname*{L^t}_{x=a} f(x) = l.$$

Ex. Prove that in the example at the beginning of § 401
$$\operatorname*{L^t}_{x=1} \frac{x^2-3x+2}{x^2-1} = -\frac{1}{2}.$$

Consider
$$\left| \frac{x^2-3x+2}{x^2-1} \sim \frac{-1}{2} \right| \text{ or } \left| \frac{3x^2-6x+3}{2(x^2-1)} \right| = u \text{ (say)}$$

for values of x nearly $=1$.

Suppose k is a small positive quantity and let $|x-1| = k$, *i.e.* either $x = 1+k$ or $x = 1-k$.

If $\quad x = 1+k, \quad u = \left| \dfrac{3k^2}{2k(2+k)} \right| = \dfrac{3k}{2(2+k)}.$

If $\quad x = 1-k, \quad u = \left| -\dfrac{3k^2}{2k(2-k)} \right| = \dfrac{3k}{2(2-k)}.$

Of these two expressions the second is the greater, and therefore each expression is less than ϵ provided that
$$3k < 2(2-k)\epsilon, \quad i.e. \quad k < \frac{4\epsilon}{3+2\epsilon}.$$

It follows that
$$\left|\frac{x^2-3x+2}{x^2-1} - -\frac{1}{2}\right| < \epsilon \text{ if } |x-1| < \frac{4\epsilon}{3+2\epsilon}.$$
$$\therefore \underset{x=1}{L^t} \frac{x^2-3x+2}{x^2-1} = -\frac{1}{2}.$$

Special cases of limits.

The definition cannot be directly applied to the case when $f(x)$ increases indefinitely as x approaches a.

Take the case $\underset{x=1}{L^t} \frac{1}{(x-1)^2}.$

Put $x = 1 + h$. Then $\frac{1}{(x-1)^2} >$ any quantity A however large if $\frac{1}{h^2} > A$ or if $h < \sqrt{\frac{1}{A}}$. In this case the limit is said to be *infinite*.

The required modification of the above definition is seen to be the following:—

DEFINITION.—$\underset{x=a}{L^t} f(x) = \infty$ if, when A is any selected finite quantity however great, $|f(x)| > A$ for all values of x such that $|x - a| < h$, h being a finite quantity depending on A.

Similarly, some functions tend to a definite limit as the variable (x) increases indefinitely.

DEFINITION.—$\underset{x=\infty}{L^t} f(x) = l$ if $|f(x) - l| < \epsilon$ for all values of x such that $|x| > A$, if ϵ be any selected finite quantity however small, and A is some finite quantity depending on ϵ.

If we are dealing with real quantities, the notation $|f(x) - l|$ is equivalent to $f(x) \sim l$.

Whenever the value $f(a)$, which is taken by $f(x)$ when $x = a$, is finite and can be found by direct substitution, this value satisfies the ordinary definition for $\underset{x=a}{L^t} f(x)$. When $f(a)$ is infinite, then it satisfies the above definition for an infinite limit. And if $f(\infty)$ can be found by direct substitution of ∞ for x, it satisfies the definition of $\underset{x=\infty}{L^t} f(x)$.

The following statements should now be obvious:—
$$\underset{x=b}{L^t}(1+x) = 1+b, \ \underset{x=0}{L^t}(a+x) = a, \ \underset{x=0}{L^t}\frac{1}{x} = \infty.$$

402. The limit of the sum of a finite number of functions is equal to the sum of their limits.

Thus if $\operatorname*{Lt}_{x=a} f_1(x) = l_1$, $\operatorname*{Lt}_{x=a} f_2(x) = l_2 \ldots \operatorname*{Lt}_{x=a} f_n(x) = l_n$,

then $\operatorname*{Lt}_{x=a} \{f_1(x) + f_2(x) \ldots f_n(x)\} = l_1 + l_2 \ldots l_n$.

For if ϵ be any finite quantity however small we can choose h, so that

$$f_1(a+h) - l_1 < \frac{\epsilon}{n}, \quad f_2(a+h) - l_2 < \frac{\epsilon}{n} \ldots f_n(a+h) - l_n < \frac{\epsilon}{n},$$

and $\therefore$ $\{f_1(a+h) + f_2(a+h) \ldots + f_n(a+h)\}$
$$- (l_1 + l_2 \ldots + l_n) < \epsilon,$$

which proves the theorem.

This we may point out does not necessarily hold in the case of an infinite number of quantities. Thus for example we have

$$\frac{n}{n+1} > \frac{1}{n+1} + \frac{1}{n+2} + \ldots + \frac{1}{2n} > \frac{n}{2n},$$

where n is a positive integer,

$\therefore \operatorname*{Lt}_{n=\infty} \left(\frac{1}{n+1} + \frac{1}{n+2} + \ldots + \frac{1}{2n} \right)$ lies between $\frac{1}{2}$ and 1,

whereas the limit of each of the terms is manifestly zero.

403. The limit of the product of a finite number of functions is equal to the product of their limits, *if these limits are finite.*

First take two functions $f_1(x)$ and $f_2(x)$, and suppose that

$$\operatorname*{Lt}_{x=a} f_1(x) = l_1 \text{ and } \operatorname*{Lt}_{x=a} f_2(x) = l_2,$$

then we shall have

$$\operatorname*{Lt}_{x=a} \{f_1(x) \cdot f_2(x)\} = l_1 l_2.$$

For we have

$f_1(x) \sim l_1 < \epsilon_1$, whenever $x \sim a < h_1$

and $f_2(x) \sim l_2 < \epsilon_2$, ,, $x \sim a < h_2$.

Let therefore $f_1(x) - l_1 = k_1$, $f_2(x) - l_2 = k_2$
where $|k_1| < \epsilon_1$, $|k_2| < \epsilon_2$.

Hence $f_1(x) = l_1 + k_1$, $f_2(x) = l_2 + k_2$,
$$\therefore f_1(x) f_2(x) = (l_1 + k_1)(l_2 + k_2)$$
or $f_1(x) f_2(x) - l_1 l_2 = k_1 l_2 + k_2 l_1 + k_1 k_2$,
$$\therefore |f_1(x) f_2(x) - l_1 l_2| < |l_1 \epsilon_2| + |l_2 \epsilon_1| + |\epsilon_1 \epsilon_2|$$
if only $x \sim a < h$.

Now, by taking ϵ_1 and ϵ_2 small enough we can make the right-hand side less than any finite quantity ϵ, consequently
$$f_1(x) f_2(x) \sim l_1 l_2 < \epsilon, \text{ if only } x \sim a < h,$$
however small ϵ may be; i.e.
$$L^t\{f_1(x) f_2(x)\} = l_1 l_2.$$

If we have three quantities f_1, f_2, f_3, then
$$L^t\{f_1 . f_2 . f_3\} = L^t(f_1) \times L^t(f_2 f_3) = L^t(f_1) \times L^t(f_2) \times L^t(f_3)$$
and so on for any number. Q. E. D.

Here again the result does not necessarily hold for an infinite number of functions.

For the limit of each of the quantities
$$\frac{n}{n+1}, \frac{n+1}{n+2}, \cdots \frac{2n-1}{2n},$$
when n is infinite is unity while the limit of the product is $\frac{1}{2}$.

COR. I. $\underset{x=a}{L^t}\{f(x)\}^r = \{\underset{x=a}{L^t} f(x)\}^r$.

This follows by taking $f_1(x) = f_2(x) = \ldots$
$= f_r(x) = f(x)$ in the main proposition.

COR. II. If $f_1(x)$ has a definite *value* k when $x = a$ and if $\underset{x=a}{L^t} f_2(x) = l$, then
$$\underset{x=a}{L^t}[f_1(x) . f_2(x)] = f_1(a) \underset{x=a}{L^t} f_2(x) = kl.$$

In this both k and l must be finite.

LIMITING VALUES AND THEORY OF INFINITE SERIES. 427

404. We shall now give some examples of limits.

Ex. 1. To show that
$$\operatorname*{L^t}_{n=\infty} \frac{1 \cdot 3 \cdot 5 \ldots 2n-1}{2 \cdot 4 \cdot 6 \ldots 2n} = 0.$$

Here let
$$v = \tfrac{1}{2} \cdot \tfrac{3}{4} \cdot \tfrac{5}{6} \ldots \frac{2n-1}{2n} \quad\ldots\ldots\ldots\ldots\ldots\ldots (1),$$

then
$$v < \tfrac{2}{3} \cdot \tfrac{4}{5} \cdot \tfrac{6}{7} \ldots \frac{2n}{2n+1} \quad\ldots\ldots\ldots\ldots\ldots\ldots (2),$$

for each factor in (2) is greater than the corresponding one of (1).

Multiplying these results together we have
$$v^2 < \frac{1}{2n+1}.$$

Consequently v^2, and therefore v, can be made to differ from zero by less than any quantity however small if n be sufficiently increased, and
$$\therefore \operatorname*{L^t}_{n=\infty} v = 0.$$

Ex. 2.
$$\operatorname*{L^t}_{n=\infty} \frac{3 \cdot 5 \cdot 7 \ldots 2n+1}{2 \cdot 4 \cdot 6 \ldots 2n} = \infty.$$

Here
$$v = \tfrac{3}{2} \cdot \tfrac{5}{4} \cdot \tfrac{7}{6} \ldots \frac{2n+1}{2n},$$

and
$$v > \tfrac{4}{3} \cdot \tfrac{6}{5} \cdot \tfrac{8}{7} \ldots \frac{2n+2}{2n+1},$$

hence on multiplication,
$$v^2 > \frac{2n+2}{2} \text{ or } n+1;$$

$\therefore$ by sufficiently increasing n we can make v^2 and therefore v greater than any finite quantity however large, or
$$\operatorname*{L^t}_{n=\infty} v = \infty.$$

Ex. 3. $\operatorname*{L^t}_{n=\infty} \left(x - \frac{a_1}{n}\right)\left(x - \frac{a_2}{n}\right)\left(x - \frac{a_3}{n}\right) \ldots \left(x - \frac{a_r}{n}\right) = x^r;$

this follows at once from § 403 since the limit of each term of the product is x, the a's being all finite.

We shall examine more closely the case in which the a's are all positive.

Taking the first two factors we have

$$\left(x - \frac{a_1}{n}\right)\left(x - \frac{a_2}{n}\right) > x^2 - \frac{a_1+a_2}{n}x;$$

then multiplying by another factor,

$$\left(x - \frac{a_1}{n}\right)\left(x - \frac{a_2}{n}\right)\left(x - \frac{a_3}{n}\right) > x^3 - \frac{a_1+a_2+a_3}{n}x^2,$$

and so on.

Thus the continued product is less than x^r and greater than

$$x^r - \frac{a_1+a_2 \ldots +a_n}{n}x^{r-1},$$

a result which will be useful to us hereafter.

Ex. 4. $\qquad \underset{n=\infty}{\mathrm{L}^t} \dfrac{x^n}{n!} = 0.$

For $\qquad \dfrac{x^n}{n!} < \dfrac{x^r}{r!}\left(\dfrac{x}{r}\right)^{n-r}$ if $n > r$.

Now $\dfrac{x^r}{r!}$ is finite if r be finite, and if r be greater than x the other factor can be made as small as we please;

$$\therefore \underset{n=\infty}{\mathrm{L}^t} \frac{x^n}{n!} = 0.$$

Ex. 5. If $x > 1$ $\underset{n=\infty}{\mathrm{L}^t} x^n = \infty$, n being an integer.

Let $x = 1 + \xi$ then $x^2 = 1 + 2\xi + \xi^2 > 1 + 2\xi$,

$\qquad x^3 > 1 + 3\xi, \ldots x^n > 1 + n\xi$.

Thus $\qquad x^n > r$ if $1 + n\xi > r$, i.e. if $n > \dfrac{r-1}{\xi}$;

$\therefore$ by sufficiently increasing n we can make x^n greater than any quantity however large.

Ex. 6. If $x < 1$ $\underset{n=\infty}{\mathrm{L}^t} x^n = 0$, n being an integer.

Let $x = \dfrac{1}{y}$ then by taking n large enough we can make $y^n > r$;

$$\therefore x^n = \frac{1}{y^n} < \frac{1}{r} \text{ however large } r \text{ may be;}$$

$$\therefore \underset{n=\infty}{\mathrm{L}^t} x^n = 0.$$

LIMITING VALUES AND THEORY OF INFINITE SERIES. 429

Both these results are true when n is not an integer for, when $x>1$, x^n increases continually with n; but we only need the cases given above.

Ex. 7. If n be a positive integer
$$\operatorname*{Lt}_{x=1} \frac{x^n - 1}{x - 1} = n.$$

The student will notice that if we put $x=1$ at once we get $\frac{0}{0}$, a result quite indeterminate, so that we can *only* assign a value to this by the method of limits.

We have $\quad \dfrac{x^n - 1}{x - 1} = x^{n-1} + x^{n-2} + x^{n-3} \ldots + 1,$

and $\quad x^{n-1} + x^{n-2} + x^{n-3} \ldots + 1 - n$
$\quad = (x^{n-1} - 1) + (x^{n-2} - 1) \ldots + (x - 1)$

contains $x-1$ as a factor, and can therefore be made as small as we please by bringing x sufficiently near unity; hence as the other factor is finite we have
$$\operatorname*{Lt}_{x=1} \frac{x^n - 1}{x - 1} = n.$$

405. Theorem. Let $a_0 + a_1 x + a_2 x^2 + \ldots + a_p x^p$ be any function of x. Choose any term $a_r x^r$ of this expression. Then the sum of all the terms following $a_r x^r$ can be made $< a_r x^r$ provided x is taken small enough.

First let x, and all the coefficients, be real and positive. Then suppose K to be the value of the greatest of the coefficients $\quad a_{r+1} \quad a_{r+2} \quad a_{r+3} \ldots a_p$

Then $\quad (a_{r+1} x^{r+1} + a_{r+2} x^{r+2} \ldots a_p x^p) \ldots (a)$
$\quad < K(x^{r+1} + x^{r+2} + \ldots x^p)$
$\quad < K x^{r+1} \left[\dfrac{1 - x^{p-r}}{1 - x} \right].$

If $x < 1$ and positive the above expression is
$$< \frac{K x^{r+1}}{1 - x}.$$

Hence the object required will be attained if $a_r x^r$ can be made $< \dfrac{K x^{r+1}}{1-x}$ i.e. if $a_r > \dfrac{Kx}{1-x}$ . . . . (1)

This inequality may be multiplied up by $1-x$ which is positive.

$$\therefore a_r(1-x) > Kx$$

or $\qquad x < \dfrac{a_r}{a_r + K}.$

Hence if any value $< \dfrac{a_r}{a_r+K}$ be given to x the result is secured.

Next, if x or some of the coefficients become negative, it is evident that some terms will be negative, and the expression (a) is diminished numerically, whilst the next line remains unchanged and hence the result will still hold.

COR. If it is required to make
$$a_r x^r > q(a_{r+1} x^{r+1} + a_{r+2} x^{r+2} + \ldots a_p x^p)$$
where q is any number however great, similar reasoning applies except that (1) above becomes

$$a_r > \dfrac{Kq \cdot x}{1-x}$$

and $\qquad x < \dfrac{a_r}{a_r + Kq}.$

A proof similar in principle shows that by taking x sufficiently large $a_r x^r$ can be made as large as is desired compared with the sum of all the terms which *precede* it. For to make
$$a_r x^r > q(a_0 + a_1 x + a_2 x^2 \ldots a_{r-1} x^{r-1})$$
where q is any quantity however large, put $x = \dfrac{1}{y}$. Then a value of y must be chosen such that

$$\dfrac{a_r}{y^r} > q\left(a_0 + \dfrac{a_1}{y} + \ldots \dfrac{a_{r-1}}{y^{r-1}}\right)$$

or $\qquad a_r > q(a_0 y^r + a_1 y^{r-1} + \ldots a_{r-1} y).$

LIMITING VALUES AND THEORY OF INFINITE SERIES. 431

This can be done (Cor. above) by making y sufficiently *small*, *i.e.* by making x sufficiently large.

Examples on Limits. The preceding theorem of § 405 gives a short method for evaluating many limits.

Ex. 1. Find the limit when $x = 0$ of
$$\frac{Ax^p + Bx^{p+1} + Cx^{p+2} + \ldots Kx^{p+r}}{A_1 x^q + B_1 x^{q+1} + C_1 x^{q+2} + \ldots K_1 x^{q+r}}.$$

By the theorem of § 405 when x is made very small, Ax^p may be made as large as we please compared with the rest of the numerator; and $A_1 x^q$ as large as we please compared with the rest of the denominator.

Hence the limit required is that of $\dfrac{A}{A_1} x^{p-q}$ when $x = 0$.

(1) Suppose $p > q$, $\underset{x=0}{\text{Lt}} \cdot x^{p-q} = 0$.

∴ required limit $= 0$.

(2) Suppose $q > p$, $\underset{x=0}{\text{Lt}} \dfrac{1}{x^{q-p}} = \infty$.

∴ required limit $= \infty$.

(3) Suppose $p = q$, $x^0 = 1$, ∴ required limit $= \dfrac{A}{A_1}$.

Ex. 2. Find $\underset{x \quad a}{\text{Lt}} \dfrac{x^4 - a^2 x^2 - 2a^3 x + 2a^4}{x^4 + 4a^3 x - 11 a^2 x^2 + 6ax}$.

Put $x = a + h$. Expression becomes

$$\underset{h=0}{\text{Lt}} \frac{h^2 [5a^2 + 4a^2 h + h^2]}{h^2 [7a^2 + 8ah + h^2]}$$

$$= \underset{h=0}{\text{Lt}} \frac{5a^2 + 4a^2 h + h^2}{7a^2 + 8ah + h^2}$$

$= \tfrac{5}{7}$ neglecting terms in h and h^2 by use of § 405 above.

EXERCISES XXVII.

1. Find the limits when $x = 0$ of

(i) $\dfrac{ax^{\frac{3}{2}} + px^{\frac{5}{2}} + qx^8}{lx^{\frac{3}{2}} + mx^{\frac{5}{2}}}$.

[Put $x = z^2$, $x = 0$ when $z = 0$]   [Answer, a/l.]

(ii) $\dfrac{(1+x)^3 - 3(1+x)^2 + 2(1+x)}{(1+x)^4 - (1-x)^3}$.

[Answer, $-\frac{1}{7}$.]

2. Prove $\underset{x=1}{\mathrm{Lt}} \dfrac{x^p - x^q}{x^r - x^s} = \dfrac{p - q}{r - s}$

p, q, r, s being positive integers.

3. Find $\underset{}{\mathrm{Lt}} \dfrac{x^2 - 3x + 2}{x^2 - 2x + 1}$ when $x = 0, 1, \infty$.

[Answers, 2, ∞, 1.]

4. Prove that $\underset{x=a}{\mathrm{Lt}} \dfrac{f(x)}{\phi(x)} = \underset{x=a}{\mathrm{Lt}} f(x) \div \underset{x=a}{\mathrm{Lt}} \phi(x)$ if both are finite

and $\underset{x=a}{\mathrm{Lt}} \phi(x)$ not equal to zero.

5. Evaluate $\underset{x=0}{\mathrm{Lt}} \dfrac{\sqrt{a+x} - \sqrt{a-x}}{\sqrt{a^2 - x^2} + x - a}$.

$\left[\text{Answer, } \dfrac{1}{\sqrt{a}}\right]$

6. Find $\underset{x=1}{\mathrm{Lt}} \dfrac{x - (n+1)x^{n+1} + nx^{n+2}}{(1-x)^2}$, n being a positive integer.

[Answer, $\frac{1}{2}n(n+1)$.]

Irrational numbers. These are best treated as examples of limits. A complete theory is beyond the range of this work, but the following examples may be studied:—

Ex. 1. The number universally known as π can be proved (by advanced methods) to be equal to 3·14159265 ... where the decimal neither recurs nor terminates. Cutting off all the figures except the first, the first two, the first three, etc., this may be interpreted as meaning that π lies between

$\left.\begin{array}{c} 3 \\ \text{and} \\ 4 \end{array}\right\}$ $\left.\begin{array}{c} 3\cdot1 \\ \text{and} \\ 3\cdot2 \end{array}\right\}$ $\left.\begin{array}{c} 3\cdot14 \\ \text{and} \\ 3\cdot15 \end{array}\right\}$ $\left.\begin{array}{c} 3\cdot141 \\ \text{and} \\ 3\cdot142 \end{array}\right\}$ and so forth.

Suppose all figures after the xth decimal place are rejected: and let the number thus obtained be called X. Then it is known that π lies between X and $X + \dfrac{1}{10^{x+1}}$ and since x can be increased indefinitely (for the decimal does not terminate), the numbers X and $X + \dfrac{1}{10^{x+1}}$ can be made as nearly equal as is desired. Therefore two numbers can be obtained between which π must lie, and whose difference can be made as small as required.

LIMITING VALUES AND THEORY OF INFINITE SERIES.

This furnishes some justification for asserting that π has a **definite** finite value.

[The decimal form of π has been obtained by W. Shanks to 707 places, so that two rational numbers whose difference is less than $\frac{1}{10^{707}}$ are known, between which π must lie.]

Ex. 2. It may be shown that the square of no integer or rational fraction can equal 2. Let x be some integer: $x^2 < 2$ if $x = 1$ and $x^2 > 2$ if $x > 1$. Again, no fraction p/q (which will be supposed in its lowest terms) can make $\frac{p^2}{q^2} = 2$, since p being prime to q, p^2 is prime to q^2 and p^2/q^2 is essentially fractional.

By the usual method the decimal 1·4142 . . . (which, by the above, cannot terminate) is obtained. The only meaning that can be given to this is that by taking a sufficient number of decimal places we can make the difference between 2 and (1·4142 . . .)² as small as we please. Thus L^t 1·4142 . . . $= \sqrt{2}$, when the number of decimal places increases indefinitely.

ELEMENTARY THEORY OF SERIES.

406. Series.

From the consideration of certain particular cases, the reader will be familiar with the use of the word series in Algebra, but as we intend in this chapter to discuss some more general cases, it may be as well to lay down a formal definition as follows :—

A series is a succession of quantities formed according to some definite law.

Thus, for example, if $f(n)$ be any function of n, then the system of quantities

$$f(1), f(2), f(3)\ldots f(n)\ldots$$

forms a series whose nth term is $f(n)$.

NOTE 1. A rather more rigorous definition is as follows :—

A series is set of quantities or *terms* arranged according to a definite rule, or definite rules, such that
(1) There is a definite first term.

(2) There is no last term, i.e. *any* term in the series is conceived as being followed by another.

Ex. 1. The set of terms
$$\ldots -n, -(n-1) \ldots -1, 0, 1, \ldots, +n, +(n+1) \ldots$$
does not form a series because there is no definite first term.

Ex. 2. The set of all possible proper fractions arranged *in ascending order of magnitude* does not form a series. For suppose we start with $\frac{1}{n}$ as the first term (n being any integer) this is not the *least* proper fraction for *e.g.* $\frac{1}{2n}$ should precede it in the set of terms.

NOTE 2. The terms of a series need not all follow *the same law.* Thus
$$a + \beta^2 + a^3 + \beta^4 + \ldots$$
is a series in which the nth term is a^n if n is odd and is β^n if n is even. Using the symbol u_n for the nth term, the series may be defined by the *two* laws

(1) $u_n = a^n$ (n odd),
(2) $u_n = \beta^n$ (n even).

Generally in what follows a single rule only will be needed, so that if the nth term is $f(n)$ the series will be
$$f(1) + f(2) + f(3) \ldots + f(n) +$$

407. Before proceeding to the discussion of infinite series, it is necessary to bear in mind that, since an infinite series of operations can never be carried out in their totality, there is at present no *definition* of the sum of an infinite series. It will be found that *in certain cases only* a meaning can be given to the term *sum* in this connection. The question then arises, Can series undergo the four elementary operations of Algebra in the same way as if such series was a single term? The answer to this is even more restricted: if the sums of certain series can be defined, it is not always possible to add or subtract, multiply or divide them by the rules of elementary algebra.

Definition of Sum of a Series.

Let $u_1 + u_2 + u_3 + \ldots + u_n + \ldots$ denote an infinite series, or briefly, a series.

Form the sums
$$S_1 = u_1$$
$$S_2 = u_1 + u_2$$
$$S_3 = u_1 + u_2 + u_3$$
$$\dots\dots\dots\dots$$
$$S_n = u_1 + u_2 + u_3 + \dots u_n$$

When n is increased indefinitely one of three things must happen.

(1) S_n may approach a *definite finite limit* S so that $|S - S_n| < \epsilon$ for all values of n greater than some fixed value. [ϵ is as usual in this chapter any arbitrarily small quantity.]

In this case the series is said to **converge** and S is called the **limiting sum** or **sum to infinity**.

(2) S_n may increase (numerically) beyond any finite limit. The series is said to **diverge**. Sometimes its sum is loosely said to be infinite.

(3) S_n may assume *different* values, finite or infinite, according to the character of n, e.g. S_n may be $+2$ for all even values of n, -2 for all odd values, etc. No *definite* limit is attained and the series is said to **oscillate**.

Examples.

If $r < 1$, the series
$$1 + r + r^2 + r^3 \dots + r^n + \dots$$
is convergent, and the sum to infinity is $\dfrac{1}{1-r}$.

If $r > 1$, the same series is divergent, since
$$S_n = \frac{r^n - 1}{r - 1}$$
may be made as large as we please by sufficiently increasing n.

It may be mentioned here that series of this class possess little interest, and in divergent series we are only concerned with individual terms and their law of succession, whereas a convergent series represents a definite function when considered in its entirety, and such series are thus of the utmost importance in analysis.

436 LIMITING VALUES AND THEORY OF INFINITE SERIES.

As an example of an oscillating series take
$$2 - \tfrac{1}{2} + \tfrac{5}{4} - \tfrac{7}{8} + \tfrac{17}{16} - \tfrac{31}{32} + \cdots$$
which may be written
$$(1+1) + (\tfrac{1}{2} - 1) + (\tfrac{1}{4} + 1) + (\tfrac{1}{8} - 1) + (\tfrac{1}{16} + 1) + \cdots.$$
Here $S_n = (1-1+1-1$ for n terms$)+(1+\tfrac{1}{2}+\tfrac{1}{4}+\ldots$ for n terms$)$, the sum in the first bracket is $+1$ or 0 according as n is odd or even, and in the second bracket the sum is $2 - \dfrac{1}{2^{n-1}}$.

$\therefore S_n = 3 - \dfrac{1}{2^{n-1}}$ or $2 - \dfrac{1}{2^{n-1}}$ according as n is odd or even. Thus the series oscillates, tending to the limit 3 for odd values of n and the limit 2 for even values of n.

As further examples take the series
(1) $1 - 2 + 3 - 4 + \ldots + (-1)^{n-1} n$.
Here
$$\operatorname*{Lt}_{n=\infty} S_{2n} = \operatorname*{Lt}_{n=\infty}(-n) = -\infty$$
$$\operatorname*{Lt}_{n=\infty} S_{2n+1} = \operatorname*{Lt}_{n=\infty}(n+1) = +\infty$$

(2) $1 - 1 + 1 - 1 + 1 - 1 \ldots$
Here
$$\operatorname*{Lt}_{n=\infty} S_{2n} = 0 \qquad \operatorname*{Lt}_{n=\infty} S_{2n+1} = 1$$

In Example 1, $+\infty$ and $-\infty$ are sometimes called the *limits of indeterminacy* of the series: similarly for 0 and 1 in the second example.

408. Before proceeding to tests which enable us to ascertain whether a given series is convergent or not, it may not be out of place to offer a few remarks as to the use of infinite series, and an explanation of the necessity of the rest of this chapter.

I. A convenient way of calculating an irrational quantity (say $\sqrt{2}$ for example) is to have it expressed as the sum of a number of rational quantities. Now the sum of a finite number of rational fractions is clearly rational itself, so to express an irrational quantity we must use an infinite number. Even then the result *may* be rational, *e.g.*,
$$2 = 1 + \tfrac{1}{2} + \tfrac{1}{4} \ldots \text{ ad inf.,}$$
but we shall see afterwards how to express irrational quantities in terms of rational ones (Chap. XXVIII.)

LIMITING VALUES AND THEORY OF INFINITE SERIES. 437

II. The reader will do well always to bear in mind the caution, "An argument valid for a finite number of quantities may be fallacious for an infinite number."

For instance, it would be fallacious to argue that there are as many even numbers as there are odd and even numbers together, because the double of every number is even and therefore an even number corresponds to every number odd or even.

409. In a convergent series the sum of any number of terms after the nth may be made as small as we please by sufficiently increasing n.

For, by the definition of a limit, we can find a number n, such that

$$s \sim s_m < \frac{\epsilon}{2}, \text{ whenever } m > n - 1,$$

however small ϵ may be.

Hence s_m and s_n both lie between

$$s + \frac{\epsilon}{2} \text{ and } s - \frac{\epsilon}{2},$$

$$\therefore s_m \sim s_n < \epsilon;$$

i.e. $u_{n+1} + u_{n+2} \ldots + u_m < \epsilon$ ....(A),

whatever m may be, provided n be large enough.   Q. E. D.

The limit towards which $u_{n+1} + u_{n+2} \ldots u_m$ tends, when m is infinite, is called *the remainder after n terms*, and is denoted by R_n.

COR. I. $\underset{n=\infty}{L^t} R_n = 0$.

Conversely the condition (A) is *sufficient* for convergence. For if $s_m \sim s_n < \epsilon$, s_m lies between $s_n + \epsilon$ and $s_n - \epsilon$. Keep n fixed and increase m indefinitely. Then $\underset{m=\infty}{L^t} s_n$ lies between $s_n + \epsilon$ and $s_n - \epsilon$, and hence tends to a definite finite limit [see definition of limit, § 401].

A proof that the series cannot oscillate is really redundant, but may be given in the following form. Suppose $s_m = a$, $s_{m'} = \beta$, where a and β are different finite quantities. $s_m - s_{m'} = a - \beta$ and is *finite*. But (m and m' being greater than n)

$$|s_m - s_n| < \epsilon \text{ and } |s_{m'} - s_n| < \epsilon,$$
$$\therefore |s_m - s_n - (s_{m'} - s_n)| < 2\epsilon,$$
i.e. $$|s_m - s_{m'}| < 2\epsilon.$$

But 2ϵ may be made *as small as we please* and also by supposition $(s_m - s_{m'}) = a - \beta$, which is *finite*. The two results are contradictory: therefore oscillation is impossible.

COR. II. *The terms of a convergent series diminish without limit when n becomes infinite.*

For making $m = n + 1$ in (A) we see that $u_{n+1} < \epsilon$.

To guard the student against supposing that this is a sufficient test, we consider the series

$$1 + \frac{1}{2} + \frac{1}{3} \ldots + \frac{1}{n} + \ldots$$

which satisfies it;

since however

$$u_{n+1} + u_{n+2} \ldots + u_m = \frac{1}{n+1} + \frac{1}{n+2} \ldots + \frac{1}{m}$$
$$> \left(\frac{1}{m} + \frac{1}{m} \ldots + \frac{1}{m}\right)$$
$$> \frac{m-n}{m};$$
$$> 1 - \frac{n}{m}.$$

Now if $m = 2n$ (say)
$$u_{n+1} + u_{n+2} + \ldots u_{2n} > 1 - \tfrac{1}{2},$$
and hence however far we go in the series a sufficient number of terms can be taken beyond this point, so that the sum of these is $> \tfrac{1}{2}$. This shows that no finite limit can be obtained.

LIMITING VALUES AND THEORY OF INFINITE SERIES. 439

The single criterion. The criterion of § 409 $\mathrm{L}^t_{m=\infty} R_n = 0$, or the practically equivalent one $\mathrm{L}^t_{n=\infty} (s - s_n) = 0$ is the *only* criterion yet devised which is both *necessary* and *sufficient*. Its inconvenience lies in the fact that it is usually impossible to get an expression for the sum to n terms of a series as a closed expression; as can be done, *e.g.* in the case of a geometric series. Hence the necessity arises for other tests of convergency. A large number of these have been devised, the few very important ones will be given later.

Graphical illustration. Suppose the usual rectangular axes are drawn and values of n are plotted as abscissae and corresponding values of S_n as ordinates. Of course s_n is not defined for *fractional* values of n, but the points obtained may be joined, each to the next, either by straight lines or by a smooth curve.

This curve will illustrate the properties of a series as to convergency or divergency.

(**A**) *Suppose the series convergent.* Then from and after some value n, s_n lies between $s - \epsilon$ and $s + \epsilon$. All the points of the curve after the value n will lie between the lines $y = s - \epsilon$ and $s + \epsilon$. The curve *will approach a single horizontal asymptote.*

(**B**) *In the case of divergency* the curve will trend away from the x-axis and ultimately go to an infinite distance from it.

(**C**) *An oscillatory series* will give a wavy curve, the distance between a "trough" and a succeeding "crest" either remaining finite or becoming infinite as $x(n)$ increases.

Illustrations—

(A) (1) $S = 2 + 1 + \frac{1}{2} + \frac{1}{4} + \frac{1}{8} + \frac{1}{16} + \dots$
(2) $S = 1 - \frac{1}{2} + \frac{1}{3} - \frac{1}{4} + \frac{1}{5} - \frac{1}{6} + \dots$

The graphs are easily drawn and are marked (1) (2) in Fig. 38.

(B) $S = 1 + \dfrac{1}{\sqrt{2}} + \dfrac{1}{\sqrt{3}} + \dfrac{1}{\sqrt{4}} + \dots$

The graph is given in Fig. 39.

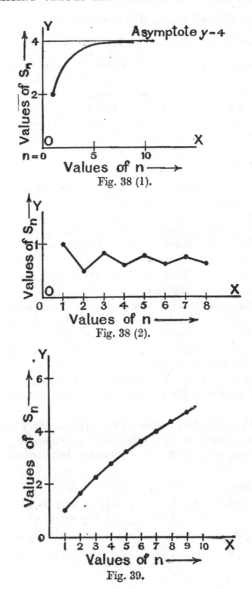

Fig. 38 (1).

Fig. 38 (2).

Fig. 39.

LIMITING VALUES AND THEORY OF INFINITE SERIES. 441

(C) (1) $S = 1 + 1 - 2 + 1 + 1 - 2 + 1 + 1 - 2 + \ldots$
(2) $S = 1 - 2 + 3 - 4 + 5 - 6 + \ldots$

The graphs are given in Figs. 40 and 41.

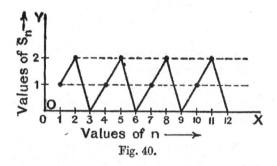

Fig. 40.

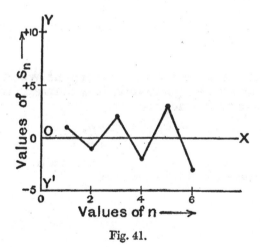

Fig. 41.

442 LIMITING VALUES AND THEORY OF INFINITE SERIES.

Examples. Draw the graphs of the following series :—

(α) $1 + \dfrac{1}{1!} + \dfrac{1}{2!} + \dfrac{1}{3!} + \dfrac{1}{4!} + \ldots \quad \dfrac{1}{n!}\ldots$

(β) $1 + \dfrac{1}{2^2} + \dfrac{1}{3^2} + \dfrac{1}{4^2} + \dfrac{1}{5^2} + \ldots \quad \dfrac{1}{n^2} + \ldots$

(γ) $\dfrac{1}{1.2} + \dfrac{1}{2.3} + \dfrac{1}{3.4} + \ldots \quad + \dfrac{1}{n.n+1}$

$\left(\text{Here } s_n = 1 - \dfrac{1}{n+1}\right).$

(δ) $\log 1 + \log 2 + \log 3 \ldots \quad + \log n + \ldots$
where the logs are taken to base 10.

(ϵ) $2 - 1\tfrac{1}{2} + 1\tfrac{1}{3} - 1\tfrac{1}{4} + 1\tfrac{1}{5} - 1\tfrac{1}{6} + \ldots$

The results indicate that (α) (β) (γ) are convergent, (δ) divergent, and (ϵ) oscillatory.

410. The convergency or divergency of any series is not altered if each of a finite number of terms is changed in any way. For let S denote the original series, Σ the altered series. Suppose the altered terms are all included in the first r terms of S. Then if $n > r$, the sum $u_n + u_{n+1} + \ldots + u_m$ is the same for S as for Σ: and by § 409, Cor. I., the convergence or divergence depends only on this sum.

411. If a series be convergent when all its terms are taken positively, it is convergent when some are taken negatively.

Let $\qquad u_1 + u_2 + u_3 \ldots + u_n + \ldots$
$\qquad\qquad v_1 + v_2 + v_3 \ldots + v_n + \ldots$
be two series, such that $v_n = u_n$ numerically, while the u's are all positive and the v's are not.

Then, $\quad v_{n+1} + v_{n+2} + \ldots + v_m \not> u_{n+1} + u_{n+2} + \ldots + u_m$
$\qquad\qquad\qquad \not> \epsilon,$
however small ϵ may be,

$\therefore$ the v series is convergent.

LIMITING VALUES AND THEORY OF INFINITE SERIES. 443

The converse is not necessarily true, thus
$$1 - \tfrac{1}{2} + \tfrac{1}{3} - \tfrac{1}{4} + \ldots$$
is convergent, while
$$1 + \tfrac{1}{2} + \tfrac{1}{3} + \tfrac{1}{4} + \ldots$$
is divergent.

DEFINITION I.—A series which is convergent when all its terms are made positive is said to be **absolutely, or essentially, convergent.**

DEFINITION II.—A series such as $1 - \tfrac{1}{2} + \tfrac{1}{3} - \tfrac{1}{4} + \tfrac{1}{5}\ldots$ which does not converge if all the signs are made positive is said to be **conditionally, or accidentally, convergent.**

Remark.—Any series which contains positive and negative terms can be arranged so that the terms are alternatively positive and negative, *e.g.* if the series is
$$u_1 + u_2 + u_3 - u_4 - u_5 + u_6 + u_7 - \ldots$$
it can be written $\quad v_1 - v_2 + v_3 - \ldots$
where $v_1 = u_1 + u_2 + u_3,\ v_2 = u_4 + u_5,\ v_3 = u_6\ u_7,$
and so on.

412. A series is **convergent if, from and after some fixed term, each of its terms bears to the corresponding term of a given convergent series, having all its terms positive, a ratio which is numerically less than some finite quantity.**

Taking the series to be $v_1 + v_2 \ldots + v_n + \ldots,$

and the given convergent series as $u_1 + u_2 + \ldots u_n + \ldots,$

and supposing the $(n+1)$th term to be beyond the fixed term mentioned, then
$$\frac{v_{n+1} + v_{n+2} + \ldots + v_m}{u_{n+1} + u_{n+2} + \ldots + u_m}$$
lies between the greatest and least of the ratios
$$\frac{v_{n+1}}{u_{n+1}},\ \frac{v_{n+2}}{u_{n+2}},\ \ldots\ \frac{v_m}{u_m},$$

and since all these are less than some finite quantity it is equal to some smaller finite quantity t, say;

$$\therefore\ v_{n+1} + v_{n+2} + \ldots + v_m = t(u_{n+1} + u_{n+2} + \ldots + u_m) < t\epsilon,$$

however small ϵ may be, if n be large enough.

Thus since t is finite the v series is convergent. Q. E. D.

If the ratio u_m/v_m approaches a limit, when m is infinite, which is neither 0 nor infinite, and if the terms of both series be positive, then the u series and the v series are alike in respect of convergence.

For let $\underset{n=\infty}{\mathrm{L}^t}\ u_m/v_m = k$ then, if n be taken large enough, u_m/v_m will be as nearly equal to k as we like when $m > n$. Hence also v_m/u_m is as near as we like to $1/k$. Consequently v_m/u_m is always less than some finite quantity, and u_m/v_m is always less than another finite quantity, if $m > n$. Hence if the u series converges, the v series does also; and conversely if the v series converges the u **series converges** also.

COR. I. *If the u series be divergent so also is the v series.*

For if possible let the v series be convergent, then since each of the ratios u_n/v_n is finite, the u series would be convergent, which is contrary to hypothesis.

COR. II. *If the u series have some of its terms negative but be such that it would not lose its convergency by taking all its terms positively, then the v series is convergent.*

Let $\Sigma u_n'$ be the series derived from Σu_n by taking all its terms positively, and $\Sigma v_n'$ the series derived in like manner from Σv_n.

Then $\Sigma u_n'$ is convergent by hypothesis, $\therefore$ by the theorem of this article so also is $\Sigma v_n'$, and hence Σv_n is convergent.

Ex. Show that

$$f(1).x + \frac{f(2).x^2}{2^r} + \frac{f(3)x^3}{3^r} + \ldots \frac{f(n)x^n}{n^r}$$

where $f(n) = An^r + Bn^{r-1} + \ldots K$.

A, B, C, etc., being constants, converges if $x < 1$.

LIMITING VALUES AND THEORY OF INFINITE SERIES. 445

Compare with the series $x + x^2 + \ldots x^{n-1} + x^n +$ which converges when $x < 1$. Let $u_n = x^n$, $v_n = \dfrac{f(n)}{n^r} \cdot x^n$.

$$\frac{v_n}{u_n} = \frac{f(n)}{n^r} = A + \frac{B}{n} + \ldots \frac{K}{n^r}.$$

Hence $\dfrac{v_n}{u_n}$ is always finite [its limit for $n = \infty$ is A], and the theorem of this article applies.

EXERCISES XXVII.

Show by comparison with the Geometric Series that the following series are convergent:—

7. $1 + \dfrac{1}{2} \cdot \dfrac{1}{2} + \dfrac{1}{3} \cdot \dfrac{1}{2^2} + \dfrac{1}{4} \cdot \dfrac{1}{2^3} + \ldots + \dfrac{1}{n} \cdot \dfrac{1}{2^{n-1}} + \ldots$

8. $1 + \dfrac{1}{2} \cdot x + \dfrac{2}{3} x^2 + \dfrac{3}{4} x^3 + \ldots \dfrac{n-1}{n} \cdot x^n$, if < 1.

9. $1 + x + \dfrac{2^2}{1 \cdot 3} x^2 + \dfrac{3^2}{2 \cdot 4} x^3 + \ldots \dfrac{(n-1)^2}{(n-2)n} x^{n-1} + \ldots$
if $x < 1$.

10. $x + \dfrac{x^2}{2} + \dfrac{x^3}{3} + \ldots \dfrac{x^n}{n} \quad (x < 1)$.

Comparison Series. The following series

$$\Sigma \equiv \frac{1}{1^r} + \frac{1}{2^r} + \frac{1}{3^r} + \ldots + \frac{1}{n^r}$$

is very useful for comparison, and enables the convergency of many other series to be determined.

Theorem. The series Σ converges if $r > 1$ and diverges for all other values of r.

CASE I. $r = 1$. The series now becomes

$$1 + \frac{1}{2} + \frac{1}{3} + \ldots + \frac{1}{n},$$

and has already (§ 409) been shown to be divergent.

A proof may be given that the sum of n terms of this series can actually be made greater than any quantity A by taking n sufficiently large.

For since each term is greater than the following, we have
$$\frac{1}{3}+\frac{1}{4} > 2.\frac{1}{4} > \frac{1}{2}$$
$$\frac{1}{5}+\frac{1}{6}+\frac{1}{7}+\frac{1}{8} > 4.\frac{1}{8} > \frac{1}{2}$$
$$\frac{1}{9}+\ldots+\frac{1}{16} > 8.\frac{1}{16} > \frac{1}{2}$$

Proceeding in this way let m such groups be taken, the number of terms in each group being doubled at each step. Also let these groups contain $n-2$ terms.

Then $1+\frac{1}{2}+\left(\frac{1}{3}+\frac{1}{4}\right)+\left(\frac{1}{5}+\frac{1}{6}+\frac{1}{7}+\frac{1}{8}\right)+\ldots$

to n terms $> m \times \frac{1}{2}+1+\frac{1}{2}$.

Hence this is $> A$ if
$$\frac{3}{2}+\frac{m}{2} > A \text{ or if } m > 2A-3.$$

CASE II. $r < 1$, i.e. $1 > r > -\infty$.

Here $n^r < n$ and $\therefore \frac{1}{n^r} > \frac{1}{n}$.

Thus each term of the series is greater than the corresponding term in the series
$$1+\frac{1}{2}+\frac{1}{3}+\ldots+\frac{1}{n}+\ldots$$
and so the series **diverges**.

CASE III. $r > 1$.

Here $\frac{1}{2^r}+\frac{1}{3^r} < 2.\frac{1}{2^r} < \frac{1}{2^{r-1}}$,

$\frac{1}{4^r}+\frac{1}{5^r}+\frac{1}{6^r}+\frac{1}{7^r} < 4.\frac{1}{4^r} < \frac{1}{4^{r-1}}$,

and so on : the number of terms in each group being

LIMITING VALUES AND THEORY OF INFINITE SERIES. 447

double that in the preceding. Thus the sum of m groups is less than
$$\frac{1}{2^{r-1}} + \frac{1}{4^{r-1}} + \frac{1}{8^{r-1}} + \frac{1}{16^{r-1}} + \ldots$$
i.e. less than
$$\frac{1}{2^{r-1}} + \frac{1}{2^{2(r-1)}} + \frac{1}{2^{3(r-1)}} + \frac{1}{2^{4(r-1)}} \ldots$$

Hence each *group* is less than the corresponding *term* of a G.P. of common ratio $\frac{1}{2^{r-1}}$. Since $r > 1$, $2^{r-1} > 1$, $\frac{1}{2^{r-1}} < 1$ and the G.P. is convergent. Hence the original series **converges**.

413. A series is **convergent** if, from and after some fixed term, **the ratio of each term to the preceding term is numerically less than some quantity which is itself numerically less than unity.**

First consider the case in which all the terms are positive.

Suppose the pth to be the fixed term, then when $n > p$
$$\frac{u_{n+1}}{u_n} < r, \quad \frac{u_{n+2}}{u_{n+1}} < r \ldots \frac{u_m}{u_{m-1}} < r, \text{ where } r < 1.$$

Hence $u_{n+1} + u_{n+2} + u_{n+3} + \ldots + u_m$
$$= u_{n+1}\left(1 + \frac{u_{n+2}}{u_{n+1}} + \frac{u_{n+3}}{u_{n+2}} \cdot \frac{u_{n+2}}{u_{n+1}} + \ldots + \frac{u_m}{u_{m-1}} \cdot \frac{u_{m-1}}{u_{m-2}} \ldots \frac{u_{n+2}}{u_{n+1}}\right)$$
$$< u_{n+1}\left(1 + r + r^2 + r^3 + \ldots + r^{m-n-1}\right)$$
$$< u_{n+1} \frac{1 - r^{m-n}}{1 - r} < u_{n+1} \frac{1}{1 - r}.$$

Now $u_{n+1} < u_p r^{n+1-p}$. Hence since $\underset{n=\infty}{\mathrm{L}^t} r^{n+1-p} = 0$ and $\frac{1}{1-r}$ is finite, we can make
$$u_{n+1} + u_{n+2} + \ldots + u_m$$
as small as we please by taking n large enough, ∴ the series is convergent.

448 LIMITING VALUES AND THEORY OF INFINITE SERIES.

Further, since it is convergent when all its terms are taken positively it is convergent when some of them are taken negatively.

Note that the theorems of § 412 and of this article give conditions *sufficient* for convergence but not *necessary*.

Thus take the series
$$a + \beta^2 + a^3 + \beta^4 + \ldots \beta^{2n} + a^{2n+1} \ldots \qquad (a < \beta < 1)$$
$$\frac{u_{2n+1}}{u_{2n}} = \frac{a^{2n+1}}{\beta^{2n}} = a\left(\frac{a}{\beta}\right)^{2n}.$$

This is certainly < 1 for all values of u: but
$$\frac{u_{2n}}{u_{2n-1}} = \frac{\beta^{2n}}{a^{2n-1}} = \beta\left(\frac{\beta}{a}\right)^{2n-1}$$
and this *increases indefinitely* as n increases. The series is, however, convergent, for each term is not greater than the corresponding term of the series $\beta + \beta^2 + \beta^3 + \ldots \beta^{2n} + \beta^{2n+1} \ldots$, and the latter series is convergent.

The meaning of the wording "from and after some fixed term" in the enunciation of these theorems should be noted.

Consider the series
$$1 + \frac{x^2}{1^2} + \frac{x^4}{1^2 \cdot 2^2} + \ldots \frac{x^{2n}}{(n!)^2}.$$

Here
$$\frac{u_{n+1}}{u_n} = \frac{x^2}{n^2}.$$

This ratio is > 1 if $x > n$ (numerically).

Let $|x|$ lie between the integers m and $m+1$. Then if $n \geqslant m+1$, $\frac{x^2}{n^2} < 1$, and so the condition for convergency is satisfied from and after the $(m+1)$th term. (Cf. § 430 post.)

414. If in a series $\underset{n=\infty}{L^t} \frac{u_{n+1}}{u_n} < 1$ numerically, the series is convergent.

Suppose
$$\underset{n=\infty}{L^t} \frac{u_{n+1}}{u_n} = r,$$

and first let r be positive, then by taking n large enough we can make
$$\frac{u_{n+1}}{u_n} \sim r < \epsilon,$$

LIMITING VALUES AND THEORY OF INFINITE SERIES. 449

where ϵ is as small as we please, and $m > n$. Now, when ϵ is sufficiently small $r + \epsilon < 1$, $\therefore$ from and after the nth term (n being finite) $u_{m+1}/u_m < 1$, hence the series is convergent by the last article.

If r be negative, the series is convergent because, by the above, it would be convergent with all its terms taken positively.

Ex. 1. A series having all its terms positive, is divergent if, from and after some fixed term, the ratio of each term to the one preceding it is equal to or greater than unity.

{For $u_{n+1} + u_{n+2} + \ldots + u_m \nless (m-n) u_{n+1}$.}

Ex. 2. If $\underset{n=\infty}{\mathrm{L}^t} \dfrac{u_{n+1}}{u_n} > 1$ the series is divergent. {Proof as in this article.}

Ex. 3. The series $1 + 1 + \dfrac{1}{2!} + \dfrac{1}{3!} + \dfrac{1}{4!} + \ldots$

is convergent, since $u_2/u_1 = 1$; $u_3/u_2 = \tfrac{1}{2}$; $u_4/u_3 = \tfrac{1}{3}, \ldots$, and therefore each ratio after the second is less than $\tfrac{1}{2}$, which is less than unity.

Ex. 4. Prove that the series $a + (a+d)r + \ldots + (a+nd)r^n + \ldots$ is convergent if $r < 1$.

Here $\dfrac{u_{n+1}}{u_n} = \dfrac{(a+nd) r^n}{(a + \overline{n-1} . d) r^{n-1}} = \dfrac{(a/n + d) r}{a/(n-1) + d} \cdot \dfrac{n}{n-1}$

$\qquad = \dfrac{(a/n + d) r}{a/(n-1) + d} \cdot \left(1 + \dfrac{1}{n-1}\right).$

Hence $\underset{n=\infty}{\mathrm{L}^t} \dfrac{u_{n+1}}{u_n} = r$, which is less than 1.

Hence the series is convergent if $r < 1$.

Ex. 5. Prove that the infinite series
$$1^r x + 2^r x^2 + 3^r x^3 + \ldots$$
is convergent if $x < 1$ and r is a finite positive integer.

Here $\dfrac{u_{n+1}}{u_n} = \dfrac{(n+1)^r x^{n+1}}{n^r x^n} = \left(\dfrac{n+1}{n}\right)^r x = \left(1 + \dfrac{1}{n}\right)^r x$

$\qquad = \left\{1 + r \cdot \dfrac{1}{n} + \dfrac{r(r-1)}{1 \cdot 2} \cdot \dfrac{1}{n^2} + \ldots + \dfrac{1}{n^r}\right\} x$ [§ 50.]

$\qquad = x$, when n infinitely increased.

Hence the series is convergent when $x < 1$.

TUT. ALG. II.

EXERCISES XXVII.

11. Show that $u_1 + u_2 + u_3 \ldots + u_n + \ldots$ converges if

$$\underset{n=\infty}{\text{Lt}} \sqrt[n]{u_n} < 1$$

and diverges if

$$\underset{n=\infty}{\text{Lt}} \sqrt[n]{u_n} > 1.$$

Show that the following series converge under the conditions stated.

12. $x - \dfrac{x^3}{3!} + \dfrac{x^5}{5!} \ldots + (-1)^{n-1} \dfrac{x^{2n-1}}{\underline{|2n-1|}}$.

13. $1 + \dfrac{x^2}{\underline{|2}} + \dfrac{x^4}{\underline{|4}} + \dfrac{x^6}{\underline{|6}} \ldots + \dfrac{x^{2n}}{\underline{|2n}}$.

14. $1 - z + \dfrac{z^2}{1^2 \cdot 2^2} - \dfrac{z^3}{1^2 \cdot 2^2 \cdot 3^2} + \ldots + \dfrac{(-1)^{n-1} z^{n-1}}{\{(n-1)!\}^2}$.

These converge for all finite values of the variable x (or z).

15. $1 + \dfrac{a}{\beta} x + \dfrac{a \cdot (a+1)}{\beta (\beta + 1)} x^2 + \ldots$

$\qquad + \dfrac{a \cdot (a+1)(a+2) \ldots (a+n-1)}{\beta(\beta+1)(\beta+2) \ldots (\beta+n-1)} \cdot x^n$,

provided $|x| < 1$ and β is not a negative integer.

16. $1 + \dfrac{a \cdot \beta \cdot x}{1 \cdot \gamma} + \dfrac{a \cdot \overline{a+1} \cdot \beta \cdot \overline{\beta+1}}{1 \cdot 2 \cdot \gamma \cdot \overline{\gamma+1}} \cdot x^2$

$\qquad + \ldots + \dfrac{a(a+1) \ldots (a+n-1)\beta(\beta+1) \ldots (\beta+n-1) x^n}{\underline{|n} \, \gamma \cdot (\gamma+1)(\gamma+2) \ldots (\gamma+n-1)}$

provided $|x| < 1$ and γ is not a negative integer.

17. $\dfrac{1}{x-a} + \dfrac{1}{x-2a} + \dfrac{1}{x-3a} + \ldots \dfrac{1}{x-na}$

diverges for all finite values of x.

$$\left[\text{Compare with } 1 + \frac{1}{2} + \frac{1}{3} + \ldots + \frac{1}{n}. \right]$$

LIMITING VALUES AND THEORY OF INFINITE SERIES. 451

18. $\dfrac{1}{(x-a)^2} + \dfrac{1}{(x-2a)^2} + \dfrac{1}{(x-3a)^2} + \ldots \dfrac{1}{(x-na)^2}$

converges for all values of x except the values $x = a, 2a, 3a, 4a \ldots ra$ (r any positive integer).

$$\left[\text{Compare with } 1 + \frac{1}{2^2} + \frac{1}{3^2} + \ldots + \frac{1}{n^2}. \right]$$

19. If in the series $a_1 x + a_2 x^2 + \ldots a_n x^n + \ldots$

$$\underset{n=\infty}{\mathrm{Lt}} \left| \frac{a_{n+1}}{a_n} \right| = \frac{1}{\lambda}$$

the series converges *absolutely* so long as $|x| < \lambda$.

Theorem I. The series $u_1 - u_2 + u_3 - u_4 + u_5 \ldots$ is at least conditionally convergent if (1) $u_1 > u_2 > u_3 > u_4 > u_5 \ldots$ and (2) $\underset{n=\infty}{\mathrm{Lt}}\, u_n = 0$.

For $S_{2n} = (u_1 - u_2) + (u_3 - u_4) \ldots + (u_{2n-1} - u_{2n})$, and is *positive* by first condition.

Also $S_{2n+1} = u_1 - (u_2 - u_3) - (u_4 - u_5) \ldots$
$\ldots - (u_{2n} - u_{2n+1})$,

and is **less than** u_1.

$S_{2n+1} - S_{2n} = u_{2n+1}$, and this difference vanishes when n approaches ∞.

Since $S_{2n} > 0$ and $S_{2n+1} < u_1$, and since their difference vanishes when n increases indefinitely, they must both tend to a common limit between 0 and u_1; and so the series converges.

Theorem II. If a series arranged in ascending powers of a variable x be absolutely convergent when $x = a$ it is convergent for all values of x such that $|x| < a$.

Let the series be $a_0 + a_1 x + a_2 x^2 + \ldots a_r x^r + \ldots$

Let $|a| = \alpha$. Then by hypothesis

$$|a_{n+1} \alpha^{n+1}| + |a_{n+2} \alpha^{n+2}| + \ldots + |a_m \alpha^m| < \epsilon$$

for a suitably chosen n and for any value of m.

Now put $x = b$ where $|b| = \beta < \alpha$.

Then $\quad |a_{n+1} b^{n+1}| + |a_{n+2} b^{n+2}| + \ldots + |a_m b^m|$
$= |a_{n+1}| \beta^{n+1} + |a_{n+2}| \beta^{n+2} + \ldots |a_m| \beta^m$
$< |a_{n+1}| \alpha^{n+1} + |a_{n+2}| \alpha^{n+2} + \ldots |a_m| \alpha^m$
$< \epsilon.$

Hence by the general criterion the series converges absolutely when $x = b$.

415. The Binomial Series

$$1 + nx + \frac{n(n-1)}{2!}x^2 + \frac{n(n-1)(n-2)}{3!}x^3 + \ldots$$
$$+ \frac{n(n-1)\ldots(n-r+1)}{r!}x^r + \ldots$$

is always convergent when x is numerically less than unity.

Here $\quad u_r = \dfrac{n(n-1)(n-2)\ldots(n-r+2)}{(r-1)!} x^{r-1}$,

$\therefore \dfrac{u_{r+1}}{u_r} = \dfrac{n-r+1}{r} x = -x\left(1 - \dfrac{n+1}{r}\right);$

$\therefore \underset{r=\infty}{\mathrm{L}^t} \dfrac{u_{r+1}}{u_r} = -x,\ \text{since } \underset{r=\infty}{\mathrm{L}^t} \dfrac{n+1}{r} = 0;$

consequently when x is numerically less than unity the series is convergent.

When n is a positive integer, of course the series terminates and the question of convergency does not arise.

Cor. When x is positive and $r > n+1$, u_{r+1}/u_r is negative and the terms are alternately positive and negative after $r > n+1$; but even if all the terms were taken positively the series would still be convergent, since we should still have

$$\mathrm{L}^t\, u_{r+1}/u_r < 1.$$

If x be negative the terms are, after a certain limit, all positive.

416. The Exponential Series

$$1 + x + \frac{x^2}{2!} + \frac{x^3}{3!} + \ldots + \frac{x^n}{n!} + \ldots$$

is convergent for all values of x.

Here $\quad u_n = \dfrac{x^{n-1}}{(n-1)!};\quad \therefore \dfrac{u_{n+1}}{u_n} = \dfrac{x}{n};$

$\therefore \underset{n=\infty}{\mathrm{L}^t} \dfrac{u_{n+1}}{u_n} = 0,\ \text{since } \underset{n=\infty}{\mathrm{L}^t} \dfrac{x}{n} = 0;$

$\therefore$ the series is convergent for all values of x.

LIMITING VALUES AND THEORY OF INFINITE SERIES. 453

If x be positive the terms are all positive, if x be negative they are alternately positive and negative.

In the latter case the series remains convergent when all its terms are taken positively, for this only amounts to changing the sign of x.

417. The Logarithmic Series

$$x + \frac{x^2}{2} + \frac{x^3}{3} + \ldots + \frac{x^n}{n} + \ldots$$

is convergent when x is positive and less than unity.

Here $u_n = x^n/n$;

$$\therefore \frac{u_{n+1}}{u_n} = \frac{x^{n+1}}{n+1} \Big/ \frac{x^n}{n} = x\,\frac{n}{n+1} = x\left(1 - \frac{1}{n+1}\right);$$

$$\therefore \operatorname*{L^t}_{n=\infty} \frac{u_{n+1}}{u_n} = x,$$

and when $x < 1$ the series is convergent.

COR. If $x < 1$ the logarithmic series

$$x - x^2/2 + x^3/3 - x^4/4 + \ldots$$

is convergent, for it is convergent when all its terms are taken positively.

418. Product of two absolutely convergent series. *If a series*

$$S = u_0 v_0 + (u_0 v_1 + u_1 v_0) x + (u_0 v_2 + u_1 v_1 + u_2 v_0) x^2 + \ldots$$

be formed, the coefficient of any power of x in which is the same as in the product of the two series

$$A = u_0 + u_1 x + u_2 x^2 + u_3 x^3 + \ldots$$
$$B = v_0 + v_1 x + v_2 x^2 + v_3 x^3 + \ldots,$$

then the series S is convergent and equals the product of the two series A and B, supposing that these series (1) *are convergent, and* (2) *have either all their terms positive or would not lose their convergency when all the negative terms in them have their signs changed.*

Let A_{2n}, B_{2n} and S_{2n} denote the sums of the first $2n+1$ terms of the series A, B and S respectively.

I. Let the terms of A and B be all positive. Then $A_{2n} \times B_{2n} = S_{2n} +$ terms containing $x^{2n+1}, x^{2n+2}, \ldots x^{4n}$.

Hence $\qquad S_{2n} < A_{2n} \times B_{2n}$.

Again, since $S_{2n} = A_n \times B_n +$ some other terms*,
$$S_{2n} > A_n \times B_n.$$

Hence S_{2n} has a value between $A_{2n} \times B_{2n}$ and $A_n \times B_n$.

Now, however small ϵ may be we can find a value of n, such that $A_n > A - \epsilon$ and $B_n > B - \epsilon$, also $A_{2n} < A$, $B_{2n} < B$.

Therefore, as $\qquad A_{2n} B_{2n} > S_{2n} > A_n B_n$,

we have $\quad AB > S_{2n} > (A - \epsilon)(B - \epsilon) > AB - \epsilon(A + B)$,

and since A **and** B are finite we can make $\epsilon(A+B) < \epsilon'$, however small the latter may be,

$$\therefore \ AB - S_{2n} < \epsilon',$$

hence the series S is convergent and its sum is AB.

II. Let the terms in A and B be not all of the same sign. Suppose A', B' and S' are the series derived from A, B and S respectively by making all their terms positive. Then A', B' are convergent by hypothesis, and hence by the foregoing we can choose n so large that

$$A_n' B_n' - S_n' < \epsilon,$$

however small ϵ may be.

Hence $A_n B_n - S_n$, which cannot be numerically greater than $A_n' B_n' - S_n'$ of which all the terms are positive, can be made as small as we please by increasing n, consequently

$$S = A \times B \text{ and the series } S \text{ is convergent.}$$

---

* For example, the coefficient of x^{2n} in $A_n \times B_n$ is $u_n v_n$, while that of x^{2n} in S_{2n} is $u_{2n} v_0 + u_{2n-1} v_1 + \ldots + u_n v_n + \ldots + u_0 v_{2n}$.

419. Uniform Convergence.

Suppose the terms of a series are each functions of a single variable x, continuous between a and b, then we have seen that, when the series is convergent, we can always choose a number n such that
$$R_n < \epsilon,$$
where ϵ is any finite quantity, however small.

DEFINITION. If we can choose the value of n so that for **all values of x**, between a and b, $R_n < \epsilon$, then the series is said to be **uniformly convergent** in this interval.

420. Continuity of Convergent Series.
If a series be uniformly convergent in a given interval and each term be continuous in that interval, then, within the same limits, it represents a continuous function.

Suppose $S(x)$ is the sum of the series and S_n the sum of the first n terms, then
$$S(x) = S_n(x) + R_n, \text{ where } R_n < \epsilon,$$
and $\quad S(x+h) = S_n(x+h) + R_n'$ and $R_n' < \epsilon$;
$$\therefore\ S(x+h) - S(x) = S_n(x+h) - S_n(x) + R_n' - R_n,$$
and since R_n' and R_n are each numerically $< \epsilon$, $R_n' - R_n < 2\epsilon$. Also $S_n(x)$, being the sum of a finite number of continuous functions, is itself continuous, and hence by sufficiently diminishing h we can make $S_n(x+h) - S_n(x) < \epsilon$;
$$\therefore\ S(x+h) - S(x) < 3\epsilon,$$
which can be made as small as we please;
$$\therefore\ S(x) \text{ is a continuous function of } x. \quad \text{Q. E. D.}$$

421. Application to Power Series.

We proceed to an important application of the foregoing, viz. to series in which the nth term is $a_n x^n$ when a_n is a constant. Such series are known as **power series**, and may be written

$$a_0 + a_1 x + a_2 x^2 + \ldots + a_n x^n + \ldots.$$

When
$$x < \operatorname*{L^t}_{n=\infty} \frac{a_n}{a_{n+1}},$$

the power series $a_0 + a_1 x + a_2 x^2 + \ldots$ is *not only convergent, but uniformly convergent*.

The series is convergent if only
$$\operatorname*{L^t}_{n=\infty} \frac{a_n x^n}{a_{n+1} x^{n+1}} > 1,$$

i.e. if only
$$\operatorname*{L^t}_{n=\infty} \frac{a_n}{a_{n+1}} > x,$$

which proves the first part.

Let ρ be the value of $\operatorname*{L^t}_{n=\infty} \dfrac{a_n}{a_{n+1}}$.

Suppose in the first place that all the a's are positive, then, clearly R_n diminishes with x, consequently if we choose n so that $R_n < \epsilon$ when x has the value $\rho - \eta$, where η is a small finite quantity, this value of n will serve for the whole interval in which the series converges, and our result follows at once.

If the a's have not all the same sign, R_n can in no case be greater than it would be if they all had the same sign, and x had the value $\rho - \eta$; consequently the value of n, selected previously, will serve for the whole interval in this case, and the theorem is completely demonstrated.

Thus, for example, the series
$$1 + x + \frac{x^2}{2!} + \ldots + \frac{x^r}{r!} + \ldots$$

converges for all values of x; $\therefore$ it is uniformly convergent throughout.

The series $1 + mx + \dfrac{m(m-1)}{2!} x^2 + \ldots$ converges if $x < 1$, and thus in any interval lying entirely between $+1$ and -1 the series is uniformly convergent.

EXERCISES XXVII.

20. Show that $\underset{x=1}{\mathrm{L^t}} \dfrac{x^2 - 3x + 2}{x-1} = -1$.

21. Show that $\underset{n=\infty}{\mathrm{L^t}} \left(x - \dfrac{1}{n}\right)\left(x - \dfrac{2}{n}\right)\dots\left(x - \dfrac{r}{n}\right) = x^r$, r being a positive integer.

Show also that the above expression lies between $x^r - \dfrac{r(r+1)}{2n} x^{r-1}$ and x^r for positive values of n.

22. Prove very carefully that the limit of the difference of two quantities is equal to the difference of their limits when these limits are finite.

23. Prove that the series
$$1 + x + \dfrac{2^r x^2}{2!} + \dfrac{3^r x^3}{3!} + \dfrac{4^r x^4}{4!} + \dots$$
is convergent for all finite values of x.

24. The series
$$1 + x + 2^r x^2 + 3^r x^3 + 4^r x^4 + \dots$$
is convergent when $x < 1$ but otherwise it is divergent, r being a positive integer.

25. If $f(n)$ be a rational integral function of n then the series whose nth term is
$$f(n) \dfrac{x^n}{n!}$$
is convergent for all values of x; and the series whose nth term is
$$f(n)\, x^n$$
is convergent when x is less than unity.

CHAPTER XXVIII.

THE BINOMIAL THEOREM.

FRACTIONAL AND NEGATIVE INDICES.

422. It was shown in §§ 327, 328, 329, that, when n is a positive integer, the series

$$1 + nx + \frac{n(n-1)}{1 \cdot 2} x^2 + \ldots$$
$$+ \frac{n(n-1)(n-2) \ldots (n-r+1)}{1 \cdot 2 \cdot 3 \ldots r} x^r + \ldots$$

terminates at the $(n+1)$th term, and that its value is $(1+x)^n$.

When n is not a positive integer, none of the factors $n-1$, $n-2$, $n-3$, ... vanish, and the series is therefore infinite.

It was proved in § 415 that this infinite series has a finite value when x is numerically less than 1, and we propose to find this value after stating and proving Vandermonde's Theorem, which we require in the investigation, in its most general form. [See § 339, Ex. 7.]

423. Vandermonde's Theorem.

If r be a positive integer, and m and n have any values whatever,

$$(m+n)_r = m_r + {}_rC_1 \cdot m_{r-1}n_1 + {}_rC_2 \cdot m_{r-2}n_2 + \ldots + n_r,$$

THE BINOMIAL THEOREM.

where the notation m_r stands for
$$m(m-1)(m-2)\ldots(m-r+1),$$
that is, for the continued product of the r factors
$$m,\ m-1,\ m-2,\ \ldots\ m-r+1,$$
m being integral or fractional.

[According to this notation, $(4\tfrac{1}{2})_3 = 4\tfrac{1}{2} \cdot 3\tfrac{1}{2} \cdot 2\tfrac{1}{2} = 39\tfrac{3}{8}$.]

We have proved in § 339, Ex. 7, that, when m, n, and r are positive integers, such that $m + n > r$,
$$_{m+n}C_r = {}_mC_r + {}_mC_{r-1} \cdot {}_nC_1 + {}_mC_{r-2} \cdot {}_nC_2 + \ldots + {}_nC_r;$$
that is, that
$$\frac{(m+n)_r}{r!} = \frac{m_r}{r!} + \frac{m_{r-1}}{(r-1)!} \cdot \frac{n_1}{1!} + \frac{m_{r-2}}{(r-2)!} \cdot \frac{n_2}{2!} + \ldots + \frac{n_r}{r!}.$$

Multiplying both sides by $r!$, we have
$$(m+n)_r = m_r + r \cdot m_{r-1} n_1 + \frac{r(r-1)}{1 \cdot 2} m_{r-2} n_2 + \ldots + n_r.$$

This equality has been proved on the supposition that m, n and r are positive integers, such that $m + n$ is not less than r. We shall now show that the equality will be true, if all these restrictions are removed, except that r is a positive integer.

When m takes any integral value greater than r, the equality is true for *all* positive integral values of n, and therefore for more than r values of n. That is, when m takes any integral value greater than r, the expressions on the two sides of the sign of equality which are of the rth degree are equal for more than r values of n. Hence by § 373, the equality is true for that particular value of m and any value whatever of n. It is thus true for *any* integral value of m greater than r, and for any value whatever of n. It is therefore true for more than r values of m, and for any value whatever of n. Hence, again, by § 373, the theorem is true for any value whatever of m and any value whatever of n.

424. Cayley's Proof of Vandermonde's Theorem.

When $r = 1$ the theorem reduces to the obvious identity
$$(m + n) = m + n.$$

Multiply both sides by $(m + n - 1)$, then the left-hand side becomes $(m + n)_2$.

We then have
$$(m + n)_2 = m\,\overline{(m - 1 + n)} + n\,\overline{(n - 1 + m)}$$
$$= m_2 + m_1 n_1 + m_1 n_1 + n_2$$
$$= m_2 + 2m_1 n_1 + n_2,$$
which establishes the result for $r = 2$.

Multiplying by $(m + n - 2)$ we get
$$(m + n)_3 = m_2\,\overline{(m - 2 + n)} + 2m_1 n_1\,\overline{(m - 1 + n - 1)} + n_2\,(n - 2 + m)$$
$$= m_3 + m_2 n_1 + 2m_2 n_1 + 2m_1 n_2 + n_3 + m_1 n_2$$
$$= m_3 + 3m_2 n_1 + 3m_1 n_2 + n_3.$$

This suggests that
$$(m + n)_r = m_r + {}^rC_1 m_{r-1} n_1 + {}^rC_2 m_{r-2} n_2 + \ldots + n_r.$$

Assuming that this is so we shall show that the same holds for $(m + n)_{r+1}$.

Multiplying both sides by
$$(m + n - r)$$
we have in fact
$$(m + n)_{r+1} = m_r\,\overline{(m - r + n)} + {}^rC_1\, m_{r-1} n_1\,\overline{(m - r + 1 + n - 1)}$$
$$+ {}^rC_2 m_{r-2} n_2\,\overline{(m - r + 2 + n - 2)} \ldots n_r\,\overline{(n - r + m)}$$
$$= (m_{r+1} + m_r n_1) + {}^rC_1\,(m_r n_1 + m_{r-1} n_2) + {}^rC_2\,(m_{r-1} n_2 + m_{r-2} n_3) + \ldots n_{r+1}$$
$$= m_{r+1} + m_r n_1\,(1 + {}^rC_1) + m_{r-1} n_2\,({}^rC_1 + {}^rC_2)$$
$$+ \ldots + m_{r-s} n_{s+1}\,({}^rC_s + {}^rC_{s+1}) + \ldots + n_{r+1},$$
and since ${}^rC_s + {}^rC_{s+1} = {}^{r+1}C_{s+1}$ this becomes
$$(m + n)_{r+1} = m_{r+1} + {}^{r+1}C_1 m_r n_1$$
$$+ {}^{r+1}C_2 m_{r-1} n_2 + \ldots + {}^{r+1}C_{s+1} m_{r-s} n_{s+1} \ldots + n_{r+1}.$$

Thus if the result holds for any value of r it holds for the next higher value of r, but it does hold when $r = 1, 2, \ldots$; $\therefore$ it is true universally. [Q.E.D.]

The student will notice that by splitting up the multiplier $(m + n - r)$ suitably for each term, as we have done, the coefficients on the right for $(m + n)_{r+1}$ are formed from those for $(m + n)_r$ in just the same way as those of $(a + b)^{n+1}$ are from those of $(a + b)^n$.

425. The Binomial Theorem for a positive fractional or negative index.

To prove that when x is numerically less than unity,

$$(1+x)^n = 1 + nx + \frac{n(n-1)}{1.2} x^2 + \dots$$
$$+ \frac{n(n-1)(n-2)\dots(n-r+1)}{1.2.3\dots r} x^r + \dots.$$

Let $f(m)$ denote the series on the right side of

$$f(m) = 1 + \frac{m_1}{1!} x + \frac{m_2}{2!} x^2 + \dots + \frac{m_r}{r!} x^r + \dots.$$

Then $\quad f(n) = 1 + \frac{n_1}{1!} x + \frac{n_2}{2!} x^2 + \dots + \frac{n_r}{r!} x^r + \dots,$

and $\quad f(m+n) = 1 + \frac{(m+n)_1}{1!} x$

$$+ \frac{(m+n)_2}{2!} x^2 + \dots + \frac{(m+n)_r}{r!} x^r + \dots.$$

Now the coefficient x^r in the product of $f(m)$ and $f(n)$

$$= \frac{m_r}{r!} + \frac{m_{r-1}}{(r-1)!} \cdot \frac{n_1}{1!} + \frac{m_{r-2}}{(r-2)!} \cdot \frac{n_2}{2!} + \dots + \frac{n_r}{r!}$$

$$= \frac{1}{r!} \left\{ m_r + r \cdot m_{r-1} n_1 + \frac{r(r-1)}{1.2} \cdot m_{r-2} n_2 + \dots + n_r \right\}$$

$$= \frac{1}{r!} \cdot (m+n)_r \quad \text{[By Vandermonde's Theorem, § 423.]}$$

= the coefficient of x^r in $f(m+n)$*.

---

* By the older writers this used to be established by the principle of Fixity of Algebraic Form in virtue of which it was asserted that since the series has the same form whether m is a positive integer or not, and since moreover when m and n are positive integers $f(m) \times f(n) = f(m+n)$; ∴ for all values of m and n this holds.
Without going deeply into the question of the validity of the principle we may point out the fatal objection that it apparently establishes the Binomial Theorem for all values of x, a ludicrous result, when we remember that the series is divergent if $x > 1$.

And when x is numerically less than 1, $f(m)$ and $f(n)$ are convergent series which have all their terms positive or are such as would not lose their convergency when all their terms are made positive. [§ 415.]

Hence, by the theorem for the distribution of infinite series, § 418, $f(m+n)$ is a convergent series whose sum is equal to the product of the sums of the series $f(m)$ and $f(n)$.

Thus, when x is numerically less than 1,
$$*f(m) \times f(n) = f(m+n) \quad \ldots \ldots \ldots \ldots (i),$$
whatever m and n may be;
$$\therefore \ f(m) \times f(n) \times f(p) = f(m+n) \times f(p) = f(m+n+p).$$
And $\ f(m) \times f(n) \times f(p) \ldots = f(m+n+p+\ldots)$
for all values of $m, n, p \ldots$.

Now take s factors on the left side of this equality, and as $m, n, p \ldots$ may take any values whatever, let
$$m = n = p = \ldots = \frac{r}{s},$$
where r and s are positive integers.

Then $\quad f\left(\dfrac{r}{s}\right) \times f\left(\dfrac{r}{s}\right) \times f\left(\dfrac{r}{s}\right) \ldots$ to s factors
$$= f\left(\frac{r}{s} + \frac{r}{s} + \frac{r}{s} + \ldots \text{ to } s \text{ terms}\right).$$

That is, $\left\{f\left(\dfrac{r}{s}\right)\right\}^s = f(r)$
$$= 1 + rx + \frac{r(r-1)}{1 \cdot 2} x^2 + \frac{r(r-1)(r-2)}{1 \cdot 2 \cdot 3} x^3 + \ldots$$
$= (1+x)^r$, since r is a positive integer.
$$\therefore \ f\left(\frac{r}{s}\right) = (1+x)^{\frac{r}{s}}.$$

---

* Or from this point we may proceed as in the next Article.

THE BINOMIAL THEOREM.   463

That is, $(1+x)^{\frac{r}{s}} = 1 + \frac{r}{s}x + \frac{\frac{r}{s}\left(\frac{r}{s}-1\right)}{1.2}x^2 + \ldots,$

which proves the binomial theorem for a positive fractional exponent, x being numerically less than 1.

And as the binomial theorem is true for any positive integral exponent, it is true for any positive index.

We shall now prove it to be true for any negative exponent.

In $\quad f(m) \times f(n) = f(m+n),$
let $m = -n$, where n is positive.

Then $f(-n) \times f(n) = f(0) = 1 + 0x + \frac{0(0-1)}{1.2}x^2 + \ldots$

$\qquad\qquad\qquad = 1.$

$\therefore\ f(-n) = \frac{1}{f(n)} = \frac{1}{(1+x)^n},$ since n is positive,

$\qquad\qquad = (1+x)^{-n};$

$\therefore\ (1+x)^{-n} = 1 + (-n)x + \frac{(-n)(-n-1)}{1.2}x^2 + \ldots,$

and the theorem is completely proved when x is numerically < 1.

426. Alternative Proof. We shall now give a slightly different proof of the result of the last Article. It is more complete in itself than the foregoing, inasmuch as Vandermonde's Theorem is proved in the course of the work, but it depends on the relation

$$f(m) \times f(n) = f(m+n)$$

in just the same manner. The demonstration is as follows.

When $x < 1$ the series

$$1 + mx + \frac{m\cdot(m-1)}{2!}x^2 + \ldots \frac{m(m-1)\ldots(m-r+1)}{r!}x^r + \ldots \text{ad inf.}$$

is convergent (§ 415). Denote it by $f(m)$ and by $f(n)$ the corresponding series with n written in the place of m.

THE BINOMIAL THEOREM.

Then since neither series loses its convergency when all its terms are taken positively $f(m) \times f(n)$, i.e. the product of the two series, is equal to the sum of a convergent series in which the coefficient of x^r is

$$A_r = \frac{m(m-1)\ldots(m-r+1)}{r!} + n \cdot \frac{m(m-1)\ldots(m-r+2)}{(r-1)!}$$
$$+ \frac{n(n-1)}{2!} \frac{m(m-1)\ldots m-r+3}{(r-2)!} + \ldots + \frac{n(n-1)\ldots(n-r+1)}{r!} \quad (\S 418)..(1).$$

Now whenever m and n are positive integers, the series whose sums are $f(m)$ and $f(n)$ represent $(1+x)^m$ and $(1+x)^n$ respectively (§ 328).

Consequently A_r becomes the coefficient of x^r in

$$(1+x)^m \times (1+x)^n,$$

i.e. in $(1+x)^{m+n}$.

Hence in all such cases, i.e. when m and n are integers greater than r,

$$A_r = \frac{(m+n)(m+n-1)\ldots(m+n-r+1)}{r!} \quad \ldots\ldots\ldots\ldots (2).$$

The equation obtained by equating these two values of A_r is of the rth degree in both m and n, for each factor in the numerator in (2) contains m and n and there are r factors. Also this equation is satisfied by more than r values of m and similarly for n, for we can assign more than r positive integral values each greater than r to either m or n, and all these satisfy the equation. Hence it is an identity (§ 372) and is satisfied by all values of m and n, therefore the coefficient of x^r in $f(m)$, $f(n)$ is *always*

$$\frac{(m+n)(m+n-1)\ldots(m+n-r)}{r!}.$$

Consequently the product of $f(m)$ and $f(n)$ is a convergent series similar in form to $f(m)$ with $(m+n)$ written in the place of m,

$$\therefore f(m) \times f(n) = f(m+n) \quad \ldots\ldots\ldots\ldots\ldots\ldots (3).$$

From this point we may proceed as from equation (i) in § 425 or thus:

This relation is identical in form with the Index Law

$$a^m a^n = a^{m+n}. \quad (\S 2.)$$

Now the Index Law, combined with the property that when m^r is a positive integer,

$$a^m = a \times a \times a \ldots \text{ to } m \text{ factors},$$

enables us to assign meanings to a^m when m is negative and fractional. (§§ 8, 10.)

THE BINOMIAL THEOREM.

Bearing in mind that when m is a positive integer
$$f(m) = (1+x)^m = (1+x)(1+x) \ldots \text{to } m \text{ factors,}$$
we see by analogy that (3) affords interpretations of $f(m)$ for negative and fractional values of m, which must be identical with the corresponding interpretations of $(1+x)^m$.

Hence by the Properties of Indices,

$f\left(\dfrac{p}{q}\right)$ represents the qth root of $f(p)$ or $\sqrt[q]{(1+x)^p}$,

$f(-n)$ represents the reciprocal of $f(n)$ or $\dfrac{1}{(1+x)^n}$.

Hence $\quad f\left(\dfrac{p}{q}\right) = (1+x)^{\frac{p}{q}}$, and $f(-n) = (1+x)^{-n}$.

427. There are one or two points in connection with the theorem of §§ 425, 426 of which the student should take particular notice. In the first place, it is absolutely essential for the validity of our proof that x should be numerically less than unity, for otherwise we do not know that the series for $f(m)$ is convergent, § 415, and hence cannot apply the result of § 418. We may remark that under certain conditions the theorem is true when x is numerically equal to unity, but the proper treatment of these cases is beyond the scope of the present work. When x is greater than unity the series for $f(m)$ is divergent, and so the theorem is never true. Thus, for example, we can only apply the theorem to $(1+2x)^{\frac{1}{2}}$ when $2x$ is numerically less than unity, i.e. when x lies between $-\frac{1}{2}$ and $+\frac{1}{2}$.

Again, when n is a positive integer we can expand $(a+x)^n$ in just the same way as $(1+x)^n$, but if n be not a positive integer we must write $(a+x)^n$ in the form

$$a^n\left(1+\frac{x}{a}\right)^n,$$

and we can now expand it provided $\dfrac{x}{a}$ is numerically less than unity, i.e. if x is less than a. If $x > a$ we can expand

$$x^n\left(1+\frac{a}{x}\right)^n.$$

428. Illustrative Examples.

Ex. 1. Find the first four terms in the expansion of $(8+x)^{1\frac{1}{3}}$,

$$(8+x)^{1\frac{1}{3}} = \left\{8\left(1+\frac{x}{8}\right)\right\}^{1\frac{1}{3}} = 8^{1\frac{1}{3}}\left(1+\frac{1}{8}x\right)^{1\frac{1}{3}}$$

$$= 16\left\{1 + 1\tfrac{1}{3}\cdot\tfrac{1}{8}x + \frac{1\tfrac{1}{3}(1\tfrac{1}{3}-1)}{1\cdot 2}(\tfrac{1}{8}x)^2 + \frac{1\tfrac{1}{3}(1\tfrac{1}{3}-1)(1\tfrac{1}{3}-2)}{1\cdot 2\cdot 3}\cdot(\tfrac{1}{8}x)^3 + \ldots\right\}$$

$$= 16\left\{1 + \frac{1}{6}x + \frac{2}{9}\cdot\frac{1}{64}x^2 - \frac{2}{9}\cdot\frac{1}{64}\cdot\frac{1}{36}x^3 + \ldots\right\};$$

$\therefore$ the first four terms $= 16 + 2\frac{2}{3}x + \dfrac{1}{18}x^2 - \dfrac{1}{648}x^3$.

Ex. 2. Find the cube root of $(8+x)^4$ to four terms.

The question is identical with that of Example 1, since

$$\sqrt[3]{(8+x)^4} = (8+x)^{\frac{4}{3}} = (8+x)^{1\frac{1}{3}}.$$

Ex. 3. Expand $(1-x)^{-1}$ to four terms and find the general term.

$(1-x)^{-1}$

$$= 1 + (-1)(-x) + \frac{(-1)(-2)}{1\cdot 2}(-x)^2 + \frac{(-1)(-2)(-3)}{1\cdot 2\cdot 3}(-x)^3 + \ldots$$

$$= 1 + x + x^2 + x^3 + \ldots$$

The general term $= \dfrac{(-1)(-2)\ldots\ldots(-r)}{1\cdot 2\cdot 3\ldots\ldots r}(-x)^r$

$$= (-1)^r\frac{1\cdot 2\cdot 3\ldots\ldots r}{1\cdot 2\cdot 3\ldots\ldots r}\cdot(-1)^r\cdot x^r = (-1)^{2r}\cdot x^r = x^r,$$

since $(-1)^{2r} = 1$.

Ex. 4. Find the first five terms of the expansion of $(1-x)^{-2}$.

In some cases it will be advisable to find the general term first and to deduce therefrom the particular terms required.

$$T_{r+1} = \frac{(-2)(-3)(-4)\ldots\ldots(-r-1)}{1\cdot 2\cdot 3\ldots\ldots r}(-x)^r$$

$$= (-1)^r\cdot\frac{2\cdot 3\cdot 4\ldots\ldots(r+1)}{1\cdot 2\cdot 3\ldots\ldots r}(-1)^r\cdot x^r$$

$$= (-1)^{2r}(r+1)x^r = (r+1)x^r.$$

$\therefore T_1 = (0+1)x^0 = 1$; $T_2 = (1+1)x^1 = 2x$; $T_3 = (2+1)x^2 = 3x^2$; and so on. Thus the first five terms are

$$1 + 2x + 3x^2 + 4x^3 + 5x^4.$$

THE BINOMIAL THEOREM.

Ex. 5. Find the general term of $(1-x)^{-3}$, and thence deduce the first four terms.

$$T_{r+1} = \frac{(-3)(-4)(-5)\ldots\ldots(-r-2)}{1.2.3\ldots\ldots r}(-x)^r$$

$$= (-1)^r \cdot \frac{1.2.3\ldots\ldots r}{1.2.3\ldots\ldots r} \cdot \frac{(r+1)(r+2)}{1.2}(-1)^r x^r = \frac{(r+1)(r+2)}{1.2} x^r.$$

$\therefore$ the first four terms $= 1 + 3x + 6x^2 + 10x^3$.

NOTE. The expansions of $(1-x)^{-1}$, $(1-x)^{-2}$, and of $(1-x)^{-3}$ are *worthy of note*.

$$(1-x)^{-1} = 1 + x + x^2 + x^3 + \ldots\ldots + x^r + \ldots\ldots$$

$$(1-x)^{-2} = 1 + 2x + 3x^2 + 4x^3 + \ldots\ldots + (r+1)x^r + \ldots\ldots$$

$$(1-x)^{-3} = 1 + 3x + 6x^2 + 10x^3 + 15x^4 + \ldots\ldots$$

$$+ \frac{(r+1)(r+2)}{1.2} x^r + \ldots\ldots$$

The terms of $(1+x)^{-1}$, $(1+x)^{-2}$ and of $(1+x)^{-3}$ will be numerically those of $(1-x)^{-1}$, $(1-x)^{-2}$ and of $(1-x)^{-3}$ respectively, but will be alternately positive and negative.

Ex. 6. Find the general term of $(3-x)^{-25}$.

$$(3-x)^{-25} = 3^{-25}(1-\tfrac{1}{3}x)^{-25}.$$

$\therefore$ the general term $= T_{r+1} = (3)^{-25} \times \dfrac{-25(-26)\ldots(-25-r+1)}{1.2.3\ldots\ldots r}(-\tfrac{1}{3}x)^r$

$$= 3^{-25} \cdot (-1)^r \cdot \frac{25.26.27\ldots\ldots(r+24)}{1.2.3\ldots\ldots r}(-1)^r \frac{x^r}{3^r}$$

$$= 3^{-r-25}(-1)^{2r} \cdot \frac{(r+1)(r+2)\ldots\ldots(r+24)}{1.2.3\ldots\ldots 24} x^r, \text{ when common factors are struck out}$$

$$= \frac{1}{3^{r+25}} \cdot \frac{(r+1)(r+2)\ldots\ldots(r+24)}{1.2.3\ldots\ldots 24} x^r.$$

Ex. 7. Find the first five terms and the general term of the expansion of $(1-x)^{\frac{5}{2}}$.

$$(1-x)^{\frac{5}{2}} = 1 + \frac{5}{2}(-x) + \frac{\frac{5}{2}\left(\frac{5}{2}-1\right)}{1.2}(-x)^2 + \frac{\frac{5}{2}\left(\frac{5}{2}-1\right)\left(\frac{5}{2}-2\right)}{1.2.3}(-x)^3$$

$$+ \frac{\frac{5}{2}\left(\frac{5}{2}-1\right)\left(\frac{5}{2}-2\right)\left(\frac{5}{2}-3\right)}{1.2.3.4}(-x)^4 + \ldots\ldots;$$

∴ the first five terms $= 1 - 2\tfrac{1}{2} x + 1\tfrac{7}{8} x^2 - \dfrac{5}{16} x^3 - \dfrac{5}{128} x^4$.

And the general term $= T_{r+1} = \dfrac{\tfrac{5}{2}\left(\tfrac{5}{2} - 1\right)\left(\tfrac{5}{2} - 2\right)\ldots\ldots\left(\tfrac{5}{2} - r + 1\right)}{1 \,.\, 2 \,.\, 3 \ldots\ldots r}(-x)^r$

$= \dfrac{\tfrac{5}{2} \cdot \tfrac{3}{2} \cdot \tfrac{1}{2} \cdot \left(-\tfrac{1}{2}\right) \cdot \left(-\tfrac{3}{2}\right) \ldots\ldots \left(-\tfrac{2r-7}{2}\right)}{1\,.\,2\,.\,3\ldots\ldots r}(-x)^r \ldots\ldots$ (i).

As there are $r - 3$ negative factors in the numerator, the general term

$= (-1)^{r-3} \cdot \dfrac{5\,.\,3\,.\,1\,.\,1\,.\,3\,.\,5\,.\,7\ldots\ldots(2r-7)}{2^r\,.\,1\,.\,2\,.\,3\ldots r}(-1)^r (x)^r$

$= (-1)^{2r-3} \cdot \dfrac{5\,.\,3\,.\,1\,.\,1\,.\,3\,.\,5\ldots\ldots(2r-7)}{1\,.\,2\,.\,3\ldots\ldots r}\left(\dfrac{x}{2}\right)^r$

$= -\dfrac{5\,.\,3\,.\,1\,.\,1\,.\,3\,.\,5\ldots\ldots(2r-7)}{1\,.\,2\,.\,3\ldots\ldots r} \cdot \left(\dfrac{x}{2}\right)^r \ldots\ldots\ldots\ldots\ldots$ (ii),

[since $(-1)^{2r-3} = -1$, $2r - 3$ being odd.]

NOTE. The expression in the line marked (ii) **has a peculiarity** which we shall do well to notice. It seems to indicate that all the terms in the expansion are negative, which is clearly incorrect, since the first and third terms have been obtained above as positive quantities.

That the first four terms are alternately positive and negative, and that the terms from the fourth are all negative will be evident from the form of the general expression in the line marked (i) which may be exhibited in the form:—

$$\dfrac{\tfrac{5}{2}(-x) \times \tfrac{3}{2}(-x) \times \tfrac{1}{2}(-x) \times -\tfrac{1}{2}(-x) \times -\tfrac{3}{2}(-x)\ldots\ldots -\left(\dfrac{2r-7}{2}\right)(-x)}{1\,.\,2\,.\,3\ldots\ldots r}.$$

The expression which appears in the line marked (ii), though it is called the general term, must be carefully taken to apply to the terms of the expansion, only after negative factors begin to appear in the numerator; this we shall be able to remember if we notice that the factors of the numerator first decrease and afterwards increase, and also that the last factor $2r - 7$ shows that r must have a value greater than $3\tfrac{1}{2}$, if the expression is to be applicable.

Ex. 8. Expand $(1 + x)^{\frac{5}{2}}$ to five terms, and find the general term.

Proceeding as in Example 7, we find that the first five terms of

$$(1 + x)^{\frac{5}{2}} = 1 + 2\tfrac{1}{2} x + 1\tfrac{7}{8} x^2 + \dfrac{5}{16} x^3 - \dfrac{5}{128} x^4.$$

THE BINOMIAL THEOREM.

The general term $= T_{r+1} = \dfrac{2\frac{1}{2}\,(2\frac{1}{2}-1)\ldots\ldots(2\frac{1}{2}-r+1)}{1\,.\,2\,.\,3\ldots\ldots r}\, x^r$

$\qquad\qquad\qquad = (-1)^{r-3} \cdot \dfrac{5\,.\,3\,.\,1\,.\,1\ldots\ldots(2r-7)}{1\,.\,2\,.\,3\ldots\ldots r\,.\,2^r} \cdot x^r$

for values of r greater than 3.

NOTE. The first four terms are all positive and the terms from the fourth are alternately positive and negative.

Ex. 9. Expand $\dfrac{\sqrt{1+x}-\sqrt{1-x}}{\sqrt{1+x}+\sqrt{1-x}}$ in ascending powers of x.

Rationalising the denominator, we have

$\dfrac{\sqrt{1+x}-\sqrt{1-x}}{\sqrt{1+x}+\sqrt{1-x}} = \dfrac{(1+x)+(1-x)-2\sqrt{1-x^2}}{(1+x)-(1-x)}$

$\qquad\qquad\qquad = \dfrac{2\{1-(1-x^2)^{\frac{1}{2}}\}}{2x}$

$\qquad\qquad\qquad = x^{-1}\{1-(1-x^2)^{\frac{1}{2}}\}$

$\qquad\qquad\qquad = x^{-1} - x^{-1}(1-x^2)^{\frac{1}{2}}$

$= x^{-1} - x^{-1}\left\{1 - \dfrac{1}{2}x^2 + \dfrac{\frac{1}{2}\left(\frac{1}{2}-1\right)}{1\,.\,2}x^4 - \dfrac{\frac{1}{2}\left(\frac{1}{2}-1\right)\left(\frac{1}{2}-2\right)}{1\,.\,2\,.\,3}x^6 + \ldots\ldots\right\}$

$= \dfrac{1}{2}x + \dfrac{1}{8}x^3 + \dfrac{1}{16}x^5 + \ldots\ldots + \dfrac{1\,.\,3\,.\,5\ldots\ldots(2n-3)}{2\,.\,4\,.\,6\ldots\ldots 2n}x^{2n-1} + \ldots\ldots$

429. Modification of the general term. The ultimate sign of the terms.

Ex. 1. Find the $(r+1)$th term of $(1+x)^{-n}$.

$T_{r+1} = \dfrac{-n(-n-1)(-n-2)\ldots\ldots(-n-r+1)}{1\,.\,2\,.\,3\ldots\ldots r}\, x^r$

$\qquad = (-1)^r \dfrac{n(n+1)(n+2)\ldots\ldots(n+r-1)}{1\,.\,2\,.\,3\ldots\ldots r}\, x^r.$

NOTE. The factors of the numerator of this expression go on increasing by unity from n until we come to the last factor which is $n+r-1$. The factor $(-1)^r$ shews that the terms are alternately positive and negative from the beginning, x and n being positive.

THE BINOMIAL THEOREM.

Ex. 2. Find the $(r+1)$th term of $(1-x)^{-n}$.

$$T_{r+1} = \frac{-n(-n-1)\ldots(-n-r+1)}{1.2.3\ldots r}(-x)^r$$

$$= (-1)^{2r}\frac{n(n+1)(n+2)\ldots(n+r-1)}{1.2.3\ldots r}x^r$$

$$= \frac{n(n+1)(n+2)\ldots(n+r-1)}{1.2.3\ldots r}x^r.$$

NOTE 1. The last expression indicates that every term in the expansion of $(1-x)^{-n}$ is positive, if x and n are positive.

NOTE 2. Though the general term of the expansion of any binomial may be derived from the formula,

$$T_{r+1} = \frac{n(n-1)\ldots(n-r+1)}{1.2\ldots r}x^r,$$

it will be more convenient in practice to use the formula,

$$T_{r+1} \text{ of } (1-x)^{-n} = \frac{n(n+1)\ldots(n+r-1)}{1.2.3\ldots r}x^r.$$

when the index is **negative**.

Ex. 3. Find the general term of $(1+x)^{-\frac{p}{q}}$.

$$T_{r+1} = \frac{-\frac{p}{q}\left(-\frac{p}{q}-1\right)\ldots\left(-\frac{p}{q}-r+1\right)}{1.2.3\ldots r}x^r$$

$$= (-1)^r \frac{p(p+q)(p+2q)\ldots\{p+(r-1)q\}}{1.2\ldots r \cdot q^r}x^r.$$

NOTE 1. The general term of $(1-x)^{-\frac{p}{q}}$ is obtained by substituting $-x$ for x in the foregoing expression.

Hence T_{r+1} of $(1-x)^{-\frac{p}{q}} = \frac{p(p+q)(p+2q).\{p+(r-1)q\}}{1.2\ldots r \cdot q^r}x^r$.

NOTE 2. The terms of $(1+x)^{-\frac{p}{q}}$ will be alternately positive and negative from the very beginning, while those of $(1-x)^{-\frac{p}{q}}$ will be all positive, p, q and x being positive. [See Note 1, Examples 1 and 2.]

Ex. 4. Find the general term of $(1+x)^{\frac{p}{q}}$.

$$T_{r+1} = \frac{\frac{p}{q}\left(\frac{p}{q}-1\right)\ldots\left(\frac{p}{q}-r+1\right)}{1.2\ldots r}x^r$$

$$= \frac{p(p-q)(p-2q)\ldots\ldots\{p-(r-1)q\}}{1 \cdot 2 \ldots\ldots r \cdot q^r} x^r.$$

NOTE 1. The general term of $(1-x)^{\frac{p}{q}}$ is obtained by substituting $-x$ for x in the foregoing expression.

Hence

$$T_{r+1} \text{ of } (1-x)^{\frac{p}{q}} = (-1)^r \frac{p(p-q)(p-2q)\ldots\ldots\{p-(r-1)q\}}{1 \cdot 2 \cdot 3 \ldots\ldots r \cdot q^r} x^r.$$

NOTE 2. As has been explained in the preceding article, if x, p and q are positive, the terms of $(1+x)^{\frac{p}{q}}$ will be all positive until a negative factor begins to appear in the numerator of the coefficient, and will then be alternately positive and negative; while the terms of $(1-x)^{\frac{p}{q}}$ will, until a negative factor begins to enter the numerator, be alternately positive and negative, and will then be all positive or all negative.

Ex. 5. Find the first negative term in $(1+x)^{\frac{11}{2}}$, x being positive.

$$T_{r+1} = T_r \times \frac{(n-r+1)}{r} x$$

$$= T_r \times \left(\frac{5\frac{1}{2} - r + 1}{r}\right) x$$

$$= T_r \times \left(\frac{6\frac{1}{2} - r}{r}\right) x.$$

Since x is positive, T_{r+1} will *first* be negative, as soon as $6\frac{1}{2} - r$ is negative; that is, as soon as $r > 6\frac{1}{2}$, that is, when $r = 7$.

Hence the term which involves x^7 or the 8th term is the first negative term.

Ex. 6. From what term will the terms of $(1-x)^{10\frac{1}{4}}$ have the same sign, x being positive? And what is that sign?

Since $$T_{r+1} = T_r \times \frac{n-r+1}{r}(-x),$$

two successive terms will *first* have the same sign as soon as $n - r + 1$ begins to be negative, that is, as soon as $10\frac{1}{4} - r + 1$, that is, $11\frac{1}{4} - r$ is negative, that is, as soon as $r > 11\frac{1}{4}$; that is, when $r = 12$.

∴ the 13th term has the same sign as the 12th.

Thus all the terms from the 12th term are of the same sign; this sign being of course *negative*, since the terms are at first alternately positive and negative, odd terms being positive and even terms negative.

EXERCISES XXVIII.

1. Show that *if* x *be positive, and* n *unrestricted in value, the terms of* $(1+x)^n$ *will be alternately positive and negative from the term containing* x^r, *where* r *is the least integer greater than* n.

2. Hence deduce that the terms of $(1+x)^{-n}$, where x and n are positive, are alternately positive and negative from the first term forwards.

3. Deduce also that in the expansion of $(1+x)^n$ where x is positive and n a positive fraction, the first negative term will be that containing x^{r+1}, where r is the least integer which exceeds n.

4. Show that *if* x *be positive and* n *unrestricted in value, the terms of* $(1-x)^n$ *from and after the term containing* x^r *will be of the same sign, where* r *is the least integer which exceeds* n.

5. Hence deduce that the terms of $(1-x)^{-n}$ where x and n are positive are *all* positive.

6. Show also that in the expansion of $(1-x)^n$, where x is positive and n a positive fraction, the terms will have the same sign from and after the term containing x^r, r being the least integer which exceeds n; and show that this sign will be positive or negative according as r is even or odd.

Expand each of the following expressions to four terms:—

7. $(1+x)^{\frac{1}{4}}$.   **8.** $(1+x)^{-\frac{1}{2}}$.   **9.** $(1-\frac{1}{2}x)^{\frac{1}{2}}$.

10. $(1+x)^{\frac{2}{3}}$.   **11.** $(a^2-x^2)^{-1}$.   **12.** $\sqrt{(a^2-x^2)}$.

13. $(a^2-ax)^{-\frac{3}{10}}$.   **14.** $(a-a^{-1}x^{-1})^{-3}$.   **15.** $\dfrac{m}{\sqrt{(b^2+c^4)}}$.

16. $(c^2-x^2)^{\frac{3}{4}}$.   **17.** $\dfrac{1}{(\sqrt[3]{a}-\sqrt[3]{x})^6}$.   **18.** $(a^5-x^5)^{-\frac{1}{5}}$.

Find and simplify the terms named in the following expansions:—

19. The 4th and the $(r+1)$th term in the expansion of $(1+3x)^{-2}$.

20. The 8th and the $(r+1)$th term in the expansion of $(1+x)^{-4}$.

21. The 7th and the $(r+1)$th term in the expansion of $(1-x)^{-6}$.

22. The $(r+1)$th term in $(1+2x)^{\frac{1}{2}}$.

23. The 13th term in the expansion of $(256+64x)^{\frac{11}{2}}$.

24. The $(r+1)$th term in the expansion of $3a/(a^3-x^2)^{\frac{1}{3}}$.

25. The general term in the expansion of $(xy-\sqrt{yz})^{\frac{17}{3}}$.

26. Show that the nth terms of $(1-x)^{-n}$ and $(1+x)^{2n-2}$ are equal.

THE BINOMIAL THEOREM. 473

27. Find the coefficient of x^{4r} in $(1-x^2)^{-3}$.

28. If p and q be the rth terms of $(1-x)^{-\frac{1}{2}}$ and $(1-x)^{-\frac{3}{2}}$, then $q = (2r-1)p$.

29. Show that all the coefficients of $(1+x)^{-n}$ are integers, if n be an integer.

What terms in the following expansions are first negative, x being positive?—

30. $(1-x)^{\frac{3}{2}}$. **31.** $(1+x)^{\frac{19}{2}}$. **32.** $(1+x)^{\frac{171}{10}}$.

33. $(1+x)^{\frac{20}{3}}$. **34.** $(1+x)^{-3\frac{1}{4}}$.

430. To find the numerically greatest term in a binomial series.

In the expansion of $(1+x)^n$ we shall suppose x positive, since only the *numerical* value of the greatest term has to be attended to.

Case I. *Let n be a positive integer.*

Here the $(r+1)$th term is obtained by multiplying the rth term by $\dfrac{n-r+1}{r}x$; that is, by $\left(\dfrac{n+1}{r}-1\right)x$.

Hence the terms continue to increase, each term upon the one preceding, so long as

$$\left(\frac{n+1}{r}-1\right)x > 1;$$

that is, $\dfrac{n+1}{r}-1 > \dfrac{1}{x}$;

or $\dfrac{n+1}{r} > 1+\dfrac{1}{x}$; or $\dfrac{r}{n+1} < \dfrac{x}{x+1}$;

or $r < \dfrac{(n+1)x}{x+1}$, which expression we shall call p.

Arguing as in § 334 we find that if p be an integer, two adjacent terms of the expansion are equal, *viz.*, the pth and the $(p+1)$th; and these are greater than any other term.

If p be not an integer, suppose q to be the integral part of p; then the $(q+1)$th term will be the greatest.

Case II. *Let n be a positive fraction, x being less than unity.*

Here, as in the preceding case, the $(r+1)$th term may be obtained by multiplying the rth term by $\left(\dfrac{n+1}{r}-1\right)x$.

THE BINOMIAL THEOREM.

Hence here, as in the first case, if $p\left[=\dfrac{(n+1)}{x+1}x\right]$ be an integer, the pth term is equal to the $(p+1)$th term, and these are greater than any other term. If p be not an integer, let q be the integral part of p; then the $(q+1)$th term will be the greatest.

It may be remarked that the multiplying factor $\left(\dfrac{n+1}{r}-1\right)x$ indicates that the terms continue to be positive only so long as $r<n+1$; when $r>n+1$, the multiplying factor becomes negative, but is numerically less than 1, since both x and $\left(\dfrac{n+1}{r}-1\right)$ are less than unity.

Hence the greatest term will be found between the first term and the term containing x^r, where r is the integral part of $n+1$.

Case III. *Let n be negative, x being less than unity.*

Let $n=-m$, so that m is positive. The numerical value of the $(r+1)$th term may be obtained by multiplying that of the rth term by $\left(\dfrac{m+r-1}{r}\right)x$, that is $\left(\dfrac{m-1}{r}+1\right)x$.

(a) If $m<1$, $\dfrac{m-1}{r}$ is a negative quantity less than unity, and $\left(\dfrac{m-1}{r}+1\right)$ is therefore a proper fraction; and as x is also less than 1, the multiplying factor is less than unity for all values of r. Hence each term is less than the preceding, and the first term is therefore the greatest.

(b) If $m=1$, then $\left(\dfrac{m-1}{r}+1\right)x=x$ for all values of r, and as x is less than unity, the terms go on decreasing, the first term being the greatest as in subdivision (a).

(c) If $m>1$, then $\left(\dfrac{m-1}{r}+1\right)x$ will be greater than unity, so long as $\left(\dfrac{m-1}{r}+1\right)x>1$;

that is $\dfrac{(m-1)x}{r}>1-x$; that is, $\dfrac{r}{(m-1)x}<\dfrac{1}{1-x}$;

that is, $r<\dfrac{(m-1)x}{1-x}$; which expression we shall call p.

If p be a positive integer, the pth term is equal to the $(p+1)$th term, and these are greater than any other term. If p be a positive fraction, suppose q the integral part of p, then the $(q+1)$th term is the greatest.

THE BINOMIAL THEOREM.

Ex. 1. Find the greatest term in the expansion of $(1+x)^{-n}$, when $x=\frac{1}{5}$ and $n=60$.

Here $\quad T_{r+1} = (-1)^r \dfrac{n(n+1)(n+2)\ldots\ldots(n+r-1)}{1.2.3\ldots\ldots r} x^r$

$\qquad = T_r \times \dfrac{n+r-1}{r} x \text{ numerically.}$

$\qquad = T_r \times \dfrac{60+r-1}{r} \times \tfrac{1}{5} = T_r \times \dfrac{59+r}{5r}\,;$

$\therefore T_{r+1} > T_r$, so long as $(59+r)/5r > 1$,

that is, $\qquad\qquad\qquad 59+r > 5r\,;$

that is, $\qquad\qquad\qquad r < 14\tfrac{3}{4}$.

Hence for values of r up to 14 inclusive, $T_{r+1} > T_r$ and for values of r from 15, $T_{r+1} < T_r$.

Hence the 15th term is the greatest.

Ex. 2. Find the greatest term in the expansion of $(1-x)^{-30}$, where $x=\tfrac{5}{6}$.

$T_{r+1} = \dfrac{n(n+1)\ldots\ldots(n+r-1)}{1.2.3\ldots\ldots r} x^r$

$\qquad = T_r \cdot \dfrac{n+r-1}{r} x = T_r \dfrac{30+r-1}{r} \cdot \tfrac{5}{6} = \dfrac{(29+r)5}{6r} \cdot T_r\,;$

$\therefore T_{r+1} > T_r$, so long as $(29+r)5 > 6r$; that is, $r < 145$.

Hence for values of r up to 144 inclusive,

$\qquad\qquad T_{r+1} > T_r\,;\ \text{when}\ r = 145,\ T_{r+1} = T_r\,;$

and for values of r greater than 145, $T_{r+1} > T_r$. Hence the 145th and 146th terms are the greatest.

EXERCISES XXVIII.

35. From what term in each of the following expansions are all the terms of the same sign?

$$(1-x)^{\frac{40}{3}};\ (1-x)^{\frac{51}{2}};\ (1-x)^{-3}.$$

36. If r be the greatest whole number contained in p/q, show that $(a+x)^{\frac{p}{q}}$ has the first $r+2$ terms of its expansion positive, and the $(r+m)^{\text{th}}$ of the same sign as $(-1)^m$, but $(a-x)^{\frac{p}{q}}$ has the first $(r+1)$ terms alternatively positive and negative, and all the rest of the same sign as $(-1)^{r+1}$.

THE BINOMIAL THEOREM.

Find the greatest term in the following expansions:—

37. $(1+x)^{\frac{3}{2}}$, when $x = \frac{5}{6}$.   **38.** $\left(1 + \frac{2}{3}\right)^{\frac{31}{5}}$.

39. $(a^2 + ax)^{\frac{3}{2}}$, when $x = 2$ and $a = 3$.

40. $(1 - x/a)^{-\frac{4}{3}}$, when $x = 12$ and $a = 13$.

41. At what term does the series for $\left(1 + \frac{9}{10}\right)^{\frac{3}{2}}$ begin to converge? [See Art. 430.]

42. Show that the greatest term of $\left(\frac{1}{4} + \frac{1}{5}\right)^{\frac{31}{3}}$ is $\frac{11}{900}\sqrt[3]{4}$.

43. Show that if x be less than 2, $(2+x)^{\frac{1}{2}}$ has the first term the greatest.

$\left[x \text{ being less than 2}, \left\{\left(\frac{3}{2} - r\right)\middle/ r\right\}\frac{1}{2}x \text{ is less than 1 for all values of } r.\right]$

44. What is the meaning of $(1+x)^{\sqrt{3}}$? Has the binomial theorem been proved in any sense to extend to such a quantity?

45. Find the greatest term in $(\sqrt{2}+1)^{\sqrt{2}}$.

46. Find which is the greatest coefficient in $(1+x)^{\sqrt[11]{2}}$.

431. Summation of Series.

Sometimes a series of numbers can be summed by showing that its terms, after suitable modification where necessary, are those of the expansion of a binomial.

Ex. 1. Sum the infinite series

$$1 + 2 \cdot \frac{1}{3^2} + \frac{2 \cdot 5}{1 \cdot 2} \cdot \frac{1}{3^4} + \frac{2 \cdot 5 \cdot 8}{1 \cdot 2 \cdot 3} \cdot \frac{1}{3^6} + \ldots.$$

Taking the expansion of $(1-x)^{-n}$, we have

$$(1-x)^{-n} = 1 + nx + \frac{n(n+1)}{1 \cdot 2}x^2 + \ldots + \frac{n(n+1)\ldots(n+r-1)}{1 \cdot 2 \cdot 3 \ldots r}x^r + \ldots.$$

Here (1) the indices of the powers of x increase regularly by unity, the index of x in any term being the number of the factors in the numerator or denominator of its coefficient; (2) the numerator of the coefficient is the product of an A.P. of which the first term is n, and the common difference is 1, and (3) the denominator of the coefficient is also the product of an A.P. of which the first term is 1, and the common difference is 1.

In the series given for summation, the first and second conditions do not appear to be satisfied. We therefore modify it, by dividing by 3 each factor of the numerator of every coefficient, the requisite number of 3's for this division being taken from the powers of $\frac{1}{3}$.

THE BINOMIAL THEOREM.

Thus, the series after modification

$$=1+\frac{2}{3}\cdot\frac{1}{3}+\frac{\frac{2}{3}\cdot\frac{5}{3}}{1.2}\cdot\left(\frac{1}{3}\right)^2+\frac{\frac{2}{3}\cdot\frac{5}{3}\cdot\frac{8}{3}}{1.2.3}\cdot\left(\frac{1}{3}\right)^3+\ldots$$

$$=\left(1-\frac{1}{3}\right)^{-\frac{2}{3}}=\left(\frac{2}{3}\right)^{-\frac{2}{3}}=\left(\frac{3}{2}\right)^{\frac{2}{3}}=\frac{1}{2}\sqrt[3]{18}.$$

Ex. 2. Sum to infinity:

$$1+\frac{1}{10^2}+\frac{1.3}{1.2}\cdot\frac{1}{10^4}+\frac{1.3.5}{1.2.3}\cdot\frac{1}{10^6}+\ldots$$

The series $=1+\frac{1}{10^2}+\frac{\frac{1}{2}\cdot\frac{3}{2}}{1.2}\cdot\frac{1}{2^2.5^4}+\frac{\frac{1}{2}\cdot\frac{3}{2}\cdot\frac{5}{2}}{1.2.3}\cdot\frac{1}{2^3.5^6}+\ldots$

$$=1+\frac{1}{2}\left(\frac{1}{2}\cdot\frac{1}{5^2}\right)+\frac{\frac{1}{2}\cdot\frac{3}{2}}{1.2}\left(\frac{1}{2}\cdot\frac{1}{5^2}\right)^2+\frac{\frac{1}{2}\cdot\frac{3}{2}\cdot\frac{5}{2}}{1.2.3}\left(\frac{1}{2}\cdot\frac{1}{5^2}\right)^3+\ldots$$

$$=\left(1-\frac{1}{2}\cdot\frac{1}{25}\right)^{-\frac{1}{2}}=\left(1-\frac{1}{50}\right)^{-\frac{1}{2}}=\left(\frac{49}{50}\right)^{-\frac{1}{2}}=\left(\frac{50}{49}\right)^{\frac{1}{2}}=\frac{5}{7}\sqrt{2}.$$

Ex. 3. Sum to infinity:

$$2\tfrac{1}{5}+\frac{6.2}{5.10}-\frac{6.2.2}{5.10.15}+\frac{6.2.2.6}{5.10.15.20}-\ldots$$

The series $=2\tfrac{1}{5}+\frac{6.2}{1.2}\cdot\left(\frac{1}{5}\right)^2-\frac{6.2.2}{1.2.3}\cdot\left(\frac{1}{5}\right)^3+\frac{6.2.2.6}{1.2.3.4}\cdot\left(\frac{1}{5}\right)^4-\ldots$

$$=1+\frac{6}{5}+\frac{\frac{6}{4}\cdot\frac{2}{4}}{1.2}\cdot\left(\frac{4}{5}\right)^2+\frac{\frac{6}{4}\cdot\frac{2}{4}\cdot\left(-\frac{2}{4}\right)}{1.2.3}\cdot\left(\frac{4}{5}\right)^3$$

$$+\frac{\frac{6}{4}\cdot\frac{2}{4}\cdot\left(-\frac{2}{4}\right)\left(-\frac{6}{4}\right)}{1.2.3.4}\cdot\left(\frac{4}{5}\right)^4+\ldots$$

$$=1+\frac{3}{2}\left(\frac{4}{5}\right)+\frac{\frac{3}{2}\cdot\frac{1}{2}}{1.2}\cdot\left(\frac{4}{5}\right)^2+\frac{\frac{3}{2}\cdot\frac{1}{2}\cdot\left(-\frac{1}{2}\right)}{1.2.3}\cdot\left(\frac{4}{5}\right)^3+\ldots$$

$$=\left(1+\frac{4}{5}\right)^{\frac{3}{2}}=\left(\frac{9}{5}\right)\times\left(\frac{9}{5}\right)^{\frac{1}{2}}=\frac{27}{25}\sqrt{5}.$$

THE BINOMIAL THEOREM.

Ex. 4. Sum to infinity:

$$\frac{1}{12} - \frac{1.4}{12.18} + \frac{1.4.7}{12.18.24} - \ldots$$

The series $= \dfrac{\frac{1}{3}}{4} - \dfrac{\frac{1}{3} \cdot \frac{4}{3}}{4.6} + \dfrac{\frac{1}{3} \cdot \frac{4}{3} \cdot \frac{7}{3}}{4.6.8} - \ldots$

$= \dfrac{\frac{1}{3}}{2} \cdot \left(\dfrac{1}{2}\right) - \dfrac{\frac{1}{3} \cdot \frac{4}{3}}{2.3} \cdot \left(\dfrac{1}{2}\right)^2 + \dfrac{\frac{1}{3} \cdot \frac{4}{3} \cdot \frac{7}{3}}{2.3.4} \cdot \left(\dfrac{1}{2}\right)^3 - \ldots$

The factors of the denominators are in A.P.; but they do not begin with 1. Hence one additional factor, namely unity, has to be introduced into the denominator of each coefficient; and as the number of factors in the numerator is the same as that of the factors in the denominator, we have to introduce an additional factor in the numerator also, which factor is clearly, $-\frac{2}{3}$. Let S denote the series.

Now, $-\frac{2}{3} S = \dfrac{(-\frac{2}{3}) \cdot \frac{1}{3}}{1.2} \cdot \frac{1}{2} - \dfrac{(-\frac{2}{3}) \cdot \frac{1}{3} \cdot \frac{4}{3}}{1.2.3} \cdot (\frac{1}{2})^2 + \dfrac{(-\frac{2}{3}) \cdot \frac{1}{3} \cdot \frac{4}{3} \cdot \frac{7}{3}}{1.2.3.4} (\frac{1}{2})^3 - \ldots$

Again, since the index of x in every term must be the same as the number of factors in the numerator or denominator of the coefficient, we have

$$-\frac{2}{3} S \times \frac{1}{2} = \dfrac{\frac{2}{3} \cdot \left(-\frac{1}{3}\right)}{1.2} \cdot \left(\frac{1}{2}\right)^2 + \dfrac{\frac{2}{3} \cdot \left(-\frac{1}{3}\right)\left(-\frac{4}{3}\right)}{1.2.3} \cdot \left(\frac{1}{2}\right)^3 + \ldots$$

Here the first two terms are evidently wanting. When we supply these, we find that

$$-\frac{2}{3} S \times \frac{1}{2} + 1 + \frac{2}{3}\left(\frac{1}{2}\right) = 1 + \frac{2}{3}\left(\frac{1}{2}\right) + \dfrac{\frac{2}{3}\left(\frac{2}{3} - 1\right)}{1.2}\left(\frac{1}{2}\right)^2 + \ldots$$

$$\therefore \quad -\frac{1}{3} S + 1\frac{1}{3} = \left(1 + \frac{1}{2}\right)^{\frac{2}{3}} = \left(\frac{3}{2}\right)^{\frac{2}{3}} = \frac{1}{2}\sqrt[3]{18}.$$

$$\therefore \quad S = 3\left(1\frac{1}{3} - \frac{1}{2}\sqrt[3]{18}\right) = 4 - 1\frac{1}{2}\sqrt[3]{18}.$$

NOTE:—Other series connected with the expansions of binomials may be similarly summed up. But if the student should be at a loss for a clue for the identification of the expansion of which the given series is a disguised form or an incomplete portion, the method of Example 1, of § 429 will always be useful, four adjacent terms in order, or these terms multiplied by a constant k, being taken to be four consecutive terms of a binomial expansion.

THE BINOMIAL THEOREM. 479

432. Sum of Coefficients.

Ex. 1. Find the sum of the first $r+1$ coefficients in the expansion of $(1-x)^n$.

$$(1-x)^n = 1 - nx + \frac{n(n-1)}{1.2}x^2 + \ldots$$
$$+ (-1)^r \frac{n(n-1)\ldots(n-r+1)}{1.2.3\ldots r} x^r + \ldots.$$

And $(1-x)^{-1} = 1 + x + x^2 + x^3 + \ldots + x^r + \ldots$.

Here the coefficient of x^r in the product of the two series is equal to the coefficient of x^r in $(1-x)^n \times (1-x)^{-1}$, that is, in the expansion of $(1-x)^{n-1}$. [§ 407.]

The coefficient of x^r in the product of the two series is the sum of the first $r+1$ coefficients in the expansion of $(1-x)^n$. And the coefficient of x^r in the expansion of $(1-x)^{n-1}$ is

$$(-1)^r \frac{(n-1)(n-2)(n-3)\ldots(n-r)}{1.2.3\ldots r};$$

∴ the required sum $= (-1)^r \dfrac{(n-1)(n-2)(n-3)\ldots(n-r)}{1.2.3\ldots r}$.

Ex. 2. Find the sum of the coefficients of the first $r+1$ terms in the expansion of $(1-x)^{-3}$.

$$(1-x)^{-3} = 1 + 3x + 6x^2 + 10x^3 + \ldots + \frac{(r+1)(r+2)}{1.2} x^r + \ldots.$$

And $(1-x)^{-1} = 1 + x + x^2 + x^3 + \ldots + x^r + \ldots$.

Now, the coefficient of x^r in the product of the two series
$= 1 + 3 + 6 + \ldots + (r+1)(r+2)/1.2$
$=$ the sum of the first $r+1$ coefficients in the expansion of $(1-x)^{-3}$.
And this is equal to the coefficient of x^r in $(1-x)^{-3} \times (1-x)^{-1}$.

∴ the required sum $=$ the coefficient of x^r in the expansion of

$$(1-x)^{-4} = \frac{1}{6}(r+1)(r+2)(r+3).$$

Ex. 3. Find the sum of n terms of the series

$$1.2.3.4 + 2.3.4.5 + 3.4.5.6 + \ldots.$$

The $(r+1)^{\text{th}}$ term of the series $= (r+1)(r+2)(r+3)(r+4)$

$$= 1.2.3.4 \times \frac{(r+1)(r+2)(r+3)(r+4)}{1.2.3.4}.$$

$$= 24 \times \frac{5.6.7\ldots r(r+1)(r+2)(r+3)(r+4)}{(1.2.3.4).5.6.7\ldots r}$$

$= 24 \times$ the $(r+1)^{\text{th}}$ coefficient of $(1-x)^{-5}$.

∴ the 1st term of the given series $= 24 \times$ the 1st coefficient of $(1-x)^{-5}$.

The 2nd term of the given series $= 24 \times$ the 2nd coefficient of $(1-x)^{-5}$; and so on.

Hence the sum of the 1st n terms of the given series

$= 24 \times \{$sum of the 1st n coefficients of $(1-x)^{-5}\}$

$= 24 \times \{$coefficient of x^{n-1} in the expansion of $(1-x)^{-6}\}$
[See Example 1]

$= 24 \times \dfrac{n(n+1)(n+2)(n+3)(n+4)}{1.2.3.4.5}$

$= \dfrac{1}{5} n(n+1)(n+2)(n+3)(n+4).$

433. Coefficient of x^r in expressions other than direct expansions of binomials, &c.

Ex. 1. Find the coefficient of x^r in $\dfrac{4+2x+3x^2}{(1-x)^2}$.

The expansion $= (4+2x+3x^2)(1-x)^{-2}$

$= (4+2x+3x^2)\{1+2x+3x^2+\ldots$
$\qquad + (r-1)x^{r-2} + rx^{r-1} + (r+1)x^r + \ldots\}.$

In the product of these two factors, the terms involving x^r are obtained by multiplying $(r+1)x^r$ by 4, rx^{r-1} by $2x$ and $(r-1)x^{r-2}$ by $3x^2$.

∴ the coefficient of $x^r = 4(r+1) + 2r + 3(r-1) = 9r+1.$

Ex. 2. Find the expansion in ascending powers of x of

$$\dfrac{a+bx}{p+qx}.$$

$\dfrac{a+bx}{p+qx} = (a+bx)(p+qx)^{-1} = (a+bx) \cdot p^{-1}\left(1+\dfrac{qx}{p}\right)^{-1}$

$= \dfrac{a+bx}{p}\left\{1 - \dfrac{qx}{p} + \dfrac{q^2x^2}{p^2} - \dfrac{q^3x^3}{p^3} + \ldots\right\}$

$= \dfrac{a}{p} + \dfrac{x}{p}\left(b - \dfrac{aq}{p}\right) - \dfrac{qx^2}{p^2}\left(b - \dfrac{aq}{p}\right) + \ldots.$

THE BINOMIAL THEOREM. 481

Ex. 3. Expand $\dfrac{(1+x)^4}{(1-x^2)^3}$ in a series of ascending powers of x.

The expression $= (1+x)^4 (1+x)^{-3} (1-x)^{-3} = (1+x)(1-x)^{-3}$
$= (1+x)(1 + p_1 x + p_2 x^2 \ldots + p_{r-1} x^{r-1} + p_r x^r + \ldots)$, suppose.

∴ the coefficient of x^r

$$= p_r + p_{r-1} = \tfrac{1}{2}(r+1)(r+2) + \tfrac{1}{2} r (r+1) = (r+1)^2.$$

[Art. 414, Ex. 5.]

Hence the expansion $= 1^2 + 2^2 x + 3^2 x^2 + \ldots + (r+1)^2 x^r + \ldots$.

Ex. 4. Find the coefficient of x^r in the expansion of
$$(1 - 2x + 3x^2 - 4x^3 + \ldots \text{to infinity})^{\frac{1}{2}}.$$

The quantity within the brackets $= (1+x)^{-2}$.

[See Art. 414, Ex. 5, Note.]

Hence the given expression $= \{(1+x)^{-2}\}^{\frac{1}{2}} = (1+x)^{-1}$.

Hence the required coefficient of $x^r = (-1)^r$.

[See Art. 414, Ex. 5, Note.]

Ex. 5. Find the coefficient of x^{100} and x^{101} in the expansion of
$$(1 + x + x^2)^{-2}.$$

The expression $= (1-x)^2 \times \dfrac{(1+x+x^2)^{-2}}{(1-x)^2}$

$= (1-x)^2 (1-x)^{-2} (1+x+x^2)^{-2} = (1-x)^2 \{1 - x^3\}^{-2}$

$= (1 - 2x + x^2)\{1 + 2x^3 + 3x^6 + 4x^9 + \ldots$
$\hspace{4cm} + (r+1) x^{3r} + \ldots\}.$

In the expansion of $(1-x^3)^{-2}$, the coefficients of x^{98}, x^{100} and x^{101} are evidently nothing; and the coefficient of x^{99} is 34.

∴ the coefficient of $x^{100} = 1 \times 0 - 2 \times 34 + 1 \times 0 = -68$;

and the coefficient of $x^{101} = 1 \times 0 - 2 \times 0 + 1 \times 34 = 34$.

Ex. 6. Show that the coefficient of x^{11} in the expansion of
$$\sqrt{\dfrac{(1+x)^3}{(1-x)^5}}$$

in ascending powers of x is

$$\dfrac{7 \cdot 9 \cdot 11 \cdot 23}{8 \lfloor 4}.$$

TUT. ALG. II. 31

THE BINOMIAL THEOREM.

The expression $=\dfrac{(1+x)^{\frac{3}{2}}}{(1-x)^{\frac{5}{2}}}$.

To escape the difficulty of determining the required coefficient by dealing with two infinite series, multiply the numerator and the denominator of the expression by $(1+x)^{\frac{5}{2}}$ which operation will give us an integral power of $1+x$ for the numerator, leaving the denominator still a binomial.

Thus the expression $=\dfrac{(1+x)^{\frac{3}{2}}(1+x)^{\frac{5}{2}}}{(1+x)^{\frac{5}{2}}(1-x)^{\frac{5}{2}}} = \dfrac{(1+x)^4}{(1-x^2)^{\frac{5}{2}}} = (1+x)^4(1-x^2)^{-\frac{5}{2}}$

$= (1+4x+6x^2+4x^3+x^4)\left(1+\dfrac{5}{2}x^2+\ldots\right)$.

Since even powers of x alone appear in the expansion of $(1-x^2)^{-\frac{5}{2}}$, the coefficient of x^{11} required in the question can be got by multiplying the coefficients of $(x^2)^4$ and $(x^2)^5$ in this expansion by the coefficients of x^3 and x respectively in the expansion of $(1+x)^4$, and therefore the coefficient of x^{11}

$= 4 \cdot \dfrac{\frac{5}{2} \cdot \frac{7}{2} \cdot \frac{9}{2} \cdot \frac{11}{2}}{1.2.3.4} + 4 \cdot \dfrac{\frac{5}{2} \cdot \frac{7}{2} \cdot \frac{9}{2} \cdot \frac{11}{2} \cdot \frac{13}{2}}{1.2.3.4.5}$

$= \dfrac{7.9.11}{\lfloor 4} \cdot \left(\dfrac{5}{4} + \dfrac{13}{8}\right) = \dfrac{7.9.11.23}{8\lfloor 4}$.

Ex. 7. Expand $(1+2x+4x^2)^{-\frac{1}{2}}$ as far as x^2 inclusive.

$(1+2x+4x^2)^{-\frac{1}{2}} = \{1+2x(1+2x)\}^{-\frac{1}{2}}$

$= 1 - \dfrac{1}{2} \cdot 2x(1+2x) + \dfrac{\left(-\dfrac{1}{2}\right)(-1\frac{1}{2})}{1.2} \cdot (2x)^2(1+2x)^2$

  + terms involving powers higher than x^2.

$= 1 - x - 2x^2 + \dfrac{3}{8} \cdot 4x^2 = 1 - x - \dfrac{1}{2}x^2 + \ldots$.

Ex. 8. Find the first four terms of the expansion of
$(1+ax+bx^2)^n$.

$(1+ax+bx^2)^n = \{1+(ax+bx^2)\}^n$

$= 1 + n(ax+bx^2) + \dfrac{n(n-1)}{1.2}(ax+bx^2)^2$

$\quad + \dfrac{n(n-1)(n-2)}{.1.2.3}(ax+bx^2)^3$

  + terms not involving powers of x lower than x^4.

$= 1 + nax + \left\{nb + \dfrac{n(n-1)}{1.2}a^2\right\}x^2$

$\quad + \left\{n(n-1)ab + \dfrac{n(n-1)(n-2)}{1.2.3}a^3\right\}x^3 + \ldots$.

THE BINOMIAL THEOREM.

Ex. 9. Find the coefficient of x^r in the expansion of
$$\frac{3-x}{(1-x)(2-x)}.$$

Resolving the given expression into partial fractions
$$\frac{3-x}{(1-x)(2-x)} = \frac{2}{1-x} - \frac{1}{2-x} = \frac{2}{1-x} - \frac{1}{2\left(1-\frac{1}{2}x\right)}$$
$$= 2(1-x)^{-1} - \frac{1}{2}\left(1-\frac{1}{2}x\right)^{-1}$$
$$= 2(1+x+x^2+\ldots+x^r+\ldots)$$
$$- \frac{1}{2}\left(1+\frac{1}{2}x+\frac{1}{4}x^2+\ldots+x^r/2^r+\ldots\right).$$

Hence the coefficient of $x^r = 2 - 1/2^{r+1}$.

EXERCISES XXVIII.

Show that

47. $\sqrt[3]{\dfrac{10}{7}} = 1 + \dfrac{1}{10} + \dfrac{1 \cdot 4}{10 \cdot 20} + \dfrac{1 \cdot 4 \cdot 7}{10 \cdot 20 \cdot 30} + \ldots$ to infinity.

48. $\sqrt{2} = 1 + \dfrac{1}{2^2} + \dfrac{1 \cdot 3}{\lfloor 2 \cdot 2^4} + \dfrac{1 \cdot 3 \cdot 5}{\lfloor 3 \cdot 2^6} + \ldots$ to infinity.

49. $\sqrt[3]{\dfrac{3}{2}} = 1 + \dfrac{1}{3^2} + \dfrac{1 \cdot 4}{1 \cdot 2} \cdot \dfrac{1}{3^4} + \dfrac{1 \cdot 4 \cdot 7}{1 \cdot 2 \cdot 3} \cdot \dfrac{1}{3^6} + \ldots$ to infinity.

50. $2 = 1 + \dfrac{1}{2 \cdot 2} + \dfrac{1 \cdot 3}{2 \cdot 3 \cdot 2^2} + \dfrac{1 \cdot 3 \cdot 5}{2 \cdot 3 \cdot 4 \cdot 2^3} + \ldots$ to infinity.

51. $\sqrt{5} = \dfrac{5}{2}\left\{1 - \dfrac{1}{8} + \dfrac{1 \cdot 3}{8 \cdot 16} - \dfrac{1 \cdot 3 \cdot 5}{8 \cdot 16 \cdot 24} + \ldots \text{ to infinity}\right\}$.

52. $\dfrac{4}{3}\sqrt{3} - 2\dfrac{1}{4} = \dfrac{3}{4} \cdot \dfrac{1}{2^4} + \dfrac{3 \cdot 5}{4 \cdot 6} \cdot \dfrac{1}{2^6} + \dfrac{3 \cdot 5 \cdot 7}{4 \cdot 6 \cdot 8} \cdot \dfrac{1}{2^8} + \ldots$ to infinity.

53. Find the sum of the infinite series whose p^{th} term is
$$p\left(\frac{x}{1+x}\right)^{p-1}.$$

54. Find, by multiplication or otherwise, the cube of
$$1 + \frac{1}{3}x + \frac{1 \cdot 4}{3 \cdot 6}x^2 + \frac{1 \cdot 4 \cdot 7}{3 \cdot 6 \cdot 9}x^3 + \ldots \text{ to infinity},$$
as far as x^3 in its simplest form.

THE BINOMIAL THEOREM.

55. Find the sum of the first $r+1$ coefficients in $(1-x)^{-5}$.

56. Find the sum of the first 7 coefficients in $(1-x)^{\frac{1}{2}}$.

57. If $(1+x)^{-4} = 1 + p_1 x + p_2 x^2 + p_3 x^3 + \ldots$,
sum the series $1 - p_1 + p_2 - p_3 + \ldots$ to n terms.

58. Sum to n terms the series $1.2.3 + 2.3.4 + 3.4.5 + \ldots$.

59. Prove that in the expansion of $(1-x)^{-\frac{1}{n}}$, the sum of the coefficients of the first r terms bears to the coefficient of the rth term the ratio of $1 + n(r-1) : 1$.

60. In the expansion of $(1-x)^{-n}$, find the conditions that the sum of the coefficients of the first three terms may be equal to the coefficient of the fourth term. Find also the condition that the sum of the coefficients of the first r terms may be equal to the coefficient of the $(r+1)^{\text{th}}$ term.

61. Find the coefficient of x^r in $(1+2x)/(1-x)$.

62. Find the coefficient of x^{2n} in $(1-x)/(1+x)^2$.

63. Find the coefficient of x^{10} in $(1+x)/(1-x)^3$.

64. Find the coefficient of x^r in $(1-x)^2/(1-2x)^3$.

65. Find the coefficient of x^n in $(1+2x+3x^2)/(1-x)^3$.

66. Show that the coefficient of x^n in $(1+2x)/(1-2x)^2$ is $2^n(2n+1)$.

67. Find the term involving x^5 in the expansion of
$$(1 + 2x + 3x^2 + \ldots \text{ to infinity})^6.$$

68. Find the coefficient of x^r in $(1 + ax + a^2 x^2 + \ldots \text{ to infinity})^n$.

69. Show that the coefficient of x^2 in $(1 - 2x + 3x^2 - \ldots \text{ to infinity})^{-\frac{3}{2}}$ is 3.

70. Show that the coefficient of x^n in $(1+x)^n/(1-x)$ is 2^n.

71. Show that the coefficient of x^{n+1} in the expansion of
$$(1-x)\sqrt{1+x}$$
is
$$(-1)^n \cdot \frac{1}{2^{n+1}} \cdot \frac{1.3.5\ldots(2n-3)}{\lfloor n+1} \cdot (4n+1).$$

72. Find the coefficient of x^r in the expansion of
$$\{(1+x)/(1-x)\}^{\frac{1}{2}}.$$

73. Show that the coefficient of x^{2n} in $\dfrac{1}{(1-x)^3(1+x)^4}$ is
$$\frac{1}{6}(n+1)(n+2)(n+3).$$

THE BINOMIAL THEOREM.

74. Write down the first three terms in the expansion of
$$\{(1+x)/(1-x)\}^{\frac{1}{3}}.$$

75. (a) Show that the coefficient of x^4 in
$$\left(\frac{1+x}{1-x}\right)^n \text{ is } \frac{2}{3} n^2 (n^2+2).$$

(b) Show that the coefficient of x^r in the expansion of
$$(1.2 + 2.3x + 3.4x^2 + \ldots \text{ to infinity})^2$$
is
$$\frac{1}{30}(r+1)(r+2)(r+3)(r+4)(r+5).$$

76. Show that the coefficient of x^n in $(1-2x)^n/(1-3x)$ is 1 and that of x^{n+r} is 3^r.

$$\left[\frac{(1-2x)^n}{1-3x} = \frac{1}{1-3x}(\overline{1-3x}+x)^n\right.$$
$$= \frac{1}{1-3x}\{(1-3x)^n + n(1-3x)^{n-1}x + \ldots + n(1-3x)x^{n-1} + x^n\}$$
$$= (1-3x)^{n-1} + n(1-3x)^{n-2}.x + \ldots + nx^{n-1} + x^n(1-3x)^{-1}.$$
Here the only term that contains x^n and x^{n+r} is the last, viz. $x^n(1-3x)^{-1}$. Hence, etc.$\Big]$

77. Show that if t_r denote the middle term of $(1+x)^{2r}$, then will $t_0 + t_1 + t_2 + \ldots = (1-4x)^{-\frac{1}{2}}.$

78. Show that in the expansion of $(1-x)^{-n}$, no coefficient can be equal to the next following unless all the coefficients are equal.

79. Prove that $(1 + x + x^2 + \ldots \text{ to infinity})(1 + 2x + 3x^2 + \ldots \text{ to infinity})$
$$= \frac{1}{2}(1.2 + 2.3x + 3.4x^2 + \ldots \text{ to infinity}).$$

80. Show that $x^n = 1 + n\left(1 - \frac{1}{x}\right) + \frac{n(n+1)}{1.2}\left(1 - \frac{1}{x}\right)^2 + \ldots.$
Hint. $[x^n = (1/x)^{-n} = \{1-(1-1/x)\}^{-n}.]$

81. Show that
$$(a-b)^n = a^n\left\{1 - n.\frac{b}{a-b} + \frac{n(n+1)}{1.2}\left(\frac{b}{a-b}\right)^2 - \ldots\right\}.$$
Hint. $\left[(a-b)^n = a^n\left(\frac{a-b}{a}\right)^n\right.$
$$\left. = a^n\left(\frac{a}{a-b}\right)^{-n} = a^n\left(1+\frac{b}{a-b}\right)^{-n}\right].$$

THE BINOMIAL THEOREM.

82. Show that
$$\left(\frac{1+x}{1-x}\right)^n = 1 + n\cdot\left(\frac{2x}{1+x}\right) + \frac{n(n+1)}{1.2}\cdot\left(\frac{2x}{1+x}\right)^2 + \ldots$$

Hint. $\left[\left(\frac{1+x}{1-x}\right)^n = \left(\frac{1-x}{1+x}\right)^{-n} = \left(1 - \frac{2x}{1+x}\right)^{-n}\right].$

83. Prove that $1 + n\dfrac{2n}{1+n} + \dfrac{n(n+1)}{1.2}\left(\dfrac{2n}{1+n}\right)^2 + \ldots$
$$= 1 + n\frac{2n}{1-n} + \frac{n(n-1)}{1.2}\left(\frac{2n}{1-n}\right)^2 + \ldots.$$

*****84.** Find the coefficient of x^{n-1} in $(1+x)^n(1-x)^{-4}$, and thence show that the expression that you get is the sum of the first n coefficients of $(1+x)^n/(1-x)^3$.

85. Find the coefficient of x^4 in $(1+x+x^2)^{-5}$.

86. Find the coefficient of x^2 in $(1+ax+bx^2)^{-\frac{1}{2}}$.

87. Expand $(a - bx - cx^2)^{-\frac{2}{3}}$ to four terms.

88. Find the first two terms in the expansion of
$$\frac{2 + 3x + (1-3x)^{\frac{1}{2}}}{1 - \frac{1}{2}x + (4-x)^{\frac{1}{2}}}$$

according to ascending powers of x.

89. Show that
$$\frac{1}{x^2 - 5x + 6} = \frac{1}{x-3} - \frac{1}{x-2} = \frac{1}{2\left(1 - \frac{1}{2}x\right)} - \frac{1}{3\left(1 - \frac{1}{3}x\right)}$$
$$= \frac{1}{2}\left(1 - \frac{1}{2}x\right)^{-1} - \frac{1}{3}\left(1 - \frac{1}{3}x\right)^{-1};$$

and hence show that the coefficient of x^n in
$$1/(x^2 - 5x + 6) = \left(\frac{1}{2}\right)^{n+1} - \left(\frac{1}{3}\right)^{n+1}.$$

90. Prove that
$$\frac{1}{(1-x)(1-2x)(1-3x)} = \frac{1}{2}(1-x)^{-1} - 4(1-2x)^{-1} + 4\tfrac{1}{2}(1-3x)^{-1},$$

and hence show that the coefficient of x^n in $1/(1-x)(1-2x)(1-3x)$ is $\frac{1}{2}(3^{n+2} - 2^{n+3} + 1)$.

91. Observing that
$$\frac{a^{-1}/a}{(1-ax)(1-x/a)} = \frac{1}{a^2 - 1}\left(\frac{1}{1-ax} - \frac{1}{a^2}\frac{1}{1-x/a}\right),$$
show that $a^n + a^{n-2} + a^{n-4} + \ldots + a^{-n} = (a^{2n+2} - 1)/a^n(a^2 - 1)$
by obtaining the coefficient of x^n on both sides of the identity.

92. (a) Find the coefficient of x^4 and of x^7 in
$$(1 + x + x^2 + x^3 + x^4)^3.$$
(b) Show that the coefficient of x^{4m} in $(1 - x + x^2 - x^3)^{-1}$ is 1.

93. A student is examined in 3 papers with a maximum of n marks for each paper. In how many ways can he get a total of n marks?

[In each paper he may get 0, 1, 2, ... or n marks.

Hence the total number of ways in which he can get a total of n marks is the coefficient of x^n in
$$(1 + x + x^2 + \ldots + x^n)(1 + x + x^2 + \ldots + x^n)(1 + x + x^2 + \ldots + x^n),$$
since every way of forming the coefficient of x^n, say $(x^2 \times x^2 \times x^{n-4})$, corresponds to a way of getting a total of n marks (2 marks in the first paper, 2 marks in the second paper and $n - 4$ marks in the third).

Hence the number required = coefficient of x^n in
$$(1 + x + \ldots + x^n)^3 = \left(\frac{1 - x^{n+1}}{1 - x}\right)^3 = (1 - x^{n+1})^3 (1 - x)^{-3}$$
$$= (1 - 3x^{n+1} + 3x^{2n+2} - x^{3n+3})$$
$$\left(1 + 3x + 6x^2 + \ldots + \frac{(n+1)(n+2)}{1.2}x^n + \ldots\right).$$

Hence the answer required is $\frac{1}{2}(n+1)(n+2)$.]

94. A man goes in for an examination in which there are four papers with a maximum of m marks for each paper. Show that the number of ways of getting half marks on the whole is
$$\frac{1}{3}(m+1)(2m^2 + 4m + 3).$$

434. Application of the theorem to Approximations.

To obtain, by the binomial theorem, an approximate value of the nth root of a given number.

Let N be the number whose n^{th} root is required. Express N in the form of $a^n + b$, where a is so chosen that a^n differs very little from N, and b, which may be positive or negative, is thus made very small in comparison with a^n.

$$N^{\frac{1}{n}} = (a^n + b)^{\frac{1}{n}} = (a^n)^{\frac{1}{n}}\{1 + b/a^n\}^{\frac{1}{n}} = a(1+x)^{\frac{1}{n}}, \text{ where } x = b/a^n.$$

The terms in the expansion of $(1+x)^{\frac{1}{n}}$ will gradually diminish, and an approximate value of $(1+x)^{1/n}$ and therefore of $N^{1/n}$ may thus be obtained by taking a small number of these terms. The number of terms to be retained will depend upon the degree of accuracy required, since evidently the greater the number of terms that we take, the more nearly shall we approach to the exact value of the required root.

Ex. 1. Express $\sqrt[5]{31}$ as a series and find its value to four places of decimals.

$$\sqrt[5]{31} = (32-1)^{\frac{1}{5}} = \left\{32\left(1 - \frac{1}{32}\right)\right\}^{\frac{1}{5}} = 2\left(1 - \frac{1}{2^5}\right)^{\frac{1}{5}}$$

$$= 2\left\{1 + \frac{1}{5}\left(-\frac{1}{2^5}\right) + \frac{\frac{1}{5}\cdot\left(-\frac{4}{5}\right)}{1\cdot 2}\cdot\left(-\frac{1}{2^5}\right)^2 + \ldots\right\}$$

$$= 2\left\{1 - \frac{1}{5}\cdot\frac{1}{2^5} - \left(\frac{1}{5}\cdot\frac{1}{2^5}\right)\left(\frac{1}{5}\cdot\frac{1}{2^4}\right) - \ldots\right\}$$

$$= 2\left\{1 - \frac{2}{10}\cdot\frac{5^5}{10^5} - \left(\frac{2}{10}\cdot\frac{5^5}{10^5}\right)\left(\frac{2}{10}\cdot\frac{5^4}{10^4}\right) - \ldots\right\}$$

$$= 2\left\{1 - \frac{625}{10^5} - \left(\frac{625}{10^5}\right)\left(\frac{125}{10^4}\right) - \ldots\right\}$$

$$= 2\{1 - \cdot 00625 - \cdot 00625 \times \cdot 0125\},$$

other terms not being taken into account, as their value will not affect the fourth place of decimals;

$$= 1 \cdot 9873 \text{ to four places of decimals.}$$

Ex. 2. Find the cube root of $\dfrac{1}{345}$ to four places of decimals.

$$\sqrt[3]{\frac{1}{345}} = (345)^{-\frac{1}{3}} = (343+2)^{-\frac{1}{3}} = \left\{7^3\left(1 + \frac{2}{7^3}\right)\right\}^{-\frac{1}{3}} = \frac{1}{7}\left(1 + \frac{2}{7^3}\right)^{-\frac{1}{3}}.$$

THE BINOMIAL THEOREM.

The student may complete the solution by expanding the binomial, &c. The arithmetical calculation of the result will be simplified, if we proceed as follows:—

$$\begin{array}{r}7\,|\,1{\cdot}00\\ 7\,|\underline{{\cdot}1428571}=1/7\\ 7\,|\underline{{\cdot}0204081}=1/7^2\\ {\cdot}0029154=1/7^3\text{; and so on.}\end{array}$$

Ex. 3. Find, by the binomial theorem, the value of $\sqrt{2}$ to four places of decimals.

$$\sqrt{2} = 7\sqrt{\frac{2}{49}} = \frac{7}{5}\sqrt{\frac{50}{49}} = \frac{7}{5}\left(\frac{50}{49}\right)^{\frac{1}{2}} = \frac{7}{5}\left(\frac{49}{50}\right)^{-\frac{1}{2}} = \frac{7}{5}\left(1-\frac{1}{50}\right)^{-\frac{1}{2}}$$

$$= \frac{7}{5}\left\{1 + \left(-\frac{1}{2}\right)\left(-\frac{1}{50}\right) + \frac{\left(-\frac{1}{2}\right)\left(-\frac{1}{2}-1\right)}{1\,.\,2}\cdot\left(-\frac{1}{50}\right)^2\right.$$

$$\left.+ \frac{\left(-\frac{1}{2}\right)\left(-\frac{1}{2}-1\right)\left(-\frac{1}{2}-2\right)}{1\,.\,2\,.\,3}\cdot\left(-\frac{1}{50}\right)^3 + \ldots\right\}.$$

$$= \frac{7}{5}\left(1 + \frac{1}{100} + \frac{3}{20{,}000} + \frac{1}{400{,}000} + \ldots\right)$$

$$= 1{\cdot}4 \times (1 + {\cdot}01 + {\cdot}00015 + {\cdot}0000025 + \ldots)$$

$$= 1{\cdot}4142 \text{ to four places of decimals.}$$

NOTE. The value of $\sqrt{2}$ can also be found by putting it in the form

$$\sqrt{2} = \frac{10}{7}\sqrt{\frac{49}{50}} = \frac{10}{7}\left(1-\frac{1}{50}\right)^{\frac{1}{2}}.$$

Ex. 4. Show that $(a+x)^n : a^n$ is nearly $a+nx : a$, if x is small compared with a.

$$(a+x)^n : a^n = \left(\frac{a+x}{a}\right)^n = \left(1+\frac{x}{a}\right)^n = \left(1+\frac{nx}{a}\right) \text{ nearly} = \frac{a+nx}{a}.$$

Ex. 5. If x is so small that x^2 and the higher powers of x may be omitted, show that

$$\frac{\sqrt{1+x} + \sqrt[3]{(1-x)^2}}{1+x+\sqrt{1+x}} = 1 - \frac{5}{6}x \text{ nearly.}$$

The expression $= \dfrac{(1+x)^{\frac{1}{2}} + (1-x)^{\frac{2}{3}}}{(1+x) + (1+x)^{\frac{1}{2}}}$

$$= \frac{\left(1+\frac{1}{2}x+\ldots\right)+\left(1-\frac{2}{3}x+\ldots\right)}{1+x+\left(1+\frac{1}{2}x+\ldots\right)}.$$

Since x^2 and the higher powers of x may be neglected, the expression

$$=\frac{1+\frac{1}{2}x+1-\frac{2}{3}x}{1+x+1+\frac{1}{2}x}=\frac{2-\frac{1}{6}x}{2+1\frac{1}{2}x}=\frac{1-\frac{1}{12}x}{1+\frac{3}{4}x}$$

$$=\left(1-\frac{1}{12}x\right)\left(1+\frac{3}{4}x\right)^{-1}=\left(1-\frac{1}{12}x\right)\left(1-\frac{3}{4}x+\ldots\right)$$

$$=1-\frac{1}{12}x-\frac{3}{4}x; \; x^2 \text{ and the higher powers of } x \text{ being omitted}$$

$$=1-\frac{5}{6}x \text{ nearly}.$$

Ex. 6. If b be **very small** compared with a, show **that**
$$\frac{(a-b)^n}{(a+b)^n} \text{ is nearly } 1-2nb/a.$$

$$\frac{(a-b)^n}{(a+b^n)}=\frac{a^n(1-b/a)^n}{a^n(1+b/a)^n}=(1-b/a)^n(1+b/a)^{-n}$$

$$=(1-nb/a)(1-nb/a); \; \frac{b^2}{a^2} \text{ and higher powers of } \frac{b}{a} \text{ being omitted};$$

$$=1-2nb/a \text{ nearly}.$$

Ex. 7. Prove that $(1-mx)^n - (1-nx)^m = \frac{1}{2}mn(m-n)x^2$ approximately, when x is so small that x^3 and higher powers of x may be neglected.

The expression $= \left\{1-nmx+\frac{n(n-1)}{1.2}m^2x^2-\ldots\right\}$

$$-\left\{1-mnx+\frac{m(m-1)}{1.2}n^2x^2-\ldots\right\}$$

$$=\frac{1}{2}\left\{n(n-1)m^2x^2-m(m-1)n^2x^2\right\}; \text{ since } x^3 \text{ and}$$
$$\text{higher powers of } x \text{ may be omitted};$$

$$=\frac{1}{2}mn(m-n)x^2 \text{ approximately}.$$

THE BINOMIAL THEOREM. 491

Ex. 8. Prove that

$$\sqrt{(x^2+a^2)(x^2+b^2)} - \sqrt{(x^2+c^2)(x^2+d^2)} = \frac{1}{2}(a^2+b^2-c^2-d^2),$$

when x is very great.

The expression $= (x^2+a^2)^{\frac{1}{2}}(x^2+b^2)^{\frac{1}{2}} - (x^2+c^2)^{\frac{1}{2}}(x^2+d^2)^{\frac{1}{2}}$

$= x^2(1+a^2/x^2)^{\frac{1}{2}}(1+b^2/x^2)^{\frac{1}{2}} - x^2(1+c^2/x^2)^{\frac{1}{2}}(1+d^2/x^2)^{\frac{1}{2}}$

$= x^2\left(1+\frac{1}{2}\cdot\frac{a^2}{x^2}+...\right)\left(1+\frac{1}{2}\cdot\frac{b^2}{x^2}+...\right)$

$\qquad - x^2\left(1+\frac{1}{2}\cdot\frac{c^2}{x^2}+...\right)\left(1+\frac{1}{2}\cdot\frac{d^2}{x^2}+...\right)$

$= x^2\left(1+\frac{a^2}{2x^2}+\frac{b^2}{2x^2}\right) - x^2\left(1+\frac{c^2}{2x^2}+\frac{d^2}{2x^2}\right);$

reciprocals of powers of x higher than x^2 being neglected, since x is very great;

$= x^2 - x^2 + \frac{1}{2}x^2\left(\frac{a^2+b^2-c^2-d^2}{x^2}\right)$ approximately

$= \frac{1}{2}(a^2+b^2-c^2-d^2)$ approximately.

Ex. 9. If x be nearly equal to unity, show that

$$\frac{mx^m - nx^n}{m-n} = x^{m+n} \text{ nearly.}$$

Let $x = 1+y$, where y must be small, since x is nearly equal to 1.

$\therefore$ the expression $= \dfrac{m(1+y)^m - n(1+y)^n}{m-n}$

$= \dfrac{m(1+my+...) - n(1+ny+...)}{m-n}$

$= \dfrac{m-n+y(m^2-n^2)}{m-n};$

$= 1+(m+n);$

y^2 and higher powers of y being omitted.

Also $\qquad x^{m+n} = (1+y)^{m+n}$

$\qquad\qquad = 1+(m+n)y,$

under the same assumptions, which proves the identity.

Ex. 10. If a be the greatest integer contained in $N^{\frac{1}{3}}$ and the difference be so small that its cube may be neglected, prove that a nearer approximation to $\sqrt[3]{N}$ will be

$$\frac{1}{2}\left\{a+\left(\frac{4N-a^3}{3a}\right)^{\frac{1}{2}}\right\}.$$

Let $\sqrt[3]{N}=a+h$, where h is a small positive quantity.

$\therefore N=a^3+3a^2h+3ah^2$; h^3 being neglected.

Solving the quadratic in h, we have

$$h=\frac{-3a^2\pm\sqrt{9a^4-12a^4+12aN}}{6a}$$

$$=-\frac{1}{2}a\pm\frac{\sqrt{3a(4N-a^3)}}{6a}.$$

Simplifying and taking only the positive sign, since h is positive, we have

$$h=-\frac{1}{2}a+\frac{1}{2}\left(\frac{4N-a^3}{3a}\right)^{\frac{1}{2}}.$$

Hence $\sqrt[3]{N}=a+h=\dfrac{1}{2}\left\{a+\left(\dfrac{4N-a^3}{3a}\right)^{\frac{1}{2}}\right\}$.

Ex. 11. Required approximate values of the roots of

$$ax^2+bx+c=0,$$

when a is very small compared with b and c.

The roots are $=\dfrac{-b\pm\sqrt{b^2-4ac}}{2a}$.

Now $\sqrt{b^2-4ac}=(b^2-4ac)^{\frac{1}{2}}=b\left\{1-\dfrac{4ac}{b^2}\right\}^{\frac{1}{2}}$

$$=b\left\{1-\frac{1}{2}\cdot\frac{4ac}{b^2}-\frac{1}{8}\left(\frac{4ac}{b^2}\right)^2-\ldots\right\}.$$

If this value of b^2-4ac be substituted in the roots of the equation, they become

$$-\frac{c}{b}-\frac{ac^2}{b^3}-\ldots;\text{ and }-\frac{b}{a}+\frac{c}{b}+\frac{ac^2}{b^3}+\ldots.$$

And as a is given to be very small compared with b and c, the former does not differ much from $-\dfrac{c}{b}$, while the latter root is numerically very large. [See § 192.]

THE BINOMIAL THEOREM.

NOTE.—In using the Binomial Series, or indeed any other series, for purposes of approximation *to a stated degree of accuracy*, say to p decimal places, two points require special attention —

(1) To ensure that the terms retained include all those which can influence the result as far as the pth place of decimals.

(2) To calculate each term retained at least as far as $p + 2$ decimal places so that the accumulation of errors may not cause inaccuracy in the figures retained.

Under (1) a point to be noticed is that the *greatest* term in the series should be retained—

(a) For the Binomial Expansion the greatest term is determined as in § 430.

(b) For the Exponential Series, ratio of $n + 1$th term to nth term is x/n: therefore the greatest term is obtained by taking n as the integer nearest in value to x, *e.g.* if e^3 is required, the ratio is $3/n$ and so the third and fourth terms are the greatest.

(c) In the Logarithmic Series, since x is numerically < 1, it is evident that the first term is always the greatest.

It is also necessary to make some estimate of the numerical value of the remainder after n terms of the series. There is no convenient general form for this in the Binomial Series* suited to elementary work, but for the Exponential and Logarithmic Series § 440 and § 449 (Note) respectively supply the required information.

Ex. Calculate $e^{\frac{1}{5}}$ correct to 6 decimal places (see § 438).

(1) Here the ratio $\dfrac{x}{n}$ is $\dfrac{1}{5n}$: evidently the first term is the greatest.

(2) Remainder after n terms $< \dfrac{\left(\dfrac{1}{5}\right)^n}{n!} \Big/ \left(1 - \dfrac{1}{5n}\right)$.

This is found by trial to be $< \cdot 000,000,1$ if $n = 6$. *Hence only 6 terms need be retained.*

The calculation reduces to

$$1 + \frac{1}{5} + \left(\frac{1}{5}\right)^2 \Big/ 2! + \left(\frac{1}{5}\right)^3 \Big/ 3! + \left(\frac{1}{5}\right)^4 \Big/ 4! + \left(\frac{1}{5}\right)^5 \Big/ 5!$$

Calculate each term to 8 decimal places. The result is $1 \cdot 221403$.

---

* Since however (§ 430) T_{n+1}/T_n always diminishes, when $x < 1$, after we have passed the greatest term, it follows that the remainder after n terms is less than a certain G.P. whose common ratio is T_{n+1}/T_n, provided that n is taken (as above) so that the terms retained include the greatest term.

435. *To find the number of homogeneous products of r dimensions that can be formed out of n letters a, b, c ... and their powers.*

[For another solution see § 321.]

By common division, or by the Binomial Theorem, if ax, bx, cx, ... are each < 1,

$$\frac{1}{1-ax} = 1 + ax + a^2x^2 + a^3x^3 + \ldots,$$

$$\frac{1}{1-bx} = 1 + bx + b^2x^2 + b^3x^3 + \ldots,$$

$$\frac{1}{1-cx} = 1 + cx + c^2x^2 + c^3x^3 + \ldots; \text{ and so on.}$$

$$\therefore \frac{1}{1-ax} \cdot \frac{1}{1-bx} \cdot \frac{1}{1-cx} \ldots$$

$$= (1 + ax + a^2x^2 + \ldots)(1 + bx + b^2x^2 + \ldots)(1 + cx + c^2x^2 + \ldots)$$

$$= 1 + x(a + b + c + \ldots) + x^2(a^2 + ab + ac + \ldots + b^2 + bc + \ldots) + \ldots$$

$$= 1 + S_1 x + S_2 x^2 + S_3 x^3 + \ldots,$$

where S_1, S_2, S_3 ... are evidently the sums of the homogeneous products of *one, two, three* ... dimensions that can be formed out of the letters $a, b, c, \ldots$ and their powers.

To get the *number* of these homogeneous products, put $a, b, c \ldots$ each equal to 1; then each term in S_1, S_2, S_3 ... becomes 1; and the values of S_1, S_2, S_3 ... therefore become the *number* of the homogeneous products of one, two, three... dimensions which we shall denote by $_nH_1$, $_nH_2$, $_nH_3$, ...

But, by this substitution,

$$\frac{1}{1-ax} \cdot \frac{1}{1-bx} \cdot \frac{1}{1-cx} \ldots = \frac{1}{1-x} \cdot \frac{1}{1-x} \cdot \frac{1}{1-x} = \ldots = \frac{1}{(1-x)^n}$$

$$= (1-x)^{-n}$$

$$= 1 + nx + \frac{n(n+1)}{1 \cdot 2} x^2 + \ldots;$$

$$\therefore \quad _nH_1 = n; \quad _nH_2 = \frac{n(n+1)}{1 \cdot 2}; \quad _nH_3 = \frac{n(n+1)(n+2)}{1 \cdot 2 \cdot 3},$$

and generally

$$_nH_r = \frac{n(n+1)(n+2) \ldots (n+r-1)}{1 \cdot 2 \cdot 3 \ldots r}.$$

THE BINOMIAL THEOREM.

NOTE 1. Hence $_nH_r = {}_{n+r-1}C_r$, where $_nH_r$ is the number of the homogeneous products of r dimensions formed out of n letters and their powers.

NOTE 2. *The number of combinations of n letters taken r at a time with repetitions*, i.e. when each letter may occur once, twice, thrice... up to r times in any combination, is clearly the same as the number of homogeneous products of r dimensions that can be formed out of these letters and their powers, and is, therefore, equal to

$$n(n+1)\ldots(n+r-1)/r! = {}_{n+r-1}C_r,$$

as proved independently in the Chapter on Combinations.

Ex. Find the number of homogeneous products of 6 dimensions that can be formed out of the 5 letters a, b, c, d, e and their powers.

By proceeding as in the solution of the foregoing proposition, we find that the number required is the coefficient of x^6 in $(1-x)^{-5}$

$$= 5.6.7.8.9.10/(1.2.3.4.5.6) = 210.$$

436. *To find the number of terms in the expansion of a power of any multinomial, the index being a positive integer.*

Let it be required to find the number of terms in the expansion of

$$(a_1 + a_2 + a_3 + \ldots + a_r)^n.$$

The expansion contains all possible combinations of the r letters $a_1, a_2, a_3 \ldots a_r$ and of their powers. Hence the total number of terms in the expansion is the same as the number of homogeneous products of n dimensions that can be formed out of r letters and their powers, and is, therefore, by the preceding article

$$= \frac{r(r+1)(r+2)\ldots(r+n-1)}{1.2.3\ldots n}.$$

Ex. 1. Find the number of terms in $(a+b)^n$, n being a positive integer.

This is the same as the number of homogeneous products of n dimensions of 2 letters and their powers, and therefore

$$= {}_2H_n = {}_{2+n-1}C_n = {}_{n+1}C_n = n+1.$$

Ex. 2. Find the number of terms in $(a+b+c)^n$, n being a positive integer.

The required number $= {}_3H_n = {}_{3+n-1}C_n = {}_{n+2}C_n = {}_{n+2}C_2$

$$= \tfrac{1}{2}(n+1)(n+2).$$

EXERCISES XXVIII.

95. Find the fifth root of 31 to four places of decimals.

96. Find the sixth root of 719 to four places of decimals.

97. Show that $\sqrt[3]{3} = \left(\dfrac{1000}{343} \times \dfrac{1029}{1000}\right)^{\frac{1}{3}} = \dfrac{10}{7}(1 + \cdot 029)^{\frac{1}{3}}$, and hence find the cube root of 3 to four places of decimals.

98. Prove that $\sqrt[3]{2} = 1\frac{1}{4}(1 + \cdot 024)^{\frac{1}{3}}$, and hence find the cube root of 2 to four places of decimals.

99. Find the cube root of 29 to four places of decimals.

100. Find the seventh root of 108 to four places of decimals.

If x be so small that x^2 and higher powers of x may be neglected, show that

101. $(1+2x)^{\frac{1}{2}}(1-4x)^{-2\frac{1}{2}} = 1 + 11x$ nearly.

102. $\dfrac{(9+2x)^{\frac{1}{2}}(3+4x)}{\sqrt[5]{1-x}} = 9 + 14\frac{4}{5}x$ nearly.

103. $\dfrac{\sqrt{1+x}+\sqrt[3]{1+2x}}{\sqrt[4]{1+3x}+\sqrt[5]{1+4x}} = 1 - \dfrac{23}{120}x$ nearly.

104. $\dfrac{2+3x+(1-3x)^{\frac{1}{2}}}{1-\frac{1}{2}x+(4-x)^{\frac{1}{2}}} = 1 + \dfrac{3}{4}x$ nearly.

105. If c be a quantity so small that c^3 may be neglected in comparison with l^3, show that

$$\sqrt{\dfrac{l}{l+c}} + \sqrt{\dfrac{l}{l-c}}$$

is very nearly equal to $2 + 3c^2/4l^2$.

106. If x differs very little from 1, show that

$$\dfrac{ax^b - bx^a}{x^b - x^a} = \dfrac{1}{1-x} \text{ very nearly.}$$

107. When x is very nearly equal to a, prove that

$$\dfrac{(5a^2x - x^3)^{\frac{1}{2}} - 2a^{\frac{1}{2}}x}{(3ax - 2x^2)^{\frac{1}{4}} - a^{\frac{1}{2}}} = 6a \text{ very nearly.}$$

THE BINOMIAL THEOREM.

108. If N and n be nearly equal, then

$$\sqrt{\frac{N}{n}} = \frac{N}{N+n} + \frac{1}{4} \cdot \frac{N+n}{n} \text{ very nearly.}$$

$\left[\text{Let } N = m+x; \text{ and } n = m-x \text{ where } x \text{ is small. Then} \right.$

$\left. \sqrt{\frac{N}{n}} = \frac{\sqrt{(m^2-x^2)}}{m-x} = \frac{m}{m-x}\left(1 - \frac{x^2}{2m^2}\right) \text{ nearly} = \frac{m+x}{2m} + \frac{1}{4} \cdot \frac{2m}{m-x} \cdot \right]$

109. If $\sqrt{N} = a + x$ where x is very small, then

$$\sqrt{N} = a \cdot \frac{3N + a^2}{N + 3a^2} \text{ nearly.}$$

*110. If $\sqrt[3]{N} = a - x$ where x is very small, then

$$\sqrt[3]{N} = a \cdot \frac{2N + a^3}{N + 2a^3} \text{ nearly.}$$

111. Prove that if M differ from N^2 by a small quantity, the square root of M is approximately equal to

$$\frac{3}{2}N - \frac{(3N^2 - M)^2}{8N^3}.$$

112. Show that if $p - q$ be small compared with p or q, then

$$\sqrt[n]{\frac{p}{q}} = \frac{(n+1)p + (n-1)q}{(n-1)p + (n+1)q} \text{ very nearly.}$$

$$\left[\sqrt[n]{\frac{p}{q}} = \frac{\{(p+q)+(p-q)\}^{\frac{1}{n}}}{\{(p+q)-(p-q)\}^{\frac{1}{n}}} = \frac{\{1+(p-q)/(p+q)\}^{\frac{1}{n}}}{\{1-(p-q)/(p+q)\}^{\frac{1}{n}}} = \ldots \right]$$

113. Apply the foregoing rule to find to three places of decimals (a) the square root of $\dfrac{19}{20}$; (b) the cube root of $\dfrac{24}{25}$; (c) $\sqrt[7]{\dfrac{119}{121}}$.

NOTE. In connection with the last two articles, solve exercises 191—206 in the chapter on Permutations and Combinations.

CHAPTER XXIX.

EXPONENTIAL AND LOGARITHMIC SERIES.

437. We have already seen that the series
$$1 + x + \frac{x^2}{2!} + \frac{x^3}{3!} + \ldots + \frac{x^r}{r!} + \ldots$$
is convergent for all finite values of x. (§ 416.)

If we make x equal to unity the above series becomes
$$1 + 1 + \frac{1}{2!} + \frac{1}{3!} + \ldots + \frac{1}{r!} + \ldots,$$
this is denoted by the letter e.

438. Exponential Theorem. If x be any commensurable quantity, then
$$e^x = 1 + x + \frac{x^2}{2!} + \frac{x^3}{3!} + \ldots + \frac{x^r}{r!} + \ldots \ldots \ldots (1).$$

For let us denote the series on the right by $f(x)$, so that
$$f(y) = 1 + y + \frac{y^2}{2!} + \frac{y^3}{3!} + \ldots + \frac{y^r}{r!} + \ldots,$$
then since the series are both absolutely convergent we may form their product by the ordinary rules of multiplication. (§ 418.)

Now the terms of rth degree in this product are
$$\frac{x^r}{r!} + \frac{x^{r-1}y}{(r-1)!\,1!} + \frac{x^{r-2}y^2}{(r-2)!\,2!} + \ldots$$
$$+ \frac{x^{r-s}y^s}{(r-s)!\,s!} + \ldots + \frac{xy^{r-1}}{1!\,(r-1)!} + \frac{y^r}{r!},$$

or what is the same thing
$$\frac{1}{r!}\left\{x^r + \frac{r!}{(r-1)!\,1!}x^{r-1}y + \frac{r!}{(r-2)!\,2!}x^{r-2}y^2 + \cdots \right.$$
$$\left. + \frac{r!}{(r-s)!\,s!}x^{r-s}y^s + \cdots + y^r\right\}$$
$$= \frac{(x+y)^r}{r!}; \qquad (\S\ 328)$$

and therefore we have
$$f(x) \times f(y) = 1 + \frac{x+y}{1} + \frac{(x+y)^2}{2!} + \cdots + \frac{(x+y)^r}{r!} + \cdots$$
$$= f(x+y) \quad\dotfill(2)$$
for all values of x and y.

*Hence we deduce that if $a_1\,a_2\,a_3\ldots a_r$ be any r quantities, we have
$$f(a_1) \times f(a_2) \times f(a_3) \ldots f(a_r) = f(a_1 + a_2) \times f(a_3) \times \ldots f(a_r)$$
$$= \ldots\ldots\ldots$$
$$= f(a_1 + a_2 + a_3 \ldots a_r),$$
and in particular we see that
$$\{f(a)\}^r = f\{ra\} \quad\dotfill(3),$$
where a is any quantity and r a positive integer.

Putting $a = 1$ in this last equation we obtain
$$\{f(1)\}^r = f(r),$$
$\therefore f(r) = e^r$ when r is a positive integer.

Next, when $x = \dfrac{h}{\epsilon}$, where h and ϵ are positive integers, we have on putting $a = \dfrac{h}{\epsilon}$ and $r = \epsilon$ in (3),
$$\{f(x)\}^\epsilon = \left\{f\left(\frac{h}{\epsilon}\right)\right\}^\epsilon = f\left\{\frac{h}{\epsilon} \times \epsilon\right\} = f(h) = e^h,$$

---

* From this point the proof is exactly as in § 425.

since h is a positive integer. Hence, extracting the ϵth root, we get
$$f\left(\frac{h}{\epsilon}\right) = e^{\frac{h}{\epsilon}},$$
which proves the theorem when x is a positive fraction, and therefore completes the demonstration for positive commensurable quantities.

Next, if x be negative and equal to $-m$, we get
$$f(x) \times f(m) = f\{-m + m\} = f(0) = 1;$$
$$\therefore f(x) = \frac{1}{f(m)} = \frac{1}{e^m} = e^{-m} = e^x,$$
since m is a positive quantity.

Thus the theorem is true if x be any positive or negative commensurable quantity. Q. E. D.

Cor. $\{f(m)\}^x = f(mx)$ for all rational values of x whatever m may be.

For when x is integral we have seen (3) that
$$\{f(m)\}^x = f(mx).$$

If x is fractional and equal to $\dfrac{h}{\epsilon}$, then
$$\{f(m)\}^x = \{f(m)\}^{\frac{h}{\epsilon}} = \{f(mh)\}^{\frac{1}{\epsilon}} = f\left(m\frac{h}{\epsilon}\right) = f(mx),$$
since h and ϵ are both integers.

If x is negative and equal to $-y$, then
$$\{f(m)\}^x = \{f(m)\}^{-y} = \frac{1}{\{f(m)\}^y} = \frac{1}{f(my)} = f(-my) = f(mx).$$

This proof has the advantage of depending only on the fundamental equation (3). Of course, if m be rational
$$\{f(m)\}^x = \{e^m\}^x = e^{mx},$$
by the Laws of Indices.

439. Alternative Proof.

After the relation $f(m)f(n) = f(m+n)$ has been obtained we may proceed thus:—

This equation is identical in form with the Index Law
$$a^m \times a^n = a^{m+n} \quad (\S\ 2) \quad \ldots\ldots\ldots\ldots\ldots\ldots\ldots (4).$$

Now the Index Law combined with the fact that when m is a positive integer
$$a^n = a \times a \times a \ldots\ldots \text{ to } m \text{ factors} \ldots\ldots\ldots\ldots\ldots (5),$$
enables us to assign meanings to a^m when m is negative or fractional (§§ 8, 10).

Bearing in mind that when m is a positive integer
$$f(m) = e \times e \times e \ldots\ldots \text{ for } m \text{ factors} = e^m \ldots\ldots\ldots\ldots (6),$$
we see by analogy that (6) affords interpretations of $f(m)$ for negative and fractional values of m, which must be identical in form with the corresponding interpretations of e^m.

Hence by the Properties of Indices,

$f\left(\dfrac{h}{e}\right)$ represents the eth root of $f(h)$ or is $e^{\frac{h}{e}}$,

$f(-m)$ represents the reciprocal of $f(n) = \dfrac{1}{e^m} = e^{-m}$.

Hence for all values of m
$$f(m) = e^m$$
which is the Exponential Theorem.

Cor. $f(mx)$ represents $\{f(m)\}^x$ for all rational values of x.

440. Limits to the Remainder in the Exponential Series.

In the series for e^x it is useful to be able to fix limits for the error committed in taking any assigned number of terms to represent the function. For this purpose we proceed to find an upper limit for the remainder after the first n terms which we denote by R_n.

We have
$$R_n = \frac{x^n}{n!} + \frac{x^{n+1}}{(n+1)!} + \frac{x^{n+2}}{(n+2)!} + \ldots,$$
$$\therefore R_n < \frac{x^n}{n!}\left(1 + \frac{x}{n} + \frac{x^2}{n^2} + \ldots \text{ ad inf.}\right),$$

i.e. $R_n < \dfrac{x^n}{n!} \dfrac{1}{1-\dfrac{x}{n}}$,

n being supposed greater than x (§ 197);

$$\therefore R_n < \frac{x^n}{(n-x)(n-1)!}.$$

Cor. The error in taking the first n terms of the expansion lies between

$$\frac{x^n}{n!} \text{ and } \frac{x^n}{n!} \frac{1}{1-\dfrac{x}{n}}.$$

When n is very large this error is exceedingly small as of course we know from the fact that the series is convergent, but in any case the above are two very close limits for the error.

441. e is incommensurable.

We may employ the preceding result to show that the number e is incommensurable.

For if possible let $e = \dfrac{m}{n}$ where m and n are integers, then we have

$$\frac{m}{n} = 1 + 1 + \frac{1}{2!} + \frac{1}{3!} \cdots \frac{1}{n!} + R'_{n+1},$$

where R'_{n+1} is positive and $= \dfrac{1}{n(n)!}$, as we see by making $x = 1$ and changing n into $n+1$ in the preceding article.

Consequently, multiplying across by $n!$ we have

$$I_1 = I_0 + n! \, R'_{n+1},$$

where I_1 and I_0 are both positive integers.

But $n! \, R'_{n+1} < \dfrac{1}{n}$, and is therefore a proper fraction, hence we have the difference between two integers equal to a fraction which is absurd, therefore e cannot be commensurable.

As regards the actual value of e we note in the first place that it is greater than 2, and in the second that since $R_3' < \frac{1}{4}$ (§ 440) it is less than $2\frac{3}{4}$.

For practical purposes of calculation it is useful to remark that each term in the series for e is derived from the preceding one by dividing by an integer.

442. We shall now establish a very important theorem in connection with limiting values, viz.
$$\mathrm{L^t}_{n=\infty}\left(1+\frac{1}{n}\right)^{nx}=e^x.$$

To prove this, suppose x, in the first instance, positive and write $nx = m$, so that m will become infinite at the same time as n, and we have therefore to prove that
$$\mathrm{L^t}_{m=\infty}\left(1+\frac{x}{m}\right)^m=e^x.$$

First, let m be a positive integer, then, by the binomial theorem, we have

$$\left(1+\frac{x}{m}\right)^m = 1 + m\frac{x}{m} + \frac{m(m-1)}{2!}\frac{x^2}{m^2} + \ldots$$
$$+ \frac{m(m-1)(m-2)\ldots(m-r+1)}{r!}\frac{x^r}{m^r} + \ldots$$
$$+ \frac{m(m-1)(m-2)\ldots 1}{m!}\frac{x^m}{m^m} \quad (\S\ 328)$$
$$= 1 + x + \frac{x^2}{2!}\left(1-\frac{1}{m}\right) + \ldots$$
$$+ \frac{x^r}{r!}\left(1-\frac{1}{m}\right)\left(1-\frac{2}{m}\right)\ldots\left(1-\frac{r-1}{m}\right) + \ldots$$
$$+ \frac{x^m}{m!}\left(1-\frac{1}{m}\right)\left(1-\frac{2}{m}\right)\ldots\left(1-\frac{(m-1)}{m}\right).$$

EXPONENTIAL AND LOGARITHMIC SERIES.

Putting, for a moment,
$$a_m = 1 + \frac{x}{1!} + \frac{x^2}{2!} + \ldots + \frac{x^m}{m!},$$

and remembering that
$$\left(1 - \frac{1}{m}\right)\left(1 - \frac{2}{m}\right) \ldots \left(1 - \frac{r-1}{m}\right) > 1 - \frac{1+2+3\ldots+(r-1)}{m}$$
$$> 1 - \frac{r(r-1)}{2m} \text{ (§ 404, Ex. 3)},$$

we see that
$$\left(1 + \frac{x}{m}\right)^m < a_m,$$

and $> 1 + \frac{x}{1!} + \frac{x^2}{2!}\left(1 - \frac{1\cdot 2}{2m}\right) + \ldots$
$$+ \frac{x^r}{r!}\left(1 - \frac{(r-1)\cdot r}{2m}\right) + \ldots + \frac{x^m}{m!}\left(1 - \frac{m(m-1)}{2m}\right),$$

i.e. $\left(1 + \frac{x}{m}\right)^m > 1 + \frac{x}{1!} + \frac{x^2}{2!} + \ldots$
$$+ \frac{x^m}{m!} - \frac{x^2}{2m}\left(1 + x + \frac{x^2}{2!} + \ldots + \frac{x^{m-2}}{(m-2)!}\right)$$
$$> a_m - \frac{x^2}{2m} a_{m-2};$$

therefore, since $a_{m-2} < a_m,$
we find that $\left(1 + \frac{x}{m}\right)^m$ lies between a_m and $a_m\left(1 - \frac{x^2}{2m}\right)$.

Now a_m is finite for all values of m however large (§ 416) and tends to the limit e^x (§ 438), whereas by sufficiently increasing m we can make $\frac{x^2}{2m}$ as small as we please;

$$\therefore \operatorname{L^t}_{m=\infty}\left(1 + \frac{x}{m}\right)^m = \operatorname{L^t}_{m=\infty} a_m = e^x.$$

Next, if m be not an integer but lie between the positive integers h and $h+1$, we have

$$1 + \frac{x}{h} > 1 + \frac{x}{m} > 1 + \frac{x}{h+1};$$

$$\therefore \left(1 + \frac{x}{h}\right)^m > \left(1 + \frac{x}{m}\right)^m > \left(1 + \frac{x}{h+1}\right)^m,$$

or $\left(1 + \frac{x}{h}\right)^h \left(1 + \frac{x}{h}\right)^{m-h} > \left(1 + \frac{x}{m}\right)^m$

$$> \left(1 + \frac{x}{h+1}\right)^{h+1} \div \left(1 + \frac{x}{h+1}\right)^{h+1-m} \quad \ldots(A).$$

Now both $\left(1 + \frac{x}{h}\right)^h$ and $\left(1 + \frac{x}{h+1}\right)^{h+1}$ tend to the limit e^x, and since $m-h$ and $h+1-m$ are proper fractions, $\left(1 + \frac{x}{h}\right)^{m-h}$ and $\left(1 + \frac{x}{h+1}\right)^{h+1-m}$ both tend to the limit unity;

$\therefore$ the limits of the extreme expressions (A) are both e^x;

$$\therefore \mathrm{L^t}_{m=\infty}\left(1 + \frac{x}{m}\right)^m = e^x,$$

when m is positive.

Finally, if m be negative and equal to $-\mu$,

$$\mathrm{L^t}_{m=-\infty}\left(1 + \frac{x}{m}\right)^m = \mathrm{L^t}_{\mu=\infty}\left(1 - \frac{x}{\mu}\right)^{-\mu} = \mathrm{L^t}_{\mu=\infty}\left(\frac{\mu}{\mu-x}\right)^{\mu}$$

$$= \mathrm{L^t}_{\mu=\infty}\left(1 + \frac{x}{\mu-x}\right)^{\mu}$$

$$= \mathrm{L^t}_{\mu=\infty}\left(1 + \frac{x}{\mu-x}\right)^{\mu-x}\left(1 + \frac{x}{\mu-x}\right)^{x}.$$

The first factor in this expression has e^x for limit by what precedes and the second factor, since x is finite, has unity for limit; hence even when m tends to $-\infty$ we have

$$\mathrm{L^t}_{m=-\infty}\left(1 + \frac{x}{m}\right)^m = e^x,$$

which completely establishes the theorem when x is positive.

If x be negative and equal to $-y$, we have

$$\mathrm{Lt}_{m=\infty}\left(1+\frac{x}{m}\right)^m = \mathrm{Lt}_{m=\infty}\left(1-\frac{y}{m}\right)^m$$

$$= \mathrm{Lt}_{m=-\infty}\frac{1}{\left(1+\frac{y}{n}\right)^n} = \frac{1}{e^y} = e^x,$$

and accordingly the theorem is completely established.

The reader may regard the above proof as unnecessarily long, but inasmuch as this result is one of the most important in the whole range of mathematics, it is important to give a satisfactory demonstration even at the sacrifice of brevity.

Cor. $\mathrm{Lt}_{z=0}(1+xz)^{\frac{1}{z}}=e^x$ as is seen at once by putting $z=\frac{1}{m}$, and z may be either positive or negative.

A shorter but less satisfactory proof of these results will be found in § 447.

443. Existence of a logarithm corresponding to every number.

As we have already stated, the functions e^x and

$$1+x+\frac{x^2}{2!}+\frac{x^3}{3!}+\ldots$$

have been identified for all rational values of x. Now, what are we to say of incommensurable values? We remark that e^x is defined in this case by replacing x by a series of rational approximations; strictly speaking, as the reader will observe, we cannot give an elementary definition of e^x here, so we may suppose the function e^x to be defined by the series

$$1+x+\frac{x^2}{2!}+\frac{x^3}{3!}+\ldots,$$

if x be incommensurable.

Now, we showed in a previous chapter (§ 421) that this series is uniformly convergent, and therefore represents a continuous function of x for all finite values of x (§ 420); further, when $x = 0$ the value is unity, whilst by sufficiently increasing x we can make the series as great as we please; hence there is a value of x which makes the series take any assigned value $a > 1$ (§ 400). If this value be m, we have

$$a = e^m \text{ or } m = \log_e a.$$

Moreover, since $f(x)$ increases with x when x is positive, there can only be one value of m corresponding to each value of a.

If b be positive and less than unity then $b = \dfrac{1}{a}$ where $a > 1$, and

$$a = f(m); \therefore b = f(-m) = e^{-m},$$
or
$$\log_e b = -m,$$

which shows the existence of a unique logarithm for all positive members.

444. Generalised form of the Exponential Theorem.

Now since $a = e^m = f(m)$,

we have $a^x = \{f(m)\}^x = \{f(mx)\} = e^{mx}$ (§ 438, Cor.).

Consequently from the exponential theorem (§ 438)

$$a^x = e^{mx} = 1 + \frac{mx}{1!} + \frac{m^2 x^2}{2!} + \frac{m^3 x^3}{3!} + \ldots$$

$$= 1 + x (\log_e a) + \frac{x^2}{2!} (\log_e a)^2 + \frac{x^3}{3!} (\log_e a)^3,$$

which is the extended form of the Exponential Theorem. If x be incommensurable a^x may be supposed to be defined by the above equation.

This series is in like manner a continuous function of x (§ 420), and we can easily establish thus the existence of a unique logarithm of any positive number to any positive base.

EXPONENTIAL AND LOGARITHMIC SERIES.

Ex. 1. Show that $n + \dfrac{1}{n} = 2\left\{1 + \dfrac{(\log_e n)^2}{2!} + \dfrac{(\log_e n)^4}{4!} + \ldots\right\}$.

In $\quad a^x = 1 + (\log_e a)\,x + \dfrac{(\log_e a)^2 x^2}{2!} + \dfrac{(\log_e a)^3 x^3}{3!} + \ldots,$

first put $a = n$, and $x = 1$; and then put $a = n$, and $x = -1$.

Then $\quad n = 1 + \log_e n + \dfrac{(\log_e n)^2}{2!} + \dfrac{(\log_e n)^3}{3!} + \ldots,$

and $\quad n^{-1} = 1 - \log_e n + \dfrac{(\log_e n)^2}{2!} - \dfrac{(\log_e n)^3}{3!} + \ldots.$

By addition, we get the required result.

Ex. 2. Show that $\dfrac{1}{e} = \dfrac{2}{3!} + \dfrac{4}{5!} + \dfrac{6}{7!} + \ldots.$

$$e^{-1} = 1 - 1 + \dfrac{1}{2!} - \dfrac{1}{3!} + \dfrac{1}{4!} - \dfrac{1}{5!} + \ldots$$

$$= \left(\dfrac{1}{2!} - \dfrac{1}{3!}\right) + \left(\dfrac{1}{4!} - \dfrac{1}{5!}\right) + \ldots + \left\{\dfrac{1}{(2n)!} - \dfrac{1}{(2n+1)!}\right\};$$

$$\therefore \dfrac{1}{e} = \dfrac{2}{3!} + \dfrac{4}{5!} + \dfrac{6}{7!} + \ldots + \dfrac{2n}{(2n+1)!} + \ldots.$$

Ex. 3. Show that

$$\dfrac{e^2 + 1}{e^2 - 1} = \left\{1 + \dfrac{1}{2!} + \dfrac{1}{4!} + \dfrac{1}{6!} + \ldots\right\} \Big/ \left\{1 + \dfrac{1}{3!} + \dfrac{1}{5!} + \dfrac{1}{7!} + \ldots\right\}.$$

$$\dfrac{e^2+1}{e^2-1} = \dfrac{e + 1/e}{e - 1/e} = \dfrac{e + e^{-1}}{e - e^{-1}}$$

$$= \dfrac{\left(1 + 1 + \dfrac{1}{2!} + \dfrac{1}{3!} + \ldots\right) + \left(1 - 1 + \dfrac{1}{2!} - \dfrac{1}{3!} + \ldots\right)}{\left(1 + 1 + \dfrac{1}{2!} + \dfrac{1}{3!} + \ldots\right) - \left(1 - 1 + \dfrac{1}{2!} - \dfrac{1}{3!} + \ldots\right)},$$

from which the required result readily follows.

Ex. 4. Find the coefficient of x^r in the expansion of $\dfrac{2+3x}{e^{2x}}$.

$(2+3x)/e^{2x} = (2+3x)\,e^{-2x}$

$= (2+3x)\left\{1 - 2x + \dfrac{2^2 x^2}{2!} - \ldots + \dfrac{(-1)^{r-1} 2^{r-1} x^{r-1}}{(r-1)!} + \dfrac{(-1)^r 2^r x^r}{r!} + \ldots\right\};$

$\therefore$ the coefficient of $x^r = 3 \cdot \dfrac{(-1)^{r-1} 2^{r-1}}{(r-1)!} + 2 \cdot \dfrac{(-1)^r 2^r}{r!}$

$= \dfrac{(-1)^r}{r!}\{2^{r+1} - 3r \cdot 2^{r-1}\} = (-1)^r \cdot 2^{r-1}(4 - 3r)/r!.$

Ex. 5. Sum to infinity
$$\frac{1}{1!} + \frac{1+3}{2!} + \frac{1+3+3^2}{3!} + \frac{1+3+3^2+3^3}{4!} + \dots$$

Here
$$U_n = \frac{1+3+3^2+\dots+3^{n-1}}{n!}$$
$$= \frac{1}{2} \cdot \frac{3^n-1}{n!} = \frac{1}{2}\left(\frac{3^n}{n!} - \frac{1}{n!}\right).$$

Hence the required sum
$$= \frac{1}{2}\left\{\left(\frac{3}{1!} + \frac{3^2}{2!} + \frac{3^3}{3!} + \dots\right) - \left(\frac{1}{1!} + \frac{1}{2!} + \frac{1}{3!} + \frac{1}{4!} + \dots\right)\right\}$$
$$= \frac{1}{2}\left\{\left(1 + 3 + \frac{3^2}{2!} + \frac{3^3}{3!} + \dots\right) - \left(1 + 1 + \frac{1}{2!} + \frac{1}{3!} + \frac{1}{4!} + \dots\right)\right\}$$
$$= \frac{1}{2}(e^3 - e) = \frac{1}{2}e(e^2 - 1).$$

Ex. 6. Sum to infinity $\frac{2 \cdot 1^2 + 3}{2!} + \frac{2 \cdot 2^2 + 4}{3!} + \frac{2 \cdot 3^2 + 5}{4!} + \dots$

Here
$$U_n = \frac{2n^2 + n + 2}{(n+1)!}.$$

We have to express this fraction as the sum of a number of fractions with numerical constants for the numerators and factorials for the denominators. Hence we proceed as follows:—

Let $2n^2 + n + 2 = A + B(n+1) + C(n+1)n + D(n+1)n(n-1) + \dots$

Since the coefficients of n^3 and higher powers of n on the left side equal zero, the coefficients $D, E, F, \dots$ vanish;

$$\therefore 2n^2 + n + 2 = A + B(n+1) + C(n+1)n$$
$$= (A+B) + (B+C)n + Cn^2.$$

Hence by § 93, $C = 2$; $B + C = 1$; and $A + B = 2$,
which makes $C = 2$, $B = -1$, and $A = 3$;

$$\therefore 2n^2 + n + 2 = 3 - (n+1) + 2(n+1)n;$$
$$\therefore U_n = \frac{2n^2 + n + 2}{(n+1)!}$$
$$= \frac{3}{(n+1)!} - \frac{n+1}{(n+1)!} + \frac{2(n+1)n}{(n+1)!}$$
$$= \frac{3}{(n+1)!} - \frac{1}{n!} + \frac{2}{(n-1)!}.$$

510   EXPONENTIAL AND LOGARITHMIC SERIES.

Hence substituting 1, 2, 3 ... in order for n, we have the required

sum $= \left(\dfrac{3}{2!} + \dfrac{3}{3!} + \dfrac{3}{4!} + \ldots\right) - \left(\dfrac{1}{1!} + \dfrac{1}{2!} + \dfrac{1}{3!} + \ldots\right)$
$+ \left(\dfrac{2}{0!}* + \dfrac{2}{1!} + \dfrac{2}{2!} + \ldots\right)$
$= 3(e-2) - (e-1) + 2e = 4e - 5.$

Ex. 7. Show that the coefficient of x^r in the expansion of e^{e^x} is
$$\dfrac{1}{r!}\left(\dfrac{1^r}{1!} + \dfrac{2^r}{2!} + \dfrac{3^r}{3!} + \ldots\right).$$

$e^{e^x} = 1 + e^x + \dfrac{e^{2x}}{2!} + \dfrac{e^{3x}}{3!} + \ldots + \dfrac{e^{nx}}{n!} + \ldots$

$= 1 + \left(1 + x + \dfrac{x^2}{2!} + \ldots + \dfrac{x^r}{r!} + \ldots\right) + \dfrac{1}{2!}\left(1 + 2x + \dfrac{4x^2}{2!} + \ldots + \dfrac{2^r x^r}{r!} + \ldots\right)$
$+ \dfrac{1}{3!}\left(1 + 3x + \dfrac{9x^2}{2!} + \ldots + \dfrac{3^r x^r}{r!} + \ldots\right) + \ldots.$

Hence the coefficient of x^r in the expansion is
$$\dfrac{1}{r!}\left(\dfrac{1^r}{1!} + \dfrac{2^r}{2!} + \dfrac{3^r}{3!} + \ldots\right).$$

Ex. 8. Show that the coefficient of x^3 in the expansion of e^{e^x} is $5e/3!$.

$e^{e^x} = e^{1 + x + x^2/2! + \ldots} = e \cdot e^{x + x^2/2! + x^3/3! + \ldots}$

$= e\left\{1 + \left(x + \dfrac{x^2}{2!} + \ldots\right) + \dfrac{1}{2!}\left(x + \dfrac{x^2}{2!} + \ldots\right)^2 + \dfrac{1}{3!}\left(x + \dfrac{x^2}{2!} + \ldots\right)^3 + \ldots\right\};$

$\therefore$ coefficient of $x^3 = e\left(\dfrac{1}{3!} + \dfrac{1}{2!} \cdot 2 \cdot \dfrac{1}{2^2} + \dfrac{1}{3!}\right) = \dfrac{5e}{3!}.$

NOTE. Comparing the results of Examples 7 and 8, we get
$$\dfrac{1^3}{1!} + \dfrac{2^3}{2!} + \dfrac{3^3}{3!} + \ldots \text{ to infinity} = 5e.$$

Ex. 9. If x be positive, show that $\log_e(1 + x)$ is less than x; and hence prove that $\log \dfrac{n+1}{2} < \dfrac{1}{2} + \dfrac{1}{3} + \dfrac{1}{4} + \ldots + \dfrac{1}{n}.$

(i) If $\qquad y = \log_e(1 + x),\ e^y = 1 + x.$

---

* In operations like these, we occasionally get factorials like $(-1)!, (-2)!, (-3)! \ldots$ of which the values may be inferred from $n(n-1)! = n!$ to be $+\infty, -\infty, +\infty, \ldots$ alternately.

But $\qquad e^y = 1 + y + y^2/2! + y^3/3! + \ldots$
$\qquad\qquad\quad > 1 + y,$

since y is positive being the logarithm of a number greater than 1.

Hence y is less than x;
that is, $\qquad \log_e(1+x) < x.$

(ii) Putting in this inequality $\dfrac{1}{2}, \dfrac{1}{3}, \dfrac{1}{4} \ldots \dfrac{1}{n}$ in succession for x,

we have $\log \dfrac{3}{2} < \dfrac{1}{2}$; $\log \dfrac{4}{3} < \dfrac{1}{3}$; $\log \dfrac{5}{4} < \dfrac{1}{4}$; $\ldots$; $\log \dfrac{n+1}{n} < \dfrac{1}{n}$;

$\therefore \log \dfrac{3}{2} + \log \dfrac{4}{3} + \log \dfrac{5}{4} + \ldots + \log \dfrac{n+1}{n} < \dfrac{1}{2} + \dfrac{1}{3} + \dfrac{1}{4} + \ldots + \dfrac{1}{n};$

$\therefore \log \dfrac{3}{2} \cdot \dfrac{4}{3} \cdot \dfrac{5}{4} \ldots \dfrac{n+1}{n} = \log \dfrac{n+1}{2} < \dfrac{1}{2} + \dfrac{1}{3} + \dfrac{1}{4} + \ldots + \dfrac{1}{n}.$

Ex. 10. Show how the values of the constants $A, B, C \ldots$ may be found so that $e^{a+h} = Ae^a + Be^{a+rh} + Ce^{a+sh} + \ldots$ when h is very small, and prove that $e^{a+h} = 2e^a + 8e^{a+\frac{1}{2}h} - 9e^{a+\frac{1}{3}h}$ when h may be taken to be so small that h^3 and higher powers of h may be neglected.

(1) Let $e^{a+h} = Ae^a + Be^{a+rh} + Ce^{a+sh} + \ldots$, when h is very small.

Taking out the factor e^a, we have

$$e^h = A + Be^{rh} + Ce^{sh} + \ldots;$$

$$\therefore e^h - A - Be^{rh} - Ce^{sh} - \ldots = 0.$$

$$\therefore \left(1 + h + \frac{h^2}{2!} + \frac{h^3}{3!} + \ldots\right) - A - B\left(1 + rh + \frac{r^2h^2}{2!} + \frac{r^3h^3}{3!} + \ldots\right)$$
$$- C\left(1 + sh + \frac{s^2h^2}{2!} + \frac{s^3h^3}{3!} + \ldots\right) - \ldots = 0.$$

Arranging the terms of these series in powers of h, we have

$$(1 - A - B - C - \ldots) + h(1 - Br - Cs - \ldots)$$
$$+ \frac{h^2}{2!}(1 - Br^2 - Cs^2 - \ldots) - \ldots = 0.$$

As the equality must be true for all values of h, the coefficients in the expression must vanish. And as h is very small compared with the constants, it will be enough if we take as many coefficients as there are constants whose values have to be determined. These coefficients equated to zero will give us the required values of the constants.

EXPONENTIAL AND LOGARITHMIC SERIES.

(2) For example, let it be required to find the values of A, B, C, so that
$$e^{a+h} = Ae^a + Be^{a+\frac{1}{2}h} + Ce^{a+\frac{1}{3}h},$$
when h is very small.

Here $e^h = A + Be^{\frac{1}{2}h} + Ce^{\frac{1}{3}h};$
$$\therefore e^h - A - Be^{\frac{1}{2}h} - Ce^{\frac{1}{3}h} = 0.$$

$\therefore (1 - A - B - C) + h(1 - \frac{1}{2}B - \frac{1}{3}C) + \dfrac{h^2}{2!}\left(1 - \dfrac{B}{4} - \dfrac{C}{9}\right) +$ terms containing the third and higher powers of h which may be neglected as h is very small.

Equating the coefficients to zero, we have

$\left.\begin{array}{r}A + B + C = 1\\ \frac{1}{2}B + \frac{1}{3}C = 1\\ \frac{1}{4}B + \frac{1}{9}C = 1\end{array}\right\}$ from which we get $A = 2$, $B = 8$, and $C = -9$;

$\therefore e^{a+h} = 2e^a + 8e^{a+\frac{1}{2}h} - 9e^{a+\frac{1}{3}h}$ when h is small enough.

Ex. 11. Expand $e^{\frac{1}{\sqrt{1-x}}}$ up to the third power of x.

$$e^{\frac{1}{\sqrt{1-x}}} = e^{(1-x)^{-\frac{1}{2}}} = e^{1 + \frac{1}{2}x + \frac{3}{8}x^2 + \frac{5}{16}x^3 + \ldots}$$
$$= e \cdot e^{\frac{1}{2}x + \frac{3}{8}x^2 + \frac{5}{16}x^3 + \ldots}$$
$$= e\left\{1 + (\tfrac{1}{2}x + \tfrac{3}{8}x^2 + \tfrac{5}{16}x^3 + \ldots) + \dfrac{(\frac{1}{2}x + \frac{3}{8}x^2 + \ldots)^2}{2!} + \dfrac{(\frac{1}{2}x + \ldots)^3}{3!} + \ldots\right\}$$
$$= e\left(1 + \tfrac{1}{2}x + \tfrac{5}{8}x^2 + \tfrac{25}{48}x^3\right) \text{ up to the third power of } x.$$

Ex. 12. Expand $\dfrac{x}{e^x - 1}$ up to the third power of x.

Since $e^x - 1 = x + \dfrac{x^2}{2!} + \dfrac{x^3}{3!} + \dfrac{x^4}{4!} + \ldots,$

$$\dfrac{x}{e^x - 1} = \dfrac{1}{1 + \dfrac{x}{2!} + \dfrac{x^2}{3!} + \dfrac{x^3}{4!} + \ldots} = \left\{1 + \left(\dfrac{x}{2!} + \dfrac{x^2}{3!} + \dfrac{x^3}{4!} + \ldots\right)\right\}^{-1}$$
$$= 1 - \left(\dfrac{x}{2!} + \dfrac{x^2}{3!} + \dfrac{x^3}{4!} + \ldots\right) + \left(\dfrac{x}{2!} + \dfrac{x^2}{3!} + \ldots\right)^2 - \left(\dfrac{x}{2!} + \ldots\right)^3 + \ldots$$
$$= 1 - \tfrac{1}{2}x + \tfrac{1}{12}x^2 + 0 \cdot x^3 \text{ up to the third power of } x.$$

EXPONENTIAL AND LOGARITHMIC SERIES.

445. Elementary logarithmic formulæ.

Taking e^x to be defined as the series

$$f(x) = 1 + x + \frac{x^2}{2!} + \ldots + \frac{x^r}{r!} + \ldots,$$

it is quite easy to prove the elementary formulæ of logarithms thus:

(i) $\log(1) = 0$, since $f(0) = 1$.

(ii) $\log a + \log b = \log ab$ for if $\log a = m$ and $\log b = n$, then $a = f(m)$; $b = f(n)$, and therefore

$$ab = f(m) \times f(n) = f(m+n),\qquad \S\, 438.$$

or $\log ab = \log a + \log b$.

(iii) $\log a + \log 1/a = \log 1$ by (ii),

$$\therefore \log(1/a) = -\log a.$$

(iv) $\log a^r = r \log a$,

for here $\qquad a = f(m),$

and $\therefore$ if r be rational,

$$a^r = f(mr),$$

this being simply a reproduction of the extended form of the exponential theorem. $\qquad \S\, 444.$

446. To prove that

$$\mathop{\mathrm{Lt}}_{x=0} \frac{a^x - 1}{x} = \log_e a.$$

We have $a^x = 1 + x \log_e a + \dfrac{x^2}{2!}(\log_e a)^2 + \ldots;$

$$\therefore \frac{a^x - 1}{x} = \log_e a$$

$$+ x\left\{\frac{1}{2!}(\log_e a)^2 + \frac{x}{3!}(\log_e a)^3 + \frac{x^2}{4!}(\log_e a)^4 + \ldots\right\}.$$

Now the series within the brackets is convergent for all values of x, and therefore is always finite; hence by taking x sufficiently small we can make $(a^x - 1)/x$ differ from $\log_e a$ by as small a quantity as we please;

$$\therefore \mathop{\mathrm{Lt}}_{x=0} (a^x - 1)/x = \log_e a.$$

TUT. ALG. II.

447. $\operatorname*{Lt}_{m=\infty}\left(1+\dfrac{x}{m}\right)^m = e^x$ **(alternative proof).**

Putting $y = \log\left(1+\dfrac{x}{m}\right)$ in $\operatorname*{Lt}_{y=0} \dfrac{e^y-1}{y} = 1$,

we obtain $\operatorname*{Lt}_{m=\infty} \dfrac{1+\dfrac{x}{m}-1}{\log\left(1+\dfrac{x}{m}\right)} = 1$,

since $m = \infty$ when $y = 0$.

Hence $\operatorname*{Lt}_{m=\infty} \dfrac{\log\left(1+\dfrac{x}{m}\right)}{\dfrac{x}{m}} = 1$,

i.e. $\operatorname*{Lt}_{m=\infty} m \log\left(1+\dfrac{x}{m}\right) = x$; or $\operatorname*{Lt}_{m=\infty} \log\left(1+\dfrac{x}{m}\right)^m = x$,

i.e. $\operatorname*{Lt}_{m=\infty} \left(1+\dfrac{x}{m}\right)^m = e^x$.

We can easily deduce the other results of § 442.

Thus putting $m = nx$ we get $\operatorname*{Lt}_{n=\infty} \left(1+\dfrac{1}{n}\right)^{nx} = e^x$,

and writing $m = \dfrac{1}{n}$ we derive

$$\operatorname*{Lt}_{n=0} (1+nx)^{\frac{1}{n}} = e^x.$$

448. Conversely, assuming the result of § 442 we can deduce that of § 446 as follows.

Let $\dfrac{a^x-1}{x} = y$, then $a = (1+xy)^{\frac{1}{x}}$.

But we know that $\operatorname*{Lt}_{x=0} (1+xy)^{\frac{1}{x}} = e^y$, § 442, Cor.

$\therefore$ when $x = 0$, we have $a = e^y$;

$$\therefore y = \log_e a,$$

or $\operatorname*{Lt}_{x=0} \dfrac{a^x-1}{x} = \log_e a.$ Q. E. D.

EXPONENTIAL AND LOGARITHMIC SERIES.

449. To prove that if $x < 1$

$$\log_e(1+x) = x - \frac{x^2}{2} + \frac{x^3}{3} - \frac{x^4}{4} + \ldots$$

We have $\quad \log_e(1+x) = \underset{y=0}{\mathrm{L}^t} \dfrac{(1+x)^y - 1}{y}.\quad$ (§ 446.)

Now by the binomial theorem we have, since $x < 1$,

$$(1+x)^y = 1 + yx + \ldots + \frac{y(y-1)\ldots(y-r+1)}{r!} x^r + \ldots$$

$$\therefore \log_e(1+x) = \underset{y=0}{\mathrm{L}^t} \frac{1}{y} \left\{ yx + \frac{y(y-1)}{2!} x^2 + \ldots \right.$$

$$\left. + \frac{y(y-1)\ldots(y-r+1)}{r!} x^r + \ldots \right\}$$

$$= \underset{y=0}{\mathrm{L}^t} \left\{ x + \frac{y-1}{2!} x^2 + \ldots + \frac{(y-1)(y-2)\ldots(y-r+1)}{r!} x^r + \epsilon \right\},$$

where ϵ can be made as small as we please by sufficiently increasing r, since the series in the brackets is convergent when $x < 1$.

Hence since the limit of the sum of a *finite* number of quantities is equal to the sum of their limits, we have

$$\log_e(1+x) = x + \frac{-1}{2!} x^2 + \frac{(-1)(-2)}{3!} x^3 + \ldots$$

$$+ \frac{(-1)^{r-1}(r-1)!}{r!} x^r + \epsilon$$

$$= x - \frac{x^2}{2} + \frac{x^3}{3} - \ldots + (-1)^{r-1} \frac{x^r}{r} + \epsilon',$$

where ϵ' is what ϵ becomes when $y = 0$, and can therefore be made as small as we please by increasing r.

Consequently, we have when $x < 1$

$$\log_e(1+x) = x - \frac{x^2}{2} + \frac{x^3}{3} + \ldots + (-1)^{r-1} \frac{x^r}{r} + \ldots \text{ ad inf.}$$

This result holds even when $x = 1$, but we shall not give a rigorous proof of this here, merely remarking that

$$\log_e 2 = 1 - \frac{1}{2} + \frac{1}{3} - \frac{1}{4} + \dots$$

Since the above reasoning only assumes x to be numerically < 1 we may make it negative, and thus obtain

$$\boldsymbol{\log (1-x) = -x - \frac{x^2}{2} - \frac{x^3}{3} - \dots \text{ ad inf.,}}$$

when $x < 1$.

Note. Since the $(n+1)^{\text{th}}$ term of the series on the right side of

$$\log_e (1+x) = x - \frac{x^2}{2} + \frac{x^3}{3} - \frac{x^4}{4} + \dots$$

is numerically $x^{n+1}/(n+1)$, and the terms are alternately positive and negative, it is evident that the difference between $\log_e (1+x)$ and the sum of the first n terms of the series is less than $x^{n+1}/(n+1)$, for each term is less than the preceding one.

Thus, when x is positive and less than unity, $\log_e (1+x)$ is less than x and greater than $x - x^2/2$.

450. Modification of the foregoing series.

The series for $\log(1+x)$ does not hold for $x > 1$, and even when x is < 1 so many terms have to be taken for a good approximation, that the series may conveniently be replaced by others which we proceed to investigate.

We have, if x be numerically less than 1,

$$\log_e (1+x) = x - \frac{x^2}{2} + \frac{x^3}{3} - \frac{x^4}{4} + \dots$$

Putting $-x$ for x, we have

$$\log_e (1-x) = -x - \frac{x^2}{2} - \frac{x^3}{3} - \frac{x^4}{4} - \dots;$$

$$\therefore \log_e \frac{1+x}{1-x} = \log_e (1+x) - \log_e (1-x)$$

$$= 2 \left\{ x + \frac{x^3}{3} + \frac{x^5}{5} + \frac{x^7}{7} + \dots \right\}.$$

EXPONENTIAL AND LOGARITHMIC SERIES. 517

Putting $\dfrac{m}{n}$ for $\dfrac{1+x}{1-x}$, so that $x = \dfrac{m-n}{m+n}$, we have

$$\log_e \frac{m}{n} = 2\left\{\frac{m-n}{m+n} + \frac{1}{3}\left(\frac{m-n}{m+n}\right)^3 + \frac{1}{5}\left(\frac{m-n}{m+n}\right)^5 + \ldots\right\}.$$

If now we put $n+1$ for m,

$$\log_e \frac{n+1}{n} = 2\left\{\frac{1}{2n+1} + \frac{1}{3(2n+1)^3} + \frac{1}{5(2n+1)^5} + \ldots\right\}.$$

451. To construct a table of Napierian logarithms.

Since $\log_e \dfrac{n+1}{n} = \log_e (n+1) - \log_e n$, the last formula gives the logarithm of one of two consecutive integers when that of the other is known. Hence by putting $n = 1, 2, 3 \ldots$ in succession, we get the logarithms of $2, 3, 4, \ldots$ to the base e.

Example 1. Calculate the value of $\log_e 2$.

If we put $n = 1$ in the last formula of the last article,

$$\log_e 2 = 2\left\{\frac{1}{3} + \frac{1}{3 \cdot 3^3} + \frac{1}{5 \cdot 3^5} + \frac{1}{7 \cdot 3^7} + \ldots\right\}.$$

The work of calculation may be exhibited as follows :—

$$\begin{array}{lll}
 & 1/3 & = \cdot 333{,}333{,}333 \\
1/3^3 = (1/3) \div 9 = \cdot 037{,}037{,}037 & \therefore\ 1/(3 \cdot 3^3) = & 12{,}345{,}679 \\
1/3^5 = (1/3^3) \div 9 = 4{,}115{,}226 & \therefore\ 1/(5 \cdot 3^5) = & 823{,}045 \\
1/3^7 = (1/3^5) \div 9 = 457{,}247 & \therefore\ 1/(7 \cdot 3^7) = & 65{,}321 \\
1/3^9 = (1/3^7) \div 9 = 50{,}805 & \therefore\ 1/(9 \cdot 3^9) = & 5{,}645 \\
1/3^{11} = (1/3^9) \div 9 = 5{,}645 & \therefore\ 1/(11 \cdot 3^{11}) = & 513 \\
1/3^{13} = (1/3^{11}) \div 9 = 627 & \therefore\ 1/(13 \cdot 3^{13}) = & 48 \\
1/3^{15} = (1/3^{13}) \div 9 = 70 & \therefore\ 1/(15 \cdot 3^{15}) = & 5 \\
1/3^{17} = (1/3^{15}) \div 9 = 8 & \therefore\ 1/(17 \cdot 3^{17}) = & 0 \\
 & & \overline{\cdot 346{,}573{,}589} \\
 & & 2 \\
 & & \overline{\cdot 693{,}147{,}178} \text{ very nearly}
\end{array}$$

The value of $\log_e 2 = \cdot 693{,}147{,}180$ (true to the 9th place of decimals).

If, in the series for $\log_e \dfrac{n+1}{n}$, we put $n = 2$, we get the value of $\log_e 3 - \log_e 2$; and as the value of $\log_e 2$ is known, we know the value of $\log_e 3$. Similarly the values of $\log_e 4$, $\log_e 5 \ldots$ may be computed.

EXPONENTIAL AND LOGARITHMIC SERIES.

Example 2. Find the value of $\log_e 5$.

Putting $n=4$ in the series for $\log_e (n+1)/n$, we have

$$\log_e 5 - \log_e 4 = 2\left\{\frac{1}{9} + \frac{1}{3 \cdot 9^3} + \frac{1}{5 \cdot 9^5} + \ldots\right\};$$

$$\therefore \log_e 5 = 2\log_e 2 + 2\left\{\frac{1}{9} + \frac{1}{3 \cdot 9^3} + \frac{1}{5 \cdot 9^5} + \ldots\right\}.$$

The process of calculation may be exhibited as follows:—

$$\begin{array}{llll}
& & 1/9 & = \cdot 111{,}111{,}111 \\
1/9^2 = \cdot 012{,}345{,}679 & 1/9^3 = \cdot 001{,}371{,}742 & 1/(3 \cdot 9^3) = & 457{,}247 \\
1/9^4 = \phantom{000{,}}152{,}416 & 1/9^5 = \phantom{000{,}00}16{,}935 & 1/(5 \cdot 9^5) = & 3{,}387 \\
1/9^6 = \phantom{000{,}00}1{,}882 & 1/9^7 = \phantom{000{,}000{,}}209 & 1/(7 \cdot 9^7) = & 30 \\
1/9^8 = \phantom{000{,}000{,}0}23 & 1/9^9 = \phantom{000{,}000{,}00}3 & 1/(9 \cdot 9^9) = & 0 \\
\end{array}$$

$$\begin{aligned}
&\cdot 111{,}571{,}775 \\
&\phantom{\therefore\ \log_e 5 = 1\cdot 000{,}000{,}00}2 \\
&\overline{\cdot 223{,}143{,}550} \\
&2\log_e 2 = 1\cdot 386{,}294{,}360 \\
&\therefore\ \log_e 5 = \overline{1\cdot 609{,}437{,}910} \\
&\phantom{\therefore\ \log_e 5 = 1\cdot 609{,}437{,}910}\text{very nearly}
\end{aligned}$$

The value of $\log_e 5 = 1\cdot 609{,}437{,}912$ (true to the 9th place of decimals).

It is evident that in making a table of logarithms of the natural numbers, we need not employ the series for composite numbers, since composite numbers are the products of two or more prime numbers, and the logarithms of the composite numbers are thus equal to the sums of the logarithms of their factors.

For example, $\quad\log_e 10 = \log_e 2 + \log_e 5$
$$= \cdot 693{,}147{,}180 + 1\cdot 609{,}437{,}912$$
$$= 2\cdot 302{,}585{,}092 \text{ very nearly.}$$

The value of $\log_e 10 = 2\cdot 302{,}585{,}093$ (true to the 9th place of decimals).

NOTE. Hence to find the logarithms of numbers, say from 1 to 50, we have to use the series only for calculating the logarithms of the prime numbers 2, 3, 5*, 7, 11, 13, 17, 19, 23, 29, 31, 37, 41, 43, 47.

452. We shall now show how the value of $\log_e 10$ can be calculated to any required number of places from the last formula in § 450 by employing the relation

---

* In the common logarithms, it will be enough if we calculate either $\log_{10} 2$ or $\log_{10} 5$.

EXPONENTIAL AND LOGARITHMIC SERIES. 519

$\log 10 = 23 \log 10/9 - 6 \log 25/24 + 10 \log 81/80$,
which is proved as follows :—

$23 \log \dfrac{10}{9} - 6 \log \dfrac{25}{24} + 10 \log \dfrac{81}{80}$

$= 23 \log \dfrac{2 \cdot 5}{3^2} - 6 \log \dfrac{5^2}{3 \cdot 2^3} + 10 \log \dfrac{3^4}{5 \cdot 2^4}$

$= 23 (\log 2 + \log 5 - 2 \log 3) - 6 (2 \log 5 - \log 3 - 3 \log 2)$
$\qquad + 10 (4 \log 3 - \log 5 - 4 \log 2)$

$= (23 + 18 - 40) \log 2 + (- 46 + 6 + 40) \log 3$
$\qquad + (23 - 12 - 10) \log 5$

$= \log 2 + \log 5 = \log (2 \times 5) = \log 10$.

Now, if we put n in turn equal to 9, 24 and 80, in

$$\log_e \dfrac{n+1}{n} = 2 \left\{ \dfrac{1}{2n+1} + \dfrac{1}{3(2n+1)^3} + \dfrac{1}{5(2n+1)^5} + \ldots \right\},$$

we can find the values of $\log_e 10/9$, $\log_e 25/24$ and $\log_e 81/80$ to any number of places of decimals, and the relation proved above enables us to calculate $\log_e 10$ in terms of these logarithms to any required degree of accuracy.

453. The modulus of the common system of logarithms.

If we know the Napierian logarithm of any number, we can obtain the common logarithm or logarithm to the base 10 of that number by the method explained in § 239.

Since $\qquad \log_{10} a = \log_e a / \log_e 10$,

Napierian logarithms are converted into common logarithms by multiplying them by the constant multiplier

$$\dfrac{1}{\log_e 10} = \dfrac{1}{2 \cdot 302{,}585{,}093} = 0 \cdot 434{,}294{,}48 \ldots.$$

This constant multiplier is called, as has been mentioned in § 240, the *modulus* of the common system, and is usually denoted by the letter μ.

454. To construct a table of common logarithms.

It would be a laborious and tedious task to compile the Napierian logarithms of all the numbers and convert them into common logarithms by multiplying them by μ. The following direct method is simpler in practice.

To construct a table of seven figure common logarithms of all numbers from 1 to 100,000, it will be enough if we find the logarithms of the numbers from 10,000 to 100,000; for we can get the logarithms of the numbers below 10,000 from these by the rule for mantissa [see §§ 240—242]. Thus, for example, the logarithms of 53, 743 and 3751 are the same, so far as the mantissæ are concerned, as those of 53,000, 74,300, and 37,510 respectively.

Now, taking the formula

$$\log_e \frac{n+1}{n} = 2\left\{\frac{1}{2n+1} + \frac{1}{3(2n+1)^3} + \frac{1}{5(2n+1)^5} + \cdots\right\},$$

we find that the error that will arise from neglecting the terms after the first term in the series will be less than

$$2\left\{\frac{1}{3(2n+1)^2} + \frac{1}{3(2n+1)^3} + \frac{1}{3(2n+1)^4} + \cdots\right\},$$

i.e. less than $\quad 2 \cdot \dfrac{1}{3(2n+1)^2} \Big/ \left(1 - \dfrac{1}{2n+1}\right)$

or less than $\quad 2 \cdot \dfrac{1}{3(2n+1)^2} \cdot \dfrac{2n+1}{2n} = \dfrac{1}{3n(2n+1)}.$

This error, when n is not less than 10,000, will be less than $\quad \dfrac{1}{3 \cdot 10000 \cdot 20001} = \cdot 000000001\ldots,$

and cannot therefore affect the eighth place of decimals in the logarithm.

EXPONENTIAL AND LOGARITHMIC SERIES. 521

Hence, retaining only the first term in the series, we have
$$\log_e \frac{n+1}{n} = \frac{2}{2n+1};$$
$$\therefore \log_{10} \frac{n+1}{n} = \frac{2\mu}{2n+1};$$
$$\therefore \log_{10}(n+1) = \log_{10} n + \frac{2\mu}{2n+1}.$$

Hence, taking $\log 10{,}000 = 4$, we have
$$\log 10{,}001 = \log 10{,}000 + \frac{2\mu}{20{,}001},$$
$$\log 10{,}002 = \log 10{,}001 + \frac{2\mu}{20{,}003},$$

and so on, where the logarithms cannot but be correct to the seven places of decimals that we require.

455. Proof of the method of interpolation.

The method of interpolation is based on the theory of proportional parts which we have already stated in § 245 in the following terms:—

If N be any number not less than 10,000, and h and K any numbers not greater than 1; then, as far as seven places of decimals, the following relation holds good,
$$\frac{\log_{10}(N+h) - \log_{10} N}{\log_{10}(N+K) - \log_{10} N} = \frac{h}{K}.$$

We offer now a proof of this proposition.

$$\begin{aligned}
\text{Log}_{10}(N+h) - \log_{10} N &= \log_{10} \frac{N+h}{N} \\
&= \log_{10}\left(1 + \frac{h}{N}\right) \\
&= \mu \log_e\left(1 + \frac{h}{N}\right) \\
&= \mu\left(\frac{h}{N} - \frac{1}{2}\frac{h^2}{N^2} + \frac{1}{3}\frac{h^3}{N^3} - \cdots\right).
\end{aligned}$$

522   EXPONENTIAL AND LOGARITHMIC SERIES.

As the terms in this logarithmic series are continually decreasing and are alternately positive and negative, the series differs from $\mu h/N$ by a quantity which is less than

$$\frac{\mu h^2}{2N^2}.$$

Now $\mu = \log_{10} e < \log_9 3$ or $\frac{1}{2}$;

and $\frac{h}{n}$ is, according to supposition, not greater than $\frac{1}{10000}$ or 10^{-4}.

$$\therefore \frac{\mu h^2}{2N^2} < \frac{1}{4} \cdot 10^{-8}.$$

Hence $\log_{10}(N+h) - \log_{10} N = \mu \dfrac{h}{n}$, correct to 8 places of decimals.

Similarly $\log_{10}(N+K) - \log_{10} N = \mu \dfrac{K}{n} \ldots$;

$$\therefore \frac{\log_{10}(N+h) - \log_{10} N}{\log_{10}(N+K) - \log_{10} N} = \frac{\mu \dfrac{h}{n}}{\mu \dfrac{K}{n}}$$

$$= \frac{h}{K},$$

which was required to be proved.

Hence in tables of seven figure logarithms, if we require the logarithm of a fraction lying between two consecutive integers each of five digits, we may use the formula

$$\frac{\log_{10}(N+h) - \log_{10} N}{\log_{10}(N+1) - \log_{10} N} = \frac{h}{1}.$$

Note. $\log_{10}(N+1) - \log_{10} N = \dfrac{\mu}{n}$, which is the "tabular difference."
[See § 247.]

EXPONENTIAL AND LOGARITHMIC SERIES.

456. Illustrative Examples.

Ex. 1. Calculate, by the aid of the logarithmic series, $\log_{10} 1041$ to 5 places of decimals, given that $\log_{10} e = \cdot 43429$.

We have
$$\log_{10} 1041 = \log_{10} (1000 \times 1\cdot 041)$$
$$= \log_{10} 1000 + \log_{10} 1\cdot 041$$
$$= 3 + \log_{10} e \cdot \log_e 1\cdot 041.$$

Expanding $\log_e 1\cdot 041$ by the logarithmic series, we find
$$\log_{10} 1041 = 3 + \cdot 43429 \{ \cdot 041 - \tfrac{1}{2}(\cdot 041)^2 + \tfrac{1}{3}(\cdot 041)^3 - \ldots \}.$$

The calculation stands as follows, six places of decimals being sufficient:—

```
·43429
 ·041
─────
·017372                              3·000000
   434                ·43249 ×  ·041  =  ·017806
─────                 ·43249 × ⅓(·041)³ =  ·000010
·017806 = ·43429 × ·041              ─────────
 ·041                                 3·017816
─────
·000712               Subtract
   18                 ·43249 × ½(·041)² =  ·000365
─────                 ∴ log₁₀ 1041 = 3·01745(1)
·000730 = ·43429 × (·041)²
 ·041                 true to five places of decimals.
─────
·000030 = ·43429 × (·041)³
```

Ex. 2. Show that, if x be positive,
$$\log_e x = \frac{x-1}{x+1} + \frac{1}{2} \cdot \frac{x^2-1}{(x+1)^2} + \frac{1}{3} \cdot \frac{x^3-1}{(x+1)^3} + \ldots.$$

$$\mathrm{Log}_e x = \log_e \left\{ \left(1 - \frac{1}{x+1}\right) \Big/ \left(1 - \frac{x}{x+1}\right) \right\}$$

$$= \log_e \left(1 - \frac{1}{x+1}\right) - \log_e \left(1 - \frac{x}{x+1}\right)$$

$$= \left\{ -\frac{1}{x+1} - \frac{1}{2(x+1)^2} - \frac{1}{3(x+1)^3} - \ldots \right\}$$

$$\quad + \left\{ \frac{x}{x+1} + \frac{x^2}{2(x+1)^2} + \frac{x^3}{3(x+1)^3} + \ldots \right\}$$

$$= \frac{x-1}{x+1} + \frac{x^2-1}{2(x+1)^2} + \frac{x^3-1}{3(x+1)^3} + \ldots.$$

Ex. 3. Expand $\log_e (1+x+x^2)$ in ascending powers of x when $x < 1$.

$\text{Log}_e (1+x+x^2) = \log_e \{(1-x^3)/(1-x)\} = \log_e (1-x^3) - \log_e (1-x)$

$= \left(-x^3 - \dfrac{x^6}{2} - \dfrac{x^9}{3} - \ldots - \dfrac{x^{3n}}{n} - \ldots\right) + \left(x + \dfrac{x^2}{2} + \dfrac{x^3}{3} + \ldots + \dfrac{x^n}{n} + \ldots\right)$

$= x + \dfrac{x^2}{2} - \dfrac{2}{3}x^3 + \dfrac{x^4}{4} + \dfrac{x^5}{5} - \dfrac{2}{6}x^6 + \ldots,$

the coefficient of x^n being $1/n$ or $-2/n$, according as x is not or is a multiple of 3.

Ex. 4. Expand $\log_e (x^2 + 3x + 2)$ in descending powers of x.

$\text{Log}_e (x^2 + 3x + 2) = \log_e \{(x+1)(x+2)\} = \log_e \{x^2 (1+1/x)(1+2/x)\}$

$= 2 \log_e x + \log_e (1+1/x) + \log_e (1+2/x)$

$= 2 \log_e x + \left(\dfrac{1}{x} - \dfrac{1}{2x^2} + \dfrac{1}{3x^3} - \ldots\right) + \left(\dfrac{2}{x} - \dfrac{2^2}{2x^2} + \dfrac{2^3}{3x^3} - \ldots\right)$

$= 2 \log_e x + \left(\dfrac{2+1}{x} - \dfrac{2^2+1}{2x^2} + \dfrac{2^3+1}{3x^3} + \ldots\right).$

Here x must be greater than two, since $\dfrac{2}{x} < 1$.

Ex. 5. If $1/a$ and $1/b$ are the roots of $x^2 - px + q = 0$, show that

$\log_e (x^2 + px + q) = \log_e q + (a+b)x - \tfrac{1}{2}(a^2+b^2)x^2 + \tfrac{1}{3}(a^3+b^3)x^3 - \ldots,$

where x is such that neither ax nor bx is > 1.

Since $1/a$ and $1/b$ are the roots of $x^2 - px + q = 0$,

$$\dfrac{1}{a} + \dfrac{1}{b} = \dfrac{a+b}{ab} = p ; \text{ and } \dfrac{1}{ab} = q ;$$

$$\therefore\ q + px + x^2 = q\left(1 + \dfrac{p}{q}x + \dfrac{1}{q}x^2\right) = q\{1 + (a+b)x + abx^2\}$$

$$= q(1+ax)(1+bx).$$

$\therefore\ \text{Log}_e (q + px + x^2) = \log_e q + \log_e (1+ax) + \log_e (1+bx)$

$= \log_e q + \left(ax - \dfrac{a^2 x^2}{2} + \dfrac{a^3 x^3}{3} - \ldots\right) + \left(bx - \dfrac{b^2 x^2}{2} + \dfrac{b^3 x^3}{3} - \ldots\right)$

$= \log_e q + \{(a+b)x - \tfrac{1}{2}(a^2+b^2)x^2 + \tfrac{1}{3}(a^3+b^3)x^3 - \ldots\}.$

EXPONENTIAL AND LOGARITHMIC SERIES.

Ex. 6. If $a+b+c=0$, prove that
$$\tfrac{1}{7}(a^7+b^7+c^7) = \tfrac{1}{3}(a^3+b^3+c^3) \cdot \tfrac{1}{2}(a^4+b^4+c^4).$$

Let $-p = bc+ca+ab$, and $q = abc$.

Then $1 - px^2 - qx^3 = 1 - (a+b+c)x + (bc+ca+ab)x^2 - abcx^3$
$$= (1-ax)(1-bx)(1-cx);$$

$\therefore \text{Log}_e\{1 - x^2(p+qx)\} = \log_e(1-ax) + \log_e(1-bx) + \log_e(1-cx).$

Expanding these by the logarithmic theorem, we have

$x^2(p+qx) + \tfrac{1}{2}x^4(p+qx)^2 + \tfrac{1}{3}x^6(p+qx)^3 + \ldots$
$$= (a+b+c)x + \tfrac{1}{2}(a^2+b^2+c^2)x^2 + \tfrac{1}{3}(a^3+b^3+c^3)x^3 + \ldots.$$

Equating the coefficients of x^7, x^3 and x^4, we find

$$\tfrac{1}{7}(a^7+b^7+c^7) = \tfrac{1}{3} \cdot 3p^2q = p^2q\,;$$
$$\tfrac{1}{3}(a^3+b^3+c^3) \qquad = q\,;$$
and $\qquad \tfrac{1}{4}(a^4+b^4+c^4) \qquad = \tfrac{1}{2}p^2\,;$

from which the required result readily follows.

Ex. 7. Find the limit of $\sqrt[x]{\dfrac{1+x}{1-x}}$, when $x=0$.

We have $\qquad \sqrt[x]{\dfrac{1+x}{1-x}} = (1+x)^{\tfrac{1}{x}} \times (1-x)^{-\tfrac{1}{x}}.$

Now $\qquad\qquad \mathop{L^t}_{x=0} (1+x)^{\tfrac{1}{x}} = e,$

and $\qquad\qquad \mathop{L^t}_{x=0} (1-x)^{-\tfrac{1}{x}} = e,$

$\therefore \mathop{L^t}_{x=0} \sqrt[x]{\dfrac{1+x}{1-x}} = e \times e = e^2.$

Ex. 8. Show that $\log_e 2 = \dfrac{1}{1 \cdot 2} + \dfrac{1}{3 \cdot 4} + \dfrac{1}{5 \cdot 6} + \ldots$.

Putting $x=1$ in the series for $\log_e(1+x)$, we have

$$\log_e 2 = 1 - \frac{1}{2} + \frac{1}{3} - \frac{1}{4} + \ldots \quad [\text{see § 136}]$$
$$= \left(1 - \frac{1}{2}\right) + \left(\frac{1}{3} - \frac{1}{4}\right) + \left(\frac{1}{5} - \frac{1}{6}\right) + \ldots$$
$$= \frac{1}{1 \cdot 2} + \frac{1}{3 \cdot 4} + \frac{1}{5 \cdot 6} + \ldots.$$

EXPONENTIAL AND LOGARITHMIC SERIES.

Ex. 9. Show that $\log_e 2 - \dfrac{1}{2} = \dfrac{1}{1.2.3} + \dfrac{1}{3.4.5} + \dfrac{1}{5.6.7} + \cdots$

$$2\log_e 2 = 2\left(1 - \frac{1}{2} + \frac{1}{3} - \frac{1}{4} + \cdots\right)$$

$$= 1 + 1 - \frac{1}{2} - \frac{1}{2} + \frac{1}{3} + \frac{1}{3} - \frac{1}{4} - \frac{1}{4} + \frac{1}{5} + \cdots$$

$$= 1 + \left(1 - \frac{1}{2}\right) - \left(\frac{1}{2} - \frac{1}{3}\right) + \left(\frac{1}{3} - \frac{1}{4}\right) - \left(\frac{1}{4} - \frac{1}{5}\right) + \cdots$$

$$= 1 + \frac{1}{1.2} - \frac{1}{2.3} + \frac{1}{3.4} - \frac{1}{4.5} + \cdots$$

$$= 1 + \left(\frac{1}{1.2} - \frac{1}{2.3}\right) + \left(\frac{1}{3.4} - \frac{1}{4.5}\right) + \cdots$$

$$= 1 + \frac{2}{1.2.3} + \frac{2}{3.4.5} + \frac{2}{5.6.7} + \cdots,$$

from which the required result follows.

Ex. 10. Sum to infinity $\dfrac{1}{1.2.3} + \dfrac{5}{3.4.5} + \dfrac{9}{5.6.7} + \dfrac{13}{7.8.9} + \cdots$.

Here $u_n = \dfrac{4n-3}{(2n-1)\,2n\,(2n+1)} = \dfrac{2(2n+1) - 5}{(2n-1)\,2n\,(2n+1)}$

$$= \frac{2}{(2n-1)\,2n} - \frac{5}{(2n-1)\,2n\,(2n+1)}.$$

Substituting 1, 2, 3... in succession for n, we have the required sum

$$= 2\left(\frac{1}{1.2} + \frac{1}{3.4} + \frac{1}{5.6} + \cdots\right) - 5\left(\frac{1}{1.2.3} + \frac{1}{3.4.5} + \frac{1}{5.6.7} + \cdots\right)$$

$$= 2\log_e 2 - 5(\log_e 2 - \tfrac{1}{2}) \qquad\qquad \text{[Examples 8 and 9]}$$

$$= \tfrac{5}{2} - 3\log_e 2.$$

Ex. 11. Find the values of the constants A, B and C so that
$$\log(a+h) = A\log a + B\log(a + \tfrac{1}{2}h) + C\log(a + \tfrac{1}{3}h)$$
when h is very small.

We have $\log a + \log\left(1 + \dfrac{h}{a}\right)$

$$= A\log a + B\left\{\log a + \log\left(1 + \frac{h}{2a}\right)\right\} + C\left\{\log a + \log\left(1 + \frac{h}{3a}\right)\right\}.$$

EXPONENTIAL AND LOGARITHMIC SERIES. 527

$$\therefore \log a (1 - A - B - C) + \frac{h}{a}\left(1 - \frac{1}{2}B - \frac{1}{3}C\right) - \frac{h^2}{2a^2}\left(1 - \frac{1}{4}B - \frac{1}{9}C\right) + \text{terms}$$

involving powers of h higher than h^2, which may be neglected.

Then we have

$$\left.\begin{array}{r}A + B + C = 1\\ \tfrac{1}{2}B + \tfrac{1}{3}C = 1\\ \tfrac{1}{4}B + \tfrac{1}{9}C = 1\end{array}\right\} \text{ from which we get that } A = 2, B = 8 \text{ and } C = -9.$$

Ex. 12. Find the value of $\dfrac{e^x - e^{-x}}{\log(1+x)}$ when $x = 0$.

The expression $= \dfrac{1 + x + \dfrac{x^2}{2!} + \dfrac{x^3}{3!} + \ldots - \left(1 - x + \dfrac{x^2}{2!} - \dfrac{x^3}{3!} + \ldots\right)}{x - \dfrac{x^2}{2} + \dfrac{x^3}{3} - \ldots}$

$= \dfrac{2\left(x + \dfrac{x^3}{3!} + \ldots\right)}{x - \dfrac{x^2}{2} + \dfrac{x^3}{3} - \ldots} = 2\,\dfrac{1 + \dfrac{x^2}{3!} + \ldots}{1 - \dfrac{x}{2} + \ldots}$

$= 2$ when $x = 0$. [See § 146]

Ex. 13. Defining $\operatorname{Log}_e a = \operatorname*{L^t}_{x=0} \dfrac{a^x - 1}{x}$, prove from this that

$$\log_e(ab) = \log_e a + \log_e b.$$

$$\frac{(ab)^x - 1}{x} = \frac{(ab)^x - a^x}{x} + \frac{a^x - 1}{x}$$

$$= \frac{a^x(b^x - 1)}{x} + \frac{a^x - 1}{x}$$

$$= \frac{b^x - 1}{x} + \frac{a^x - 1}{x},$$

when $x = 0$, a^x being equal to 1.

$$\therefore \operatorname*{L^t}_{x=0} \frac{(ab)^x - 1}{x} = \operatorname*{L^t}_{x=0} \frac{b^x - 1}{x} + \operatorname*{L^t}_{x=0} \frac{a^x - 1}{x}.$$

That is, $\log_e(ab) = \log_e a + \log_e b.$

EXERCISES XXIX.

1. Show that $e =$ the limit of $(1+n)^{1/n}$, when $n=0$.

2. Show, by applying the binomial theorem, that
$$(1+1/a)^a > (1+1/b)^b,$$
when a and b are positive integers and $a>b$; and hence show that
$$(1\tfrac{1}{2})^2 < (1\tfrac{1}{3})^3 < (1\tfrac{1}{4})^4 < \ldots < e.$$

3. Show that $e^x =$ the limit of $(1+x/n)^n$, when $n=\infty$
$$= \text{the limit of } (1+nx)^{1/n}, \text{ when } n=0.$$

4. If x be very small, show that $ee^x = e(1+x)$ very nearly.

5. Express the equation
$$q(\log_e p + x) = r \log_e x + s$$
in an exponential form.

6. Show that $\dfrac{1}{e} = \dfrac{1}{1.3} + \dfrac{1}{1.2.3.5} + \dfrac{1}{1.2.3.4.5.7} + \ldots$
to infinity.

7. Sum to infinity $\dfrac{1}{2!} + \dfrac{2}{3!} + \dfrac{3}{4!} + \dfrac{4}{5!} + \ldots$

8. Sum to infinity $\dfrac{1.2}{1!} + \dfrac{2.3}{2!} + \dfrac{3.4}{3!} + \dfrac{4.5}{4!} + \ldots$

9. Sum to infinity $1 + \dfrac{1}{3!} + \dfrac{1}{5!} + \dfrac{1}{7!} + \ldots$

10. Sum to infinity $\dfrac{1}{1!} + \dfrac{1+2}{2!} + \dfrac{1+2+3}{3!} + \dfrac{1+2+3+4}{4!} + \ldots$

11. Sum to infinity $\dfrac{1}{1!} + \dfrac{1+2}{2!} + \dfrac{1+2+4}{3!} + \dfrac{1+2+4+8}{4!} + \ldots$

12. Show that $\sqrt{e} = 1 + \dfrac{1}{2!} + \dfrac{1.3}{4!} + \dfrac{1.3.5}{6!} + \ldots$ to infinity.

13. Sum to infinity $\dfrac{1^2}{2!} + \dfrac{2^2}{3!} + \dfrac{3^2}{4!} + \ldots$

14. Sum to infinity $\dfrac{2^3}{1!} + \dfrac{3^3}{2!} + \dfrac{4^3}{3!} + \ldots$

15. Show that the coefficient of x^n in the infinite series
$$1 + \dfrac{a+bx}{1!} + \dfrac{(a+bx)^2}{2!} + \ldots + \dfrac{(a+bx)^n}{n!} + \ldots$$
is $e^a b^n / n!$

EXPONENTIAL AND LOGARITHMIC SERIES.

16. Prove (i) by the exponential theorem and (ii) by simple multiplication with the aid of the binomial theorem, that
$$\left(1 + x + \frac{x^2}{2!} + \frac{x^3}{3!} + \ldots\right)\left(1 - x + \frac{x^2}{2!} - \frac{x^3}{3!} + \ldots\right) = 1.$$

17. Find the coefficient of x^n in the expansion of $(1 + 2x - 3x^2)/e^x$.

18. Expand $e^x(1-x)$ in a series of ascending powers of x.

19. Show that the difference of the coefficients of x^n and x^{n-1} in the expansion of $e^x/(1-x)$ is $1/n!$

20. Find the coefficient of x^4 in the expansion of
$$1/\{(1+x)\,e^x\}.$$

21. By equating the coefficients of x^n in the expansions of
$$(e^x - 1)^n = (x + x^2/2! + x^3/3! + \ldots)^n,$$
prove that $\quad n^n - n(n-1)^n + \dfrac{n(n-1)}{1 \cdot 2}(n-2)^n - \ldots = n!$

22. Prove that, if x be positive, $e > x^{1/x}$.

23. Prove that
$$2\log_e m - \log_e(m+1) - \log_e(m-1)$$
$$= 2\left\{\frac{1}{2m^2-1} + \frac{1}{3(2m^2-1)^3} + \frac{1}{5(2m^2-1)^5} + \ldots\right\}.$$

24. Prove that
$$\log_e(n+1) - \log_e(n-1) = 2\left(\frac{1}{n} + \frac{1}{3n^3} + \frac{1}{5n^5} + \ldots\right).$$

25. Prove that
$$\log_e\left(\frac{1}{1-x}\right) = x + \frac{x^2}{2} + \frac{x^3}{3} + \ldots.$$

26. Prove that, if x be positive and less than unity,
$$\log_e\{1/(1-x)\} > x.$$

27. Prove that
$$\log_e(x+n) = \log_e x + \log_e\left(1 + \frac{1}{x}\right) + \log_e\left(1 + \frac{1}{1+x}\right)$$
$$+ \log_e\left(1 + \frac{1}{2+x}\right) + \ldots + \log_e\left\{1 + \frac{1}{(n-1)+x}\right\}.$$

28. Prove that
$$\log_e(1+x)^{\frac{1}{2}(x+1)} + \log_e(1-x)^{\frac{1}{2}(1-x)} = \frac{x^2}{1\cdot 2} + \frac{x^4}{3\cdot 4} + \frac{x^6}{5\cdot 6} + \ldots.$$

TUT. ALG. II.

29. Prove that the difference between the logarithms of any two successive numbers decreases as the number increases.

30. Prove that, when n is large,
$$\left(1+\frac{1}{n}\right)^{n+\frac{1}{2}} = e^{1+\frac{1}{12n^2}} \text{ approximately.}$$

31. Show that
$$\log_v u = \frac{(u-1) - \frac{1}{2}(u-1)^2 + \frac{1}{3}(u-1)^3 - \ldots}{(v-1) - \frac{1}{2}(v-1)^2 + \frac{1}{3}(v-1)^3 - \ldots}.$$

*32. Show that the limit of
$$\frac{1}{n+1} + \frac{1}{n+2} + \frac{1}{n+3} + \ldots + \frac{1}{2n} = \log_e 2, \text{ when } n = \infty.$$

33. Show that the limit of
$$\frac{e^x - 1 - \log_e(1+x)}{x^2} = 1, \text{ when } x = 0.$$

34. Prove, by taking the logarithms of both the sides, that
$$\sqrt[x]{\frac{1+x}{1-x}} > \sqrt[y]{\frac{1+y}{1-y}},$$
if x and y are proper fractions and $x > y$.

35. Prove that the ratio $\log a : \log b$, where a and b are given numbers, is the same for all systems.

36. If α and β are the roots of $ax^2 + bx + c = 0$, show that
$$\log_e(a - bx + cx^2) = \log_e a + (\alpha + \beta)x - \tfrac{1}{2}(\alpha^2 + \beta^2)x^2 + \tfrac{1}{3}(\alpha^3 + \beta^3)x^3 - \ldots.$$

37. If m and n be the roots of $x^2 + px + q = 0$, show that
$$\log_e(1 - px + qx^2) = (m+n)x - \tfrac{1}{2}(m^2 + n^2)x^2 + \tfrac{1}{3}(m^3 + n^3)x^3 - \ldots.$$

38. If $a = x - \tfrac{1}{2}x^2 + \tfrac{1}{3}x^3 - \tfrac{1}{4}x^4 + \ldots$, prove that
$$x = a + \frac{a^2}{2!} + \frac{a^3}{3!} + \ldots.$$

39. Expand $\log_e(x^2 + 5x + 6)$ in a series of descending powers of x.

40. Expand $\log_e\left(\dfrac{1+x+x^2}{1-x+x^2}\right)$ in a series of ascending powers of x.

41. If $\log_e \dfrac{1}{1+x+x^2+x^3}$ be expanded in a series of ascending powers of x, show that the coefficient of x^n will be $3/n$ if n be a multiple of 4, and $-1/n$ if n be otherwise.

42. If $a+b+c=0$, prove that

(i) $\frac{1}{5}(a^5+b^5+c^5) = \frac{1}{3}(a^3+b^3+c^3) \cdot \frac{1}{2}(a^2+b^2+c^2) = -abc(bc+ca+ab)$.

(ii) $\frac{1}{7}(a^7+b^7+c^7) = \frac{1}{3}(a^3+b^3+c^3) \cdot \frac{1}{4}(a^2+b^2+c^2)^2 = abc(bc+ca+ab)^2$.

43. Prove, by employing the logarithmic series, that
$$\log_e(1{\cdot}01) = {\cdot}00995033 \text{ nearly};$$
and having given $\log_{10} e = {\cdot}43429448$, prove that
$$\log_{10}(1{\cdot}01) = {\cdot}0043214 \text{ nearly}.$$

44. Prove that $\log 10 = 23\log \frac{10}{9} - 6\log \frac{25}{24} + 10\log \frac{81}{80}$; and explain how this formula may be used to calculate $\log_e 10$ to any required number of places.

45. Show that, when x is positive and less than unity, $-\log_e(1-x)$ exceeds $x + \frac{1}{2}x^2 + \frac{1}{3}x^3 + \ldots + x^n/n + \ldots$ by less than $\dfrac{x^{n+1}}{n+1}(1-x)^{-1}$.

46. Given $\log_{10} 3 = {\cdot}47712$, $\log_{10} e = {\cdot}43429$, compute $\log_{10} 29$ to five places of decimals.

47. Given $\log_e 3 = 1{\cdot}098612$, and $\log_e 10 = 2{\cdot}302585$, calculate $\log_{10} 9$.

48. If $\log_e \dfrac{(2x+1)^{\frac{3}{2}}}{3^{\frac{3}{2}}x}$ be expanded in ascending powers of $x-1$, prove that the first term in the expansion will be $\frac{1}{6}(x-1)^2$.

49. Sum to infinity $\dfrac{1}{2 \cdot 3} + \dfrac{1}{4 \cdot 5} + \dfrac{1}{6 \cdot 7} + \ldots$.

50. Sum to infinity $\dfrac{1}{2 \cdot 3 \cdot 4} + \dfrac{1}{4 \cdot 5 \cdot 6} + \dfrac{1}{6 \cdot 7 \cdot 8} + \ldots$.

51. Sum to infinity $\dfrac{1}{1 \cdot 2 \cdot 3 \cdot 4} + \dfrac{1}{3 \cdot 4 \cdot 5 \cdot 6} + \dfrac{1}{5 \cdot 6 \cdot 7 \cdot 8} + \ldots$.

52. Show that
$$\left(1+\frac{1}{x}\right)\log_e(1+x) - 1 = \frac{x}{1 \cdot 2} - \frac{x^2}{2 \cdot 3} + \frac{x^3}{3 \cdot 4} - \frac{x^4}{4 \cdot 5} + \ldots \text{ to infinity}.$$

53. Deduce from example 49 that
$$\log_e \frac{4}{e} = \frac{1}{1 \cdot 2} - \frac{1}{2 \cdot 3} + \frac{1}{3 \cdot 4} - \frac{1}{4 \cdot 5} + \ldots.$$

***54.** Show, by equating the coefficients of x^n in the expansions of $2\log_e(1-x)$ and $\log_e(1-2x+x^2)$, that
$$2^n - n \cdot 2^{n-2} + \frac{n(n-3)}{1 \cdot 2} 2^{n-4} - \frac{n(n-4)(n-5)}{1 \cdot 2 \cdot 3} 2^{n-6} + \ldots = 2.$$

CHAPTER XXX.

SUMMATION OF SERIES.

457. In the present Chapter we shall explain methods of finding expressions for the sum of any number of terms of certain series. We may remark at the outset that, as there is no general process for effecting such a summation, we have to confine ourselves to the discussion of certain types of series only.

458. The first class of series we shall deal with is that in which the nth term is a rational integral function of n of finite degree, and it will be seen that a formula can always be found for the sum of the first n terms of such a series.

It is convenient to begin with a particular case, and to this we address ourselves in the next article.

459. Series of products of arithmetical progressions.

To find the sum of n terms of a series each term of which is the product of r factors in arithmetical progression, and the first factors of the successive terms are in the same arithmetical progression.

Let us denote the series by
$$u_1 + u_2 + u_3 + \ldots\ldots + u_n,$$
where
$$u_n = (a + nd)(a + \overline{n+1}.d)(a + \overline{n+2}.d)\ldots\ldots(a + \overline{n+r-1}.d);$$
and let
$$v_n = (a + \overline{n+r}.d)u_n.$$

SUMMATION OF SERIES.

$$\therefore v_n = (a+nd)(a+\overline{n+1}.d)\ldots\ldots(a+\overline{n+r-1}.d)(a+\overline{n+r}.d);$$

and

$$v_{n-1} = (a+\overline{n-1}.d)(a+nd)\ldots\ldots(a+\overline{n+r-2}.d)(a+\overline{n+r-1}.d).$$

By subtraction

$$v_n - v_{n-1} = (a+nd)(a+\overline{n+1}.d)\ldots\ldots(a+\overline{n+r-1}.d)$$
$$\{a+\overline{n+r}.d - (a+\overline{n-1}.d)\}.$$

That is, $\quad v_n - v_{n-1} = u_n \cdot (r+1)d.$

Similarly, $\quad v_{n-1} - v_{n-2} = u_{n-1} \cdot (r+1)d,$

$\ldots\ldots\ldots\ldots\ldots\ldots\ldots\ldots\ldots\ldots\ldots$

$\qquad v_2 - v_1 = u_2 \cdot (r+1)d,$

$\qquad v_1 - v_0 = u_1 \cdot (r+1)d,$

where v_0, the term preceding the first, is what the nth term becomes when in it we put $n = 0$. Hence, by addition,

$$v_n - v_0 = (r+1)d \cdot \{u_1 + u_2 + u_3 + \ldots\ldots + u_n\} = (r+1)d \cdot S_n.$$

$$\therefore S_n = \frac{v_n - v_0}{(r+1)d} = \frac{v_n}{(r+1)d} - \frac{v_0}{(r+1)d}.$$

Hence we have the following rule for summing up the series:

The required sum is the value of $A - B$, where A is a fraction having for its numerator the nth term of the given series multiplied by the factor next after its last, and for its denominator the number of factors in the numerator multiplied by the common difference of the A.P.; while B is what A becomes when in it we put $n = 0$.

COR. An important particular case of this proposition is that

$$1.2.3\ldots\ldots r + 2.3.4\ldots\ldots(r+1) + \ldots\ldots$$
$$+ n(n+1)(n+2)\ldots(n+r-1) = \frac{n(n+1)(n+2)\ldots(n+r)}{r+1}.$$

Ex. 1. Find the sum to n terms of
$$1.2.3.4+2.3.4.5+3.4.5.6+\ldots.$$

(i) By the rule, since $u_n = n(n+1)(n+2)(n+3)$; $r=4$; and $d=1$;

$$S_n = \frac{n(n+1)(n+2)(n+3)(n+4)}{5 \times 1} - \frac{0.1.2.3.4}{5 \times 1}$$
$$= \frac{n(n+1)(n+2)(n+3)(n+4)}{5},$$

a result which can be obtained at once from the corollary by putting $r=4$.

(ii) Or solving this question independently of the rule, we have
$$u_n = n(n+1)(n+2)(n+3).$$

Let $v_n = n(n+1)(n+2)(n+3)(n+4)$;
$$\therefore v_{n-1} = (n-1)n(n+1)(n+2)(n+3);$$
$$\therefore v_n - v_{n-1} = n(n+1)(n+2)(n+3)\{(n+4)-(n-1)\}.$$

That is, $v_n - v_{n-1} = 5u_n$;
$$\therefore v_{n-1} - v_{n-2} = 5u_{n-1},$$
$$\ldots\ldots\ldots\ldots\ldots\ldots\ldots$$
$$v_2 - v_1 = 5u_2,$$
$$v_1 - v_0 = 5u_1;$$

$$\therefore S_n = \frac{1}{5}(v_n - v_0) = \frac{n(n+1)(n+2)(n+3)(n+4)}{5} - \frac{0.1.2.3.4}{5}$$
$$= \frac{1}{5}n(n+1)(n+2)(n+3)(n+4) \text{ as before.}$$

Ex. 2. Find to n terms the sum of
$$2.5.8+5.8.11+8.11.14+\ldots.$$

Here $v_n = (3n-1)(3n+2)(3n+5)$; $r=3$; $d=3$.
Hence we have, by the rule,

$$S_n = \frac{(3n-1)(3n+2)(3n+5)(3n+8)}{4.3} - \frac{-1.2.5.8}{4.3}$$
$$= \frac{1}{12}n(81n^3 + 378n^2 + 459n + 42) \text{ after reduction,}$$

which may be easily verified by putting $n=1$ and $n=2$.

460. Series in which the nth term is a rational integral function of n.

This class of series can always be summed by the methods of § 459.

Suppose for example the nth term, $\phi(n)$, is a function of n of the second degree, *then if A_0, A_1, A_2 be the remainders when $\phi(n)$ is divided successively by n, $n+1$ and $n+2$ we have*

$$\phi(n) = A_2 n(n+1) + A_1 n + A_0.$$

The final quotient is zero because $\phi(n)$ is only of the second degree.

Hence the A's being numerical we have

$$\sum_1^n \phi(n) = A_2 \sum_1^n n(n+1) + A_1 \Sigma n + A_0 n$$

$$= A_2 \frac{n(n+1)(n+2)}{3} + A_1 \frac{n(n+1)}{2} + A_0 n.$$

We thus effect the summation by splitting up the nth term into others to each of which the method of § 459 applies.

Ex. 1. Suppose the nth term is $n^2 + 4n + 1$, then we write the work thus

$$\begin{array}{r} n\,)\ n^2+4n+1 \\ \overline{n+1\,)\ n+4}\ +1 \\ \overline{n+2\,)\ 1}\ +3 \\ \overline{\ 0}\ +1 \end{array}$$

Consequently $\quad n^2 + 4n + 1 = n(n+1) + 3n + 1$,

and the sum required is $\dfrac{n(n+1)(n+2)}{3} + \dfrac{3n(n+1)}{2} + n$

$$= \frac{n}{6}\{2(n+1)(n+2) + 9(n+1) + 6\}$$

$$= \frac{n}{6}\{2n^2 + 15n + 19\}.$$

Ex. 2. Find by this method the sum of the squares of the first n natural numbers.

Here $A_0 = 0$, $A_1 = -1$, $A_2 = 1$, and proceeding as before, we find the sum to be $\dfrac{n(n+1)(2n+1)}{6}$.

461. General Case. When $\phi(n)$ is of the rth degree **divide it successively by**
$$n, n+1, n+2, \ldots\ldots n+r-1, n+r,$$
and let $A_0, A_1, \ldots\ldots A_r$ be the remainders,
then $\phi(n) = A_r n \ (n+1)(n+2) \ldots\ldots (n+r-1)$
$$+ A_{r-1} n (n+1)(n+2) \ldots\ldots (n+r-2)$$
$$+ \ldots\ldots + A_2 n(n+1) + A_1 n + A_0,$$
where the A's are numbers not involving n.

It is clear that the degree of the quotient is diminished by unity on each division, therefore the quotient after dividing by $n+r-1$ is of degree $r-r = 0$, i.e. it is numerical only, and the final quotient is zero.

The relation may be formally established as follows:—

Let $\phi_1, \phi_2 \ldots\ldots \phi_r$ be the successive quotients, then
$\phi = \phi_1 n + A_0,$
$\phi_1 = \phi_2(n+1) + A_1; \quad \therefore \ \phi = \phi_2 n (n+1) + A_1 n + A_0,$
$\phi_2 = \phi_3(n+2) + A_2; \quad \therefore \ \phi = \phi_3 n(n+1)(n+2) + A_2 n(n+1) + A_1 n + A_0,$
$\ldots\ldots\ldots\ldots\ldots\ldots\ldots\ldots\ldots\ldots\ldots\ldots\ldots\ldots\ldots\ldots\ldots\ldots\ldots\ldots\ldots\ldots$
$\phi_{r-1} = \phi_{r-2}(n+r-1) + A_{r-1};$
$\therefore \ \phi = \phi_{r-2} n(n+1) \ldots (n+r-1) + A_{r-1} n(n+1) \ldots (n+r-2)$
$$+ \ldots + A_2 n(n+1) + A_1 n + A_0,$$
and ϕ_{r-2} being numerical is equal to A_r.

Hence $\overset{n}{\underset{1}{\Sigma}} \phi(n) = A_r \dfrac{n(n+1)\ldots(n+r)}{r+1} + A_{r-1} \dfrac{n(n+1)\ldots(n+r-1)}{r} + \ldots$
$$\ldots + A_2 \dfrac{n(n+1)(n+2)}{6} + A_1 \dfrac{n(n+1)}{2} + A_0 n.$$

Ex. **To find the sum of the cubes of the first n natural numbers.**

Here $\phi(n) = n^3$. As in Ex. 1 § 460
$$n^3 = 1 . n(n+1)(n+2) - 3n(n+1) + n,$$
$$\therefore \ \text{sum} = \dfrac{n(n+1)(n+2)(n+3)}{4} - \dfrac{3(n)(n+1)(n+2)}{3} + \dfrac{n(n+1)}{2}$$
$$= \dfrac{n(n+1)}{4} \{(n+2)(n+3) - 4(n+2) + 2\} = \dfrac{n(n+1)}{4} \{n^2 + n\} = \left\{\dfrac{n(n+1)}{2}\right\}^2.$$

SUMMATION OF SERIES. 537

462. Sums of any power of the first n natural numbers.

The series $\quad 1^r + 2^r + 3^r \ldots\ldots + n^r$

is so important that it may be worth while to explain another process for its evaluation. Denoting it by ${}^nS_r$, we shall show how it may be determined when we know the values of

$$ {}^nS_{r-1}, \quad {}^nS_{r-2}, \quad {}^nS_{r-3} \ldots\ldots {}^nS_2, \quad {}^nS_1. $$

We have in fact

$$(x+1)^{r+1} = x^{r+1} + {}_{r+1}C_1 x^r + {}_{r+1}C_2 x^{r-1} \ldots\ldots {}_{r+1}C_r x + 1,$$

or putting in this identity $n, n-1, n-2, \ldots\ldots 3, 2, 1$ in succession for x, we have

$$(n+1)^{r+1} = n^{r+1} + {}_{r+1}C_1 n^r + {}_{r+1}C_2 n^{r-1} + \ldots\ldots + {}_{r+1}C_r n + 1$$
$$n^{r+1} = (n-1)^{r+1} + {}_{r+1}C_1 (n-1)^r + {}_{r+1}C_2 (n-1)^{r-1} + \ldots\ldots$$
$$+ {}_{r+1}C_r (n-1) + 1$$
$$\ldots\ldots\ldots\ldots\ldots\ldots\ldots\ldots\ldots\ldots\ldots\ldots\ldots\ldots\ldots\ldots\ldots\ldots\ldots\ldots\ldots\ldots\ldots\ldots$$
$$3^{r+1} = 2^{r+1} + {}_{r+1}C_1 2^r + {}_{r+1}C_2 2^{r-1} + \ldots\ldots + {}_{r+1}C_r . 2 + 1$$
$$2^{r+1} = 1^{r+1} + {}_{r+1}C_1 1^r + {}_{r+1}C_2 1^{r-1} + \ldots\ldots + {}_{r+1}C_r . 1 + 1.$$

Hence by addition we derive

$$(n+1)^{r+1} = 1 + {}_{r+1}C_1 \, {}^nS_r + {}_{r+1}C_2 \, {}^nS_{r-1} + \ldots\ldots + {}_{r+1}C_r \, {}^nS_1 + n,$$

and this equation enables us to calculate ${}^nS_r$ when

$$ {}^nS_{r-1}, \quad {}^nS_{r-2} \ldots\ldots {}^nS_2, \quad {}^nS_1 $$

are known.

Then knowing the value of ${}^nS_1$ we can find the values of ${}^nS_2, {}^nS_3 \ldots$ in order.

Ex. Find by the method of this article ${}^nS_1$ and ${}^nS_2$.

Putting $r=1$ in the formulæ, we have

$$(n+1)^2 = 1 + {}_2C_1 \, {}^nS_1 + {}_2C_2 \, {}^nS_0; \qquad \{{}^nS_0 = n\}$$
$$\therefore (n+1)^2 = 1 + 2\,{}^nS_1 + n;$$
i.e. $\quad 2\,{}^nS_1 = n(n+1); \therefore {}^nS_1 = \tfrac{1}{2} n(n+1).$

Putting $r = 2$, we derive

$$(n+1)^3 = 1 + 3\,{}^nS_2 + \frac{3 \cdot 2}{1 \cdot 2}\,{}^nS_1 + n;$$

i.e. $3\,^nS_2 = (n+1)^3 - 1 - 3\,^nS_1 - n = (n+1)^3 - (n+1) - 3\dfrac{n(n+1)}{2}$

$= \dfrac{n+1}{2}\{2(n+1)^2 - 2 - 3n\} = \dfrac{n+1}{2}\{2n^2 + n\} = \dfrac{n(n+1)(2n+1)}{2};$

$\therefore\ ^nS_2 = \dfrac{n(n+1)(2n+1)}{6}.$

To find $^nS_3$ we should put $r=3$, and then use the values of $^nS_1$, $^nS_2$ already found.

463. If the nth term of a series be a rational integral function of n, the sum of n terms is also a rational integral function of n and of one degree higher.

This follows at once from the general method of summation explained Art. 461, for when the nth term is of degree r it there appeared that the sum is of degree $r+1$.

Alternative proof.

First, the sum is a rational integral function of n, for it may be written as the sum of a number of terms of the type ^nS_r, which are all such functions as follows from the last article.

Next, if the sum Σ_n be the degree s, we have

$\Sigma_n - \Sigma_{n-1} = u_n$, where u_n is the nth term.

Now $\Sigma_n = a_0 n^s + a_1 n^{s-1} + a_2 n^{s-2} + \ldots,$

and $\Sigma_{n-1} = a_0(n-1)^s + a_1(n-1)^{s-1} + a_2(n-1)^{s-2} + \ldots$

$\therefore\ u_n = a_0(n^s - \overline{n-1}^s) + a_1(n^{s-1} - \overline{n-1}^{s-1}) + \ldots$

$= n^{s-1} \cdot sa_0 + \text{terms involving lower powers of } n.$

$\therefore\ u_n$ is of degree $s-1$ when Σ_n is of degree s, and hence if u_n be of degree r, Σ_n is of degree $r+1$, for otherwise u_n would not be of degree r.

Using this fact we may effect summations by means of undetermined coefficients as is seen by the following examples.

Ex. 1. Find to n terms the sum of the series

$1^2 + 2^2 + 3^2 + 4^2 + \ldots.$

Here $u_n = n^2$, which is a function of n of the second degree.

Hence S_n is a function of n of the third degree.

SUMMATION OF SERIES.

∴ let $S_n{}^* = Bn + Cn^2 + Dn^3,$

where B, C, D are quantities independent of n of which the values have to be determined.

∴ $S_{n-1} = B(n-1) + C(n-1)^2 + D(n-1)^3.$

By subtraction,

$$S_n - S_{n-1} = u_n = n^2 = B + C(2n-1) + D(3n^2 - 3n + 1)$$
$$= (B - C + D) + (2C - 3D)n + 3Dn^2;$$

∴ $B - C + D = 0;\ 2C - 3D = 0;\ 3D = 1.$

∴ $D = \tfrac{1}{3};\ C = \tfrac{1}{2};$ and $B = \tfrac{1}{6}.$

Hence $S_n = \tfrac{1}{6}n + \tfrac{1}{2}n^2 + \tfrac{1}{3}n^3 = \tfrac{1}{6}n(n+1)(2n+1).$

∴ $S_n = \tfrac{1}{6}n + \tfrac{1}{2}n^2 + \tfrac{1}{3}n^3$
$= \tfrac{1}{6}n(n+1)(2n+1),$ as in § 145.

Ex. 2. Find to n terms the sum of the series

$$1.3.4 + 4.5.5 + 7.7.6 + 10.9.7 + \ldots$$

Here $u_n = \{1 + (n-1)3\}\{3 + (n-1)2\}\{4 + (n-1)1\}$
$= (3n - 2)(2n + 1)(n + 3) = 6n^3 + 17n^2 - 5n - 6,$

which is a function of n of the third degree.

∴ let $S_n = Bn + Cn^2 + Dn^3 + En^4;$

∴ $S_{n-1} = B(n-1) + C(n-1)^2 + D(n-1)^3 + E(n-1)^4.$

∴ $S_n - S_{n-1} = u_n = 6n^3 + 17n^2 - 5n - 6$
$= B + C(2n-1) + D(3n^2 - 3n + 1) + E(4n^3 - 6n^2 + 4n - 1)$
$= (B - C + D - E) + (2C - 3D + 4E)n$
$\qquad + (3D - 6E)n^2 + 4En^3.$

∴ $4E = 6;\ 3D - 6E = 17;\ 2C - 3D + 4E = -5;$

and $B - C + D - E = -6;$

∴ $E = 1\tfrac{1}{2};\ D = 8\tfrac{2}{3};\ C = 7\tfrac{1}{2}$ and $B = -5\tfrac{2}{3};$

∴ $S_n = -5\tfrac{2}{3}n + 7\tfrac{1}{2}n^2 + 8\tfrac{2}{3}n^3 + 1\tfrac{1}{2}n^4;$

∴ $S_n = 1\tfrac{1}{2}n^4 + 8\tfrac{2}{3}n^3 + 7\tfrac{1}{2}n^2 - 5\tfrac{2}{3}n$
$= \tfrac{1}{6}n(9n^3 + 52n^2 + 45n - 34),$

which may be easily verified by putting $n = 1$ and $n = 2$.

---

* The sum of n terms of a series whose n^{th} term is a rational integral function of n evidently cannot contain the absolute term. If it did S_n would not vanish for $n = 0$.

SUMMATION OF SERIES.

464. To find the sum of a series each term of which is the product of the reciprocals of r factors in arithmetical progression, the first factors of the denominators of the successive terms being in the same arithmetical progression.

Let us denote the series by
$$u_1 + u_2 + u_3 + \ldots\ldots + u_n,$$
where $u_n = 1/\{(a+nd)(a+\overline{n+1}.d)\ldots\ldots(a+\overline{n+r-1}.d)\}$;
and let $v_n = (a+nd)u_n$;

$$\therefore v_n = \frac{1}{(a+\overline{n+1}.d)(a+\overline{n+2}.d)\ldots\ldots(a+\overline{n+r-1}.d)},$$

and $v_{n-1} = \dfrac{1}{(a+nd)(a+\overline{n+1}.d)\ldots\ldots(a+\overline{n+r-2}.d)}.$

By subtraction,
$$v_{n-1} - v_n = \frac{(a+\overline{n+r-1}.d) - (a+nd)}{(a+nd)(a+\overline{n+1}.d)\ldots\ldots(a+\overline{n+r-1}.d)}.$$

That is,
$$\begin{aligned}
v_{n-1} - v_n &= u_n \ . \ (r-1)d \\
v_{n-2} - v_{n-1} &= u_{n-1} . (r-1)d \\
&\ldots\ldots\ldots\ldots\ldots\ldots \\
v_1 - v_2 &= u_2 \ . \ (r-1)d \\
v_0 - v_1 &= u_1 \ . \ (r-1)d,
\end{aligned}$$

where v_0, the term preceding the first, is what the expression for the nth term becomes when in it we put $n = 0$.

Hence, by addition, we have
$$v_0 - v_n = (r-1)d . \{u_1 + u_2 + u_3 + \ldots\ldots + u_n\}.$$

$$\therefore S_n = \frac{v_0 - v_n}{(r-1)d} = \frac{v_0}{(r-1)d} - \frac{v_n}{(r-1)d}.$$

Hence we have the following rule for summing the series:

SUMMATION OF SERIES.   541

The required sum is the value of $A - B$, where B is the nth term of the given series with the first factor of its denominator cut off divided by the product of the number of factors in each term so diminished and the common difference of the A. P., *and A is the value of B when in it we put $n = 0$.*

COR. It may be inferred from this proposition that

$$\frac{1}{1.2.3\ldots r} + \frac{1}{2.3.4\ldots(r+1)} + \ldots + \frac{1}{n(n+1)(n+2)\ldots(n+r-1)}$$
$$= \frac{1}{r-1}\left\{\frac{1}{1.2.3\ldots(r-1)} - \frac{1}{(n+1)(n+2)\ldots(n+r-1)}\right\}.$$

Ex. Sum to n terms and to infinity the series

$$\frac{1}{1.4.7.10} + \frac{1}{4.7.10.13} + \frac{1}{7.10.13.16} + \ldots.$$

Here $u_n = \dfrac{1}{(3n-2)(3n+1)(3n+4)(3n+7)}$; $r = 4$; and $d = 3$.

Hence $S_n = \dfrac{1}{(4-1)3}\left\{-\dfrac{1}{(3n+1)(3n+4)(3n+7)} + \dfrac{1}{1.4.7}\right\}$

$= \dfrac{1}{9}\left\{\dfrac{1}{28} - \dfrac{1}{(3n+1)(3n+4)(3n+7)}\right\},$

from which we deduce that

$$S_\infty = \frac{1}{9} \cdot \frac{1}{28} = \frac{1}{252}.$$

465. Series in which the nth term is
$$\frac{\phi(n)}{n(n+1)(n+2)\ldots\ldots(n+r-1)}$$
where $\phi(n)$ is not of degree higher than $r-2$ in n.

These can always be reduced to other series to which the method of last article applies.

In fact transforming $\phi(n)$ by the method of § 461, we have

$\phi(n) = A_0 + A_1 n + A_2 n(n+1) + \ldots\ldots$
$\qquad\qquad + A_{r-2} n(n+1)(n+2)\ldots\ldots(n+r-3).$

[If the degree of $\phi(n)$ be $< r - 2$ some of the A's vanish.]

Hence the nth term is

$$\frac{A_0}{n(n+1)\ldots(n+r-1)} + \frac{A_1}{(n+1)(n+2)\ldots(n+r-1)}$$
$$+ \frac{A_2}{(n+2)(n+3)\ldots(n+r-1)} + \ldots\ldots + \frac{A_{r-2}}{(n+r-2)(n+r-1)};$$

and to each of these elements the method of the last article applies.

Ex. Sum to n terms and to infinity the series whose n^{th} term is

$$\frac{n^2+1}{n(n+1)(n+2)(n+3)}.$$

Here as in Ex. 1 § 460 $n^2+1 = n(n+1) - n + 1$. Hence the nth term is

$$\frac{1}{n(n+1)(n+2)(n+3)} - \frac{1}{(n+1)(n+2)(n+3)} + \frac{1}{(n+2)(n+3)},$$

and hence by the rule of § 464, the sum of n terms is

$$\frac{1}{3}\left\{\frac{1}{1.2.3} - \frac{1}{(n+1)(n+2)(n+3)}\right\}$$
$$- \frac{1}{2}\left\{\frac{1}{2.3} - \frac{1}{(n+2)(n+3)}\right\} + \left(\frac{1}{3} - \frac{1}{n+3}\right)$$
$$= \frac{11}{36} - \frac{1}{6(n+1)(n+2)(n+3)}\{2 - 3(n+1) + 6(n+1)(n+2)\}$$
$$= \frac{11}{36} - \frac{6n^2+15n+11}{6(n+1)(n+2)(n+3)}.$$

Since the portion involving n vanishes when n is infinite the sum to infinity is $\frac{11}{36}$.

466. Some series require modification in their denominators before the method of § 464 can be applied.

Ex. 1. Consider for example the series whose n^{th} term is

$$\frac{1}{n(n+2)}.$$

Here the difference between the factors of the denominator is 2, while the first terms of the denominators form an A.P. whose difference is unity; consequently we cannot apply § 464 at once.

SUMMATION OF SERIES.

We notice that the factor $n+1$ is missing from the denominator, so we *multiply numerator and denominator by the missing factor of the latter and then proceed as in* § 464.

The nth term is

$$\frac{n+1}{n(n+1)(n+2)} = \frac{1}{n(n+1)(n+2)} + \frac{1}{(n+1)(n+2)},$$

and hence
$$S_n = \frac{1}{2}\left\{\frac{1}{1.2} - \frac{1}{(n+1)(n+2)}\right\} + \frac{1}{2} - \frac{1}{n+2}$$

$$= \frac{3}{4} - \frac{1}{n+2} - \frac{1}{2(n+1)(n+2)}.$$

Ex. 2. Sum the series

$$\frac{3}{1.2.4} + \frac{5}{2.3.5} + \frac{7}{3.4.6} \cdots$$

to n terms and to infinity.

Here the nth term is

$$u_n = \frac{2n+1}{n(n+1)(n+3)},$$

and the factor $(n+2)$ is missing from the denominator. Thus

$$u_n = \frac{(2n+1)(n+2)}{n(n+1)(n+2)(n+3)} = \frac{2n^2+5n+2}{n(n+1)(n+2)(n+3)}$$

$$\begin{array}{r} n\ )\ 2n^2+5n+2 \\ n+1\ )\ \overline{2n+5} \quad +2 \text{ rem.} \\ \overline{2} \quad\quad +3 \text{ rem.} \end{array}$$

$$= \frac{2n(n+1) + 3n + 2}{n(n+1)(n+2)(n+3)}$$

$$= \frac{2}{(n+2)(n+3)} + \frac{3}{(n+1)(n+2)(n+3)} + \frac{2}{n(n+1)(n+2)(n+3)};$$

$$\therefore S_n = 2\left\{\frac{1}{3} - \frac{1}{n+3}\right\} + \frac{3}{2}\left\{\frac{1}{2.3} - \frac{1}{(n+2)(n+3)}\right\}$$

$$+ \frac{2}{3}\left\{\frac{1}{1.2.3} - \frac{1}{(n+1)(n+2)(n+3)}\right\}$$

$$= \frac{37}{36} - \frac{2}{n+3} - \frac{3}{2(n+2)(n+3)} - \frac{2}{3(n+1)(n+2)(n+3)},$$

and $S_\infty = \dfrac{37}{36}$.

Ex. 3. Sum to n terms

$$\frac{1}{1.5} + \frac{1}{3.7} + \frac{1}{5.9} + \ldots.$$

Here the nth term is

$$u_n = \frac{1}{(2n-1)(2n+3)},$$

and the first factors of the denominators form the A.P.

$$1, 3, 5 \ldots;$$

∴ we must introduce into the denominator the factor $(2n+1)$.

$$\therefore u_n = \frac{(2n+1)}{(2n-1)(2n+1)(2n+3)}.$$

Also $2n+1 = 2n-1+2$,

$$u_n = \frac{1}{(2n+1)(2n+3)} + \frac{2}{(2n-1)(2n+1)(2n+3)};$$

$$\therefore S_n = \frac{1}{2}\left\{\frac{1}{3} - \frac{1}{2n+3}\right\} + 2\cdot\frac{1}{2.2}\left\{\frac{1}{1.3} - \frac{1}{(2n+1)(2n+3)}\right\}$$

$$= \frac{1}{3} - \frac{1}{2(2n+3)} - \frac{1}{2(2n+1)(2n+3)}.$$

467. Polygonal Numbers. If in the expression $n + \tfrac{1}{2}n(n-1)d$, which is the sum of n terms of an A.P. of which the first term is unity, and the common difference is d, we make $d = 0, 1, 2, 3 \ldots$ in succession, we get the general term

$$n, \tfrac{1}{2}n(n+1), n^2, \tfrac{1}{2}n(3n-1)$$

of the second, the third, ... order of what are called *polygonal numbers*, the first order being that in which every term is unity.

1st order ; n^{th} term 1 ;   series 1, 1, 1, 1, 1, 1...,
2nd order ; n^{th} term n ;   series 1, 2, 3, 4, 5, 6...,
3rd order ; n^{th} term $\tfrac{1}{2}n(n+1)$; series 1, 3, 6, 10, 15......,
4th order ; n^{th} term n^2 ;   series 1, 4, 9, 16, 25......,
5th order ; n^{th} term $\tfrac{1}{2}n(3n-1)$; series 1, 5, 12, 22 ....

The numbers in the second, the third, the fourth, the fifth... series have received respectively the names *linear, triangular, square, pentagonal*...; and all these numbers the general name of *polygonal* numbers from the fact, that if we represent a unit by a dot, these numbers can be arranged in the form of the corresponding polygons in the following manner.

468. Linear Numbers. Suppose the distance between two dots to be a unit, then a line containing n units will on being divided

SUMMATION OF SERIES. 545

into units contain $(n+1)$ dots, and this is the $(n+1)^{\text{th}}$ linear number. Thus we represent the linear numbers

469. Triangular Numbers. Place one dot at O; then add two more dots to form an equilateral triangle; then to the same diagram add an equilateral triangle whose side is two units; then add one whose side is three units, etc. Divide each line into units and the diagram will contain successively 1, 3, 6, 10 etc. dots, and these are the triangular numbers.

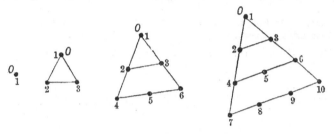

470. Square Numbers. Here we use a square instead of a triangle. As before each side is to be divided into units.

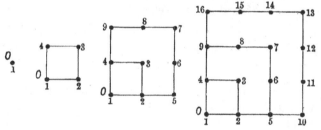

471. Pentagonal Numbers. We now use a regular pentagon.

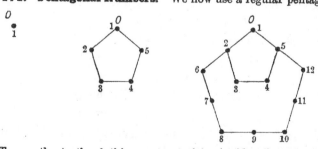

To see the truth of this representation consider the pentagonal numbers.

TUT. ALG. II.

SUMMATION OF SERIES.

When we add the nth pentagon we introduce $(n+1)$ new dots on each of three lines, but two dots are counted twice, so that the number of new dots is

$$3(n+1) - 2 = 3n + 1.$$

Hence the total number of dots when the figure contains n pentagons is

$$1 + (3.1 + 1) + (3.2 + 1) + \ldots + (3.n + 1)$$
$$= \tfrac{1}{2}(n+1)(3n+2),$$

and this is the $(n+1)^{\text{th}}$ pentagonal number.

472. To find the sum of the first n terms of the rth order of polygonal numbers.

Since the nth term of the rth order of polygonal numbers is

$$n + \tfrac{1}{2}n(n-1)(r-2);$$

$$\therefore S_n = \tfrac{1}{2}n(n+1) + \frac{r-2}{2} \cdot \frac{(n-1)n(n+1)}{3} \qquad \text{[By § 459]}$$

$$= \tfrac{1}{6}n(n+1)\{(r-2)(n-1) + 3\}.$$

473. Method of Differences.

Definition. If from each term of a series we subtract the one that immediately precedes it we obtain a new series called the *first order of differences* of the original one.

A series derived in a similar manner from the first order of differences is called the *second order of differences* and so on.

Ex. In the series

$$4, \ 10, \ 20, \ 35, \ 56, \ 84\ldots\ldots\ldots$$

the successive orders of differences are

$$6, \ 10, \ 15, \ 21, \ 28\ldots\ldots\ldots\text{first,}$$
$$4, \ 5, \ 6, \ 7, \quad\ldots\ldots\ldots\text{second,}$$
$$1, \ 1, \ 1 \qquad \ldots\ldots\ldots\text{third,}$$
$$0, \ 0, \qquad \ldots\ldots\ldots\text{fourth,}$$

wherein it will be noticed that the second order are in A.P.; the third all the same term and the fourth all zero.

SUMMATION OF SERIES.

474. If the nth term of a series be a rational integral function of n of degree r, then the $(r+1)$th order of differences vanishes.

Suppose the nth term of the series is $\phi(n)$, then the nth term of the first order of differences is

$$\phi(n+1) - \phi(n),$$

which is only of the $(r-1)$th degree in n.

Similarly in the second order of differences the degree of the nth term is one less than it is in the first, and so on.

Thus in the $(r-1)$th order of differences the degree of the nth term is $r - (r-1) = 1$, i.e. it is an A.P. In the rth all the terms are the same, and hence in the $(r+1)$th all the terms vanish.

475. To find the nth term of a series when its first order of differences is given.

Let
$$u_1, \; u_2, \; \ldots\ldots \; u_n, \; \ldots\ldots$$
be the series, and
$$v_1, \; v_2, \; \ldots\ldots \; v_n, \; \ldots\ldots$$
the first order of differences.

Then
$$v_{n-1} = u_n - u_{n-1}$$
$$v_{n-2} = u_{n-1} - u_{n-2}$$
$$\ldots\ldots\ldots\ldots\ldots\ldots$$
$$v_2 = u_3 - u_2$$
$$v_1 = u_2 - u_1$$

or on addition

$$u_n - u_1 = v_1 + v_2 + v_3 + \ldots\ldots + v_{n-1}$$

which gives u_n when u_1 and the first order of differences is known.

COR. The nth term of any order of differences may in the same way be expressed in terms of the next succeeding order.

Ex. In the series 1, 3, 6, 10, 15......... the orders of differences are
$$2, \; 3, \; 4, \; 5 \; \ldots\ldots$$
$$1, \; 1, \; 1, \; 1\ldots\ldots$$

Hence applying the above the nth term of the second order is

$$2 + (1+1+1+\ldots\text{for } \overline{n-1} \text{ terms})$$
$$= n+1 \text{ (this is evident since it is an A.P.)}.$$

Hence the nth term of the series is

$$1 + (2+3+4+\ldots\text{for } \overline{n-1} \text{ terms})$$
$$= \tfrac{1}{2} n(n+1).$$

476. If the $(r+1)$th order of differences vanishes, the nth term of the series is a rational integral function of n of the rth degree.

This is evidently true when $r=1$, for if the second order of differences vanishes the first consists of the same term, and hence the series is an A.P.

Next, if it is true that the pth differences vanish, it is when the $(p+1)$th vanish.

For let $\qquad u_1, \quad u_2, \ldots\ldots u_n, \ldots\ldots$
be the series, and $\quad v_1, \quad v_2, \ldots\ldots v_n, \ldots\ldots$
be the first order of differences.

Then for the v series the pth order of differences vanishes, and thereby by hypothesis v_n is of the $(p-1)$th degree in n.

But $\qquad u_n - u_1 = v_1 + v_2 + \ldots\ldots + v_{n-1}.$

$\therefore u_n$ is of the pth order in n.

Hence by induction the theorem is true universally.

Ex. To find the nth term of the series

$$4, \ 10, \ 20, \ 35, \ 56, \ 84\ldots\ldots$$

Here the orders of differences are

first 6, 10, 15, 21, 28………
second 4, 5, 6, 7………
third   1, 1, 1………
fourth 0, 0………

The nth term of the 2nd order of differences (an A.P.) is $n+3$.

SUMMATION OF SERIES.

Hence the nth term of the first is

$$6 + (4 + 5 + 6 + \ldots + \overline{n+2}) \qquad [\S\ 475]$$
$$= 6 + \tfrac{1}{2}(n-1)(n+6)$$
$$= \tfrac{1}{2}(n^2 + 5n + 6) = \tfrac{1}{2}(n+2)(n+3).$$

Proceeding in the same way the nth term of the series is

$$4 + \sum_{1}^{n-1} \tfrac{1}{2}(n+2)(n+3)$$
$$= 4 + \tfrac{1}{2} \cdot \tfrac{1}{3} \cdot \{(n+1)(n+2)(n+3) - 24\}$$
$$= \tfrac{1}{6}(n+1)(n+2)(n+3),$$

and this may be verified by putting $n = 1, 2, 3, 4$, etc.

EXERCISES XXX.

[In solving the following exercises, verify the answer in each case in which the nth term or the sum of n terms is required to be found, by putting $n=1$, and $n=2$.]

Sum to n terms:

1. $1^2 + 3^2 + 5^2 + \ldots$
2. $2^2 + 4^2 + 6^2 + \ldots$
3. Sum to n terms the series whose n^{th} term is $n^2 - 1$.

Find the sum of n terms of the series

4. $1 \cdot 2 + 2 \cdot 3 + 3 \cdot 4 + \ldots$
5. $1 \cdot 3 + 3 \cdot 5 + 5 \cdot 7 + \ldots$
6. $1 \cdot 2^2 + 3 \cdot 4^2 + 5 \cdot 6^2 + \ldots$
7. $1^2 \cdot 2 + 2^2 \cdot 3 + 3^2 \cdot 4 + \ldots$
8. $1 \cdot 2 \cdot 4 + 2 \cdot 3 \cdot 5 + 3 \cdot 4 \cdot 6 + \ldots$

9. Sum to n terms the series whose p^{th} term is $p^3 + 6p^2 + 8p$.

10. Sum to n terms the series whose r^{th} term is $r^2(r+2)$.

11. Find the 5th term of the series whose $2n^{\text{th}}$ term is $n^2 - n$; find also the sum to n terms.

12. If I buy n articles for 1s. each, $(n-1)$ articles at 2s. each, $(n-2)$ articles at 3s. each, and so on up to 1 article at n shillings each, show that I shall have to pay $\tfrac{1}{6}n(n+1)(n+2)$ shillings.

13. Show that $1 + 2^2 + 3 + 4^2 + 5 + 6^2 + \ldots$ to n terms is
$\tfrac{1}{12}(n+1)(2n^2 + n + 3)$ or $\tfrac{1}{12}n(n+4)(2n+1)$,
according as n is odd or even.

SUMMATION OF SERIES.

14. Find the sum to n terms and to infinity of
$$\frac{1}{1.3.4} + \frac{1}{2.4.5} + \frac{1}{3.5.6} + \ldots$$

15. Find the sum to n terms and to infinity of
$$\frac{1}{1.4} + \frac{1}{2.5} + \frac{1}{3.6} + \ldots$$

16. Find the sum to infinity of
$$\frac{3}{5.8.11} + \frac{9}{8.11.14} + \frac{15}{11.14.17} + \ldots$$

17. Find the sum to n terms of
 (i) the series whose n^{th} term is $n^4 - n^2 + 2$;
 (ii) $1(3^2 - 2^2) + 2(4^2 - 3^2) + 3(5^2 - 4^2) + \ldots$

Sum to n terms:

*18. $\dfrac{1.2}{3} + \dfrac{2.3}{3^2} + \dfrac{3.4}{3^3} + \ldots$

19. $\dfrac{1}{1.2} + \dfrac{1}{2.3} + \dfrac{1}{3.4} + \ldots$

20. $\dfrac{1}{1.3} + \dfrac{1}{3.5} + \dfrac{1}{5.7} + \ldots$

21. $\dfrac{1}{1.3.5} + \dfrac{1}{3.5.7} + \dfrac{1}{5.7.9} + \ldots$

22. Sum to infinity $\dfrac{1}{1.3} - \dfrac{1}{2.4} + \dfrac{1}{3.5} - \dfrac{1}{4.6} + \ldots$

23. Sum to infinity $\dfrac{1}{1.2.3.4} + \dfrac{1}{2.3.4.5} + \dfrac{1}{3.4.5.6} + \ldots$

24. Find the sum to n terms of
$$\frac{1}{(a+1)(a+2)} + \frac{1}{(a+2)(a+3)} + \frac{1}{(a+3)(a+4)} + \ldots$$

25. Find the sum to n terms of
$$\frac{1}{1} + \frac{1}{1+2} + \frac{1}{1+2+3} + \frac{1}{1+2+3+4} + \ldots$$

26. Find the nth term (i.e. group of terms) and the sum of n groups of
$$1 + (1+2) + (1+2+3) + (1+2+3+4) + \ldots$$
Point out the connection of this exercise with Ex. 12.

SUMMATION OF SERIES.

27. Show that the sum of n groups of terms of
$$1^2+(1^2+2^2)+(1^2+2^2+3^2)+(1^2+2^2+3^2+4^2)+\ldots\ldots$$
is
$$\tfrac{1}{12}n(n+1)^2(n+2).$$

28. Sum to n terms $8+16+36+92+256+\ldots\ldots$

29. Sum to n terms the series whose n^{th} term is $2^{n+1}+7$.

30. Write down the nth term of the series $1+3+6+10+15+\ldots$, and find the sum of n terms. [See Ex. 26.]

31. Find the sum of n terms of the series $6, 13, 22, 33, \ldots\ldots$

32. If the nth term of the series $7, 12, 19, 28, \ldots\ldots$ is 199, find n.

33. Find the sum of 15 terms of the series $1, 4, 8, 13, 19, \ldots\ldots$

34. In the series
$$1.1+3.5+5.13+7.25+\ldots\ldots$$
to n terms, prove that

(i) all the terms are odd numbers;

(ii) if unity be added to the nth term it is divisible by $2n$;

(iii) if unity be subtracted from the nth term it is divisible by $n-1$;

(iv) each term is the product of two numbers which may be taken to represent the shorter side and hypotenuse of a right-angled triangle, whose other side is represented by an integer;

(v) the sum of n terms is a square;

(vi) the square root of the sum is $1+3+5+\ldots\ldots$ to n terms.

CHAPTER XXXI.

DETERMINANTS.

477. Eliminant of three linear equations.

Taking the system of the homogeneous linear equations
$$a_1 x + b_1 y + c_1 z = 0,$$
$$a_2 x + b_2 y + c_2 z = 0,$$
$$a_3 x + b_3 y + c_3 z = 0,$$
let us eliminate x, y, z from them.

From the last two equations, by the rule of cross-multiplication, we have
$$\frac{x}{b_2 c_3 - b_3 c_2} = \frac{y}{c_2 a_3 - c_3 a_2} = \frac{z}{a_2 b_3 - a_3 b_2} = k, \text{ say.}$$

Substituting the values of x, y, z obtained from these equalities in the first equation, we have
$$k \{a_1 (b_2 c_3 - b_3 c_2) + b_1 (c_2 a_3 - c_3 a_2) + c_1 (a_2 b_3 - a_3 b_2)\} = 0.$$

The required eliminant is thus
$$a_1 (b_2 c_3 - b_3 c_2) + b_1 (c_2 a_3 - c_3 a_2) + c_1 (a_2 b_3 - a_3 b_2) = 0,$$
for if $k = 0$ we should have $x = y = z = 0$.

The expression on the left side of this result, and many other long expressions similarly formed, frequently occur in mathematics; and we shall now explain a notation by which such expressions can be expressed in a compact form.

478. Determinants.

If we take the nine quantities $a_1, a_2, a_3, b_1, b_2, b_3, c_1, c_2, c_3$ arranged in a square of three horizontal rows and three vertical columns as follows:

$$a_1 \quad b_1 \quad c_1$$
$$a_2 \quad b_2 \quad c_2$$
$$a_3 \quad b_3 \quad c_3$$

all the possible products of three quantities, such that one is taken from each column and one from each row, are

$$a_1b_2c_3, \quad a_1b_3c_2, \quad a_2b_3c_1, \quad a_2b_1c_3, \quad a_3b_1c_2, \quad a_3b_2c_1.$$

If now we consider these products as positive or negative according as there are an even or an odd number of inversions of the natural order of the figures which are suffixes, the sum of all such products will be

$$a_1b_2c_3 - a_1b_3c_2 + a_2b_3c_1 - a_2b_1c_3 + a_3b_1c_2 - a_3b_2c_1;$$

for (i) there is no inversion in the order of the suffixes in $a_1b_2c_3$ from the natural order of the figures 1, 2, 3. In this case the product is treated as positive.

(ii) there is *one* inversion in the order of the suffixes in $a_1b_3c_2$ from the arithmetical order 1, 2, 3, since 3 precedes 2—the order 123 having become 132 by a single inversion. Hence $a_1b_3c_2$ is negative.

(iii) there are *two* inversions in the order of the suffixes in $a_2b_3c_1$, since 2 and 3 both precede 1—the order 123 first becoming 213 and then becoming 231. Hence $a_2b_3c_1$ is positive.

(iv) there is *one* inversion in $a_2b_1c_3$, since 2 precedes 1—the order 123 becoming 213 by a single inversion. Hence $a_2b_1c_3$ is negative.

(v) there are *two* inversions in $a_3b_1c_2$, since 3 precedes both 1 and 2—the order 123 becoming first 132 and then becoming 312. Hence $a_3b_1c_2$ is positive.

(vi) there are *three* inversions in $a_3b_2c_1$, since 3 precedes both 2 and 1, and 2 also precedes 1—the order 123 becoming first 132, then 312, and finally 321. Hence $a_3b_2c_1$ is negative.

Now the algebraic sum

$$a_1b_2c_3 - a_1b_3c_2 + a_2b_3c_1 - a_2b_1c_3 + a_3b_1c_2 - a_3b_2c_1$$

formed as indicated above is called the **determinant** of the 9 quantities a_1, a_2, a_3, b_1, b_2, ... which are called its **constituents**.

We generally denote this determinant by writing the constituents in a square between parallel vertical lines as below, so that all the same letters stand in the same vertical column, and all the same suffixes stand in the same horizontal row.

$$\begin{vmatrix} a_1 & b_1 & c_1 \\ a_2 & b_2 & c_2 \\ a_3 & b_3 & c_3 \end{vmatrix}$$

Each of the products formed as above with the proper sign prefixed is called an **element** of the determinant.

The principal diagonal element $a_1b_2c_3$ in which the suffixes are in the natural order is called the *leading* or *principal element*, since we can obtain from it all the elements by suitable interchanges of suffixes.

The determinant whose principal element is $a_1b_2c_3$ is also expressed by the notation $\Sigma(\pm a_1b_2c_3)$, which indicates the algebraic sum of all the elements which can be got from it by proper interchanges of suffixes with the proper signs prefixed.

The determinant which we have hitherto considered is said to be of the *third* order, as it consists of three rows and three columns, and as every element of it is therefore the product of three factors.

A determinant of the *second* order is written in the following ways:

$$\begin{vmatrix} a_1 & b_1 \\ a_2 & b_2 \end{vmatrix}, \quad \Sigma(\pm a_1b_2), \quad a_1b_2 - a_2b_1.$$

DETERMINANTS. 555

This is the standard form of a determinant. Similarly if we take any nine quantities a, b, c, d, e, f, g, h, i arranged in a square of three horizontal rows and three vertical columns between parallel vertical lines as follows:

$$\begin{vmatrix} a & d & g \\ b & e & h \\ c & f & i \end{vmatrix};$$

the algebraic sum of all the possible products of three quantities taken from the columns in order, so that no two are in the same horizontal row, the products being considered positive or negative according as there are an even or an odd number of inversions in the order of the rows from which the quantities are taken, is the value of the determinant, which is therefore

$$aei - afh + bfg - bdi + cdh - ceg,$$

the order of the rows from which the quantities in these products are taken being 123, 132, 231, 213, 312, 321, and the number of inversions in them being 0, 1, 2, 1, 2, 3 respectively.

479. Determinants of the nth order.

If n^2 symbols, as in the following scheme, are arranged in a square of n horizontal rows and vertical columns between parallel vertical lines, so that all the same letters stand in the same vertical column, and all the same suffixes stand in the same horizontal row,

$$\begin{vmatrix} a_1 & b_1 & c_1 & d_1 & \ldots & k_1 \\ a_2 & b_2 & c_2 & d_2 & \ldots & k_2 \\ a_3 & b_3 & c_3 & d_3 & \ldots & k_3 \\ a_4 & b_4 & c_4 & d_4 & \ldots & k_4 \\ \cdots & \cdots & \cdots & \cdots & \cdots & \cdots \\ a_n & b_n & c_n & d_n & \ldots & k_n \end{vmatrix},$$

the determinant of the n^2 quantities is the algebraic sum of all the possible products of n quantities of which each is taken from a horizontal row and vertical column different from the rows and columns of the other $n-1$ quantities, the sign of each product being considered to be positive or negative according as there is an even or odd number of inversions in the order of the suffixes from the natural order of the figures, when in the products the quantities are written in their alphabetical order.

Each of the n^2 quantities is called a *constituent* of the determinant. Each term of the determinant is called an *element*, and the diagonal element $a_1 b_2 c_3 \ldots k_n$ is called the leading or *principal element*. The determinant itself is shortly written $\Sigma (\pm a_1 b_2 c_3 \ldots k_n)$ or even more shortly $(a_1 b_2 c_3 \ldots k_n)$.

NOTE. When the determinant is not of the standard form, the quantities for each product are taken from the successive columns, so that no two quantities are in the same row, and the signs of the products are taken to be positive or negative according as the number of inversions in the order of the rows from which they are taken is even or odd.

480. The number of terms in a determinant of the nth order of the standard form (and therefore of any determinant of the nth order) is n !, since it is obtained by writing the letters in their alphabetical order with the n suffixes permuted in all possible ways.

Thus the number of terms in a determinant of the third order is 3 ! or 6, and in a determinant of the fourth order is 4 ! or 24.

Ex. 1. Find the sign of the element $a_3 b_5 c_1 d_4 e_2$ in the determinant
$$\Sigma (\pm a_1 b_2 c_3 d_4 e_5).$$

In $a_3 b_5 c_1 d_4 e_2$, the number of inversions of the order of the suffixes from the natural order of the figures is

2 (since 3 precedes 1 and 2) + 3 (since 5 precedes 1, 4 and 2)
$\qquad\qquad\qquad\qquad$ + 1 (since 4 precedes 2) = 6.

Hence the sign of $a_3 b_5 c_1 d_4 e_2$ is positive.

DETERMINANTS. 557

Ex. 2. What is the sign of $d\,e\,l\,r$ in the determinant

$$\begin{vmatrix} a & b & c & d \\ e & f & g & h \\ k & l & m & n \\ p & q & r & s \end{vmatrix}?$$

The letters $d\,e\,l\,r$, when arranged in the order of the columns, will be $e\,l\,r\,d$, which are found in the rows of the figures in order of 2341 in which there are three inversions. The sign of $d\,e\,l\,r$ is therefore negative.

481. Expansion of a determinant. Minors.

It will now be shown how any determinant of any order, of which the constituents are numbers, is usually expanded. Let us take a determinant of the third order of the standard form

$$\begin{vmatrix} a_1 & b_1 & c_1 \\ a_2 & b_2 & c_2 \\ a_3 & b_3 & c_3 \end{vmatrix}.$$

The determinant

$$= a_1 b_2 c_3 - a_1 b_3 c_2 - a_2 b_1 c_3 + a_2 b_3 c_1 + a_3 b_1 c_2 - a_3 b_2 c_1$$
$$= a_1 (b_2 c_3 - b_3 c_2) - a_2 (b_1 c_3 - b_3 c_1) + a_3 (b_1 c_2 - b_2 c_1)$$
$$= a_1 \begin{vmatrix} b_2 & c_2 \\ b_3 & c_3 \end{vmatrix} - a_2 \begin{vmatrix} b_1 & c_1 \\ b_3 & c_3 \end{vmatrix} + a_3 \begin{vmatrix} b_1 & c_1 \\ b_2 & c_2 \end{vmatrix} \ldots\ldots\ldots(a).$$

Hence we find that the coefficient of any one of the constituents a_1, a_2, a_3 is a determinant of the lower order formed by omitting the row and column in which the constituent occurs. Such determinants are called the *minors* of the original determinant. If A_1, A_2, A_3 are the minors of a_1, a_2 and a_3 respectively, the relation (a) may be written

$$D = a_1 A_1 - a_2 A_2 + a_3 A_3.$$

It can be similarly shown that a determinant of the fourth order of the standard form is equal to

$$a_1\begin{vmatrix}b_2 & c_2 & d_2\\ b_3 & c_3 & d_3\\ b_4 & c_4 & d_4\end{vmatrix} - a_2\begin{vmatrix}b_1 & c_1 & d_1\\ b_3 & c_3 & d_3\\ b_4 & c_4 & d_4\end{vmatrix} + a_3\begin{vmatrix}b_1 & c_1 & d_1\\ b_2 & c_2 & d_2\\ b_4 & c_4 & d_4\end{vmatrix} - a_4\begin{vmatrix}b_1 & c_1 & d_1\\ b_2 & c_2 & d_2\\ b_3 & c_3 & d_3\end{vmatrix},$$

the signs + and − having to be prefixed alternately to a_1, a_2, a_3, a_4.

Hence $D = a_1A_1 - a_2A_2 + a_3A_3 - a_4A_4$,

where A_1, A_2, A_3 and A_4 are the minors of a_1, a_2, a_3, a_4 respectively in the determinant.

We may now enunciate the following rule which may be proved to be true for all determinants:

"*A determinant of any order is the sum of the products of each constituent of its first column and the minor determinant obtained by omitting the row and column in which it occurs, positive or negative sign being placed before each product according as the constituent in the first column in it comes from an odd or even horizontal row.*"

Ex. 1. $\begin{vmatrix}4 & 5 & 6\\ 3 & 7 & 4\\ 6 & 3 & 5\end{vmatrix} = 4\begin{vmatrix}7 & 4\\ 3 & 5\end{vmatrix} - 3\begin{vmatrix}5 & 6\\ 3 & 5\end{vmatrix} + 6\begin{vmatrix}5 & 6\\ 7 & 4\end{vmatrix}$

$= 4(35-12) - 3(25-18) + 6(20-42) = -61.$

Ex. 2. Find the value of

$$\begin{vmatrix}0 & 1 & 3 & 5\\ 0 & 2 & 6 & 4\\ 1 & 1 & 0 & 3\\ 0 & 5 & 3 & 3\end{vmatrix}.$$

If we denote the determinant by D,

$D = 0\begin{vmatrix}2 & 6 & 4\\ 1 & 0 & 3\\ 5 & 3 & 3\end{vmatrix} - 0\begin{vmatrix}1 & 3 & 5\\ 1 & 0 & 3\\ 5 & 3 & 3\end{vmatrix} + 1\begin{vmatrix}1 & 3 & 5\\ 2 & 6 & 4\\ 5 & 3 & 3\end{vmatrix} - 0\begin{vmatrix}1 & 3 & 5\\ 2 & 6 & 4\\ 1 & 0 & 3\end{vmatrix}$

$= \begin{vmatrix}1 & 3 & 5\\ 2 & 6 & 4\\ 5 & 3 & 3\end{vmatrix} = 1\begin{vmatrix}6 & 4\\ 3 & 3\end{vmatrix} - 2\begin{vmatrix}3 & 5\\ 3 & 3\end{vmatrix} + 5\begin{vmatrix}3 & 5\\ 6 & 4\end{vmatrix}$

$= 18 - 12 - 18 + 30 + 60 - 150 = -72.$

DETERMINANTS. 559

482. The rule of Sarrus.

To expand a determinant of the third order, we may repeat the first two columns, or imagine them to be repeated, after the third column, make products of every diagonal of three as shown below, and count the three diagonals sloping downwards positive and the other three negative.

Thus

$$\begin{vmatrix} a_1 & b_1 & c_1 \\ a_2 & b_2 & c_2 \\ a_3 & b_3 & c_3 \end{vmatrix} = \begin{matrix} a_1 & b_1 & c_1 & a_1 & b_1 \\ a_2 & b_2 & c_2 & a_2 & b_2 \\ a_3 & b_3 & c_3 & a_3 & b_3 \end{matrix}$$

$$= a_1 b_2 c_3 + b_1 c_2 a_3 + c_1 a_2 b_3 - a_3 b_2 c_1 - b_3 c_2 a_1 - c_3 a_2 b_1, \text{ as before.}$$

Ex. $\begin{vmatrix} 3 & 1 & 2 \\ 1 & 2 & 3 \\ 2 & 3 & 1 \end{vmatrix}$

$= 3.2.1 + 1.3.2 + 2.1.3 - 2.2.2 - 3.3.3 - 1.1.1$
$= -18.$

The following propositions which we prove for determinants of the third order are true for all determinants.

483. The value of a determinant (of the third order) is not altered by changing the rows into columns and the columns into rows.

Let

$$\begin{vmatrix} a_1 & b_1 & c_1 \\ a_2 & b_2 & c_2 \\ a_3 & b_3 & c_3 \end{vmatrix} \text{ and } \begin{vmatrix} a_1 & a_2 & a_3 \\ b_1 & b_2 & b_3 \\ c_1 & c_2 & c_3 \end{vmatrix}$$

be two determinants of which the rows of the second are columns of the first and the columns of the second are rows of the first.

DETERMINANTS.

We shall prove that they are equal in value by deducing the second from the first.

If we call the first determinant D, we have

$$D = a_1(b_2 c_3 - b_3 c_2) - a_2(b_1 c_3 - b_3 c_1) + a_3(b_1 c_2 - b_2 c_1)$$

$$= a_1(b_2 c_3 - c_2 b_3) - b_1(a_2 c_3 - c_2 a_3) + c_1(a_2 b_3 - b_2 a_3)$$

$$= a_1 \begin{vmatrix} b_2 & b_3 \\ c_2 & c_3 \end{vmatrix} - b_1 \begin{vmatrix} a_2 & a_3 \\ c_2 & c_3 \end{vmatrix} + c_1 \begin{vmatrix} a_2 & a_3 \\ b_2 & b_3 \end{vmatrix}$$

$$= \begin{vmatrix} a_1 & a_2 & a_3 \\ b_1 & b_2 & b_3 \\ c_1 & c_2 & c_3 \end{vmatrix}.$$

NOTE. **The value of a determinant of the standard form may therefore also be found by permuting the letters in all possible ways and keeping the suffixes in their natural order, the signs of the terms being taken to be positive or negative according as the number of inversions from the alphabetical order is even or odd.**

Alternative proof.

Without expanding the determinants this may be seen as follows:

Since the letters and suffixes are the same for both determinants any element of the one is an element of the other.

The sign of an element in the first depends on the number of transpositions required *among the suffixes*, so that a may have the suffix 1, b the suffix 2, c the suffix 3.

The sign of the same element in the second (since the leading element is the same) depends on the number of transpositions *among the letters*, so that a may have the suffix 1, b the suffix 2, and c the suffix 3.

These two numbers are the same, for any transposition between two suffixes is equivalent to one between two letters, viz. $a_p b_q c_r$ becomes $a_p b_r c_q$, either on interchanging the suffixes q and r or the letters b and c.

Hence an element has the same sign in both determinants, consequently the determinants are identical.

DETERMINANTS. 561

484. If two columns or rows of a determinant (of the third order) are interchanged, it is unaltered in absolute value, but is changed in sign.

Let the first and the third column be interchanged of the determinant

$$\begin{vmatrix} a_1 & b_1 & c_1 \\ a_2 & b_2 & c_2 \\ a_3 & b_3 & c_3 \end{vmatrix} \text{ so that it becomes } \begin{vmatrix} c_1 & b_1 & a_1 \\ c_2 & b_2 & a_2 \\ c_3 & b_3 & a_3 \end{vmatrix}.$$

Then the second determinant

$$= c_1(b_2 a_3 - b_3 a_2) - c_2(b_1 a_3 - b_3 a_1) + c_3(b_1 a_2 - b_2 a_1)$$
$$= -\{a_1(b_2 c_3 - b_3 c_2) - a_2(b_1 c_3 - b_3 c_1) + a_3(b_1 c_2 - b_2 c_1)\}$$
$$= - \begin{vmatrix} a_1 & b_1 & c_1 \\ a_2 & b_2 & c_2 \\ a_3 & b_3 & c_3 \end{vmatrix}.$$

The proposition may be similarly shown to be true for interchanges of two other columns or of any two rows.

Alternative proof.

This may be proved without expanding the determinants as follows:

Let $a_p b_q c_r$ (where p, q, r denote 1, 2, 3 written in any order) be an element of the first determinant, then it is also an element of the second, since the letters and suffixes are the same for both.

The sign of this element in the first determinant is $(-1)^n$, where n transpositions are required to bring

p, q, r to the order 1, 2, 3.

Its sign in the second is $(-1)^m$, where m transpositions are required to bring

r, q, p to the order 1, 2, 3.

Now one transposition will bring r, q, p to the order 1, 2, 3.
∴ $m = n + 1$, and the sign of the element in the second determinant is opposite to its sign in the first. The same is true for all elements, and hence the sign of the determinant is changed by interchanging two columns.

This method of proof applies to all determinants.

TUT. ALG. II.

485. If two rows or two columns of a determinant (of the third order) are identical, the determinant vanishes.

For if D be the determinant, on interchanging the two identical rows or columns, we obtain $-D$. [By § 484.]

But the interchange of two identical rows or columns cannot alter the value of the determinant.

$$\therefore D = -D; \text{ that is, } D = 0.$$

Ex. Prove that

$$\begin{vmatrix} 1 & a & a^2 \\ 1 & b & b^2 \\ 1 & c & c^2 \end{vmatrix} = (a-b)(b-c)(c-a).$$

If b were equal to a, two of the rows would become identical, and the determinant would therefore vanish. Hence $a - b$ is a factor of the determinant. Similarly $b - c$ and $c - a$ are also factors of it. The determinant can have no more factors in a, b, c, since it is of the third degree in these letters. It remains for us, therefore, only to find the constant multiplier, which is easily seen to be 1 by inspection.

486. If all the constituents of a row (or column) of a determinant be multiplied by the same quantity, the determinant is multiplied by that quantity.

In every element of the determinant, we have one constituent and only one from each row (or column). Hence if every constituent of any particular row (or column) be multiplied by any quantity, every element and therefore the determinant, which is the sum of all the elements, is multiplied by that quantity.

Thus
$$\begin{vmatrix} ma_1 & b_1 & c_1 \\ ma_2 & b_2 & c_2 \\ ma_3 & b_3 & c_3 \end{vmatrix} = \begin{vmatrix} ma_1 & mb_1 & mc_1 \\ a_2 & b_2 & c_2 \\ a_3 & b_3 & c_3 \end{vmatrix} = m \begin{vmatrix} a_1 & b_1 & c_1 \\ a_2 & b_2 & c_2 \\ a_3 & b_3 & c_3 \end{vmatrix}.$$

Ex. 1.
$$\begin{vmatrix} 8 & \tfrac{1}{2} & 16 \\ 6 & \tfrac{1}{4} & 19 \\ 16 & \tfrac{1}{8} & 27 \end{vmatrix} = 2 \begin{vmatrix} 4 & \tfrac{1}{2} & 16 \\ 3 & \tfrac{1}{4} & 19 \\ 8 & \tfrac{1}{8} & 27 \end{vmatrix} = \tfrac{2}{8} \begin{vmatrix} 4 & 4 & 16 \\ 3 & 2 & 19 \\ 8 & 1 & 27 \end{vmatrix} = \begin{vmatrix} 1 & 1 & 4 \\ 3 & 2 & 19 \\ 8 & 1 & 27 \end{vmatrix}.$$

DETERMINANTS.

Ex. 2. Express the following determinant as one of which the first column shall consist of units, and find its value.

$$\begin{vmatrix} 2 & 1 & 2 & 0 \\ 3 & 2 & 4 & 15 \\ 6 & 3 & 5 & 0 \\ 10 & 4 & 6 & 0 \end{vmatrix}.$$

Since 30 is the least common multiple of 2, 3, 6, 10, the first row can be made to consist of no number other than 30, if we multiply the rows in order by 15, 10, 5 and 3. Hence

$$D = \tfrac{1}{15} \cdot \tfrac{1}{10} \cdot \tfrac{1}{5} \cdot \tfrac{1}{3} \begin{vmatrix} 30 & 15 & 30 & 0 \\ 30 & 20 & 40 & 150 \\ 30 & 15 & 25 & 0 \\ 30 & 12 & 18 & 0 \end{vmatrix}$$

$$= \tfrac{1}{15} \cdot \tfrac{1}{10} \cdot \tfrac{1}{5} \cdot \tfrac{1}{3} \cdot 30 \cdot 75 \begin{vmatrix} 1 & 15 & 30 & 0 \\ 1 & 20 & 40 & 2 \\ 1 & 15 & 25 & 0 \\ 1 & 12 & 18 & 0 \end{vmatrix}$$

$$= \begin{vmatrix} 1 & 15 & 30 & 0 \\ 1 & 20 & 40 & 2 \\ 1 & 15 & 25 & 0 \\ 1 & 12 & 18 & 0 \end{vmatrix}$$ of which the first column consists of units,

$$= - \begin{vmatrix} 0 & 15 & 30 & 1 \\ 2 & 20 & 40 & 1 \\ 0 & 15 & 25 & 1 \\ 0 & 12 & 18 & 1 \end{vmatrix}$$

$$= +2 \begin{vmatrix} 15 & 30 & 1 \\ 15 & 25 & 1 \\ 12 & 18 & 1 \end{vmatrix}$$

$$2\{15(25-18)-15(30-18)+12(30-25)\}$$
$$= 30.$$

COR. *If each constituent of a row (or column) bears the same ratio to the corresponding constituent of another row (or column), then the determinant vanishes.*

For

$$\begin{vmatrix} ma_1 & na_1 & c_1 \\ ma_2 & na_2 & c_2 \\ ma_3 & na_3 & c_3 \end{vmatrix} = mn \begin{vmatrix} a_1 & a_1 & c_1 \\ a_2 & a_2 & c_2 \\ a_3 & a_3 & c_3 \end{vmatrix} = 0. \quad [\text{By § 485.}]$$

487. If each constituent of a row or column is resolvable into the sum of two parts, the determinant is resolvable into two determinants of the same order.

In the determinant

$$\begin{vmatrix} a_1 & b_1 & c_1 \\ a_2 & b_2 & c_2 \\ a_3 & b_3 & c_3 \end{vmatrix},$$

let* $\quad a_1 = p_1 + q_1; \quad a_2 = p_2 + q_2; \quad \text{and} \quad a_3 = p_3 + q_3.$

Then the determinant

$$= (p_1 + q_1) A_1 - (p_2 + q_2) A_2 + (p_3 + q_3) A_3,$$

where A_1, A_2 and A_3 are the minors of a_1, a_2, a_3 respectively

$$= (p_1 A_1 - p_2 A_2 + p_3 A_3)$$
$$+ (q_1 A_1 - q_2 A_2 + q_3 A_3)$$

$$= \begin{vmatrix} p_1 & b_1 & c_1 \\ p_2 & b_2 & c_2 \\ p_3 & b_3 & c_3 \end{vmatrix} + \begin{vmatrix} q_1 & b_1 & c_1 \\ q_2 & b_2 & c_2 \\ q_3 & b_3 & c_3 \end{vmatrix},$$

which establishes the proposition.

We leave it as an exercise to the student to prove this without actually expanding the determinant.

---

* A determinant is a *homogeneous linear* function of the constituents in a row or those in a column. The proposition of this article is an immediate consequence.

DETERMINANTS. 565

488. If the constituents of any row (or column) of a determinant be increased or diminished by any equimultiples of the corresponding constituents of one or more of the other rows (or columns), the determinant is not altered in value.

For

$$\begin{vmatrix} a_1 + mb_1 - nc_1 & b_1 & c_1 \\ a_2 + mb_2 - nc_2 & b_2 & c_2 \\ a_3 + mb_3 - nc_3 & b_3 & c_3 \end{vmatrix}$$

$$= \begin{vmatrix} a_1 & b_1 & c_1 \\ a_2 & b_2 & c_2 \\ a_3 & b_3 & c_3 \end{vmatrix} + \begin{vmatrix} mb_1 & b_1 & c_1 \\ mb_2 & b_2 & c_2 \\ mb_3 & b_3 & c_3 \end{vmatrix} + \begin{vmatrix} -nc_1 & b_1 & c_1 \\ -nc_2 & b_2 & c_2 \\ -nc_3 & b_3 & c_3 \end{vmatrix}$$

$$= \begin{vmatrix} a_1 & b_1 & c_1 \\ a_2 & b_2 & c_2 \\ a_3 & b_3 & c_3 \end{vmatrix} + m \begin{vmatrix} b_1 & b_1 & c_1 \\ b_2 & b_2 & c_2 \\ b_3 & b_3 & c_3 \end{vmatrix} - n \begin{vmatrix} c_1 & b_1 & c_1 \\ c_2 & b_2 & c_2 \\ c_3 & b_3 & c_3 \end{vmatrix}$$

[§ 486]

$$= \begin{vmatrix} a_1 & b_1 & c_1 \\ a_2 & b_2 & c_2 \\ a_3 & b_3 & c_3 \end{vmatrix}, \text{ since the last two vanish by § 485.}$$

Ex. 1. Find the value of

$$\begin{vmatrix} a-b & b-c & c-a \\ b-c & c-a & a-b \\ c-a & a-b & b-c \end{vmatrix}.$$

$$D = \begin{vmatrix} a-b+b-c+c-a & b-c & c-a \\ b-c+c-a+a-b & c-a & a-b \\ c-a+a-b+b-c & a-b & b-c \end{vmatrix} = \begin{vmatrix} 0 & b-c & c-a \\ 0 & c-a & a-b \\ 0 & a-b & b-c \end{vmatrix}$$

$$= 0.$$

Ex. 2. Find the value of
$$\begin{vmatrix} a-2 & 2a-7 & 13a-19 \\ b-2 & 2b-7 & 13b-19 \\ c-2 & 2c-7 & 13c-19 \end{vmatrix}.$$

$$D = \begin{vmatrix} a-2 & 2a-7-2(a-2) & 13a-19-13(a-2) \\ b-2 & 2b-7-2(b-2) & 13b-19-13(b-2) \\ c-2 & 2c-7-2(c-2) & 13c-19-13(c-2) \end{vmatrix}$$

$$= \begin{vmatrix} a-2 & -3 & 7 \\ b-2 & -3 & 7 \\ c-2 & -3 & 7 \end{vmatrix} = -21 \begin{vmatrix} a-2 & 1 & 1 \\ b-2 & 1 & 1 \\ c-2 & 1 & 1 \end{vmatrix} = 0.$$

Ex. 3.

$$\begin{vmatrix} 47 & 5 & 2 \\ 54 & 6 & 3 \\ 28 & 3 & 3 \end{vmatrix} = \begin{vmatrix} 2+9.5 & 5 & 2 \\ 0+9.6 & 6 & 3 \\ 1+9.3 & 3 & 3 \end{vmatrix} = \begin{vmatrix} 2 & 5 & 2 \\ 0 & 6 & 3 \\ 1 & 3 & 3 \end{vmatrix} + 9 \begin{vmatrix} 5 & 5 & 2 \\ 6 & 6 & 3 \\ 3 & 3 & 3 \end{vmatrix}$$

$$= \begin{vmatrix} 2 & 5 & 2 \\ 0 & 6 & 3 \\ 1 & 3 & 3 \end{vmatrix} = 2 \begin{vmatrix} 6 & 3 \\ 3 & 3 \end{vmatrix} - 0 \begin{vmatrix} 5 & 2 \\ 3 & 3 \end{vmatrix} + \begin{vmatrix} 5 & 2 \\ 6 & 3 \end{vmatrix}$$

$$= 2(18-9) + (15-12) = 21.$$

Ex. 4.
$$\begin{vmatrix} 25 & 5 & 23 \\ 33 & 6 & 28 \\ 41 & 7 & 33 \end{vmatrix}$$

$$= 25(6.33 - 7.28) - 33(5.33 - 7.23) + 41(5.28 - 6.23)$$
$$= 50 - 132 + 82 = 0.$$

We shall find the value by effecting a number of transformations to illustrate the principles of the preceding articles.

$$D = \begin{vmatrix} 5.5+0 & 5 & 4.5+3 \\ 6.5+3 & 6 & 4.6+4 \\ 7.5+6 & 7 & 4.7+5 \end{vmatrix} = \begin{vmatrix} 0 & 5 & 3 \\ 3 & 6 & 4 \\ 6 & 7 & 5 \end{vmatrix} = \begin{vmatrix} 0 & 2 & 3 \\ 3 & 2 & 4 \\ 6 & 2 & 5 \end{vmatrix}$$

$$= 2 \begin{vmatrix} 0 & 1 & 3 \\ 3 & 1 & 4 \\ 6 & 1 & 5 \end{vmatrix} = 2 \begin{vmatrix} 0 & 1 & 0 \\ 3 & 1 & 1 \\ 6 & 1 & 2 \end{vmatrix} = 0.$$

489. Simultaneous equations of the first degree.

We shall now show how the properties of determinants may be used to solve simultaneous equations of the first degree in three unknowns.

Ex. 1. Solve
$$a_1 x + b_1 y + c_1 z = d_1;$$
$$a_2 x + b_2 y + c_2 z = d_2;$$
$$a_3 x + b_3 y + c_3 z = d_3.$$

Multiply the equations in order by A_1, $-A_2$ and A_3, which are the minors of a_1, a_2, a_3 in the determinant $\Sigma(\pm a_1 b_2 c_3)$; we have, by addition,

$$(a_1 A_1 - a_2 A_2 + a_3 A_3) x + (b_1 A_1 - b_2 A_2 + b_3 A_3) y + (c_1 A_1 - c_2 A_2 + c_3 A_3) z$$
$$= d_1 A_1 - d_2 A_2 + d_3 A_3.$$

Here the coefficient of $y = b_1 A_1 - b_2 A_2 + b_3 A_3$

$$= b_1 \begin{vmatrix} b_2 & c_2 \\ b_3 & c_3 \end{vmatrix} - b_2 \begin{vmatrix} b_1 & c_1 \\ b_3 & c_3 \end{vmatrix} + b_3 \begin{vmatrix} b_1 & c_1 \\ b_2 & c_2 \end{vmatrix} = \begin{vmatrix} b_1 & b_1 & c_1 \\ b_2 & b_2 & c_2 \\ b_3 & b_3 & c_3 \end{vmatrix} = 0.$$

Similarly the coefficient of $z = c_1 A_1 - c_2 A_2 + c_3 A_3 = 0$.

$$\therefore (a_1 A_1 - a_2 A_2 + a_3 A_3) x = d_1 A_1 - d_2 A_2 + d_3 A_3;$$

$$\therefore x = \frac{d_1 A_1 - d_2 A_2 + d_3 A_3}{a_1 A_1 - a_2 A_2 + a_3 A_3} = \frac{\Sigma(\pm d_1 b_2 c_3)}{\Sigma(\pm a_1 b_2 c_3)}.$$

Similarly $y = \dfrac{\Sigma(\pm a_1 d_2 c_3)}{\Sigma(\pm a_1 b_2 c_3)}$; and $z = \dfrac{\Sigma(\pm a_1 b_2 d_3)}{\Sigma(\pm a_1 b_2 c_3)}$.

Ex. 2. Solve
$$2x + 3y + 4z = 20,$$
$$3x + 5y + 7z = 34,$$
$$x + 2y + 4z = 17.$$

Here $x = \dfrac{\begin{vmatrix} 20 & 3 & 4 \\ 34 & 5 & 7 \\ 17 & 2 & 4 \end{vmatrix}}{\begin{vmatrix} 2 & 3 & 4 \\ 3 & 5 & 7 \\ 1 & 2 & 4 \end{vmatrix}}$; $y = \dfrac{\begin{vmatrix} 2 & 20 & 4 \\ 3 & 34 & 7 \\ 1 & 17 & 4 \end{vmatrix}}{\begin{vmatrix} 2 & 3 & 4 \\ 3 & 5 & 7 \\ 1 & 2 & 4 \end{vmatrix}}$; $z = \dfrac{\begin{vmatrix} 2 & 3 & 20 \\ 3 & 5 & 34 \\ 1 & 2 & 17 \end{vmatrix}}{\begin{vmatrix} 2 & 3 & 4 \\ 3 & 5 & 7 \\ 1 & 2 & 4 \end{vmatrix}}$;

whence $x = 1$; $y = 2$; and $z = 3$.

490. Elimination.

Suppose we are required to eliminate x, y, z from the equations

$$a_1x + b_1y + c_1z = 0;$$
$$a_2x + b_2y + c_2z = 0;$$
$$a_3x + b_3y + c_3z = 0.$$

Multiplying the equations in order by $A_1, -A_2$ and A_3 which are the minors of a_1, a_2, a_3 in the determinant $\Sigma(\pm a_1b_2c_3)$ and adding, we have

$$(a_1A_1 - a_2A_2 + a_3A_3)x + (b_1A_1 - b_2A_2 + b_3A_3)y$$
$$+ (c_1A_1 - c_2A_2 + c_3A_3)z = 0.$$

Here the coefficients of y and z being zero, and the coefficient of x being the determinant $\Sigma(\pm a_1b_2c_3)$ itself, we have for the eliminant of the equations

$$\begin{vmatrix} a_1 & b_1 & c_1 \\ a_2 & b_2 & c_2 \\ a_3 & b_3 & c_3 \end{vmatrix} = 0.$$

Ex. Eliminate x, y, z from

$$bx = ay - z; \quad cy = bz - x; \quad \text{and } az = cx - y.$$

The equations are
$$bx - ay + z = 0;$$
$$x + cy - bz = 0;$$
and
$$cx - y - az = 0.$$

Hence the eliminant of the equation is

$$\begin{vmatrix} b & -a & 1 \\ 1 & c & -b \\ c & -1 & -a \end{vmatrix} = 0.$$

That is, $\quad b(-ac-b) - 1(a^2+1) + c(ab-c) = 0;$

$\therefore$ the eliminant is $\quad a^2 + b^2 + c^2 + 1 = 0.$

NOTE. We leave it as an exercise to the student to eliminate x from the equations
$$ax^2 + bx + c = 0, \text{ and } a'x^2 + b'x + c' = 0,$$
by treating these as simultaneous equations in x^2 and x.

EXERCISES XXXI.

1. Find the number of inversions from the natural order of the figures in 32541, 4325671, 532461.

2. What is the sign of $a_3 b_2 c_1 d_4 e_5$ and of $a_5 b_4 c_3 d_2 e_1$ in the determinant $\Sigma (\pm a_1 b_2 c_3 d_4 e_5)$?

3. Find the signs of the terms $b\,i\,d$ and $h\,f\,a$ in the determinant
$$\begin{vmatrix} a & b & c \\ d & e & f \\ g & h & i \end{vmatrix}.$$

4. Find the signs of the terms $c\,b\,r\,q$ and $k\,e\,d\,s$ in
$$\begin{vmatrix} a & b & k & l \\ c & d & m & n \\ e & f & p & q \\ g & h & r & s \end{vmatrix}.$$

Find the values of the following determinants:

5. $\begin{vmatrix} 2 & 3 & 2 \\ 3 & 3 & 3 \\ 3 & 2 & 3 \end{vmatrix}.$

6. $\begin{vmatrix} 2 & 3 & 3 \\ 3 & 3 & 2 \\ 3 & 2 & 3 \end{vmatrix}.$

7. $\begin{vmatrix} a & a & a \\ a & b & b \\ a & b & c \end{vmatrix}.$

8. $\begin{vmatrix} 1 & p & p^2 \\ 1 & q & q^2 \\ 1 & r & r^2 \end{vmatrix}.$

9. $\begin{vmatrix} 0 & c & b \\ c & 0 & a \\ b & a & 0 \end{vmatrix}.$

10. $\begin{vmatrix} 2 & 1 & 3 & 6 \\ 1 & 0 & 4 & 1 \\ -3 & 1 & 1 & 7 \\ 3 & 0 & 5 & 2 \end{vmatrix}.$

11. $\begin{vmatrix} a^3 & b^3 & c^3 \\ a & b & c \\ 1 & 1 & 1 \end{vmatrix}.$

12. $\begin{vmatrix} 2a+5b & 3a+7b & 4a+5b \\ a+6b & 2a+8b & 3a+6b \\ 7b & a+9b & 2a+7b \end{vmatrix}.$

13. $\begin{vmatrix} 1 & 1 & 1 \\ a & b & c \\ a^2 & b^2 & c^2 \end{vmatrix}.$

14. $\begin{vmatrix} b+c & a & a \\ b & c+a & b \\ c & c & a+b \end{vmatrix}.$

15. $\begin{vmatrix} 5 & 7 & 9 \\ 25 & 49 & 81 \\ 63 & 45 & 35 \end{vmatrix}.$

16. $\begin{vmatrix} a-x & a-y & a-z \\ b-x & b-y & b-z \\ c-x & c-y & c-z \end{vmatrix}.$

17. If ω is one of the imaginary cube roots of unity, find the value of
$$\begin{vmatrix} 1 & \omega & \omega^2 \\ \omega & \omega^2 & 1 \\ \omega^2 & 1 & \omega \end{vmatrix}.$$

Prove that

18. $\begin{vmatrix} a_1 & b_1 \\ a_2 & b_2 \end{vmatrix} \times \begin{vmatrix} c_1 & d_1 \\ c_2 & d_2 \end{vmatrix} = \begin{vmatrix} a_1c_1+b_1d_1 & a_2c_1+b_2d_1 \\ a_1c_2+b_1d_2 & a_2c_2+b_2d_2 \end{vmatrix}.$

19. $\begin{vmatrix} (b+c)^2 & a^2 & a^2 \\ b^2 & (c+a)^2 & b^2 \\ c^2 & c^2 & (a+b)^2 \end{vmatrix} = 2abc\,(a+b+c)^3.$

20. $\begin{vmatrix} 0 & ab^2 & ac^2 \\ ba^2 & 0 & bc^2 \\ ca^2 & cb^2 & 0 \end{vmatrix} = 2a^3b^3c^3.$

21. Show that $\begin{vmatrix} 11 & 12 & 8 & 1 \\ 10 & 17 & 21 & 3 \\ 15 & 38 & 19 & 2 \\ 6 & 11 & 8 & 1 \end{vmatrix} = 15.$

22. Prove that $\begin{vmatrix} 3 & 5 & 7 & 6 \\ 3 & 7 & 6 & 5 \\ 3 & 5 & 6 & 7 \\ 3 & 7 & 5 & 6 \end{vmatrix} = 0.$

23. Prove that $\begin{vmatrix} 3 & 3 & 3 & 3 \\ 3 & 4 & 3 & 3 \\ 3 & 3 & 4 & 3 \\ 3 & 3 & 3 & 4 \end{vmatrix} = 3.$

DETERMINANTS.

24. Eliminate x, y, z from $ax + by + cz = 0$,
$$bx + cy + az = 0,$$
$$cx + ay + bz = 0.$$

25. Eliminate x, y, z from
$$x + y + z = 0,$$
$$(b-c)x + (c-a)y + (a-b)z = 0,$$
$$(b+c)x + (c+a)y + (a+b)z = 0.$$

26. Solve $\quad 2x + y = z,$
$$15x - 3y = 2\tfrac{1}{4}z,$$
$$4x + 3y + 2z = 18.$$

27. Solve $\quad 7x + 10y + 5z = 42,$
$$13x + 6y + 2z = 31,$$
$$11x + 14y + 8z = 63.$$

28. Solve $\quad 5x - 6y + 4z = 15,$
$$7x + 4y - 3z = 19,$$
$$2x + y + 6z = 46.$$

29. If $x + 3y - 3z = 0$, and $2x + 9y - 8z = 0$, find the ratios $x : y : z$.

30. Find the value of a that the following equations may be consistent:
$$2x - 3y + 4 = 5x - 2y - 1 = 21x - 8y + a = 0.$$

CHAPTER XXXII.

CONTINUED FRACTIONS.

491. An expression of the form

$$a_1 + \cfrac{b_2}{a_2 + \cfrac{b_3}{a_3 + \cfrac{b_4}{a_4 + \cdots}}}$$

wherein each b is the numerator of a fraction having for denominator all that follows, is called a **Continued or Chain Fraction**, and is generally written

$$a_1 + \frac{b_2}{a_2+} \ \frac{b_3}{a_3+} \cdots \cdots$$

We shall in this treatise only deal with fractions containing a finite number of a's and b's, and following the last a we place no $+$, thus

$$\frac{1}{2+} \ \frac{3}{4+} \ \frac{4}{5}$$

means

$$\cfrac{1}{2 + \cfrac{3}{4 + \cfrac{4}{5}}},$$

and its value as calculated by the ordinary rules of arithmetic is $\frac{8}{21}$.

492. Ordinary Form of Continued Fraction.

In what follows we shall only be concerned with fractions in which the b's are all unity and the a's are all positive integers, i.e. continued fractions of the form

$$a_1 + \cfrac{1}{a_2+} \; \cfrac{1}{a_3+} \; \cdots \cdots \; \cfrac{1}{a_n},$$

and we shall give rules for facilitating the calculation of their value. In Arithmetic the usual method is to begin at the bottom (i.e. with the last letters) and gradually work up; in the rules we shall explain the process is reversed, viz. we begin at the top and work downwards.

493. Partial and Complete Quotients.

The a's are called the **partial quotients**, thus a_1 is the first partial quotient, a_2 the second partial quotient, and so on. The denominator of the part of the fraction following any given a is called the **complete quotient** corresponding to the next a. Thus in the fraction

$$a_1 + \cfrac{1}{a_2+} \; \cfrac{1}{a_3+} \; \cfrac{1}{a_4} \quad \text{or} \quad a_1 + \cfrac{1}{a_2 + \cfrac{1}{a_3 + \cfrac{1}{a_4}}}$$

the complete quotient corresponding to a_1 is the continued fraction itself; that corresponding to a_2 is $a_2 + \cfrac{1}{a_3+} \; \cfrac{1}{a_4}$, that corresponding to a_3 is $a_3 + \cfrac{1}{a_4}$, and that corresponding to a_4 is a_4, and the same applies to any continued fractions. The complete quotients corresponding to $a_1, a_2, a_3 \ldots\ldots$ are often called the *first, second, third* ...... complete quotients respectively.

NOTE. We have if x_n be the nth complete quotient

$$x_n = a_n + \frac{1}{x_{n+1}}.$$

574 CONTINUED FRACTIONS.

Ex. In the fraction $\dfrac{1}{1+}\dfrac{1}{2+}\dfrac{1}{3+}\dfrac{1}{4}$ (wherein a_1 is zero) the complete quotients are

$$\dfrac{1}{1+}\dfrac{1}{2+}\dfrac{1}{3+}\dfrac{1}{4},\ 1+\dfrac{1}{2+}\dfrac{1}{3+}\dfrac{1}{4},\ 2+\dfrac{1}{3+}\dfrac{1}{4},\ 3+\dfrac{1}{4},\ 4\ \text{respectively,}$$

and their values are $\frac{30}{43}$, $1\frac{13}{30}$, $2\frac{4}{13}$, $3\frac{1}{4}$, 4,

as the reader will most easily verify by calculating them in the reverse order to that in which they are written down.

494. Every rational number can be expressed as a Continued Fraction.

RULE. *Suppose* $\dfrac{x}{x_1}$ *is the number, x and x_1 being integers; perform the ordinary process of finding the* G.C.M. *of x and x_1 and let $a_1, a_2, a_3, \ldots\ldots a_n$ be the successive quotients, then*

$$\dfrac{x}{x_1} = a_1 + \dfrac{1}{a_2+}\dfrac{1}{a_3+}\ \ldots\ldots\ \dfrac{1}{a_n}.$$

Ex. Consider the number $\dfrac{30}{43}$. Here $x_1 > x$, so that the first partial quotient is zero; the process stands thus

$$\therefore\ \dfrac{30}{43} = \dfrac{1}{1+}\dfrac{1}{2+}\dfrac{1}{3+}\dfrac{1}{4}.$$

(See § 493)

```
        30 ) 43 ( 1
             30
             ‾‾
             13 ) 30 ( 2
                  26
                  ‾‾
                   4 ) 13 ( 3
                       12
                       ‾‾
                        1 ) 4 ( 4
                            4
```

To prove the truth of the rule let $x_2, x_3, \ldots\ldots x_n$ be the successive remainders in the process of finding the G.C.M., then

$$x = a_1 x_1 + x_2,\ \therefore\ \dfrac{x}{x_1} = a_1 + \dfrac{x_2}{x_1} = a_1 + \dfrac{1}{\dfrac{x_1}{x_2}} \quad (1)$$

CONTINUED FRACTIONS.

$$x_1 = a_2 x_2 + x_3, \quad \therefore \frac{x_1}{x_2} = a_2 + \frac{x_3}{x_2} \quad (2)$$

$$x_2 = a_3 x_3 + x_4, \quad \therefore \frac{x_2}{x_3} = a_3 + \frac{x_4}{x_3} \quad (3)$$

$$\dotsb\dotsb\dotsb\dotsb\dotsb$$

$$x_{n-2} = a_{n-1} x_{n-1} + x_n, \quad \therefore \frac{x_{n-2}}{x_{n-1}} = a_{n-1} + \frac{x_n}{x_{n-1}}$$

$$x_{n-1} = a_n x_n, \quad \therefore \frac{x_{n-1}}{x_n} = a_n$$

since the last remainder is zero.

Hence we have

$$\frac{x}{x_1} = a_1 + \frac{1}{\dfrac{x_1}{x_2}} = a_1 + \frac{1}{a_2 + \dfrac{x_3}{x_2}} \text{ by } (2),$$

$$= a_1 + \frac{1}{a_2 + } \frac{1}{\dfrac{x_2}{x_3}} = a_1 + \frac{1}{a_2 + } \frac{1}{a_3 + \dfrac{x_4}{x_3}} \text{ by } (3),$$

$$= a_1 + \frac{1}{a_2 + } \frac{1}{a_3 + } \frac{1}{\dfrac{x_3}{x_4}},$$

and so on, using the equations first found successively.

We obtain finally

$$\frac{x}{x_1} = a_1 + \frac{1}{a_2 + } \frac{1}{a_3 + } \dotsb \frac{1}{a_n}.$$

If x and x_1 have no common factor, then x_n, which is their G.C.M., is unity.

The process really amounts to this: we take $\dfrac{x}{x_1}$ an improper fraction, then taking away its integral portion we invert the fraction which becomes an improper fraction, then taking away the integral portion and inverting the rest we get another improper fraction, and so on.

Thus

$$\frac{43}{30} = 1 + \frac{13}{30}$$

$$\therefore \frac{43}{30} = 1 + \cfrac{1}{\cfrac{30}{13}}$$

$$\frac{30}{13} = 2 + \frac{4}{13}$$

$$= 1 + \cfrac{1}{2 + \cfrac{4}{13}}$$

$$\frac{13}{4} = 3 + \frac{1}{4}$$

$$= 1 + \cfrac{1}{2 + \cfrac{1}{\cfrac{13}{4}}}$$

$$4 = 4$$

$$= 1 + \cfrac{1}{2 + \cfrac{1}{3 + \cfrac{1}{4}}}$$

495. Unique Continued Fraction corresponding to a given number.

It can be shown that the continued fraction obtained in the last article is unique, i.e. that there cannot be another continued fraction equal to the number $\dfrac{x}{x_1}$.

For, if possible, let

$$\frac{x}{x_1} = b_1 + \frac{1}{b_2+} \frac{1}{b_3+} \cdots \frac{1}{b_m}.$$

Then, since b_2 is an integer, the portion added to b_1 is a proper fraction;

$\therefore b_1$ is equal to the integral part of $\dfrac{x}{x_1}$, i.e. to a_1.

Hence
$$\frac{1}{a_2+} \frac{1}{a_3+} \cdots \frac{1}{a_n} = \frac{1}{b_2+} \frac{1}{b_3+} \cdots \frac{1}{b_m},$$

and therefore the reciprocals of these are equal, i.e.

$$a_2 + \frac{1}{a_3+} \cdots \frac{1}{a_n} = b_2 + \frac{1}{b_3+} \cdots \frac{1}{b_m}.$$

CONTINUED FRACTIONS.   577

Now $\dfrac{1}{a_3+}...$ is a proper fraction, as also is $\dfrac{1}{b_3+}...$, hence a_2 being the integral portion of the left-hand side is equal to b_2 the integral portion on the right-hand side; similarly $a_3 = b_3$, and so on. Thus the two continued fractions are the same. Also, one cannot terminate before the other, for if the right-hand side did we should have

$$b_m = a_m + \dfrac{1}{a_{m+1}} + \dfrac{1}{a_{m+2}}...,$$

i.e. an integer equal to a fraction, which is absurd.

Only one exception can arise, viz. we can write a_n in the form $a_n - 1 + \dfrac{1}{1}$, so that the second continued fraction has one more partial quotient than the first, and that is unity.

496. Convergents to a Continued Fraction.

The fractions obtained by keeping only the first, the first two, the first three ...... partial quotients are called the *first, second, third* ...... **convergents** to the continued fraction.

Thus in $\quad a_1 + \dfrac{1}{a_2+} \dfrac{1}{a_3+} ...... \dfrac{1}{a_n}$

the first convergent is $\quad a_1,$

the second convergent is $\quad a_1 + \dfrac{1}{a_2} = \dfrac{a_1 a_2 + 1}{a_2},$

the third convergent is $\quad a_1 + \dfrac{1}{a_2 + \dfrac{1}{a_3}} = \dfrac{a_1 a_2 a_3 + a_1 + a_3}{a_2 a_3 + 1},$

and the nth convergent is the continued fraction itself.

The successive convergents are denoted by

$$\dfrac{p_1}{q_1},\ \dfrac{p_2}{q_2},\ \dfrac{p_3}{q_3},\ ......\dfrac{p_n}{q_n},$$

thus
$\quad p_1 = a_1, \qquad\qquad q_1 = 1,$
$\quad p_2 = a_1 a_2 + 1, \qquad q_2 = a_2,$
$\quad p_3 = a_1 a_2 a_3 + a_1 + a_3,\ q_3 = a_2 a_3 + 1,$
$\quad ....$

and $\dfrac{p_n}{q_n}$ is of course the continued fraction itself.

TUT. ALG. II.

497. Recurrence Formulae for p_n and q_n.

If p_n/q_n be the nth convergent to the continued fraction
$$a_1 + \frac{1}{a_2+} \frac{1}{a_3+} \ldots,$$
$$p_n = a_n p_{n-1} + p_{n-2}; \quad q_n = a_n q_{n-1} + q_{n-2}.$$

We have seen that

$p_1 = a_1$, $\qquad\qquad q_1 = 1$,

$p_2 = a_1 a_2 + 1$, $\qquad\qquad q_2 = a_2$,

$p_3 = a_3(a_1 a_2 + 1) + a_1 = a_3 p_2 + p_1$, $\quad q_3 = a_3 a_2 + 1 = a_3 q_2 + 1$,

and thus the formulae both hold when $n = 3$. When $n < 3$ the formulae do not exist, for $\dfrac{p_1}{q_1}$ is the first convergent.

We shall now show that if the formulae hold for p_n and q_n then they hold for p_{n+1} and q_{n+1}, i.e. we assume that
$$p_n = a_n p_{n-1} + p_{n-2}, \quad q_n = a_n q_{n-1} + q_{n-2}.$$

Now $\qquad \dfrac{p_n}{q_n} = a_1 + \dfrac{1}{a_2+}\; \dfrac{1}{a_3+} \ldots \dfrac{1}{a_n},$

and
$$\frac{p_{n+1}}{q_{n+1}} = a_1 + \frac{1}{a_2+}\;\frac{1}{a_3+}\ldots \frac{1}{a_n+}\;\frac{1}{a_{n+1}} = a_1 + \frac{1}{a_2+}\ldots\frac{1}{a_n + \dfrac{1}{a_{n+1}}},$$

thus $\dfrac{p_{n+1}}{q_{n+1}}$ differs from $\dfrac{p_n}{q_n}$ by having $a_n + \dfrac{1}{a_{n+1}}$ for its last partial quotient instead of a_n.

Hence by hypothesis
$$\frac{p_{n+1}}{q_{n+1}} = \frac{\left(a_n + \dfrac{1}{a_{n+1}}\right)p_{n-1} + p_{n-2}}{\left(a_n + \dfrac{1}{a_{n+1}}\right)q_{n-1} + q_{n-2}} = \frac{(a_n a_{n+1} + 1)p_{n-1} + a_{n+1} p_{n-2}}{(a_n a_{n+1} + 1)q_{n-1} + a_{n+1} q_{n-2}}$$

$$= \frac{a_{n+1}(a_n p_{n-1} + p_{n-2}) + p_{n-1}}{a_{n+1}(a_n q_{n-1} + q_{n-2}) + q_{n-1}} = \frac{a_{n+1} p_n + p_{n-1}}{a_{n+1} q_n + q_{n-1}}$$

by hypothesis, and hence if we take

CONTINUED FRACTIONS. 579

$$p_{n+1} = a_{n+1}p_n + p_{n-1},$$
$$q_{n+1} = a_{n+1}q_n + q_{n-1},$$

then $\dfrac{p_{n+1}}{q_{n+1}}$ is the $(n+1)$th convergent.

Thus if the relation holds for p_n it does for p_{n+1}, and it has been seen to be true for p_3; hence it is for p_4 and so on universally. Similarly for the q's.

Ex. Find the successive convergents and the value of

$$2 + \frac{1}{3+} \frac{1}{4+} \frac{1}{5+} \frac{1}{6}.$$

Here $p_1 = 2, \quad q_1 = 1, \quad p_2 = 7, \quad q_2 = 3;$
then using the formulae

$$p_3 = 4p_2 + p_1 = 30, \quad q_3 = 4q_2 + q_1 = 13,$$
$$p_4 = 5p_3 + p_2 = 157, \quad q_4 = 5q_3 + q_2 = 68,$$
$$p_5 = 6p_4 + p_3 = 972, \quad q_5 = 6q_4 + q_3 = 421,$$

and the successive convergents are

$$\frac{2}{1}, \frac{7}{3}, \frac{30}{13}, \frac{157}{68}, \frac{972}{421},$$

the last being the value of the continued fraction itself.

498. To prove that $p_n q_{n-1} - p_{n-1} q_n = (-1)^n$.

We must have $n > 1$, since p_0 and q_0 have no meanings.

We have $\quad p_1 = a_1, \qquad\qquad q_1 = 1,$
$$p_2 = a_1 a_2 + 1, \qquad q_2 = a_2,$$
$$p_3 = a_1 a_2 a_3 + a_1 + a_3, \quad q_3 = a_2 a_3 + 1,$$

and from these we see that

$p_2 q_1 - p_1 q_2 = 1 = (-1)^2,$
$p_3 q_2 - p_2 q_3 = (a_1 a_2 a_3 + a_1 + a_3) a_2 - (a_1 a_2 + 1)(a_2 a_3 + 1)$
$\hspace{6cm} = (-1) = (-1)^3,$

and the formula holds when $n = 2$ and $n = 3$.

Assuming that it holds when $n = r$ we shall show that it holds when $n = r + 1$.

580   CONTINUED FRACTIONS.

We have
$$p_{r+1}q_r - p_r q_{r+1} = (a_{r+1}p_r + p_{r-1})q_r - p_r(a_{r+1}q_r + q_{r-1}) \quad (\S\,497)$$
$$= p_{r-1}q_r - p_r q_{r-1}$$
$$= -(p_r q_{r-1} - p_{r-1}q_r) = -(-1)^r,$$
since the formula holds by hypothesis when $n = r$.
$$\therefore\ p_{r+1}q_r - p_r q_{r+1} = (-1)^{r+1}.$$

Thus if the theorem holds for any value of n it holds for the next value of n, but it has been seen to hold when $n = 2, 3$; hence by induction it is true universally.

499. The odd convergents of a continued fraction continually increase, and the even convergents continually diminish.

Suppose the continued fraction to be
$$a_1 + \frac{1}{a_2 +}\ \frac{1}{a_3 +}\ \frac{1}{a_4 +} \ldots\ldots\ \frac{1}{a_m}$$
and p_n/q_n the nth convergent, then
$$\frac{p_n}{q_n} - \frac{p_{n-2}}{q_{n-2}} = \frac{p_n q_{n-2} - q_n p_{n-2}}{q_n q_{n-2}}$$
$$= \frac{(a_n p_{n-1} + p_{n-2})q_{n-2} - p_{n-2}(a_n q_{n-1} + q_{n-2})}{q_n q_{n-2}} \quad (\S\,497)$$
$$= \frac{a_n(p_{n-1}q_{n-2} - p_{n-2}q_{n-1})}{q_n q_{n-2}} = (-1)^{n-1} \frac{a_n}{q_n q_{n-2}} \quad (\S\,498).$$

Hence if n be odd $\dfrac{p_n}{q_n} - \dfrac{p_{n-2}}{q_{n-2}}$ is positive, i.e. any odd convergent is greater than the next preceding odd convergent, or the odd convergents continually increase.

If n be even $\dfrac{p_n}{q_n} - \dfrac{p_{n-2}}{q_{n-2}}$ is negative, i.e. any even convergent is less than the next preceding even convergent, or in other words the even convergents continually diminish.

CONTINUED FRACTIONS. 581

500. Each odd convergent is less than the continued fraction, and each even convergent is greater than the continued fraction.

Let x_{n+1} be the nth complete quotient (§ 493); then the continued fraction differs from the $(n+1)$th convergent in having x_{n+1} for the nth partial quotient instead of a_{n+1}; consequently as x_1 is the continued fraction itself we have

$$x_1 = \frac{x_{n+1}p_n + p_{n-1}}{x_{n+1}q_n + q_{n-1}} \quad (\S\ 497).$$

$$\therefore \frac{p_n}{q_n} - x_1 = \frac{p_n}{q_n} - \frac{x_{n+1}p_n + p_{n-1}}{x_{n+1}q_n + q_{n-1}}$$

$$= \frac{p_n q_{n-1} - p_{n-1}q_n}{q_n(x_{n+1}q_n + q_{n-1})} = \frac{(-1)^n}{q_n(x_{n+1}q_n + q_{n-1})}.$$

Now every letter in the denominator is positive.

$\therefore \dfrac{p_n}{q_n} - x_1$ has the same sign as $(-1)^n$, i.e. $\dfrac{p_n}{q_n}$ is greater than the continued fraction if n be even and less than it if n be odd.

COR. *The continued fraction lies between the values of any two consecutive convergents.*

For one of these must be an odd convergent and the other an even convergent.

The above proof applies even for a non-terminating continued fraction. The proposition may also be proved thus:

All the odd convergents are less than the continued fraction itself except the last one, which may be equal to it; and all the even convergents are greater than the continued fraction except the last, which may be equal to it.

Suppose the continued fraction is

$$a_1 + \frac{1}{a_2+} \frac{1}{a_3+} \cdots \frac{1}{a_m},$$

then if m be odd the last convergent is an odd one and is equal to the continued fraction, while if m be even the last convergent is an even one and is equal to the continued fraction.

Now $\dfrac{p_m}{q_m} - \dfrac{p_{m-1}}{q_{m-1}} = \dfrac{p_m q_{m-1} - p_{m-1} q_m}{q_m q_{m-1}} = \dfrac{(-1)^m}{q_m q_{m-1}}$. (§ 498.)

If m be even we see that $\dfrac{p_m}{q_m} - \dfrac{p_{m-1}}{q_{m-1}}$ is positive, i.e. $\dfrac{p_{m-1}}{q_{m-1}}$ is less than $\dfrac{p_m}{q_m}$, which is the continued fraction itself.

But $\dfrac{p_{m-1}}{q_{m-1}}$, being the last odd convergent, is also the greatest odd convergent (§ 499).

Therefore in this case all the odd convergents are less than the continued fraction.

Also $\dfrac{p_m}{q_m}$ is the least even convergent.

Therefore all the even convergents are greater than the continued fraction except the last, which is equal to it.

If m be odd $\dfrac{p_m}{q_m} < \dfrac{p_{m-1}}{q_{m-1}}$ and $\dfrac{p_m}{q_m}$ is the continued fraction itself.

Therefore, now, all the odd convergents except the last are less than the continued fraction, the last is equal to it.

Also $\dfrac{p_{m-1}}{q_{m-1}}$ being the last even convergent is the least even convergent (§ 499), and as it is greater than the continued fraction, so also are all of them.

501. Any convergent is nearer in value to the continued fraction than any preceding convergent.

We have $\quad x_1 = \dfrac{x_{n+1} p_n + p_{n-1}}{x_{n+1} q_n + q_{n-1}}$,

and $\quad \dfrac{p_n}{q_n} - x_1 = \dfrac{(-1)^n}{q_n (x_{n+1} q_n + q_{n-1})}$.

Again

$\dfrac{p_{n-1}}{q_{n-1}} - x_1 = \dfrac{x_{n+1}(p_{n-1} q_n - p_n q_{n-1})}{q_{n-1}(x_{n+1} q_n + q_{n-1})} = -\dfrac{(-1)^n x_{n+1}}{q_{n-1}(x_{n+1} q_n + q_{n-1})}$.

Thus $\quad \dfrac{p_n}{q_n} \sim x_1 = \dfrac{1}{q_n (x_{n+1} q_n + q_{n-1})}$,

and $\quad \dfrac{p_{n-1}}{q_{n-1}} \sim x_1 = \dfrac{x_{n+1}}{q_{n-1}(x_{n+1} q_n + q_{n-1})}$

CONTINUED FRACTIONS. 583

Now $\quad q_n > q_{n-1}$ and $x_{n+1} > 1$;

$$\therefore \frac{1}{q_n} < \frac{1}{q_{n-1}} < \frac{x_{n+1}}{q_{n-1}},$$

hence $\quad \dfrac{p_n}{q_n} \sim x_1 < \dfrac{p_{n-1}}{q_{n-1}} \sim x_1.$

Thus any convergent is nearer the continued fraction than the next preceding convergent, and therefore than all preceding convergents.

502. Any convergent is nearer in value to the continued fraction than any fraction having a smaller denominator than that convergent.

For suppose that $\dfrac{a}{b}$ is nearer the continued fraction than $\dfrac{p_n}{q_n}$ where $b < q_n$; then $\dfrac{a}{b}$ is nearer than $\dfrac{p_{n-1}}{q_{n-1}}$ (§ 501); but the continued fraction lies between $\dfrac{p_n}{q_n}$ and $\dfrac{p_{n-1}}{q_{n-1}}$ (§ 500).

$\therefore \dfrac{a}{b}$ lies between $\dfrac{p_n}{q_n}$ and $\dfrac{p_{n-1}}{q_{n-1}}.$

Hence $\quad \dfrac{p_{n-1}}{q_{n-1}} \sim \dfrac{a}{b} < \dfrac{p_{n-1}}{q_{n-1}} \sim \dfrac{p_n}{q_n},$

i.e. $\quad \dfrac{p_{n-1} b \sim q_{n-1} a}{q_{n-1} b} < \dfrac{p_{n-1} q_n \sim p_n q_{n-1}}{q_n q_{n-1}} < \dfrac{1}{q_n q_{n-1}}.$

$$\therefore \frac{p_{n-1} b \sim q_{n-1} a}{b} < \frac{1}{q_n}.$$

Now $p_{n-1} b \sim q_{n-1} a$ is an integer and $b < q_n$.
Therefore this inequality is impossible unless

$$p_{n-1} b = q_{n-1} a, \text{ i.e. } a/b = p_{n-1}/q_{n-1}.$$

Thus a/b cannot be nearer than p_n/q_n unless it is equal to p_{n-1}/q_{n-1}, and we have already seen that it is further away in this case.

Hence the proposition is proved.

503. Limits to the error committed in taking any convergent to represent the continued fraction.

We have seen that

$$\frac{p_n}{q_n} \sim x_1 = \frac{1}{q_n(x_{n+1}q_n + q_{n-1})}.$$

Now $\qquad x_{n+1} > a_{n+1}.$

$\therefore x_{n+1}q_n + q_{n-1} > a_{n+1}q_n + q_{n-1}$, i.e. $> q_{n+1}$.

$$\therefore \frac{p_n}{q_n} \sim x_1 < \frac{1}{q_n q_{n+1}},$$

which gives an upper limit to the error.

To find a lower limit to the error we remark that

$$\frac{p_{n-1}}{q_{n-1}} \sim x_1 = \frac{x_{n+1}}{q_{n-1}(x_{n+1}q_n + q_{n-1})}$$

$$= \frac{1}{q_{n-1}\left(q_n + \dfrac{q_{n-1}}{x_{n+1}}\right)},$$

Now $\qquad x_{n+1} > 1.$

$$\therefore \frac{q_{n-1}}{x_{n+1}} < q_{n-1}.$$

Hence $\qquad \dfrac{p_{n-1}}{q_{n-1}} \sim x_1 > \dfrac{1}{q_{n-1}(q_n + q_{n-1})}.$

Changing **n** into $n+1$ in this formula we see that the error lies between

$$\frac{1}{q_n q_{n+1}} \quad \text{and} \quad \frac{1}{q_n(q_{n+1} + q_n)}.$$

COR. *The error is less than* $\dfrac{1}{a_{n+1}q_n^2}.$

For $\qquad q_{n+1} > a_{n+1}q_n.$

CONTINUED FRACTIONS. 585

504. Recapitulation. The foregoing propositions show us how to find a series of approximations in the form of fractions to any given quantity.

For (i) *We can represent the quantity as a decimal to any degree of accuracy.*

(ii) *Convert this decimal into a continued fraction.*

(iii) *Form the successive convergents to this continued fraction.*

(iv) *Then any convergent is nearer than any preceding convergent and any two consecutive ones enclose the continued fraction* (§§ 500, 501).

(v) We can find approximately the error made in taking any convergent to represent the whole fraction (§ 503).

505. Illustrative Examples.

Ex. 1. Find a fraction representing e correctly to 5 decimal places.
We have $e = 2\cdot71828\ldots = 2\frac{71828}{100000}$.
Converting this into a continued fraction in the usual way we find
$$2 + \frac{1}{1+}\frac{1}{2+}\frac{1}{1+}\frac{1}{1+}\frac{1}{4+}\frac{1}{1+}\frac{1}{1+}\frac{1}{6+}\frac{1}{10+}\frac{1}{1+}\frac{1}{1+}\frac{1}{2}.$$
The successive p's are
$$2,\ 3,\ 8,\ 11,\ 19,\ 87,\ 106,\ 193,\ 1264,\ 12833\ldots,$$
and the successive q's are
$$1,\ 1,\ 3,\ 4,\ 7,\ 32,\ 39,\ 71,\ 465,\ 4721\ldots.$$

The error must be less than $\frac{1}{100000}$, i.e. if p_n/q_n be the convergent required, we must have
$$q_n(q_n + q_{n+1}) > 100000.$$
The first value of n which satisfies this is $n = 9$, for
$$465(465 + 4721) > 100000;$$
$\therefore$ 1264/465 represents e correctly to five places of decimals.

Ex. 2. If p and q be integers with no common factor, show how to find integers x and y, such that
$$px - qy = 1.$$
Convert the fraction p/q into a continued fraction, let n be the

number of partial quotients and y/x the convergent next before the last, then since p/q and y/x are consecutive convergents,

$$px - qy = (-1)^n.$$

Thus if n be even this gives the solution. If n be odd we write the nth partial quotient in the form

$$(a_n - 1) + \frac{1}{1},$$

and now the number of partial quotients is even.

This problem is fully discussed in Chapter XXXIV.

EXERCISES XXXII.

Express as continued fractions:

1. $\dfrac{532}{1193}$. 2. $31\tfrac{76}{123}$. 3. $\dfrac{981}{114}$.

4. $\dfrac{637}{109}$. 5. $\dfrac{2345}{1234}$. 6. $\dfrac{671}{1225}$.

7. $2{\cdot}71828\,(e)$. 8. $3{\cdot}1416\,(\pi)$.

Calculate all the convergents to the following continued fractions:

9. $5 + \dfrac{1}{1+} \dfrac{1}{5+} \dfrac{1}{2+} \dfrac{1}{2+} \dfrac{1}{3}$.

10. $\dfrac{1}{1+} \dfrac{1}{1+} \dfrac{1}{1+} \dfrac{1}{1+} \dfrac{1}{1+} \dfrac{1}{1+} \dfrac{1}{1+} \dfrac{1}{1+} \dfrac{1}{3}$.

11. $\dfrac{1}{1+} \dfrac{1}{1+} \dfrac{1}{4+} \dfrac{1}{1+} \dfrac{1}{2+} \dfrac{1}{1+} \dfrac{1}{3+} \dfrac{1}{2+} \dfrac{1}{3}$.

12. $1 + \dfrac{1}{2+} \dfrac{1}{3+} \dfrac{1}{4+} \dfrac{1}{5+} \dfrac{1}{6}$. 13. $\dfrac{1}{2+} \dfrac{1}{2+} \dfrac{1}{2+} \dfrac{1}{2+} \dfrac{1}{2}$.

14. $1 + \dfrac{1}{2+} \dfrac{1}{10+} \dfrac{1}{17+} \dfrac{1}{26}$.

15. Find the first convergent to the continued fraction in Ex. 7 which will represent e correctly to four places of decimals.

16. Show that the error in taking the fourth convergent in Ex. 5 to represent the continued fraction is $< {\cdot}000008$.

17. Find limits for the difference between the continued fraction

$$3 + \frac{1}{4+} \frac{1}{5+} \frac{1}{6+} \frac{1}{7}$$

and its third convergent.

CONTINUED FRACTIONS. 587

18. Show that the second convergent to

$$3 + \frac{1}{7+} \frac{1}{15+} \frac{1}{1+} \frac{1}{292+} \frac{1}{1+} \frac{1}{2}$$

exceeds the continued fraction by less than ·01, and that the fourth exceeds it by less than ·000000267.

19. If $\frac{p_n}{q_n}$ be the nth convergent to $a_1 + \frac{1}{a_2+} \frac{1}{a_3+} \ldots \frac{1}{a_n}$, show that

$$\frac{p_n}{p_{n-1}} = a_n + \frac{1}{a_{n-1}+} \frac{1}{a_{n-2}+} \frac{1}{a_{n-3}+} \ldots \frac{1}{a_2+} \frac{1}{a_1},$$

and

$$\frac{q_n}{q_{n-1}} = a_n + \frac{1}{a_{n-1}+} \frac{1}{a_{n-2}+} \ldots \frac{1}{a_2}.$$

$$\left\{ \begin{array}{l} \text{Use the relations } \dfrac{p_n}{p_{n-1}} = a_n + \dfrac{p_{n-2}}{p_{n-1}}, \\[2mm] \dfrac{q_n}{q_{n-1}} = a_n + \dfrac{q_{n-2}}{q_{n-1}}, \text{ and so on.} \end{array} \right\}$$

CHAPTER XXXIII.

ELIMINATION.

506. Nature of the problem of Elimination.

In dealing with equations involving unknown quantities, we have hitherto been almost entirely concerned with effecting solutions of the equations—that is to say, ascertaining what value or values must be attributed to the unknown quantities in order that the equation or equations in question may be satisfied. We have seen, for instance, that the required value or values of a single unknown quantity x can be completely determined from a single linear or quadratic equation involving that unknown quantity. If we are furnished with a second equation involving the same unknown quantity, it is evident, on substituting in the second equation the value or values of x derived from the first, that there must be some relation not involving x subsisting between the various coefficients in the two equations, in order that they may be capable of being simultaneously satisfied by the same value of x.

More generally, if we have n equations sufficient for the determination of the values of n unknown quantities involved therein, together with an additional equation involving the same unknown quantities, it is evident that there must be some relation, not involving unknowns, subsisting between the various coefficients which occur in the $(n+1)$ given equations. The process of obtaining this relation, by whatever means, is known as the *elimination* of the n unknown quantities from the $(n+1)$ given equations; and the purpose of the present chapter is to give practice to the student in certain algebraical devices by which the required object can be more readily attained than by the obvious, but often cumbrous, method of effecting a solution by means of the n sufficient equations and then substituting in the $(n+1)$th.

Ex. 1. Eliminate x from the equations

$$x + \frac{1}{x} = 2a, \quad x^3 + \frac{1}{x^3} = 2b.$$

We have $\quad 8a^3 + \left(x + \frac{1}{x}\right)^3 = x^3 + \frac{1}{x^3} + 3\left(x + \frac{1}{x}\right)$

that is $\quad 8a^3 = 2b + 6a$

or $\quad 4a^3 - 3a = b.$

Note how much simpler this is than solving the first equation as a quadratic in x and substituting in the second.

[The student acquainted with Trigonometry should observe that this elimination amounts to finding the expression for $\cos 3a$ in terms of $\cos a$.]

Ex. 2. Eliminate x from the equations
$$x + \frac{1}{x} = 2a, \quad x^2 - \frac{1}{x^2} = 2b.$$

By division
$$x - \frac{1}{x} = \frac{b}{a}$$

Also
$$\left(x + \frac{1}{x}\right)^2 - \left(x - \frac{1}{x}\right)^2 = 4$$

$$\therefore\ 4a^2 - \frac{b^2}{a^2} = 4.$$

Ex. 3. Eliminate x and y from the equations
$$ax + by = c, \quad bx - ay = d, \quad x^2 + y^2 = 1.$$
Squaring the first two and adding, we have
$$(a^2 + b^2)(x^2 + y^2) = c^2 + d^2$$
or
$$a^2 + b^2 = c^2 + d^2.$$

It should be noted that whenever expressions of the form $(ax + by)$ and $(bx - ay)$ occur in an elimination, it is always advisable to try the effect of squaring the expressions and adding the results.

[The student acquainted with Trigonometry should observe that the above method furnishes the elimination of θ between the equations $a \cos \theta + b \sin \theta = c$ and $b \cos \theta - a \sin \theta = d$.]

Ex. 4. Eliminate x and y from the equations
$$ax - by = c, \quad bx - ay = d, \quad x^2 - y^2 = 1.$$
Squaring the first two and subtracting, we have
$$(a^2 - b^2)(x^2 - y^2) = c^2 - d^2$$
or
$$a^2 - b^2 = c^2 - d^2.$$

[Note that this is equivalent trigonometrically to the elimination of θ between the equations $a \sec \theta - b \tan \theta = c$, $b \sec \theta - a \tan \theta = d$.]

507. Simultaneous quadratic equations in one unknown quantity.

The student should first refresh his memory with regard to the important result of Art. 68. The process of elimination in this case is to solve the equations as simultaneous equations in x^2 and x, when the elimination follows at once.

590   ELIMINATION.

Ex. 1. Eliminate x from the equations
$$ax^2 + bx + c = 0, \quad a'x^2 + b'x + c' = 0.$$
By the rule of cross-multiplication we have
$$\frac{x^2}{bc' - b'c} = \frac{x}{ca' - c'a} = \frac{1}{ab' - a'b},$$
whence $\quad (bc' - b'c)(ab' - a'b) = (ca' - c'a)^2.$

The problem may be stated in the equivalent form "Find the condition that the two quadratics should have a common root."

Ex. 2. Eliminate x from the equations
$$ax^2 + bx + c = 0, \quad cx^2 + bx + a = 0.$$
We have as above,
$$\frac{x^2}{b(a-c)} = \frac{x}{c^2 - a^2} = \frac{1}{b(a-c)}.$$
Hence either $a = c$, (in which case the two equations are identical), or
$$b^2 = (c + a)^2, \quad \text{that is, } a + c = \pm b.$$
The student should verify the fact that whichever sign is taken the equations have a common root.

508. Manipulation of Ratios.

Before proceeding to eliminations involving the use of determinants, it is desirable to remind the student of the importance of acquiring facility in the algebraical processes involved in questions of ratio and proportion. The result established in Art. 63 and generalised in Art. 64 should be revised, and especial attention paid to the artifice by which the proof of Art. 63 is effected.

Ex. 1. If $\quad \dfrac{x}{a^2(b-c)} = \dfrac{y}{b^2(c-a)} = \dfrac{z}{c^2(a-b)}$

show that $\quad \dfrac{cy + bz}{b - c} = \dfrac{az + cx}{c - a} = \dfrac{bx + ay}{a - b}.$

We have $\quad \dfrac{cy + bz}{cb^2(c-a) + bc^2(a-b)} = \dfrac{az + cx}{ac^2(a-b) + ca^2(b-c)}$
$$= \dfrac{bx + ay}{ba^2(b-c) + ab^2(c-a)}$$
as the student may see on writing
$$x = ka^2(b-c), \quad y = kb^2(c-a), \quad z = kc^2(a-b).$$
Hence $\quad \dfrac{cy + bz}{abc(c-b)} = \dfrac{az + cx}{abc(a-c)} = \dfrac{bx + ay}{abc(b-a)}$

or $\quad \dfrac{cy + bz}{b - c} = \dfrac{az + cx}{c - a} = \dfrac{bx + ay}{a - b}.$

ELIMINATION. 591

Ex. 2. Eliminate x and y from the equations

$$\left. \begin{array}{l} \dfrac{a^4}{x^2}+\dfrac{b^4}{y^2}=(a+b)^2, \ldots\ldots\ldots\ldots \\[6pt] \dfrac{x^2}{a^2}+\dfrac{y^2}{b^2}=1, \ldots\ldots\ldots\ldots\ldots\ldots \\[6pt] \dfrac{1}{p^2}=\dfrac{x^2}{a^4}+\dfrac{y^2}{b^4} \ldots\ldots\ldots\ldots\ldots\ldots \end{array} \right\}$$

From the first two by multiplication we have

$$a^2+b^2+\frac{a^4}{b^2}\cdot\frac{y^2}{x^2}+\frac{b^4}{a^2}\cdot\frac{x^2}{y^2}=(a+b)^2$$

or
$$\left(\frac{a^2}{b}\cdot\frac{y}{x}-\frac{b^2}{a}\cdot\frac{x}{y}\right)^2=0,$$

that is
$$\frac{x^2}{a^3}=\frac{y^2}{b^3}.$$

Hence
$$\frac{x^2}{a^3}=\frac{y^2}{b^3}=\frac{\dfrac{x^2}{a^2}+\dfrac{y^2}{b^2}}{\dfrac{a^3}{a^2}+\dfrac{b^3}{b^2}}=\frac{\dfrac{x^2}{a^4}+\dfrac{y^2}{b^4}}{\dfrac{a^3}{a^4}+\dfrac{b^3}{b^4}},$$

that is
$$\frac{1}{a+b}=\frac{\dfrac{1}{p^2}}{\dfrac{1}{a}+\dfrac{1}{b}}$$

or
$$p^2=ab.$$

Ex. 3. Eliminate λ from the equations

$$\frac{\dfrac{h}{a^2+\lambda}}{l}=\frac{\dfrac{k}{b^2+\lambda}}{m}=\frac{1}{n}.$$

We have
$$\frac{n}{1}=\frac{a^2+\lambda}{\dfrac{h}{l}}=\frac{b^2+\lambda}{\dfrac{k}{m}}=\frac{a^2-b^2}{\dfrac{h}{l}-\dfrac{k}{m}},$$

$$\therefore \frac{h}{l}-\frac{k}{m}=\frac{a^2-b^2}{n}.$$

[Note the application to the locus of the pole of a given line with regard to a system of confocal conics.]

Ex. 4. Eliminate λ from the equations

$$\frac{\dfrac{h}{a^2+\lambda}-\dfrac{k}{b^2+\lambda}}{l\quad\quad m}\quad\ldots\ldots\ldots\ldots\ldots(1),$$

$$\frac{h^2}{a^2+\lambda}+\frac{k^2}{b^2+\lambda}=1\quad\ldots\ldots\ldots\ldots\ldots(2).$$

The second equation may be written

$$k^2(a^2+\lambda)+h^2(b^2+\lambda)=(a^2+\lambda)(b^2+\lambda)\ \ldots\ldots\ldots(3).$$

From the first we have

$$\frac{a^2+\lambda}{\dfrac{h}{l}}=\frac{b^2+\lambda}{\dfrac{k}{m}}=\frac{k^2(a^2+\lambda)+h^2(b^2+\lambda)}{\dfrac{k^2h}{l}+\dfrac{h^2k}{m}}$$

$$=\frac{\sqrt{\{(a^2+\lambda)(b^2+\lambda)\}}}{\sqrt{\dfrac{hk}{lm}}}$$

$$=\frac{a^2-b^2}{\dfrac{h}{l}-\dfrac{k}{m}}$$

$$\therefore\ \frac{(a^2+\lambda)(b^2+\lambda)}{\dfrac{hk}{lm}}=\frac{\{k^2(a^2+\lambda)+h^2(b^2+\lambda)\}\{a^2-b^2\}}{\left(\dfrac{k^2h}{l}+\dfrac{h^2k}{m}\right)\left(\dfrac{h}{l}-\dfrac{k}{m}\right)}.$$

Or, from (3), $\ lm\left(\dfrac{k}{l}+\dfrac{h}{m}\right)\left(\dfrac{h}{l}-\dfrac{k}{m}\right)=a^2-b^2$

that is, $\quad\quad (km+hl)(hm-kl)=(a^2-b^2)\,lm.$

[Note the application in Coordinate Geometry to the locus of the points of contact of parallel tangents to a system of confocals.]

509. Elimination by means of determinants.

We have seen in Art. 477 that the result of eliminating x, y, and z from the equations

$$a_1x+b_1y+c_1z=0,\quad a_2x+b_2y+c_2z=0,\quad a_3x+b_3y+c_3z=0$$

is $\quad a_1(b_2c_3-b_3c_2)+b_1(c_2a_3-c_3a_2)+c_1(a_2b_3-a_3b_2)=0,$

which is more usually written in the form

$$\begin{vmatrix} a_1 & b_1 & c_1 \\ a_2 & b_2 & c_2 \\ a_3 & b_3 & c_3 \end{vmatrix}=0.$$

Here we are given three equations apparently involving three unknown quantities; but since the equations are homogeneous we are in reality only concerned with the *two unknown ratios* $x:z$ and $y:z$, as may be seen on dividing the equations by z throughout. Hence the system of equations is sufficient for the elimination.

ELIMINATION.

A similar result obtains for n homogeneous linear equations involving n unknown quantities.

Ex. 1. Eliminate a, b, c from the equations
$$ax_1 + by_1 + c = 0, \quad ax_2 + by_2 + c = 0, \quad ax_3 + by_3 + c = 0.$$
The result is obviously $\begin{vmatrix} x_1 & y_1 & 1 \\ x_2 & y_2 & 1 \\ x_3 & y_3 & 1 \end{vmatrix} = 0.$

[These two eliminations express the conditions that three straight lines should meet at a point, and that three points should lie on a straight line.]

Ex. 2. Eliminate x' and y' from the equations
$$ax' + by' + g = 0, \quad \ldots\ldots\ldots\ldots\ldots\ldots(1),$$
$$hx' + by' + f = 0, \quad \ldots\ldots\ldots\ldots\ldots\ldots(2),$$
$$ax'^2 + 2hx'y' + by'^2 + 2gx' + 2fy' + c = 0 \ldots\ldots\ldots(3).$$

Equation (3) may be written
$$x'(ax' + by' + g) + y'(hx' + by' + f) + gx' + fy' + c = 0,$$
whence $\quad gx' + fy' + c = 0 \ldots\ldots\ldots\ldots\ldots\ldots\ldots\ldots(4)$

Eliminating x' and y' from (1), (2), and (4), we have
$$\begin{vmatrix} a & h & g \\ h & b & f \\ g & f & c \end{vmatrix} = 0.$$

[The condition that the general equation of the second degree should represent two straight lines.]

Ex. 3. Eliminate x' and y' from the equations
$$\frac{ax' + by' + g}{l} = \frac{hx' + by' + f}{m} = \frac{gx' + fy' + c}{n},$$
and $\quad lx' + my' + n = 0.$

Let each fraction be equal to k.

Then
$$ax' + hy' + g - lk = 0$$
$$hx' + by' + f - mk = 0$$
$$gx' + fy' + c - nk = 0$$
$$lx' + my' + n = 0.$$

Treating these as four simultaneous equations in x', y', unity, and k, and eliminating, we have
$$\begin{vmatrix} a, & h, & g, & l \\ h, & b, & f, & m \\ g, & f, & c, & n \\ l, & m, & n, & o \end{vmatrix} = 0.$$

TUT. ALG. II.

On expanding from the last row and last column, the above determinant proves to be
$$-(Al^2 + Bm^2 + Cn^2 + 2\,Fmn + 2\,Gnl + 2\,Hlm)$$
where $A \equiv bc - f^2$, $F \equiv gh - af$, etc.

Hence the required eliminant is
$$Al^2 + Bm^2 + Cn^2 + 2\,Fmn + 2\,Gnl + 2\,Hlm = 0.$$

[An example of considerable importance as illustrating the use of determinants. It expresses the condition that the straight line $lx + my + n = 0$ should touch the conic given by the general equation.]

Ex. 4. Eliminate x from the equations
$$ax^2 + bx + c = 0, \quad a'x + b' = 0.$$

We have
$$ax^2 + bx + c = 0 \dots\dots\dots\dots(1),$$
$$a'x^2 + b'x = 0 \dots\dots\dots\dots(2),$$
$$a'x + b' = 0 \dots\dots\dots\dots(3).$$

Treating these as simultaneous equations in x^2, x, and unity, and eliminating, we have
$$\begin{vmatrix} a, & b, & c \\ a', & b', & 0 \\ 0, & a', & b' \end{vmatrix} = 0.$$

The student should verify the result by solving for x in the linear equation and substituting in the quadratic. It is evident that this method can be extended to any two simultaneous equations in integral powers of the same variable, as in the following example.

Ex. 5. Eliminate x from the equations
$$ax^3 + bx^2 + cx + d = 0, \quad a'x^2 + b'x + c' = 0$$

We have
$$ax^4 + bx^3 + cx^2 + dx = 0$$
$$ax^3 + bx^2 + cx + d = 0$$
$$a'x^4 + b'x^3 + c'x^2 = 0$$
$$a'x^3 + b'x^2 + c'x = 0$$
$$a'x^2 + b'x + c' = 0$$

and the eliminant is therefore
$$\begin{vmatrix} a, & b, & c, & d, & 0 \\ 0, & a, & b, & c, & d \\ a', & b', & c', & 0, & 0 \\ 0, & a', & b', & c', & 0 \\ 0, & 0, & a' & b', & c' \end{vmatrix} = 0.$$

ELIMINATION. 595

EXERCISES XXXIII.

1. Eliminate x from the equations
$$x + \frac{1}{x} = 2a, \quad x^2 + \frac{1}{x^2} = 2b.$$

2. Eliminate x from the equations
$$x - \frac{1}{x} = 2a, \quad x^2 - \frac{1}{x^2} = 2b.$$

3. Eliminate m from the equations
$$y = mx + \frac{a}{m}, \quad y = -\frac{x}{m} - am.$$

4. Eliminate m from the equations
$$y = mx + \frac{a}{m}, \quad y = -\frac{1}{m}(x - a).$$

5. Eliminate x from the equations
$$x - \frac{1}{x} = 2a, \quad x^3 - \frac{1}{x^3} = 2b.$$

6. Eliminate x from the equations
$$x - \frac{1}{x} = 2a, \quad x^4 + \frac{1}{x^4} = 2b.$$

7. Eliminate m from the equations
$$y = mx + \sqrt{a^2m^2 + b^2}, \quad y = -\frac{x}{m} + \sqrt{\frac{a^2}{m^2} + b^2}.$$

8. Eliminate l and m from the equations
$$lx + my = \sqrt{a^2l^2 + b^2m^2}, \quad mx - ly = \sqrt{a^2m^2 + b^2l^2}.$$

9. Eliminate l and m from the equations
$$lx + my = \sqrt{a^2l^2 + b^2m^2}, \quad m(x - \sqrt{a^2 - b^2}) = ly.$$

Eliminate x, y, z from the following systems of equations :

10. $ax + by + cz = 0.$
    $bx + cy + az = 0.$
    $cx + ay + bz = 0.$

11. $ax + by - cz = 0.$
    $-bx + cy + az = 0.$
    $cx - ay + bz = 0.$

12. $(b - c)x + (c - a)y + (a - b)z = 0.$
    $(c - a)x + (a - b)y + (b - c)z = 0.$
    $(a - b)x + (b - c)y + (c - a)z = 0.$

13. Eliminate x and y from the equations
$$\frac{x+2y+3}{l} = \frac{2x+4y+5}{m} = \frac{3x+5y+6}{n},$$
$$lx + my + n = 0.$$

14. Eliminate x from the equations
$$ax^2 + 2bx + c = 0, \qquad ax + b = 0.$$

15. Eliminate x from the equations
$$ax^2 + bx + c = 0, \qquad cx^2 - bx + a = 0;$$
the coefficients being real.

16. Eliminate x and y from the equations
$$2x^2 + 14xy + 3y^2 + 12x + 10y + c = 0,$$
$$2x + 7y + 6 = 0, \quad 7x + 3y + 5 = 0.$$

17. Eliminate λ from the equations
$$\frac{l}{-2\lambda} = \frac{m}{k} = \frac{n}{2\lambda(h+2\lambda)}.$$

CHAPTER XXXIV.

INDETERMINATE EQUATIONS.

510. Nature of the problem.

Consider the equation $ax+by=c$ with integral coefficients a, b, c. It clearly admits of an infinite number of solutions, for corresponding to any given value a which may be assigned to x there is a solution $\dfrac{c-a\alpha}{b}$ for y, and corresponding to any given value β which may be assigned to y there is a solution $\dfrac{c-b\beta}{a}$ for x. If, however, we decide to regard only such solutions as afford positive integral values for x and y, the number of possible solutions may be limited, or there may be no solution of the required type possible at all.

511. Cases to be considered.

(1) If a and b are of like sign in the equation $ax+by=c$, it must be of the type $lx+my=n$ or $lx+my=-n$, where l, m, n are all positive integers. It is clear that no positive integral values of x and y can satisfy the equation $lx+my=-n$; hence this class may be considered as consisting exclusively of the type $lx+my=n$, where l, m, n are all positive integers.

(2) If a and b are of unlike sign the equation must be of the type $lx-my=n$ or $my-lx=n$, where l, m, n are all positive integers as before. It is readily seen that these two last are really of the same type, subject to an interchange of x and y; we may therefore reduce this class to the type $lx-my=n$, where l, m, n are all positive integers as before.

We shall assume the equation first reduced to its lowest terms as regards the coefficients. If after this operation l and m have a common factor which is not a factor of n, it is clear, on division throughout by such factor, that no positive integral values of x and y can satisfy either of the equations $lx \pm my = n$.

In effect, then, we need only investigate the equations $lx+my=n$, $lx-my=n$, where l, m, n are positive integers and l and m are prime to each other. These conditions will be postulated during the remainder of this chapter.

512. To find integral values of p and q satisfying the equation $lq - mp = 1$.

This is the first step, and the only step presenting any difficulty, in the solution of the indeterminate equations $lx \pm my = n$. The student acquainted with the theory of continued fractions will perform the step by converting $\dfrac{l}{m}$ into a continued fraction.

Let $\dfrac{p}{q}$ be the last convergent; then by Art. 498, $lq - mp = 1$ or $lq - mp = -1$; and since in the latter case $l(m-q) - m(l-p) = 1$, two integers are found satisfying the condition required.

The student who wishes to perform the step without the assistance of the theory of continued fractions may perform the same operation in another way; he will gather the process from the following illustration :—

Suppose it is required to find integral values of p and q satisfying the equation $8q - 19p = 1$.

By successive divisions (as if finding the H.C.F. by the usual rule) we have

$$8 \overline{\smash{)}\ 19\ } (2$$
$$\underline{16}$$
$$3 \overline{\smash{)}\ 8\ } (2$$
$$\underline{6}$$
$$2 \overline{\smash{)}\ 3\ } (1$$
$$\underline{2}$$
$$1$$

arriving at last at the remainder 1, since l and m are prime to each other.

Working from the bottom upwards we see
$$1 = 3 - 2,\quad 2 = 8 - 3 \times 2,\quad 3 = 19 - 8 \times 2.$$
$$\therefore\ 1 = 3 - (8 - 3 \times 2) = 3 \times 3 - 8$$
$$= 3 \times (19 - 8 \times 2) - 8$$
$$= 3 \times 19 - 7 \times 8$$
$$\therefore\ 7 \times 8 - 3 \times 19 = -1.$$

Change all the signs and introduce the expression
$$19 \times 8 - 19 \times 8;\quad \therefore\ (19 - 7) \times 8 - (8 - 3) \times 19 = 1$$
$$12 \times 8 - 5 \times 19 = 1.$$

Hence, whenever l and m are prime to each other, we can find a pair of integers p and q satisfying the equation $lq - mp = 1$; and it is clear that if the values of these integers are found as above p will be less than l and q than m.

513. To find the general solution in positive integers of $lx + my = n$.

Choosing the integers p and q as above to satisfy $lq - mp = 1$
we have
$$lx + my = n = n(lq - mp)$$
$$\therefore l(nq - x) = m(y + np)$$

Since l and m have no common factor, $nq - x$ must be divisible by m.

$$\therefore \frac{nq - x}{m} = \frac{y + np}{l} = t \text{ (an integer)}.$$

Hence
$$x = nq - mt,$$
$$y = lt - np,$$

is the general solution, t being an integer. Since t cannot be greater than $\frac{nq}{m}$ or less than $\frac{np}{l}$ the number of solutions is limited.

Ex. 1. Find the positive integral solutions of
$$7x + 5y = 44.$$

We have
$$\frac{7}{5} = 1 + \frac{1}{2+} \frac{1}{2}$$

The last convergent previous to $\frac{7}{5}$ is $\frac{3}{2}$, and
$$7 \cdot 2 - 3 \cdot 5 = -1 \text{ or } 7 \cdot 3 - 4 \cdot 5 = 1*$$
$$\therefore 7x + 5y = 44 \, (7 \cdot 3 - 4 \cdot 5).$$
$$\therefore 7(132 - x) = 5(y + 176).$$
$$\therefore \frac{132 - x}{5} = \frac{y + 176}{7} = t, \text{ an integer.}$$
$$\therefore x = 132 - 5t,$$
$$y = 7t - 176.$$

From the expression for y it is seen that t must be at least 26; from that for x it is seen that t cannot be greater than 26. Hence there is only one solution, corresponding to $t = 26$, namely $x = 2$, $y = 6$.

Ex. 2. Show that the equation $7x + 5y = 18$ admits of no solution in positive integers.

We find by the above method
$$x = 54 - 5t, \; y = 7t - 72.$$

From the expression for y we see that t must be at least 11; from that for x we see that it cannot be greater than 10; hence there is no solution.

---

* Or obtain this equation without the use of Continued Fractions, as in Second Method of § 512.

Ex. 3. Find all the solutions in positive integers of $7x + 5y = 111$. We find by the above method

$$x = 333 - 5t, \quad y = 7t - 444.$$

Hence, t admits of the values 66, 65, and 64, corresponding to which we have the solutions $x = 3$, $y = 18$; $x = 8$, $y = 11$; and $x = 13$, $y = 4$.

514. To find the general solution in positive integers of $lx - my = n$.

With the notation of the last article, if $lq - mp = 1$, we have

$$lx - my = n(lq - mp),$$
$$l(x - nq) = m(y - np),$$

and, as before, $\quad \dfrac{x - nq}{m} = \dfrac{y - np}{l} = t, \quad$ an integer.

Hence $x = mt + nq$, $y = lt + np$, and it is plain that the number of solutions is unlimited, since t may increase without limit.

Ex. 1. Find the positive integral solutions of

$$7x - 5y = 41.$$

We have, as above, $\quad 7x - 5y = 41\,(7.3 - 5.4)$,

$$\therefore 7(x - 123) = 5(y - 164),$$

$$\therefore \dfrac{x - 123}{5} = \dfrac{y - 164}{7} = t, \text{ an integer.}$$

$$\therefore x = 5t + 123, \quad y = 7t + 164.$$

From the value of y, we see that t must be at least -23, and may have any value from -23 upwards.

Hence the solutions are

$$x = 8, \quad y = 3; \quad x = 13, \quad y = 10, \text{ etc.}$$

Ex. 2. Solve in positive integers the equation

$$9x - 11y = 48.$$

We find by the above method

$$x = 11t + 240, \quad y = 9t + 192.$$

Hence t may have any value from -21 upwards, and the solutions are $\quad x = 9$, $y = 3$; $x = 20$, $y = 12$, etc.

515. Given one solution in positive integers of $lx + my = n$, **to find the general solution.**

Let $x = a$, $y = b$ be one solution. It is obvious from the preceding work that the general solution must be $x = a + mt$, $y = b - lt$ (t being an integer), but a strict proof may be given as follows:—

INDETERMINATE EQUATIONS.

$$lx + my = n = la + mb,$$
$$\therefore l(x - a) = m(b - y);$$

and since l is prime to m, m must divide $x - a$;

$$\therefore \frac{x-a}{m} = \frac{b-y}{l} = t, \text{ an integer.}$$

Hence the general solution is $x = a + mt$, $y = b - lt$.

516. Similarly if $x = a$, $y = b$ is one solution of $lx - my = u$, the general solution is

$$x = a + mt,$$
$$y = b + lt.$$

517. Number of solutions of $lx + my = n$.

Adopting the previous notation, and considering the case where $lq - mp = 1$, the general solution is

$$x = nq - mt,$$
$$y = lt - np. \hspace{2em} \text{(Art. 513).}$$

In order that x and y may be positive integers, t must not be greater than $\frac{nq}{m}$ or less than $\frac{np}{l}$.

Let
$$\frac{nq}{m} = \lambda + f_1, \hspace{2em} \frac{np}{l} = \mu + f_2,$$

where λ, μ are integers and f_1, f_2 proper fractions.

Then the greatest possible value of t is λ, and its least possible value $\mu + 1$; hence the number of solutions is

$$\lambda - \mu \hspace{2em} \text{or} \hspace{2em} \frac{nq}{m} - \frac{np}{l} - f_1 + f_2,$$

that is, $\frac{n}{lm} - f_1 + f_2$, since $lq - mp = 1$;

in other words, the integer next above or next below $\frac{n}{lm}$ according as the fractional part of $\frac{np}{l}$ or $\frac{nq}{m}$ is the greater.

[We leave as an exercise to the student the discussion of the special cases when either $\frac{n}{l}$, $\frac{n}{m}$, or $\frac{n}{lm}$ is integral, as well as the similar results which obtain when $lq - mp = -1$.]

Ex. Find the number of solutions of
$$9x + 11y = 302.$$

Here the last convergent preceding $\frac{9}{11}$ is $\frac{4}{5}$, and
$$lq - mp = 9.5 - 11.4 = 1$$

602   INDETERMINATE EQUATIONS.

Also
$$\frac{np}{l} = \frac{302 \times 4}{9} = \frac{1208}{9} = 134\tfrac{2}{9},$$
$$\frac{nq}{m} = \frac{302 \times 5}{11} = \frac{1510}{11} = 137\tfrac{3}{11}.$$

Hence the fractional part of $\frac{nq}{m}$ is greater than that of $\frac{np}{l}$, and the number of solutions is the integer next below $\tfrac{302}{99}$, namely 3.

EXERCISES XXXIV.

Find a solution in positive integers of each of the equations 1-3 :

1. $11x - 18y = 1.$ **2.** $11x - 18y = -1.$ **3.** $43x - 20y = 1.$

4. Find the number of solutions of $8x + 13y = 725$.

5. Find the least solution in positive integers and the general solution of $17x - 12y = 35$.

6. Find all the solutions of $9x + 19y = 533$.

7. Find the number of solutions of $12x + 7y = 599$.

8. Find the least solution in positive integers of $33x - 24y = 189$.

9. Find all the solutions of $15x + 8y = 271$.

10. Find two proper fractions with denominators 13 and 17 such that their difference is $\tfrac{4}{221}$.

11. Find the least solution in positive integers and the general solution of $11x - 16y = 1$.

12. Find the general form of all positive integers which when divided by 19 and 17 leave remainders 8 and 2 respectively.

13. Find the number of solutions of $13x + 5y = 780$.

14. In how many ways can a rope 19 feet long be made up of pieces of a yard and 4 feet in length respectively ?

15. Find the number of solutions of $9x + 14y = 784$.

16. Find all the solutions of $20x + 36y = 484$.

17. A football team scored 19 points out of tries (3 points each) and tries converted into goals (5 points each). How many tries were converted ?

18. A. races at the mile and half-mile. The former he can run in 4 min. 30 sec., and the latter in 2 min. In a series of such races his total time is 17 min. What total distance did he run in the series ?

19. A number consists of two digits. On reversing the digits the new number obtained is greater by 7 than twice the original number. Find the original number.

CHAPTER XXXV.

PARTIAL FRACTIONS.

518. Partial Fractions.—In Elementary Algebra, given a set of fractions, each of which is prefixed by a positive or negative sign, it is easy to find that single fraction which is equivalent to the algebraic sum of these fractions. For example,

$$\frac{2}{x-1} + \frac{3}{x-2} - \frac{4}{x-3} = \frac{x^2 - 10x + 13}{x^3 - 6x^2 + 11x - 6}.$$

It will be noticed that the least common denominator of the given fractions is the denominator of the resultant fraction.

The present chapter deals with the converse problem. That is, given a rational fraction such as $\dfrac{x^2 - 10x + 13}{x^3 - 6x^2 + 11x - 6}$, required to express it as the algebraic sum of a set of simple fractions.

These are called the **Partial Fractions** of the given fraction.

519. Applications.—If it is required to expand $\dfrac{3 - 4x}{1 - 3x + 2x^2}$ in powers of x it is possible to obtain the first few terms of the expansion by expanding $(1 - 3x + 2x^2)^{-1}$ by the Binomial Theorem. We then pick out the coefficients of x and its higher powers in the product $(3 - 4x)(1 - 3x + 2x^2)^{-1}$, and thus obtain as many terms of the expansion as we please; but if it is required to find the *general term* in the expansion this method obviously fails.

If, however, $\dfrac{3 - 4x}{1 - 3x + 2x^2}$ is replaced by its equivalent

$$\frac{1}{1-x} + \frac{2}{1-2x},$$

or $(1 - x)^{-1} + 2(1 - 2x)^{-1}$, it is easily expanded by the Binomial Theorem (*Ex.* 3, § 428) and the general term can be written down at once, viz. $x^r + 2 \cdot (2x)^r$, or $(2^{r+1} + 1)x^r$.

Again, suppose it is required to sum a series in which the general term is a rational fraction, and which cannot be summed by any of the methods explained in Chapter XXX. It may be then necessary to split up each term of the given series into its simple or partial fractions. A new series of terms is now formed, and some of the methods of the above chapter may be applicable for its summation.

604 PARTIAL FRACTIONS.

For example, take the series whose n^{th} term is $\dfrac{x^{n-1}}{(1+x^n)(1+x^{n+1})}$.

This cannot be summed by any known method in its present form. If, however, we express the nth term in the equivalent form

$$\frac{1}{x(1-x)}\left\{\frac{1}{1+x^{n+1}} - \frac{1}{1+x^n}\right\}$$

the series can now be easily summed by the method of Differences.

520. Case I.—Where the denominator of the given fraction has only linear factors, none of which are repeated.

Let $\dfrac{\phi(x)}{f(x)}$ denote the given fraction, and suppose for the present that $\phi(x)$ is of lower degree than $f(x)$.

Let $f(x)$ be resolved into its n linear factors

$$(a_1x - b_1)(a_2x - b_2)(a_3x - b_3)\ldots(a_nx - b_n).$$

Then *assume* that

$$\frac{\phi(x)}{f(x)} = \frac{A_1}{a_1x - b_1} + \frac{A_2}{a_2x - b_2} + \ldots + \frac{A_n}{a_nx - b_n},$$

where A_1, A_2, etc., are constants, independent of x, which are to be determined.

Next clear the above equation from fractions by multiplying by $f(x)$, and equate the coefficients of like powers of x on both sides of the resulting equation. It will be found that this gives exactly n equations to determine the n unknown constants A_1, A_2, $\ldots A_n$. Hence the initial assumption is justified.

Ex. 1. Resolve $\dfrac{5x+4}{x^2-2x-8}$ into partial fractions.

Here the denominator $= (x-4)(x+2)$.

Assume that $\dfrac{5x+4}{x^2-2x-8} = \dfrac{A_1}{x-4} + \dfrac{A_2}{x+2}$.

Clearing of fractions

$$(5x + 4) = A_1(x+2) + A_2(x-4).$$

Equating the coefficients of like powers of x,

$$5 = A_1 + A_2; \quad 4 = 2A_1 - 4A_2.$$

Whence $\qquad A_1 = 4, \quad A_2 = 1.$

Thus $\qquad \dfrac{5x+4}{x^2-2x-8} = \dfrac{4}{x-4} + \dfrac{1}{x+2}$.

Ex. 2. Resolve $\dfrac{1}{3+x-2x^2}$ into partial fractions.

The denominator $= (3-2x)(1+x)$.

PARTIAL FRACTIONS.

Let
$$\frac{1}{3+x-2x^2} = \frac{A}{(3-2x)} + \frac{B}{(1+x)}.$$

Thus
$$1 = A(1+x) + B(3-2x)$$

Equating coefficients
$$A + 3B = 1; \quad A - 2B = 0;$$

whence
$$B = \tfrac{1}{5}, \quad A = \tfrac{2}{5}.$$

Thus
$$\frac{1}{3+x-2x^2} \equiv \frac{2}{5(3-2x)} + \frac{1}{5(1+x)}.$$

521. Case II.—When the numerator of the given fraction is of higher degree than the denominator.

If the numerator of the given fraction is of higher degree than the denominator divide the numerator by the denominator until a remainder is obtained which is of lower degree than the denominator.

If this remainder be denoted by $\phi(x)$, and the denominator by $f(x)$, it is only required to resolve $\dfrac{\phi(x)}{f(x)}$ into its partial fractions.

Ex. Resolve $\dfrac{2x^3 - 3x^2 - 8x - 26}{2x^2 - 5x - 12}$ into partial fractions.

Divide the numerator by the denominator; thus

$$\frac{2x^3 - 3x^2 - 8x - 26}{2x^2 - 5x - 12} \equiv x + 1 + \frac{9x - 14}{2x^2 - 5x - 12}.$$

Now
$$2x^2 - 5x - 12 = (2x + 3)(x - 4).$$

Let
$$\frac{9x - 14}{(2x + 3)(x - 4)} = \frac{A_1}{2x + 3} + \frac{A_2}{x - 4}.$$

Clear of fractions, and equate the coefficients of the like powers of x; thus
$$9 = A_1 + 2A_2; \quad -14 = 3A_2 - 4A_1;$$
whence
$$A_1 = 5, \quad A_2 = 2.$$

Therefore
$$\frac{2x^3 - 3x^2 - 8x - 26}{2x^2 - 5x - 12} \equiv x + 1 + \frac{5}{2x + 3} + \frac{2}{x - 4}.$$

522. An Alternative Method.

Resolve $\dfrac{9x^2 + 7x - 20}{x^3 + 2x^2 - 5x - 6}$ into its partial fractions.

$$x^3 + 2x^2 - 5x - 6 = (x + 1)(x - 2)(x + 3).$$

Let
$$\frac{9x^2 + 7x - 20}{x^3 + 2x^2 - 5x - 6} = \frac{A_1}{x + 1} + \frac{A_2}{x - 2} + \frac{A_3}{x + 3}.$$

Clear of fractions,
$$9x^2 + 7x - 20 = A_1(x - 2)(x + 3) + A_2(x + 1)(x + 3) + A_3(x + 1)(x - 2).$$

PARTIAL FRACTIONS.

Now we might, as in the three preceding examples, equate the coefficients of the like powers of x, and thus find the values of A_1, A_2, A_3. But the following method is shorter, and sometimes can be done mentally.

The constants A_1, A_2, A_3 in the above equation are all independent of x, *and will therefore have the same values whatever value we give to x.*

If we put $x+1 = 0$, that is $x = -1$, the last two terms on the right-hand side of the equation vanish, and we thus obtain a simple equation to find A_1, viz. $9(-1)^2 + 7(-1) - 20 = -3 \cdot 2 \cdot A_1$,
whence $\qquad A_1 = 3$.

So to find A_2, put $x - 2 = 0$, that is $x = 2$,
$$\therefore 9 \cdot 2^2 + 7 \cdot 2 - 20 = 3 \cdot 5 \cdot A_2,$$
whence $\qquad A_2 = 2$.
Similarly by putting $x + 3 = 0$, that is $x = -3$, we find that
$$A_3 = 4,$$
$$\therefore \frac{9x^2 + 7x - 20}{x^3 + 2x^2 - 5x - 6} \equiv \frac{3}{x+1} + \frac{2}{x-2} + \frac{4}{x+3}.$$

523. Case III.—Where the denominator has repeated linear factors.

If the linear factor $(a_1x - b_1)$ occurs twice in $f(x)$ there are two corresponding partial fractions of the form $\dfrac{A_1}{a_1x - b_1} + \dfrac{B_1}{(a_1x - b_1)^2}$, and if $(a_1x - b_1)$ occurs three times as a factor of $f(x)$ the corresponding partial fractions are $\dfrac{A_1}{a_1x - b_1} + \dfrac{B_1}{(a_1x - b_1)^2} + \dfrac{C_1}{(a_1x - b_1)^3}$, and so on.

Clearing of fractions as before, by multiplying throughout by $f(x)$, and equating the coefficients of the like powers of x, we get as many equations as there are unknown constants.

524. The following examples will make this case more familiar to the student:—

Ex. 1. Resolve $\dfrac{x^2 + 7x + 11}{(x + 2)^2(x + 3)}$ into partial fractions.

Let $\qquad \dfrac{x^2 + 7x + 11}{(x + 2)^2(x + 3)} = \dfrac{A}{(x + 2)^2} + \dfrac{B}{x + 2} + \dfrac{C}{x + 3}$,
$$\therefore x^2 + 7x + 11 = A(x+3) + B(x+2)(x+3) + C(x+2)^2.$$
Put $x + 2 = 0$, or $x = -2$; thus $A = 1$.
Put $x + 3 = 0$, or $x = -3$; thus $C = -1$.
To find B, equate the coefficients of x^2.
Then $\qquad 1 = B + C, \quad \therefore B = 1 - C = +2$.
Hence $\qquad \dfrac{x^2 + 7x + 11}{(x+2)^2(x+3)} \equiv \dfrac{1}{(x+2)^2} + \dfrac{2}{x+2} - \dfrac{1}{x+3}$.

Ex. 2. Resolve $\dfrac{9}{(x-1)(x+2)^2}$ into partial fractions.

Let $\dfrac{9}{(x-1)(x+2)^2} = \dfrac{A}{x-1} + \dfrac{B}{x+2} + \dfrac{C}{(x+2)^2}$.

Thus $9 = A(x+2)^2 + B(x-1)(x+2) + C(x-1)$.

Put $x = 1$; then $A = 1$. Put $x = -2$; then $C = -3$.

Equating the coefficients of x^2, $A + B = 0$, $\therefore B = -1$.

Thus $\dfrac{9}{(x-1)(x+2)^2} \equiv \dfrac{1}{x-1} - \dfrac{1}{x+2} - \dfrac{3}{(x+2)^2}$.

Ex. 3. Resolve $\dfrac{10x^3 - 13x^2 - 6x + 4}{x^2(x-1)^2}$ into partial fractions.

Here both the factors x and $(x-1)$ occur twice in the denominator.

Let $\dfrac{10x^3 - 13x^2 - 6x + 4}{x^2(x-1)^2} = \dfrac{A}{x} + \dfrac{B}{x^2} + \dfrac{C}{x-1} + \dfrac{D}{(x-1)^2}$.

Clear of fractions,

$10x^3 - 13x^2 - 6x + 4 = Ax(x-1)^2 + B(x-1)^2 + Cx^2(x-1) + Dx^2$.

Put $x = 0$; then $B = 4$.

Put $x - 1 = 0$, or $x = +1$; then $D = -5$.

To find A and C equate the coefficients of x^3 and x^2.

Then $10 = A + C$,

$-13 = -2A + B - C + D$, *i.e.* $-12 = -2A - C$;

$\therefore A = 2$, and $C = 8$.

Thus $\dfrac{10x^3 - 13x^2 - 6x + 4}{x^2(x-1)^2} \equiv \dfrac{2}{x} + \dfrac{4}{x^2} + \dfrac{8}{x-1} - \dfrac{5}{(x-1)^2}$.

525. Case IV.—Where the denominator has quadratic factors, none of which is repeated.

If the denominator $f(x)$ has a quadratic factor of the form

$$a_1 x^2 + b_1 x + c_1,$$

then the corresponding partial fraction is

$$\dfrac{A_1 x + B_1}{a_1 x^2 + b_1 x + c_1}.$$

Thus $\dfrac{\phi(x)}{f(x)} = \dfrac{A_1 x + B_1}{a_1 x^2 + b_1 x + c_1} + \dfrac{A_2}{a_2 x + b_2} + \ldots$

The constants are found as before by clearing of fractions, and equating the coefficients of like powers of x.

Ex. Resolve $\dfrac{3x^2 + 2x - 2}{x^3 - 1}$ into partial fractions.

$$x^3 - 1 = (x-1)(x^2 + x + 1).$$

PARTIAL FRACTIONS.

Let $$\frac{3x^2 + 2x - 2}{x^3 - 1} = \frac{Ax + B}{x^2 + x + 1} + \frac{C}{x - 1}.$$

Clear of fractions,
$$(3x^2 + 2x - 2) = (Ax + B)(x - 1) + C(x^2 + x + 1).$$
Let $x = 1$, then $C = 1$.
Equate coefficients of x^2; $3 = A + C$, $\therefore A = 2$.
Equate the constant terms; $-2 = -B + C$, $\therefore B = 3$.
Hence $$\frac{3x^2 + 2x - 2}{x^3 - 1} \equiv \frac{2x + 3}{x^2 + x + 1} + \frac{1}{x - 1}.$$

526. Case V.—Where the denominator has repeated quadratic factors.

If the factor $(a_1x^2 + b_1x + c_1)$ occurs twice in the denominator, the corresponding partial fractions are
$$\frac{A_1x + B_1}{(a_1x^2 + b_1x + c_1)^2} + \frac{A_2x + B_2}{(a_1x^2 + b_1x + c_1)},$$
and if the same factor occurs three times the corresponding partial fractions are
$$\frac{A_1x + B_1}{(a_1x^2 + b_1x + c_1)^3} + \frac{A_2x + B_2}{(a_1x^2 + b_1x + c_1)^2} + \frac{A_3x + B_3}{a_1x^2 + b_1x + c_1}.$$
The constants A_1, B_1, etc., are found in the usual way.

Ex. Resolve $\dfrac{2x^4 + 2x^2 + x + 1}{x(x^2 + 1)^2}$ into partial fractions.

Here the factor $(x^2 + 1)$ occurs twice in the denominator.

Let $$\frac{2x^4 + 2x^2 + x + 1}{x(x^2 + 1)^2} = \frac{A}{x} + \frac{Bx + C}{(x^2 + 1)^2} + \frac{Ex + F}{x^2 + 1}.$$

Clear of fractions
$$2x^4 + 2x^2 + x + 1 = A(x^2 + 1)^2 + (Bx + C)x + (Ex + F)(x^2 + 1)x.$$
Put $x = 0$, then $A = 1$.
Equating the coefficients of x^4, x^3, etc.,
$$2 = A + E, \quad 0 = F, \quad 2 = 2A + B + E, \quad 1 = C + F,$$
$$\therefore E = 1, \quad F = 0, \quad B = -1, \quad C = 1.$$
Hence $$\frac{2x^4 + 2x^2 + x + 1}{x(x^2 + 1)^2} \equiv \frac{1}{x} + \frac{1 - x}{(1 + x^2)^2} + \frac{x}{(1 + x^2)}.$$

527. The method of applying the use of Partial Fractions to the expansion of a rational fraction in ascending powers of x will be most easily understood from the following examples:—

Ex. 1. Find the coefficient of x^n in the expansion of $\dfrac{5 - x}{1 - x - 2x^2}$.

PARTIAL FRACTIONS.

Resolving the given fraction into its partial fractions,
$$\frac{5-x}{1-x-2x^2} \equiv \frac{2}{1+x} + \frac{3}{1-2x}$$
$$\equiv 2(1+x)^{-1} + 3(1-2x)^{-1}.$$
$\therefore$ (§ 428) The coefficient of $x^n = 2(-1)^n + 3.2^n$.

Ex. 2. Find the general term in the expansion of $\dfrac{2x+1}{(x-1)(1+x^2)}$ in ascending powers of x.

Here put $\quad\dfrac{2x+1}{(x-1)(1+x^2)} = \dfrac{A}{x-1} + \dfrac{Bx+C}{(1+x^2)},$
$$\therefore (2x+1) = A(x^2+1) + (Bx+C)(x-1).$$
By the usual methods, $A = \frac{3}{2}$, $B = -\frac{3}{2}$, and $C = \frac{1}{2}$,
and thus the given fraction $\equiv \frac{3}{2}\dfrac{1}{x-1} + \dfrac{1-3x}{2(1+x^2)}$
$$\equiv -\tfrac{3}{2}(1-x)^{-1} + \tfrac{1}{2}(1-3x)(1+x^2)^{-1}$$
$$\equiv -\tfrac{3}{2}(1+x+x^2+x^3+\ldots) + \tfrac{1}{2}(1-3x)(1-x^2+x^4-x^6+\ldots).$$
Now pick out the terms in x^r.

If r is even the coefficient of x^r is
$$-\tfrac{3}{2} + \tfrac{1}{2}(-1)^{\frac{r}{2}} \quad \text{or} \quad \tfrac{1}{2}\{(-1)^{\frac{r}{2}} - 3\},$$
and if r is odd the coefficient of x^r is
$$-\tfrac{3}{2} + \tfrac{1}{2}.3.(-1)^{\frac{r+1}{2}} \quad \text{or} \quad -\tfrac{3}{2}\left[1 + (-1)^{\frac{r-1}{2}}\right].$$

528. Miscellaneous Examples.

Ex. 1. Resolve $\dfrac{1}{(x-1)^4(x+1)}$ into partial fractions.

Assume that $\quad\dfrac{1}{(x-1)^4(x+1)} = \dfrac{A}{x+1} + \dfrac{f(x)}{(x-1)^4}$
where A is constant, and $f(x)$ a function of x to be determined.

Clear of fractions,
$$1 = A(x-1)^4 + (x+1)f(x).$$
Put $x = -1$; then $A = \tfrac{1}{16}$,
$\therefore (x+1)f(x) = 1 - \tfrac{1}{16}(x-1)^4 = \tfrac{1}{16}\{15 + 4x - 6x^2 + 4x^3 - x^4\},$
and $\quad\therefore f(x) = \tfrac{1}{16}(15 - 11x + 5x^2 - x^3).$

We must next find the partial fractions corresponding to
$$\frac{15 - 11x + 5x^2 - x^3}{(x-1)^4}.$$
To do this put $x - 1 = y$, or $x = 1 + y$.

TUT. ALG. II.

Then $\dfrac{15-11x+5x^2-x^3}{(x-1)^4} = \dfrac{15-11(1+y)+5(1+y)^2-(1+y)^3}{y^4}$

$= \dfrac{8-4y+2y^2-y^3}{y^4} = \dfrac{8}{y^4} - \dfrac{4}{y^3} + \dfrac{2}{y^2} - \dfrac{1}{y}$,

$\therefore \dfrac{15-11x+5x^2-x^3}{(x-1)^4} = \dfrac{8}{(x-1)^4} - \dfrac{4}{(x-1)^3} + \dfrac{2}{(x-1)^2} - \dfrac{1}{x-1}$.

Thus $\dfrac{1}{(x-1)^4(x+1)}$

$= \dfrac{1}{16(x+1)} + \dfrac{1}{2(x-1)^4} - \dfrac{1}{4(x-1)^3} + \dfrac{1}{8(x-1)^2} - \dfrac{1}{16(x-1)}$.

Ex. 2. Sum to n terms the series whose n^{th} term is $\dfrac{n}{1+n^2+n^4}$.

Let u_n denote the nth term.

Resolve into partial fractions, thus

$$u_n = \tfrac{1}{2}\left[\dfrac{1}{1-n+n^2} - \dfrac{1}{1+n+n^2}\right].$$

Now it is easily seen that this may be written

$$u_n = \tfrac{1}{2}\left[\dfrac{1}{1+(n-1)+(n-1)^2} - \dfrac{1}{1+n+n^2}\right].$$

Similarly

$$u_{n-1} = \tfrac{1}{2}\left[\dfrac{1}{1+(n-2)+(n-2)^2} - \dfrac{1}{1+(n-1)+(n-1)^2}\right],$$

$$u_{n-2} = \tfrac{1}{2}\left[\dfrac{1}{1+(n-3)+(n-3)^2} - \dfrac{1}{1+(n-2)+(n-2)^2}\right],$$

. . . .
. . . .

$$u_1 = \tfrac{1}{2}\left[\dfrac{1}{1} - \dfrac{1}{3}\right].$$

Thus by addition $u_1 + u_2 + \ldots + u_n = \tfrac{1}{2}\left[1 - \dfrac{1}{1+n+n^2}\right]$.

Ex. 3. Prove that the coefficient of x^n in the expansion of

$$\dfrac{x}{1+bx+ax^2}$$

in ascending powers of x is $\dfrac{a^n - \beta^n}{a - \beta}$ where a and β are the roots of the quadratic equation $t^2 + bt + a = 0$.

$$\dfrac{x}{1+bx+ax^2} = \dfrac{x}{(1-ax)(1-\beta x)}$$

where $a\beta = a$, $a + \beta = -b$, *i.e.* a, β are the roots of $t^2 + bt + a = 0$.

By partial fractions,

$$\frac{x}{(1-ax)(1-\beta x)} \equiv \frac{1}{(a-\beta)}\left\{\frac{1}{1-ax} - \frac{1}{1-\beta x}\right\}$$

$$\equiv \frac{1}{(a-\beta)}[(1-ax)^{-1} - (1-\beta x)^{-1}].$$

Thus by the Binomial Theorem (§ 428) the coefficient of x^n is

$$\frac{1}{a-\beta}(a^n - \beta^n) \text{ or } \frac{a^n - \beta^n}{a - \beta}.$$

529. Explanation of the preceding rules.—It remains to show that the assumptions made in the preceding articles will always give one and only one set of values for the unknown coefficients.

This follows from the following two facts:—

If the denominator of the given fraction is of the nth degree, then

(1) the assumptions made always involve n unknown coefficients, and

(2) there are always n linear equations to determine these coefficients.

(1) To prove this it is sufficient to point out that in all the various cases the number of unknown coefficients corresponding to any factor in the denominator is equal to the degree of that factor.

For example,

for each factor of the form	$(ax+b)$	we assume 1 unknown coefficient,
,, ,, ,, ,,	$(ax+b)^2$	,, 2 ,, ,,
,, ,, ,, ,,	$(ax+b)^3$	,, 3 ,, ,,
,, ,, ,, ,,	(ax^2+bx+c)	,, 2 ,, ,,
,, ,, ,, ,,	$(ax^2+bx+c)^2$	,, 4 ,, ,,

and so on.

It follows that the total number of these unknown coefficients is equal to the degree of the product of these factors.

(2) When the equation is cleared of fractions the left-hand side is always of the $(n-1)$th degree or lower. Also the right-hand side is *always* of the $(n-1)$th degree; for the degree of each numerator on the right hand is less than the degree of its denominator, and there is always at least one fraction in which the degree of the numerator is exactly 1 less than the degree of its denominator.

Since the equation when cleared of fractions is of the $(n-1)$th degree, therefore by equating coefficients we always obtain n linear equations.

EXERCISES XXXV.

Resolve into partial fractions—

1. $\dfrac{2x+5}{(x+2)(x+3)}$.

2. $\dfrac{x+5}{x^2+7x+12}$.

3. $\dfrac{1}{x^2-9x+20}$.

4. $\dfrac{2x+12}{x^2-4x-12}$.

5. $\dfrac{1}{x(x+1)(2x+1)}$.

6. $\dfrac{2x^3+5x^2-3x-1}{2x^2+5x+2}$.

7. $\dfrac{4(x-1)}{(x+2)(x-2)^2}$.

8. $\dfrac{2x-5}{(x+3)(x+1)^2}$.

9. $\dfrac{23x-11x^2}{(2x-1)(9-x^2)}$.

10. $\dfrac{4x^4+4x^3+5x-1}{(x-1)(2x+1)^2}$.

11. $\dfrac{9(x+1)}{(x-1)^2(x+2)^2}$.

12. $\dfrac{4}{1-x^4}$.

13. $\dfrac{x-1}{(x+1)(x^2+1)}$.

14. $\dfrac{1}{(x^2+1)(x^2-x+1)}$.

15. $\dfrac{x^5}{(x^2+1)(x-4)}$.

16. $\dfrac{2}{x(2+x^2)^2}$.

17. $\dfrac{x^3+x+1}{(1-x)^4}$.

Find the general term in the expansion of the following expressions in ascending powers of x—

18. $\dfrac{3-5x}{1-4x+3x^2}$.

19. $\dfrac{1}{(1-x)(1-2x)(1-3x)}$.

20. $\dfrac{x}{(x-a)(x-b)}$.

21. $\dfrac{2x^2}{(x-1)^2(x^2+1)}$.

22. $\dfrac{3-x}{(2-x)(1-x)^2}$.

23. $\dfrac{1}{(1+x)(1+x^2)}$.

24. Sum to n terms the series
$$\dfrac{1}{(1+x)(1+x^2)}+\dfrac{x}{(1+x^2)(1+x^3)}+\dfrac{x^2}{(1+x^3)(1+x^4)}+\ldots$$

25. Find the sum to n terms of the series whose n^{th} terms are

(1) $\dfrac{2n+1}{n^2(n+1)^2}$, (2) $\dfrac{a^{n-1}}{(1+a^{n-1}x)(1+a^n x)}$.

PARTIAL FRACTIONS.

26. Find the sum to infinity of the series whose n^{th} term is
$$\frac{(n+1)(-x)^{n+1}}{n(n+2)}.$$

***27.** Prove that the coefficient of x^n in the expansion of
$$\frac{2-ax}{1-ax+bx^2}$$
is $\alpha^n + \beta^n$, where α and β are the roots of the quadratic
$$t^2 - at + b = 0.$$

Hence obtain a series for expressing $\alpha^n + \beta^n$ in terms of $\alpha\beta$ and $\alpha + \beta$.

APPENDIX.

ON THE SOLUTION OF EQUATIONS.

1. Reversible and irreversible operations. Throughout the book methods have been given for the solution of equations of various type. Some points in the theory of such methods require a little closer scrutiny.

The chief difficulties arise in the case of what have been called *extraneous* roots.

Consider a trivial example: the equation $x = a$, on squaring each side [which is sometimes called "squaring the equation"], becomes $x^2 = a^2$.

The original equation has the single root $x = a$: the new equation has the *two* roots $x = \pm a$; so that the extraneous root $x = -a$ is introduced. This will illustrate the meaning of the terms **"reversible"** and **"irreversible"** operations. The process of squaring the above equation is irreversible, the result being *not solely equivalent* to the original equation. The usual operations may now be classified as **reversible** or **irreversible**.

In what follows there is a single variable x, and capital letters P, Q, R, etc., will be used to represent functions of x.

I. *Addition of a constant to both sides of an equation is a reversible step.*

For if the equation is $P = Q$ and a any constant, the equation $P + a = Q + a$ involves $P + a - Q - a = 0$ or $P = Q$ only.

II. *Multiplication (or division) by any constant is reversible.*

APPENDIX. 615

Let $P = Q$ be the equation: then $aP = aQ$ is equivalent to $a(P - Q) = 0$, *i.e.* either $a = 0$ or $P = Q$. But a is not $= 0$; $\therefore aP = aQ$ is equivalent solely to $P = Q$.

The above two operations therefore cannot introduce any extraneous roots.

III. *Multiplication by a function of x (say S) is irreversible.*

For with same equation as before $SP = SQ$ is equivalent to $S(P - Q) = 0$; $\therefore$ either $S = 0$ or $P = Q$.

Hence the new equation may also possess the roots given by $S = 0$.

IV. *Raising each side of an equation to the nth power is an irreversible operation.*

For $P^n - Q^n = 0$ is equivalent to
$(P - Q)(P^{n-1} + P^{n-2}Q + \ldots + PQ^{n-2} + Q^{n-1}) = 0$:
and its roots are those of $P = Q$ *and* of
$$P^{n-1} + P^{n-2}Q + \ldots + Q^{n-1} = 0.$$

Hence *if in solving an equation either of operations III., IV. has been performed,* the results may include extraneous roots, and therefore should be substituted in the original equation to test whether they really satisfy it; or this question may sometimes be settled in other ways.

2. Number of Solutions of a system of equations.

A few general results will be indicated here and illustrated by examples.

Consider first the system of equations
$$lx + my = n \quad \ldots\ldots\ldots\ldots\ldots\ldots\ldots\ldots(\text{i})$$
$$ax^2 + 2hxy + by^2 + 2gx + 2fy + c = 0 \quad \ldots\ldots(\text{ii})$$

The method of solution is well known: from the first equation y may be obtained in terms of x and then substituted in (ii), when a quadratic in x is obtained.

Now this equation has been obtained by using *both* the

given equations. Hence if one value of x deduced from it be substituted in (i), and the corresponding value of y obtained, these values of x and y must satisfy *both* equations : and these are the *only* values which can satisfy both equations.

Consider next the system
$$ax^2 + 2hxy + by^2 = c \quad\text{...................(i)}$$
$$a_1x^2 + 2h_1xy + b_1y^2 = c_1 \quad\text{..............(ii)}$$

Let y be eliminated between these by the method of § 180 (x being considered as a mere constant during the elimination). The result will be found to be an equation of the fourth degree in x.

Call its four roots α, β, γ, δ. [The method of finding these is not required : that the equation has four roots follows from § 371.]

Suppose any root α for x is substituted in (i). Apparently *two* values of y will be deduced and thus there would be 4×2 solutions altogether. But by the method of deduction, the value α of x will make equations (i) and (ii) have one *common* root in y, and they will in general have only *one* common root.

Hence only 4 solutions altogether will be obtained. This is equal to the product of the degrees in the variables of the two equations.

A general rule may be stated.

Given any system of rational, integral equations, to determine the same number of variables, then the number of solutions to be expected is equal to the product of the degrees in the variables of all the equations.

In some cases, however, the **number** of solutions obtained apparently falls short of this, as in the following example :—
$$x^2 + xy = 8$$
$$xy + y^2 = 1$$
Adding $\quad (x+y)^2 = 9 \quad \therefore\ x+y = \pm 3$.

Dividing each side of the original equations by this result, the solutions (two in number) $x = \pm \frac{8}{3}$, $y = \pm \frac{1}{3}$ are obtained.

APPENDIX.

But from the equations (each being of the second degree) 2×2 or 4 solutions would be expected.

These may be obtained by solving in another way.

Multiply 2nd equation by 8 and subtract from the first

$$x^2 - 7xy - 8y^2 = 0$$
$$x = -y \text{ or } 8y$$

$x = 8y$ gives the former set of solutions, $x = -y$ substituted in the first equation gives

$$x^2 = \frac{8}{1-1} = \infty \qquad x = \pm \infty.$$

Hence two further solutions

$$x = \pm \infty, \quad y = \pm \infty \text{ are obtained.}$$

Infinite values as roots are usually rejected in practice: they are, however, included in the general rule given above.

[The student acquainted with analytical geometry will recognise that the hyperbolas represented by the two given equations have a common asymptote, and therefore have two common consecutive points at infinity.]

As another example take the case—

$$x + y = 2xy \quad \ldots\ldots\ldots\ldots\ldots\ldots\ldots\ldots (1)$$
$$x - y = \tfrac{1}{2}xy \quad \ldots\ldots\ldots\ldots\ldots\ldots\ldots\ldots (2)$$

Here 2×2 or 4 solutions are expected.

A convenient method (disregarding the theory of such equations) is to divide each equation by xy and solve for $\dfrac{1}{x}, \dfrac{1}{y}$.

Thus

$$\frac{1}{x} = \frac{5}{4}, \quad \frac{1}{y} = \frac{3}{4}, \quad x = \frac{4}{5}, \quad y = \frac{4}{3},$$

a single solution.

Hence 3 solutions remain to be accounted for. Eliminate y between the equations [generally a troublesome, but often the only thoroughly satisfactory, method].

We write the equations—

$$x = (2x - 1)y, \quad 2x = (x + 2)y$$

whence $\quad x(x + 2) = 2x(2x - 1), \quad 3x^2 - 4x = 0 \ldots\ldots\ldots\ldots\ldots (a)$

$$x = 4/3 \text{ or } 0.$$

The solution $x = \tfrac{4}{3}$ gives $y = \tfrac{4}{5}$ as before. $x = 0$ gives $y = 0$, a *new* solution

But the equation (a) is only a quadratic, whereas an equation of the 4th degree was to be expected (see application of § 180 above).

The other two solutions are $x = \infty$, y being indeterminate, and $y = \infty$, x being indeterminate. Each of these satisfies the equations by making both sides infinite.

[The rectangular hyperbolas represented meet at (0, 0) and ($\frac{4}{3}$, $\frac{5}{3}$) and have asymptotes parallel to the axes of x and of y. The latter account for the solutions for which x or y is infinite.]

As a last example consider the symmetrical system—
$$\left. \begin{array}{c} x + y + z = a + b + c \\ x^2 + y^2 + z^2 = a^2 + b^2 + c^2 \\ xyz = abc \end{array} \right\}$$

Here the degrees of the equations are 1, 2, 3 respectively, hence 6 solutions should be obtained.

One obvious solution is $x = a$, $y = b$, $z = c$, verifying all three equations. But since any two of the three x, y, z may be interchanged without altering the *equations*, so any two of the three may be interchanged *in the solution* and the equations will still be satisfied. Thus the set of solutions—

$x =$	a	a	b	b	c	c
$y =$	b	c	c	a	a	b
$z =$	c	b	a	c	b	a

is obtained.

[*E.g.* by interchanging x and y in the first solution above, from $x = a$, $y = b$, $z = c$ is obtained $y = a$, $x = b$, $z = c$, and similarly for all possible permutations.]

Hence 6 solutions have been obtained, and this is the number expected: the equations are therefore fully solved.

ANSWERS.

EXERCISES I.

THE THEORY OF INDICES.

1. (i) $\sqrt[5]{(\sqrt{x} \cdot \sqrt[3]{y})}$; $(x^{\frac{1}{2}}y^{\frac{1}{3}})^{\frac{1}{5}}$. (ii) $\sqrt[m]{\left(\dfrac{\sqrt{m}}{\sqrt[3]{n}}\right)^p}$; $\left(\dfrac{m^{\frac{1}{2}}}{n^{\frac{1}{3}}}\right)^{\frac{p}{m}}$.

2. $\dfrac{bc}{a} + \dfrac{ac^2}{b^3} + \dfrac{1}{abc}$.   3. $a^{-1} + 2a^{-2} + 3a^{-3}$.

4. $\dfrac{1}{a^{\frac{1}{2}} \cdot b^3}$;   5. a;   6. 2;   7. $x^{\frac{38}{13}} y^{\frac{24}{13}} z^{\frac{10}{13}}$;   8. 2^p;

9. $x(x-a)$;   10. $2c\sqrt[3]{a^2 b}$;   11. $\left(\dfrac{p}{q}\right)^{p+q}$.

12. 8.   13. (i) $x^n + 1 + x^{-n}$; (ii) $a - b^2$.

14. (i) $a^2 - b$; (ii) $x^{\frac{2}{3}} + 2x^{\frac{1}{3}} + 3$; (iii) $x^{2m} + y - z$.

15. $2a^{\frac{1}{2}} - 3b^{-\frac{1}{2}} c^{-\frac{1}{3}}$.   16. $\dfrac{x^{\frac{2}{3}} y^{\frac{2}{3}}}{(x^{\frac{2}{3}} - y^{\frac{2}{3}})^2}$.   18. 64.

19. 1.   20. $m = n^{\frac{1}{n-1}}$.

25. $5^{\frac{2}{3}} \cdot 3^{\frac{1}{3}}$; $5^{\frac{1}{2}} \cdot 3^{\frac{1}{4}} \cdot 2^{\frac{1}{4}}$.

EXERCISES II.

SURDS.

1. $\sqrt[3]{(a^{12} b^{12})}$.   2. $\sqrt[n]{(a^m b^{2m} c^{3m})}$.   3. $\sqrt{175}$.

4. $\sqrt[3]{24}$.   5. $\sqrt[5]{(a^{10} b^5 c)}$.   6. $\sqrt[5]{486}$.

7. $\sqrt{\dfrac{45}{272}}$.   8. $\sqrt[m]{\dfrac{a^m c}{b^m d}}$.   9. $\sqrt[s]{\dfrac{b^{m-rs}}{a^{2n-sn}}}$.

ANSWERS.

10. $\sqrt{\dfrac{x+y}{(x-y)(x^2-xy+y^2)}}$. 11. $5\sqrt{5}$.

12. $4\sqrt[3]{12}$.   13. $ab^3\sqrt{(bc)}$.   14. $\tfrac{1}{5}\sqrt[3]{25}$.

15. $\sqrt{6}$.   16. $3\sqrt[3]{25}$.   17. $6\sqrt[5]{48}$.

18. $\tfrac{3}{5}\sqrt{2}$.   19. $\tfrac{2}{3}\sqrt[5]{2}$.   20. $-40\sqrt{3}$.

21. $(x-1)\sqrt{m}$.   22. $\sqrt[5]{a^5 b^{\tfrac{10}{3}}}$.   23. $\sqrt[5]{a^{45}c^5}$.

24. $\sqrt[5]{\dfrac{c^{20}}{a^{\tfrac{5}{3}} b^{\tfrac{10}{3}}}}$.   25. $\dfrac{1}{\sqrt[7]{(a^{-7}b^{-\tfrac{14}{3}})}}$; $\sqrt[7]{\dfrac{a^{-14}}{b^{-14}c^{-\tfrac{7}{3}}}}$.

26. $\sqrt[m]{(a^{\tfrac{1}{2}m}b^{\tfrac{3}{2}m})}$.   27. $\dfrac{1}{\sqrt[m]{x^{mn}}}$.   28. $\dfrac{1}{\sqrt[n]{(x^{2m}y^{m/3}z^m)}}$.

29. $\sqrt[m]{(a^{\tfrac{2m}{3}} c^{\tfrac{m}{3}})}$.   30. $\{\sqrt[5]{(a+b)^4}\}^{\tfrac{5}{3}}$.   31. $6\sqrt[5]{\tfrac{2}{3}\tfrac{5}{6}}$.

32. $abc\sqrt{\dfrac{a-b}{bc}}$.   33. $2a\sqrt[5]{\dfrac{2b}{a^3}}$.

38. $\sqrt[6]{a^2}$; $\sqrt[6]{b^3}$.   39. $\sqrt[12]{64}$; $\sqrt[12]{81}$; $\sqrt[12]{64}$.

40. $\sqrt[6]{5}$; $\sqrt[6]{1331}$; $\sqrt[6]{49}$.   41. $\sqrt[12]{15{,}625}$; $\sqrt[12]{2{,}401}$; $\sqrt[12]{27}$.

42. $\sqrt[18]{531{,}441}$; $\sqrt[18]{125}$; $\sqrt[18]{49}$.   43. $\sqrt{5}>\sqrt[3]{11}$.

44. $3\sqrt[3]{2}>\sqrt{14}>2\sqrt[3]{6}$.   45. $\tfrac{1}{2}\sqrt{3}>\tfrac{1}{3}\sqrt{5}>\tfrac{1}{4}\sqrt{7}$.

46. $\sqrt{5}<2\sqrt[3]{\tfrac{3}{2}}<3(4\tfrac{1}{2})^{-\tfrac{1}{6}}$.   47. $\sqrt{6}$.   48. $7\sqrt[3]{2}$.

49. $8\tfrac{1}{2}\sqrt[3]{2}$.   50. $45\tfrac{13}{36}\sqrt{3}$.   51. $14\tfrac{1}{2}\sqrt{3}$.

52. $26\tfrac{11}{30}\sqrt{6}$.   53. $a(2ax\sqrt{ax}-5\sqrt[3]{ax^2})$.   54. 1680.

55. $\sqrt[6]{1960}$.   56. $288\sqrt[12]{72}$.   57. $\sqrt[12]{10125}$.

58. $-5\sqrt{3}-11$.   59. 4.   60. $x-y$.

61. x^8-1.   62. $\tfrac{1}{5}(\sqrt{5}+1)$.   63. $a^{\tfrac{2}{n}}, a^{\tfrac{3}{n}}, \ldots\ldots a^{\tfrac{n-1}{n}}$.

64. $a^{\tfrac{1}{6}}b^{\tfrac{1}{4}}c^{\tfrac{7}{6}}$ and $a^{\tfrac{3}{6}}b^{\tfrac{1}{2}}$.   65. 8·660.

66. 9·797.   67. 2·683.   68. 1·224.   69. ·204.

70. ·057.   71. ·819.   72. 2·339.

73. $a+b-c^2$.   74. 10.   75. $5+2\sqrt{6}$.

76. $\sqrt{5}+1$.   77. $5-2\sqrt{3}$.   78. $\tfrac{1}{5}(2\sqrt{2}+\sqrt{3})$.

79. $\tfrac{1}{11}(7\sqrt{14}-13)$.   80. $\dfrac{2\sqrt{a^2-x^2}}{x^2}$.   81. $2\sqrt{3}+2\sqrt{2}+3$.

82. $\dfrac{1}{1-x^2}$.   83. $\dfrac{2\sqrt{3}+3\sqrt{2}-\sqrt{30}}{12}$.

84. $\sqrt{6}+\sqrt{2}+\sqrt{5}$.   85. $\tfrac{1}{12}(\sqrt{2}+\sqrt{3}-\sqrt{5})$.

86. $1+\sqrt{3}+\sqrt{5}$.   87. $\sqrt{5}$.

ANSWERS.

88. 3.
89. $\frac{1}{2}(x+\sqrt{x^2-4})$.
90. (i) $\sqrt[3]{9}-\sqrt[3]{12}+\sqrt[3]{16}$; 7; (ii) $a^{\frac{5}{4}}-ab^{\frac{1}{4}}+a^{\frac{1}{2}}b^{\frac{1}{2}}-b^{\frac{3}{4}}$; a^2-b;
(iii) $9+3\sqrt[3]{4}+\sqrt[3]{16}$; 23;
(iv) $a^{\frac{5}{3}}+a^{\frac{4}{3}}b^{\frac{1}{2}}+ab+a^{\frac{2}{3}}b^{\frac{3}{2}}+a^{\frac{1}{3}}b^2+b^{\frac{5}{2}}$; a^2-b^3.
91. $\sqrt{(4+2\sqrt{2})}-\sqrt{2}-1$. **92.** $2\frac{5}{4}$. **93.** $\frac{1}{36}$.
94. ·08392. **100.** $\frac{1}{4}(\sqrt{2}+2\sqrt{3}+3\sqrt{6}-4)$.
101. $\frac{1}{2}(\sqrt{6}+1-\sqrt{3}-\sqrt{2})$. **102.** $\frac{1}{8}(1+\sqrt{15}+\sqrt{3}+\sqrt{5})$.
109. $2+\sqrt{3}$. **110.** $5+\sqrt{11}$. **111.** $3-\sqrt{5}$.
112. $\sqrt{3}+\sqrt{5}$. **113.** $\frac{1}{2}\sqrt{5}+1$. **114.** $7-2\sqrt{3}$.
115. $3\sqrt{2}+2\sqrt{3}$. **116.** $2\sqrt{3}-\sqrt{7}$. **117.** $\frac{1}{2}(4+\sqrt{2})$.
118. $\frac{2}{3}(\sqrt{3}-1)$. **119.** $\sqrt[4]{24}-\sqrt[4]{6}$. **120.** $\sqrt[4]{27}+\sqrt[4]{3}$.
121. $\frac{1}{2}\sqrt{3}-\frac{1}{3}\sqrt{2}$. **122.** $\sqrt[4]{3}+\sqrt[4]{12}$. **123.** $\sqrt[4]{2}+\sqrt[4]{\frac{1}{2}}$.
124. $\sqrt{\frac{1}{2}(1+m)}+\sqrt{\frac{1}{2}(1-m)}$. **125.** $\sqrt{a+x}+\sqrt{a-x}$.
126. $\sqrt{ax-a^2}-a$. **127.** $\sqrt{1+2x-x^2}+1-x$.
128. $\sqrt{6}+\sqrt{3}-1$. **129.** $2+\sqrt{3}-\sqrt{2}$. **130.** $\sqrt{\frac{5}{2}}+\sqrt{\frac{3}{2}}-1$.
131. $5-\sqrt{7}$. **132.** $\sqrt{7}-1$. **133.** $\sqrt{6}-2$.
134. $\sqrt{3}+\sqrt{2}$. **135.** $1+\sqrt{3}$. **136.** $1+\sqrt{5}$.
137. $4-\sqrt{3}$. **138.** $1-\frac{1}{2}\sqrt{2}$. **139.** $1+\sqrt{2}$.
140. $\frac{1}{2}(\sqrt{5}+1)$. **141.** $\sqrt{5}+\frac{1}{2}\sqrt{3}$.
144. $\sqrt[4]{\frac{1+c}{4(1-c)}}+\sqrt[4]{\frac{1-c}{4(1+c)}}$. **151.** $\sqrt{3}-\sqrt{2}$.
152. m. **153.** $\frac{1}{2}(1+2p)$.
154. $\frac{1}{2}(1\pm\sqrt{29})$. **155.** $2a(bcd)^{\frac{1}{2}}=bc+bd+cd$.
156. x^2+x+1. **157.** $3-2x+x^2$.
158. $2a^2-a-2$. **159.** $1-2ax-a^2x^2$.
160. $1-6x+12x^2-8x^3$. **161.** $a^3-2a^2b+2ab^2-b^3$.
163. $m=9$; $n=12$. **164.** $m=-4$; $n=1$.
165. $25x^2-24x+16$. **167.** $p=96$; $q=64$.
168. $p=15$; $q=-6$; $r=1$; $n=-20$.
169. $2a-3x$.

EXERCISES III.
INEQUALITIES.

3. $x>-\frac{4}{7}$. **4.** $x=-2$.
5. If in addition x is greater than -1.
24. 10 or 9; 11, 10 or 9. **25.** 3 or 4; 10, 9, 8 or 7.
26. 12, 11, 10 or 9. **27.** 15. **28.** 9 and 10.

EXERCISES IV.

Ratio.

1. $1\frac{1}{3} : 1$.   2. $1 : 1$.   3. $4 : 1$.
4. $(a+x)^2 : 1$.   5. $4 : 1$.   7. $1 : 27$.
9. (i) The latter greater than the former;
   (ii) The last, the first and the second.   10. 181.
11. 27.   13. 3.   14. The latter.
16. $(ad - bc)/(c - d)$.   17. $ab(a+b)$.   18. cd.   19. $\frac{3}{8}$.
40. $1 : 2$ or $2 : 1$.   41. $3 : 5$ or $4 : 1$.   42. $1 : 3$ or $2 : 1$.
43. $x = y = z$.   44. $\frac{1}{5}x = \frac{1}{4}y = \frac{1}{11}z$.   46. $\frac{1}{13}x = \frac{1}{2}y = \frac{1}{3}z$.
47. $x = 3$; $y = 4$.   48. $x = 5$; $y = 7$.
49. $x = \dfrac{am(bcq - npd)}{dq(bm - an)}$; $y = \dfrac{bn(acq - dmp)}{dq(an - bm)}$.
50. $x = 1$; $y = 2$; $z = 4$.   51. $x = y = z = \pm 2$.
52. $x = a$; $y = b$; $z = c$.   53. $a^2 + b^2 + c^2 = ab + ac + bc$.
54. $a^2 + b^2 + c^2 + 1 = 0$.
55. $a(b'c'' - b''c') + b(c'a'' - a'c'') + c(a'b'' - a''b') = 0$.
59. $\cdot 00008$; $\cdot 00012$.

EXERCISES V.

Proportion.

2. (i) $7\frac{1}{2}$; (ii) $4\frac{4}{7}$.   3. (i) $a - 1 : b + c = b^2 - bc + c^2 : a + 1$;
   (ii) $x : x + 1 = x + 2 : y$;   (iii) $x + 4 : y - 8 = y - 12 : x + 3$.
4. $x = 0$.   5. $x = -3\frac{1}{5}$.   6. $x = \sqrt{ab}$.   8. 2.   9. 1.
20. 8 or 0.   25. $x = 20$.
29. $x = -2$ or 1; $y = 1$.
30. $2bd - ad - bc : ad - 2ac + bc$.
31. $6, 9, 10, 15$.   32. $m^2pr : n^2qs$.   33. ± 30.
34. $6\frac{1}{4}$.   35. ± 3.   36. $-\frac{1}{2}a(1 \pm \sqrt{5})$.   37. $305 + 174\sqrt{3}$.
38. $x = 2\sqrt{2}$, $y = 4\sqrt{2}$.   39. $\sqrt[3]{45}$ and $\sqrt[3]{75}$.
40. 6 and 18.   41. $\sqrt{20}$ and $\sqrt[3]{50}$.
42. (i) $2, 4$ and 8; (ii) $2\sqrt{2}, 4$, and $4\sqrt{2}$.   44. $x = \dfrac{b^2 - ac}{a + c - 2b}$.
58. $a_2 = \sqrt[n-1]{a_1^{n-2}a_n}$; $a_3 = \sqrt[n-1]{a_1^{n-3}a_n^2}$.   60. $\frac{32}{28}$ and $\frac{40}{28}$.

ANSWERS. 623

EXERCISES VI.

Variation.

1. (i) Inverse, 40 secs. (ii) Direct, 1s. $0\frac{1}{2}d$. (iii) Inverse, $3\frac{1}{3}$ hours.
4. $y^2 = \dfrac{b^2}{a^2}(a^2 - x^2)$. 6. $xy^2 = 36a^3$. 7. $\frac{3}{5}$. 8. $3\frac{1}{3}$.
9. $1\frac{1}{5}$. 10. $x = 3y^2 + \dfrac{4}{y}$. 11. $a = \dfrac{b^2 c}{12}$.
12. $\dfrac{14}{4-5x}$. 13. $q = 2$ or 3. 15. 84.
16. $x = 1 + 2y + 3y^2$. 17. $az^{mn} = c^{mn}x$. 19. $\sqrt[3]{r^3 + r'^3}$.
21. $2(n-1)$ hours. 22. 224 carriages. 23. $8(\sqrt{2} - 1)$ inches.
24. £$\dfrac{mn^2 c}{(m+1)a^2}$; £$\dfrac{cn^{\frac{3}{2}}}{(m+1)b^{\frac{3}{2}}}$. 25. £113. 11s. $10\frac{28}{41}d$.
26. ·2385... of a year, or about 87 days.
27. 13 : 3. 28. 9·45 seconds.
41. No, unless $(a^2 + b^2)/ab = (c^2 + d^2)/cd$.
42. When $\dfrac{z}{u} = \dfrac{x+y}{x-y}$, or $\dfrac{x}{u+z}\left(\dfrac{z}{x+y} + \dfrac{u}{x-y}\right) = 1$, y need not vary as x.
44. $s = \frac{1}{2}ft^2$. 45. 9 sq. feet. 46. 420 men.
47. $7·643d$. nearly. 48. 9 : 4.
50. Length = 115·47 ft.; diameter = ·013674 inch.

EXERCISES VII.

Equations reducible to Quadratics.

(In the following answers all the values of x obtained in the solution are given. The student should decide by trial which are extraneous.)

1. $x = \pm 3$ or $\pm \sqrt{3}$. 2. $x = 16$ or $\frac{1}{16}$. 3. $x = 125$, or 20.
4. $x = 4$ or 9. 5. $y = \pm 1$ or $\pm \sqrt{\frac{5}{6}}$. 6. $x = 2$ or $\frac{1}{2}$.
7. $x = \pm a$ or $\pm a^{-1}$. 8. $x = 2$ or $-6\frac{3}{9}$.
9. $x = \pm 5$ or $\pm 3\sqrt{2}$. 10. $x = 1, -\frac{2}{5}$, or $\frac{1}{11}(-18 \pm 2\sqrt{15})$.
11. $x = 1$ or -7. 12. $x = 2, -\frac{1}{2}$, or $\frac{1}{4}(-17 \pm \sqrt{305})$.
13. $x = 0, 3\frac{1}{2}$, or $\frac{1}{4}(7 \pm 3\sqrt{5})$. 14. $x = -1, 1\frac{1}{2}$, or $\frac{1}{4}(1 \pm \sqrt{21})$.
15. $x = 1, 0, -3$, or -4. 16. $x = 8, -1$, or $\frac{1}{2}(7 \pm \sqrt{53})$.
17. $x = 2, 1\frac{1}{2}$, or $\frac{1}{4}(7 \pm \sqrt{33})$. 18. $x = 3, -\frac{1}{2}$, or $\frac{1}{4}(5 \pm \sqrt{1329})$.

ANSWERS.

19. $x = -7, 2$, or $\frac{1}{2}(-5 \pm \sqrt{53})$.

20. $x = \dfrac{b \pm \sqrt{b^2 + 2a(1 \pm \sqrt{1+4c})}}{2a}$.

21. $x = 0, a$, or $\dfrac{a}{2}\left\{1 \pm \sqrt{5 - \dfrac{8b}{a}}\right\}$.

22. $x = 2, -3$, or $\frac{1}{2}(-1 \pm \sqrt{5})$.   23. $x = 1$ or $\frac{1}{5}(-7 \pm 2\sqrt{6})$.

24. $x = \frac{1}{10}(1 \pm \sqrt{101}$ or ± 1.   25. $x = 1$ or $\frac{1}{2}(-5 \pm \sqrt{21})$.

26. $x = 2, \frac{1}{2}$, or $\frac{1}{4}(-9 \pm \sqrt{65})$.   27. $x = 1, -5$, or $-2 \pm 2\sqrt{2}$.

28. $x = -4$ or $-4 \pm \sqrt{10}$.

29. $x = -a, -2a$, or $\frac{1}{2}\{-3a \pm \frac{1}{5}a\sqrt{76}\}$.

30. $x = \frac{1}{2} \pm \frac{1}{4}\sqrt{\{\frac{1}{2} \pm \frac{1}{2}\sqrt{33}\}}$.

31. $x = 2 \pm \sqrt{10}$ or 2.   32. $x = \pm \sqrt{5}$ or 0.

33. $x = 0, -1$, or $\frac{1}{2}(-1 \pm \sqrt{-7})$.

34. $x = 3, -1$, or $1 \pm \sqrt{-28}$.

35. $x = \pm 2$ or $\pm \sqrt{-10}$.

36. $x = -2$ or $-2 \pm \sqrt{\{-5 \pm 2\sqrt{5}\}}$.

37. $x = 1$ or $1 \pm \sqrt{\{-20 \pm 8\sqrt{5}\}}$.

38. $x = -3$ or $-3 \pm \sqrt{\frac{1}{2}(-15 \pm \sqrt{165})}$.

39. $x = 3$ or $\frac{43}{6}$.   40. $x = 5$ or $\frac{4}{3}$.

41. $x = 4$ or $\frac{484}{169}$.   42. $x = 1$ or $1\frac{196}{1175}$.

43. $x = 8$.   44. $x = 2$.

45. $x = \pm \frac{1}{12}\sqrt{385}$.   46. $x = 8, 1$, or -5.

47. $x = 5$ or 6.   48. $x = 0$ or $2\frac{2}{5}$.   49. $x = 1$.

50. $x = -1, \pm 4$.   51. $x = 0$, or 5.   52. $x = 0$ or a.

53. $x = 0$, or $\pm 2\sqrt{ab}$; the three values satisfy the equation
$$\sqrt{a(a+b+x)} - \sqrt{a(a+b-x)} = x.$$

54. $x = 2$, or $-\frac{46}{3}$.   55. $x = 2$ or $-\frac{2}{21}$.

56. $x = 0$, or $4(a+b)$.   57. $x = \pm \sqrt{3}$.

58. $x = \pm \frac{1}{2}\sqrt{3} \cdot a$.   59. $x = \pm \frac{1}{2}\sqrt{13}$.   60. $x = 4$.

61. $x = \pm \frac{1}{2} p \sqrt{5}$.   62. $x = 4$ or $-1\frac{11}{14}$.

63. $x = \frac{2}{3}\{\pm \sqrt{a^2+b^2-ab} - (a+b)\}$.   64. $x = 2, 3$ or a.

65. $x = a, 3$, or $\frac{1}{2}a$.   66. $x = -2 \pm \sqrt{3}$ or -2.

67. $x = 2 \pm \sqrt{3}$ or 2.   68. $x = \pm \sqrt{6} - 1, 3$, or -1.

69. $x = \frac{1}{3}(1 \pm \sqrt{10})$ or $-\frac{2}{3}$.

70. $x = b$ or $\dfrac{-b^2 \pm \sqrt{b^4 + 4ab}}{2b}$.   71. $x = 1$, or $\dfrac{-1 \pm \sqrt{281}}{2}$.

72. $x = m + n + p$; or $\pm \sqrt{-(mn + mp + np)}$.

ANSWERS. 625

73. $x = p+q+r$; or $\frac{1}{2}(p+q+r) \pm \frac{1}{2}\sqrt{(p^2+q^2+r^2-2qr-2rp-2pq)}$.
74. $x=0$ or $\frac{1}{2}(-\Sigma a \pm \sqrt{\Sigma a^2 - 2\Sigma ab})$.
75. $x=a$ or b. **76.** $x=3, 4,$ or $3\frac{1}{2}$.
77. $x=a$ or b. **78.** $x=7$ or 5.
79. $x=2, 3,$ or 5. **80.** $x=-3, -2$ or $\frac{1}{2}(-5 \pm \sqrt{17})$.
81. $x=-1$ or $(a-b \pm \sqrt{b^2 - 2ab - 3a^2})/2a$.
82. $x=-1, -5,$ or $-\frac{1}{5}$.
83. $x=-1, +1, +1, \frac{1}{2}(3 + \sqrt{5}),$ or $\frac{1}{2}(3 - \sqrt{5})$.
84. $x=\frac{1}{2}(-3 \pm \sqrt{5})$ or 1 (two equal roots). **85.** $x=2, \frac{1}{2}, 7,$ or $\frac{1}{7}$.
86. $x=-3, -\frac{1}{3}, 5 + \sqrt{24},$ or $5 - \sqrt{24}$.
87. $x=+1, +1, -1, -1,$ or -1.
88. $x=4, -\frac{1}{4}, 2,$ or $-\frac{1}{2}$. **89.** $x=\frac{1}{3}, -3, \frac{1}{4},$ or -4.
90. $x=0$ or $1\frac{5}{34}$. **91.** $x=-34$ or 2.
92. $x=\frac{1}{2}(1 \pm \sqrt{13})$ or $\frac{1}{2}(-1 \pm \sqrt{5})$. **93.** $x = \pm\sqrt{(1 \pm \frac{1}{2}\sqrt{2})}$.
94. $x=4, -6,$ or $-1 \pm 4\sqrt{2}$. **95.** $x = \pm \dfrac{m}{2}\sqrt{\dfrac{4-m^2}{1-m^2}}$.
96. $x=1$. **97.** $x=0$.
98. $x=1$. **99.** $x=4$ or -1.

EXERCISES VIII.

SIMULTANEOUS QUADRATIC EQUATIONS WITH TWO UNKNOWN QUANTITIES.

1. $x=2$ or 3; $y=3$ or 2. **2.** $x=2$ or $\frac{1}{2}$; $y=1$ or 2.
3. $x=1$ or $-1\frac{3}{9}$; $y=2$ or $-3\frac{7}{9}$. **4.** $x=y=1$.
5. $x=a-b$ or 0; $y=-b$ or $-a$.
6. $x=5$ or -3; $y=3$ or -5. **7.** $x=3$ or -2; $y=2$ or -3.
8. $x= \pm 1$ or ± 2; $y= \pm \sqrt{2}$ or $\pm \frac{1}{2}\sqrt{2}$.
9. $x= \pm 2$ or ± 3; $y= \pm 3$ or ± 2. **10.** $x=7y$ or $-5y$.
11. $x=7y$ or $-\frac{1}{2}y$. **12.** $x=-y$ or $-\frac{1}{3}y$.
13. $x=3, -2\frac{1}{8},$ or $\frac{1}{4}(1 \pm \sqrt{105})$; $y=1, -1\frac{3}{18}$ or $\frac{1}{4}(1 \pm \sqrt{105})$.
14. $x= \pm 1$; $y= \pm 1$ or ∓ 2.
15. $x=1, -1\frac{1}{5}$ or $\frac{1}{34}(-1 \pm \sqrt{409})$;
  $y=2, -2\frac{2}{5}$ or $\frac{1}{17}(-2 \pm 2\sqrt{409})$.
16. $x=1, -2\frac{1}{2},$ or $-1 \mp \sqrt{21}$; $y=1, -2\frac{1}{2},$ or $\frac{1}{2}(1 \pm \sqrt{21})$.
17. $x= \pm 2$ or $\pm \frac{1}{5}\sqrt{3}$. **18.** $x= \pm 4$ or $\pm \frac{7}{5}\sqrt{3}$.
  $y= \pm 1$ or $\mp \frac{5}{8}\sqrt{3}$. $y= \pm 3$ or $\mp \frac{5}{8}\sqrt{3}$.

TUT. ALG. II.

ANSWERS.

19. $x = \pm 2$.
    $y = \pm 3$.
20. $x = \pm 2$ or $\pm \tfrac{1}{5}\sqrt{10}$.
    $y = \pm \tfrac{1}{2}$ or $\mp \tfrac{3}{5}\sqrt{10}$.
21. $x = \pm 6$ or $\pm \tfrac{1}{2}\sqrt{2}$.
    $y = \pm 5$ or $\mp \tfrac{11}{2}\sqrt{2}$.
22. $x = \pm 5$ or $\pm 9\sqrt{3}$.
    $y = \pm 2$ or $\mp \tfrac{11}{3}\sqrt{3}$.
23. $x = y = 1$; $x = y = 0$;
    or $x = \tfrac{3}{5}$, $y = \tfrac{1}{5}$.
24. $x = y = 0$; $x = 1$, $y = 2$.
    or $x = \tfrac{289}{64}$, $y = -\tfrac{221}{64}$.
25. $x = 4$, $y = 1$; $x = y = 0$.
    or $x = -\tfrac{244}{401}$; $y = \tfrac{128}{401}$.
26. $x = y = 2$; $x = y = 0$ or $x = -\tfrac{1}{2}$, $y = \tfrac{3}{4}$.
27. $x = y = 0$; $x = y = 1$; $x = \tfrac{1}{2}$, $y = 0$; or $x = -\tfrac{1}{2}$, $y = 1$.
28. $x = 0$ or 1; $y = 0$ or ± 2.
29. $x = y = 1$; $x = y = 0$.
30. $x = 0$, $y = 2$, or
    $x = 2$, $y = 0$.
31. $x = 3, 1$, or $2 \pm \sqrt{11\tfrac{3}{7}}$.
    $y = 1, 3$, or $2 \mp \sqrt{11\tfrac{3}{7}}$.
32. $x = 1, 0$, or $\tfrac{1}{4}(1 \pm \sqrt{19})$.
    $y = 0, 1$, or $\tfrac{1}{4}(1 \mp \sqrt{19})$.
33. $x = \tfrac{7}{5}$, or $-\tfrac{1}{5}$.
    $y = -\tfrac{1}{5}$, or $\tfrac{7}{5}$.
34. $x = \tfrac{1}{3}$ or $\tfrac{1}{6}$.
    $y = \tfrac{1}{6}$ or $\tfrac{3}{10}$.
35. $x = 0, 11$, or $\tfrac{1}{2}(-1 \pm \sqrt{-23})$.
    $y = 0, 11$, or $\tfrac{1}{2}(-1 \mp \sqrt{-23})$.
36. $x = y = 0$,
    or $x = y = \pm 3$.
37. $x = 2$ or 3; $y = 3$ or 2.
38. $x = 5$ or -2; $y = 2$ or -5.
39. $x = 9$ or 1; $y = 1$ or 9.
40. $x = y = 1$.
41. $x = 8$ or 1; $y = 1$ or 8.
42. $x = 1$ or 3; $y = 3$ or 1.
43. $x = \pm 3$ or ± 2.
    $y = \pm 2$ or ± 3.
44. $x = 8$ or 2; $y = 2$ or 8.
45. $x = 4$ or $\tfrac{1}{2}$; $y = 2$ or $\tfrac{1}{4}$.
46. $x = 1$; $y = 2$.
47. $x = \pm 6\tfrac{3}{8}$; $y = \pm 3\tfrac{3}{4}$.
48. $x = 4$ or 2; $y = 2$ or 4.
49. $x = 9$ or 4; $y = 4$ or 9.
50. $x = \pm \tfrac{1}{4}\sqrt{26}$; $y = \pm \sqrt{26}$.
51. $x = 1$ or -2.
    $y = 2$ or -4.
52. $x = 3, 2$, or $\tfrac{1}{2}(-5 \pm \sqrt{41})$.
    $y = 2, 3$, or $\tfrac{1}{2}(-5 \mp \sqrt{41})$.
53. $x = 2$ or $\tfrac{6}{5}$.
    $y = 1$ or $-\tfrac{1}{5}$.
54. $x = \tfrac{3}{2}$ or -1.
    $y = \tfrac{1}{2}$ or -2.
55. $x = \pm 2$, or $\mp 2\sqrt{3}$.
    $y = \pm 1$, or $\pm \sqrt{3}$.
56. $x = \pm \sqrt{5}$;
    $y = \pm 1$.
57. $x = 9$ or 28.
    $y = 7$ or 26.
58. $x = 2$ or 3.
    $y = 3$ or 2.
59. $x = \dfrac{1+a}{1-a}$, $y = \dfrac{1+c}{1-c}$; or $x = \dfrac{1-a}{1+a}$, $y = \dfrac{1-c}{1+c}$.
60. $x = 1$, $y = 2$, or
    $x = -\tfrac{1}{3}$, $y = -\tfrac{2}{3}$.
61. $x = 3$, $y = 2$, or
    $x = -\tfrac{3}{4}$, $y = -\tfrac{1}{2}$.

ANSWERS. 627

EXERCISES IX.

SIMULTANEOUS QUADRATICS WITH THREE UNKNOWNS.

1. $x=y=z=0$ or $x=1$, $y=2$, $z=3$.
2. $x=y=z=0$ or $x=4\frac{4}{5}$, $y=24$, $z=3\frac{3}{7}$.
3. $x=y=z=0$ or $x=\dfrac{2pq}{p+q-r}$, $y=\dfrac{2qr}{q+r-p}$, $z=\dfrac{2pr}{p+r-q}$.
4. $x=\pm\frac{1}{2}$, $y=\pm 1$, $z=\pm 2$.
5. $x=1$ or -1.
   $y=2$ or -2.
   $z=3$ or -3.
6. $x=\pm 3$, $y=\pm 4$, $z=\pm 6$.
7. $x=\pm 1\frac{1}{2}$, $y=\pm\frac{1}{2}$, $z=0$.
8. $x=\pm\dfrac{bc}{a}$, $y\pm\dfrac{ca}{b}$, $z=\pm\dfrac{ab}{c}$.
9. $x=y=z=0$ or $x=\pm\dfrac{qr}{p}$, $y=\pm\dfrac{rp}{q}$, $z=\pm\dfrac{pq}{r}$.
10. $x=y=z=0$;
    or $x=y=z=\pm 1$.
11. $x=y=z=0$;
    or $x=\pm 5$, $y=\pm 6$, $z=\pm 7$.
12. $x=\pm 3$, $y=\pm 5$, $z=\pm 7$.
13. $x=\pm 2$, $y=\pm 3$, $z=\pm 4$.
14. $x=\pm 1$, $y=\pm 2$, $z=\pm 3$.
15. $x=\pm 2$, $y=\pm 3$, $z=\pm 4$.
16. $x=\pm 4$, $y=\pm 3$, $z=\pm 2$.
17. $x=y=z=1$, or $x=-7$, $y=-11$, $z=-15$.
18. $x=y=z=1$, or $x=-7$, $y=-11$, $z=-15$.
19. $x=6$ or -8.
    $y=5$ or -7.
    $z=7$ or -9.
20. $x=3$ or -1.
    $y=4$ or -2.
    $z=5$ or -3.
21. $x=1$ or -7.
    $y=2$ or -8.
    $z=3$ or -9.
22. $x=4$ or 0.
    $y=2$ or 4.
    $z=1$ or 7.
23. $x=-1$ or 5.
    $y=-2$ or 3.
    $z=-\frac{1}{3}$ or 1.
24. $x=a\pm\sqrt{\left\{\dfrac{(q^2+ac)(r^2+ab)}{p^2+bc}\right\}}$,
    $y=b\pm\sqrt{\left\{\dfrac{(r^2+ba)(p^2+bc)}{q^2+ca}\right\}}$,
    $z=c\pm\sqrt{\left\{\dfrac{(p^2+cb)(q^2+ca)}{r^2+ab}\right\}}$
25. $x=0$ or $\frac{2}{5}$.
    $y=0$ or $\frac{1}{18}$.
    $z=1$ or $1\frac{3}{18}$.
26. $x=4$ or -7.
    $y=4$ or $-\frac{43}{3}$.
    $z=0$ or $-\frac{11}{3}$.
27. $x=2$, $y=1$, $z=1$.
28. $x=0$ or 3.
    $y=-1$ or 2.
    $z=-2$ or 1.
29. $x=3$, $y=4$, $z=5$.

ANSWERS.

30. $x = 3$ or -4.
    $y = 4$ or $-5\frac{1}{3}$.
    $z = 5$ or $-6\frac{2}{3}$.

31. $x = y = z = 2$; or $x = y = z = -2\frac{1}{2}$; or
    $x = \pm\sqrt{\frac{5}{19}}, y = \mp 2\sqrt{\frac{5}{19}}, z = \mp 3\sqrt{\frac{5}{19}}$.

32. $x = \pm 1, y = \pm 2, z = \pm 3$.

33. $x = \mp 5$.
    $y = \pm 16$.
    $z = \mp 9$.
    or $x = y = z = 0$.

34. $x = 3$ or 2.
    $y = 2$ or 3.
    $z = 1$.

35. $x = 1$ or -2.
    $y = 1$ or $-\frac{1}{2}$.
    $z = 1$.

36. $x = \pm 4, y = \pm 2, z = \pm 1$.

37. $x = 4$.
    $y = 2$.
    $z = 3$.

38. $x = 8$ or 2.
    $y = 4$.
    $z = 2$ or 8.

39. $x = \pm(4 + \sqrt{2})$.
    $y = \pm\sqrt{14}$.
    $z = \pm(4 - \sqrt{2})$.

40. $x = 1$ or 2,
    $y = 2$ or 1.
    $z = 3$.

41. $x = \pm 1$.
    $y = \pm 2$.
    $z = \mp 1$.

42. $x = \frac{1}{2}(-1 \pm \sqrt{6})$.
    $y = \frac{1}{2}(-1 \mp \sqrt{6})$.
    $z = 2$.

43. $x = \mp 2, y = 0, z = \pm 1$.

44. $x = 2\frac{2}{8}, y = 1\frac{5}{7}, z = -12$.

45. $x = 0, y = 3, z = -2$; or the solutions obtained from this by interchanging the values of x, y, z in every possible way.

EXERCISES X.

Problems leading to Quadratics.

1. 10.
2. 50s. The negative answer made positive is the solution of "what would be given per hundred loads to have ashes *removed*, when eight loads less removed for a sovereign would raise the price 1d. per load."
3. £80.
4. 12.
5. 3 miles an hour.
6. 15 days and 10 days.
7. 361.
8. 220 yds. and 165 yds.
9. 6.
10. 1296.
11. 17 and 19 or -17 and -19.
12. 7 and 17.
13. 75.
14. 7 and 2.
15. Distance walked $= 24$ miles; rate $= 3$ miles an hour.
16. $\frac{1}{2}(3 \pm \sqrt{5}), \frac{1}{2}(1 \pm \sqrt{5})$.
17. $-\frac{1}{2}\{1 \mp \sqrt{4a+1}\}$ and $\frac{1}{2}\{2a + 1 \mp \sqrt{4a+1}\}$.
18. $67\frac{1}{2}$ in.
19. 12 yds.
20. £125; £95.
21. 20 poles by 8 poles; 16 poles by 10 poles.
22. 912.

ANSWERS.   629

23. Bases, $\dfrac{a\sqrt{b}}{\sqrt{b}+\sqrt{c}}$, $\dfrac{a\sqrt{c}}{\sqrt{b}+\sqrt{c}}$,
   altitudes, $\dfrac{\sqrt{c}\,(\sqrt{b}+\sqrt{c})}{a}$, $\dfrac{\sqrt{b}\,(\sqrt{b}+\sqrt{c})}{a}$.
24. 6 miles an hour.   25. 10 per cent.
26. 3 or $4\frac{1}{4}$ miles after the nearest point.
27. $2x^2 - bx = 4a^2 - \frac{1}{4}b^2$; b must lie between $4a$ and $4\sqrt{2}a$; 144 and 130.

EXERCISES XI.

Zero and Infinity.

1. -5.   2. $\frac{6}{11}$.   3. 0.   4. ∞.
5. $\frac{1}{2}$.   6. 1.   9. $x = \infty$.
10. $x = \infty$ unless $a = b$ and then x is indeterminate.

EXERCISES XII.

The Theory of Quadratic Equations.

1. $-(a+b)$, ab.   2. $-1, -2$.   3. $2, 0$.
7. $2(x-6)\{x-(-4\frac{1}{2})\} = 0$;   8. $67(x-4)(x+\frac{191}{67}) = 0$;
9. $x(x-5) = 0$.   10. $x = a$ or $-1/a$.
11. $x = 1$ or $\dfrac{c-a}{a-b}$.   12. $x = a$ or $-\dfrac{1}{a}$.
13. $x = a$ or $1/a$;   14. $x = a$ or $15 - a$ 15. $x = 3 - m$.
16. Real, irrational, positive and negative.
17. Real, rational, positive and positive.
18. Imaginary.   19. Real, rational, negative and negative.
20. $m = \frac{16}{25}$.   21. $10p^2 = 49q$.   22. $p = \pm 4\frac{2}{7}$.
24. $n = \pm 2\sqrt{10}$.   25. $a = -1\frac{1}{4}$.   28. $a^2 = 2m^2$.
30. When $a^2c + c^2b + b^2a = 0$.   42. $2x^2 - x - 15 = 0$.
43. $5x^2 + 8x - 36 = 0$.   44. $x^2 - 6x + 4 = 0$.
45. $pqx^2 + (q^2 - p^2)x - pq = 0$.   46. $x^2 - 4x + 8 = 0$.
47. $x^2 - 2x + 2 = 0$.   48. $9x^2 - 24x - 4 = 0$.
49. $x^2 - 4x - 1 = 0$.   50. $x^2 - 24x + 137 = 0$.

ANSWERS.

51. $x^3 - 10x^2 + 23x - 14 = 0$. 52. $x^3 - 6x^2 + 7x + 4 = 0$.
53. $x^2 + 14x - 101 = 0$.
54. (i) $m^2 - 2n$; (ii) $\dfrac{m^2 - 2n}{n}$; (iii) $3mn - m^3$;
    (iv) $\dfrac{3mn - m^3}{n}$; (v) $\dfrac{m^4 - 4m^2n + 2n^2}{n^2}$.
55. 20.   56. $m^2 - 2n$.
58. (i) m; (ii) $\dfrac{m}{n}$; (iii) $\dfrac{m^2 - 2n}{n^2}$; (iv) $\dfrac{m^3 - 3mn}{n^3}$.
60. (i) $q + s + pr$; (ii) $2\{p^2 - pr + r^2 - 2(q+s)\}$.
61. $x^2 + (2a - 53)x + (a + 14)^2 = 0$.
62. (i) $a^4x^2 - 2a^2x(b^2 - ac) + b^2(b^2 - 2ac) = 0$;
    (ii) $acx^2 - x(b^2 - 2ac) + ac = 0$;
    (iii) $a^2x^2 + abx(m+1) + ac(m^2+1) = m(2ac - b^2)$.
63. $r = 2q - p^2$, $s = q^2$.
65. (i) $17x^2 - 20x + 5 = 0$; (ii) $rx^2 - qx + p = 0$; (iii) $cx^2 + bx + a = 0$.
66. $x^2 - x(p^2 - 2q) + q^2 = 0$.   67. $x^2 - 7x + 8 = 0$.
68. $2x^2 - 2mx + m^2 = n^2$.   69. $x^2 - mpx + q(m^2 - 2n) = n(2q - p^2)$.
70. -6 and -5.   71. -6 and -6.   72. $x^2 - 5x + 6 = 0$.
73. $m^2 + 2m - 19$ must be a square; m may have the values $5, -7, -11\frac{1}{2}, 9\frac{1}{2}$, etc.   74. $(x + y + 1)(2x + y + 1)$.
75. $m = 7$.   77. $y = 1$.
78. $n - m + 1 = 0$, or $n - 4m + 16 = 0$.
79. $m = 3$ or 4.   80. $(bc' - b'c)(ab' - a'b) = (ac' - a'c)^2$.
81. $(2x + y + 7)(x - 3y - 4)$.   82. $(x + 2y + 5z)(x - y - 3z)$.
83. $(x + y + z)(2x - 2y - z)$.
86. $b = c = 3a$.   91. $qs = 1$, $qr = p$.

EXERCISES XIII.

Maxima and Minima.

1. Positive, when $x > -1$, negative when $x < -1$.
2. Positive when $x < a^2$, negative when $x > a^2$.
3. Negative when x is positive. Positive when x is negative.
4. Positive when $x < 4$. Negative when $x > 4$.
5. Positive when x lies beyond 9 and 11.
   Negative when x lies between 9 and 11.

ANSWERS. 631

6. Positive for all values of x except -8 when it vanishes.
7. Positive when x lies between 9 and 11.
   Negative when x lies beyond 9 and 11.
8. Always negative.   9. Always positive.
10. Positive when x lies between 4 and 1. Negative when x lies beyond 4 and 1.
11. Positive when x lies beyond 1 and 2. Negative when x lies between 1 and 2.
12. Positive when x lies between 1 and 2. Negative when x lies beyond 1 and 2.
13. $+$ when $x>3$; $-$ when $x<3$, and ∞ when $x=3$.
14. $+$ when $x>-4$; $-$ when $x<-4$; 0 when $x=-3$ or -4.
15. $+$ when x lies between ∞ and 3 or 2 and 1; $-$ when x lies between 3 and 2 or 1 and $-\infty$; 0 when $x=1$ or 2; ∞ when $x=3$.
16. $-$ when x lies between ∞ and 8, 4 and 2, or 1 and $-\infty$; $+$ when x lies between 8 and 4, or 2 and 1; 0 when $x=1$ or 2; and ∞ when $x=4$ or 8.
17. $-$ when $x>2$ or $<-\frac{1}{2}$. $+$ when x lies between 2 and $-\frac{1}{2}$. 0 when $x=2\frac{1}{2}$, 2 or $-\frac{1}{2}$.
18. Always positive.   24. Can take all values.
25. From ∞ to 1.   26. From $-\infty$ to $-8\frac{1}{4}$.
27. From $+\infty$ to $-\frac{1}{2}$.   28. From $-\infty$ to $+\frac{5}{4}$.
29. From $-\infty$ to $-5\frac{7}{8}$.   30. From $-\infty$ to $\frac{1}{8}$.
31. From $+\infty$ to 0.   32. From 0 to $\frac{1}{4}$.
33. From 0 to $-\infty$.   34. Not beyond 0 and $\frac{4}{3}$.
35. Not beyond 0 and 1.   36. All values.
37. From 0 to ∞.   38. Beyond 2 and $-\frac{10}{9}$.
39. Between 16 and $\frac{4}{15}$.   40. All values.
42. $x=1$, greatest value is 1.
43. x between $3\frac{1}{2}$ and $-1\frac{1}{2}$. y between $\frac{3}{8}+\frac{5}{4}\sqrt{2}$ and $\frac{3}{8}-\frac{5}{4}\sqrt{2}$.
44. x and y must lie between $-\frac{1}{2}+\frac{1}{2}\sqrt{2}$ and $-\frac{1}{2}-\frac{1}{2}\sqrt{2}$.
45. x must lie beyond 2 and -2. y may have any real value.
46. x between $\frac{1}{2}a\sqrt{13}$ and $-\frac{1}{2}a\sqrt{13}$.
    y between $\frac{1}{4}a\{-1+\sqrt{13}\}$ and $\frac{1}{4}a\{-1-\sqrt{13}\}$.
49. 6 and 6.   50. 9 and 3.   51. 4 and 4.   52. $\frac{1}{2}$.
53. The vertices bisect the sides of the original square.
54. $-x^2+4x+21$; 25.   55. $x=b^3/2c^2$.

56. $x = a$. **57.** A square. **60.** $\dfrac{x^2+4}{x}$.

61. $\dfrac{x^2+x+1}{x^2-x+1}$ and $\dfrac{x^2-x+1}{x^2+x+1}$. **62.** 384 sq. yds.

EXERCISES XIV.

Arithmetical and Geometrical Progression.

8. $\frac{1}{2}\{(a-\frac{1}{2}d)-(a+nd-\frac{1}{2}d)(-1)^n\}$.

9. $\frac{5}{7}, 1, 1\frac{2}{7}\ldots\ldots$ **10.** $15:8$.

12. $a, 3a, 5a\ldots\ldots$ where $a =$ first term.

13. $2ab + 2bc + 2ca$. **22.** $\frac{1}{2}$.

25. $(a^4b)^{\frac{1}{11}}$. **26.** $\frac{1}{5}$.

27. $a, \dfrac{a}{1+p}, \dfrac{a}{(1+p)^2}\ldots\ldots$ **28.** $2, 1, \frac{1}{2}, \frac{1}{4}, \ldots\ldots$

30. $\dfrac{n}{1-r} - \dfrac{r(1-r^n)}{(1-r)^2}$. **31.** $\frac{40}{81}(10^n-1) - \frac{4}{9}n$.

32. $n + \dfrac{r^2(r^{4n}-1)}{r^{4n}(r^4-1)} + \dfrac{2r(r^{2n}-1)}{r^{2n}(r^2-1)}$. **34.** $\dfrac{a^{3n}-b^{3n}}{(a^6-b^6)b^{3n-3}}$.

35. $1\frac{4}{5}$. **36.** $4\frac{1361}{4096}$.

37. $1\frac{4333}{12383}$. **47.** $9:1$ or $1:9$.

49. $m = \sqrt{b}\,\dfrac{2\sqrt{ab}}{\sqrt{a}+\sqrt{b}},\ n = \sqrt{a}\,\dfrac{2\sqrt{ab}}{\sqrt{a}+\sqrt{b}}$.

EXERCISES XV.

Harmonical Progression.

1. $\frac{1}{4}, \frac{2}{5}; \frac{2}{13}, \frac{1}{7}$. **2.** $\frac{1}{15}$. **3.** $\dfrac{a^2-b^2}{6ab-a^2-b^2}$.

4. $\frac{6}{13}, \frac{3}{7}, \frac{2}{5}, \frac{3}{8}, \frac{6}{17}$. **5.** ∞ and 4. **6.** $-\frac{24}{11}, -1\frac{2}{5}, -\frac{3}{5}$.

7. 3. **8.** 9 and 1, or $\frac{4}{5}$ and $-7\frac{1}{5}$. **9.** $\frac{1}{2}, \frac{1}{3}, \frac{1}{4}$; or $\frac{1}{2}, \frac{1}{7}, -\frac{1}{7}$.

10. $2, 3, 6$. **14.** $-2c/b$. **22.** $\sqrt{(1+\sqrt{2})} : \sqrt[4]{2} : 1$.

28. a, b, c. **29.** $5, \pm 4, 3\frac{1}{3}, \pm 2\frac{1}{25}, 2\frac{9}{175}$.

30. 6 and 24. **31.** $\dfrac{1}{2a}\left(\dfrac{a^2+1}{a-1}\right)^2$. **54.** $4, 6, 12$.

ANSWERS. 633

EXERCISES XVI.

Some other Simple Series.

1. $(3n-2)x^{n-1}$, $\dfrac{1-(3n-2)x^n}{1-x} + \dfrac{3x(1-x^{n-1})}{(1-x)^2}$.

2. $(2n-1)\cdot 3^{n-1}$, $3^n(n-1)+1$.   3. $\dfrac{n}{3^{n-1}}$, $\dfrac{3^{n+1}-3-2n}{4\times 3^{n-1}}$.

4. $(a+n-1)(-x)^{n-1}$, $\dfrac{a+(-1)^{n-1}(a+n-1)x^n}{1+x} - \dfrac{x+(-x)^n}{(1+x)^2}$.

5. $(2n-1)(-1)^{n-1}$, $-n(-1)^n$.   6. $\dfrac{a}{r-1} + \dfrac{b}{(r-1)^2}$.

7. 3.   8. $\frac{2}{9}$.   9. $\dfrac{ar}{(1-r)^2}$.   10. $\dfrac{4a-3a^2}{(1-a)^2}$.

11. 4.   12. $\dfrac{1+22x+9x^2}{(1-x)^3}$.   13. $\dfrac{n(4n^2-1)}{3}$.

14. $\dfrac{2n(n+1)(2n+1)}{3}$.   15. $\dfrac{n}{6}(2n^2+3n-5)$.

16. $\frac{1}{3}n(n+1)(n+2)$.   17. $\frac{1}{3}n(4n^2+6n-1)$.

18. 364.   19. 2,870.   20. 880.   21. 12,700.
22. 1,484.   23. 2,185.   24. 355.   25. 1595.
26. 7.   27. 9.   28. 410.   29. 844.
30. 595.   31. 1410.   32. 1165.   33. 5; 120.

EXERCISES XVII.

Logarithms.

1. ·7781513, 1·3222193, 1·8920947, 2·1553361, 2·6646420, 3·0004341.
2. ·8037053, ·6734160, 1·6782148, 1·3102381, 2·2222828.
3. ·9030900, 1·3802113, 1·4313639, 1·6901960, 1·9822713, 2·2278868, 3·1241781.
4. 4, 5.   5. 5, 3, $\bar{4}$, $\bar{2}$.
6. $\bar{3}$·39794, $\bar{3}$·79588.   7. ·6532126, ·8293039, 1·0053952.
8. $\bar{2}$·6197887, $\bar{4}$·4524714, $\bar{2}$·7891466.   9. 12.

10. 1·00113.
11. 1·812913, $\bar{3}$·812913.
12. The 26th and 27th.
14. 3.
15. 3·7201593, $\bar{3}$·6690067, $\bar{1}$·6911864.
16. $\bar{1}$·7799827.
17. 11, 48.
18. 1·407412.
19. 4·0266150, 2·511034.
20. 5862·057.
21. 1·930698.
22. $\bar{2}$·5895301.
23. 1·048 cm.
24. (i) 1·837; (ii) −2·619.
26. 63·4 per cent.
28. 3; $\bar{3}$.
30. 435.
31. $\bar{1}$·6577393.
32. ·0334477.
33. $\frac{2}{3}$.
34. $1\frac{1}{2}$.
35. $\frac{1}{2}$.
36. ·4771213, ·6020600, ·6989700, ·7781513, ·8450980, ·9030900, ·9542425.
37. $\bar{3}$·6989700, ·2184267.
38. 1·5360405, 5·5355473, $\bar{4}$·5365584.
39. 343·56, ·003439, 34321000.
40. 5·5354536, 2·5365678, $\bar{3}$·5370363.
41. 34355·08, ·03432065, 3·440857.
42. Diff. 293. $\begin{cases} 1 & 2 & 3 & 4 & 5 & 6 & 7 & 8 & 9 \\ 29 & 59 & 88 & 117 & 147 & 176 & 205 & 234 & 264 \end{cases}$
6·1710724, $\bar{3}$·1709914, 14825180000, ·1482592.
43. $\bar{1}$·3010491 $\begin{cases} 1 & 2 & 3 & 4 & 5 & 6 & 7 & 8 & 9 \\ 22 & 43 & 65 & 87 & 108 & 130 & 152 & 174 & 195 \end{cases}$
Diff. 217
44. ·9824394.
45. 1·888175; 4·956952 × 10²¹.
46. 1·389494.
47. 38·7.
48. ·4692964.
49. 1·6268.
50. $\log_{10}m = 3\log_{10}n - \frac{3}{2}\log_{10}2 - \log 3 = 3\log_{10}n + \bar{1}·0791813$.
51. ·563119 ft.
52. $x = 5$.
53. $x = 2·4998$.
54. $x = \frac{22}{17}$.
55. $x = -·8899$.
56. ·5 and 1·292.
57. $x = 2, y = 3$ or $x = -2, y = \frac{1}{3}$.

EXERCISES XVIII.

INTEREST AND DISCOUNT.

1. £179. 1s. 8d.
2. £69. 3s. 8¼d.
3. £640. 16s. 6d.
4. £488. 17s. 4d.
5. 41000.
6. 14·2 years nearly
7. 58·7 years.
8. 54·4 years.
9. 3·5265.

ANSWERS.   635

10. 2·427.
11. £747. 5s. 2d.
12. £65. 8s. 1½d.
13. £2145.
14. £1577. 19s. 8d.
15. 1s. 7¾d. approx.
16. 1·27443 : 1 or at simple interest 5 : 4.
17. £5. 3s. 6d.
18. £11,434·105.
19. £625.
20. £2,500.
21. 22.
22. £12,151. 16s. 9d.
23. 36 years.
24. Between 139 and 140.
25. £7. 1s. 9d.
26. £5. 19s. 2d.
27. Amount £131,501. 5s. 8d. difference £125,501. 5s. 8d.
28. £6768. 7s. 10½d.

EXERCISES XIX.

ANNUITIES.

1. 40, 33⅓, 28⁴⁄₇, 25, 18$\frac{2}{11}$, 16⅔.   2. £2000.
3. (i) 19·188. (ii) 19·1425.   4. (i) £5·437. (ii) £12·329.
5. (i) £209·348. (ii) £50·3544.
6. £285⅝ + £272$\frac{8}{11}$ + £260$\frac{20}{23}$ = £819·311 approx.
7. £2977. 16s.   8. £1858. 16s.   9. £1409. 8s.
10. £2243. 8s.   11. £846. 3s. 1½d.   12. £104. 3s. 2d.
13. £49 10s. 7d.   14. £3417.   15. £140615. 12s.
16. £2219.   17. £875.   18. £650.   19. £4200.
20. Each half-yearly instalment would be a little under 4 per cent. of the debt.   21. £376. 15s. 6d.   22. £1250.
23. £19049. 3s., difference of income £253. 8s. 9d.   24. 13 years.
25. 30 years.   26. 22 years.   27. £62 8s. 5d.
28. 16·6 years.   29. £23 3s. 7d.   30. £144·966 (gross rent).
31. Amount of £15 in 10 years at 5 per cent = £24·434; amount of 10 years' subscription at same time plus £10 = £23·207, hence gain by latter method = £1. 4s. 6½d.   32. £26·2908.
33. £179·65.   34. 49.   35. £1127·406.
36. £206·612.   37. 30 years.   38. 20·15 years.

EXERCISES XX.

PERMUTATIONS.

1. 90. 3. 210. 4. 24. 5. 120. 6. 34.
14. 2520; 120. 15. 720. 16. 40,319; 719. 17. 5.
18. 5,040. 19. 120. 20. 116,280. 21. $n=9$.
22. 24. 23. 1,663,200. 24. 6. 25. 15.
26. 8. 27. 2. 28. 15. 29. 4.
30. 7. 31. 1,956. 32. 4. 33. 6 and 2.
35. 325. 36. 24. 37. 18, 48. 38. 6.
39. 60. 40. 120; 24. 41. 240. 42. 600.
43. 24. 44. 5,040. 45. 6; 2. 46. 240.
47. 36. 49. 2,880. 50. 3,600.
51. 480. 54. 139,986; 3. 55. 24. 57. 720.
58. 60. 59. 2,520. 60. 12. 61. 288.
62. 8,000. 63. 100,000. 64. 243. 65. 1,024.
66. 16. 67. 9,000,000. 68. 3125.
71. 1,080; 720; 360. 72. 120. 73. 1,023.
74. 6198. 75. 2,520. 76. 45,360. 77. 1,120.
78. 151,200. 79. 1,801,800. 80. 12,600. 81. 60; 36.
82. 277,200; 15,120; 1,680. 83. 60.
84. 119,750,400; 1,814,400. 85. 5. 86. 4.
88. 120. 89. 9. 90. 120; 60. 91. 15,120.
92. 24.

EXERCISES XXI.

COMBINATIONS.

1. 20; 136; 41. 2. 56. 3. 190. 4. 66.
5. 120. 6. 403 days 18 hrs. 7. 55.
8. 8. 9. 4 or 6. 10. 16.
11. (i) 12; (ii) 12; (iii) 10. 13. 20; 27; $\frac{1}{2}n(n+3)$.

ANSWERS.

14. 12.   **15.** 12.   **16.** $n=7$ and $r=4$.   **17.** 7.
18. $_{15}C_3=455$; $_{15}C_2=105$.   **19.** 175.   **21.** $\frac{1}{6}m(m-1)(m-2)$.
22. 90.   **23.** 126.   **24.** 252.   **28.** 70; 35.
29. 10; 92,378.   **30.** 3; 20.   **32.** 10; 3.   **33.** 14; 4.
34. 84.   **35.** 75,582; 38,760; 3,003.   **36.** 165.
39. $\dfrac{100!}{95!\,5!} - \dfrac{85!}{80!\,5!}$.   **40.** 3.   **41.** 330.
42. 120.   **43.** 100.   **44.** 140.
45. 1,900.   **46.** 1,596,000.   **47.** 96.
48. 15; 18.   **49.** 43,200.   **51.** 950,400; 158,400; 316,800.
52. 176,400.   **53.** 144.   **54.** (i) 12; (ii) 42.
56. $_{51}C_{20} \times 50! \times 20!$ **57.** 30.   **58.** 86,400.
59. $\dfrac{(2n-p-q)!}{(n-p)!\,(n-q)!} \times (n!)^2$.   **60.** 18.   **61.** 55.
62. 344.   **63.** 15.   **64.** 31.   **65.** 63.
66. 120.   **67.** 26.   **68.** 247.   **69.** 511.
70. 31; £28. 17s. 4d.   **72.** 2.   **74.** 2^p.
76. 1,953.   **77.** 3,255.   **79.** 649.   **80.** 35.
81. 5,498.   **82.** 959.
84. 8 or 10; 10 or 12, according as 1 and the given number are or are not taken as factors.
85. 3.   **89.** 22.   **90.** 22.
91. $\dfrac{52!}{(13!)^4}$.   **92.** $\dfrac{52!}{4!\,(13!)^4}$.   **93.** $\dfrac{120!}{6!\,(20!)^6}$.
94. 105.   **95.** 70; 35.   **96.** $(mn)!/(n!)^m$.
97. 20.   **98.** 126.   **99.** 126.   **100.** 120; 8.
101. 16; 10.   **102.** 330.   **103.** 20; 16.   **104.** 1,001; 126.
105. 20; 120.   **106.** 10.   **108.** 26; 22.   **109.** 399.
110. 160.   **111.** 19.   **112.** 1,638; 39.   **113.** 2,111.
114. 11.   **115.** 395.

EXERCISES XXII.

MATHEMATICAL INDUCTION.

12. Use the identity $x^n + y^n = (x+y)(x^{n-1} + y^{n-1}) - xy(x^{n-2} + y^{n-2})$.

13. Use the identity $x^n - y^n = (x+y)(x^{n-1} - y^{n-1}) - xy(x^{n-2} - y^{n-2})$.

18. This easily follows from the fact that
$(n+1)^3 - n^3 = 3n(n+1) + 1 =$ (a multiple of 6) $+ 1$, since $n(n+1)$ is divisible by 2.

EXERCISES XXIII.

THE BINOMIAL THEOREM FOR POSITIVE INTEGRAL INDEX.

1. -2.   **2.** $-39; 187$.

3. $32x^5 + 240x^4y + 720x^3y^2 + 1080x^2y^3 + 810xy^4 + 243y^5$.

4. $a^6 + 18a^5x + 135a^4x^2 + 540a^3x^3 + 1215a^2x^4 + 1458ax^5 + 729x^6$.

5. $\frac{1}{64}x^6 + \frac{3}{8}x^5y + \frac{15}{4}x^4y^2 + 20x^3y^3 + 60x^2y^4 + 96xy^5 + 64y^6$.

6. $625 + 2000x^2 + 2400x^4 + 1280x^6 + 256x^8$.

7. $x^7 + 7x^5 + 21x^3 + 35x + 35/x + 21/x^3 + 7/x^5 + 1/x^7$.

8. $x^5(243y^5 - 810y^3 + 1080y - 720/y + 240/y^3 - 32/y^5)$.

9. $x^6y^3 + 4x^{\frac{11}{2}}y^{\frac{7}{2}} + 6\frac{2}{3}x^5y^4 + 5\frac{5}{7}x^{\frac{9}{2}}y^{\frac{9}{2}} + 2\frac{22}{7}x^4y^5 + \frac{64}{81}x^{\frac{7}{2}}y^{\frac{11}{2}} + \frac{64}{729}x^3y^6$.

10. $1 - 5x + \frac{45}{4}x^2 - 15x^3 + \frac{105}{8}x^4 - \frac{63}{8}x^5 + \frac{105}{32}x^6 - \frac{15}{16}x^7 + \frac{45}{256}x^8 - \frac{5}{256}x^9 + \frac{1}{1024}x^{10}$.

11. $\frac{1}{4096}a^{12} - \frac{1}{512}a^{11}b + \frac{11}{1536}a^{10}b^2 - \frac{55}{3456}a^9b^3; \; -\frac{2}{59049}ab^{11} + \frac{1}{531441}b^{12}$.

12. $941{,}192; \; 996{,}005{,}996{,}001$.   **13.** $70x^3y^4$.

14. $30618x^{17}y^5$.   **15.** $-792a^{14}b^{15}$.   **16.** $1365a^{19}x^{11}$.

17. $\frac{105}{32}x^6$.   **18.** $-366{,}080b^5/a^{14}$.

19. $325\dfrac{y^{36}}{x^{11}}$.   **20.** $\dfrac{(n+3)!}{r!(n-r+3)!} \cdot \dfrac{2^{2r-n-3}}{3^r} \cdot x^r$.

21. 70.   **22.** $12870y^4/x^4$.   **23.** $462a^6b^5; \; 462a^5b^6$.

24. $3\frac{15}{16}x^5y^4; \; -7\frac{7}{8}x^4y^5$.   **26.** $2(x^2 + 6x + 1)$.

27. $2a(a^4 + 40a^2b + 80b^2)$.   **28.** $2(64a^3 + 2160a^2 + 4860a + 729)$.

ANSWERS.

29. 210.   **30.** $-120a^7$.   **32.** $210a^{16}b^6$.
34. $\dfrac{(2n)!}{(n-\frac{1}{2}r)!(n+\frac{1}{2}r)!}$.   **35.** 180.
36. $\dfrac{(6r)!}{(2r)!(4r)!}$.   **40.** 7.   **41.** 20.
42. $(1+2)^7$.   **43.** $(1+\frac{1}{2})^8$.   **44.** 5.
46. 8.   **49.** 14 or 23.
51. (i) The 38th and 39th; (ii) the 17th and 18th; (iii) the 12th.
52. The 3rd $= 5\frac{1}{4}$.
53. The 6th $= -1792$ and the 7th $= +1792$.
54. The 5th $= 70$.   **55.** The 2nd $= -\frac{1}{5}$, and the 3rd $= \frac{1}{7}$.
56. The 8th and the 9th $= 86,507,520$.
57. The 7th $= 5580130\frac{1}{2}$.   **58.** The 3rd $= 4860$.
59. $\frac{36}{37}$ and $\frac{37}{36}$.   **60.** 1.   **61.** 64.

EXERCISES XXIV.

Imaginary and Complex Quantities.

1. $1+8i$.   **2.** $-6(i^{\frac{1}{3}}-i^{-\frac{1}{3}})$.   **3.** $\dfrac{c^2-d^2+2cdi}{c^2+d^2-2cdi}$.
4. $2(1+i)$.   **5.** $(a^2-b^2)+2abi$; $(a^3-3ab^2)+(3a^2b-b^3)i$;
$(a^4-6a^2b^2+b^4)+4abi(a^2-b^2)$.   **6.** 56.   **7.** -36.
9. $(4+i)(4-i)$.   **10.** x^3-1.
11. $(x+\frac{1}{2}+\iota\sqrt{\frac{3}{4}})(x+\frac{1}{2}-\iota\sqrt{\frac{3}{4}})(x-\frac{1}{2}+\iota\sqrt{\frac{3}{4}})(x-\frac{1}{2}-\iota\sqrt{\frac{3}{4}})$.
12. $(x+y)(x-y)(x+yi)(x-yi)$.   **16.** $\frac{1}{3}(7+4\sqrt{2}i)$.
17. $\frac{1}{2}(5-3\sqrt{3}i)$.   **18.** $-\frac{1}{25}(3+4i)$.   **19.** $\dfrac{4abi}{a^2+b^2}$.
20. $-1+i$.   **22.** $\pm(1+i\sqrt{2})$.   **23.** $\pm(2+i\sqrt{5})$.
24. $\pm(1+i)$.   **25.** $\pm(1+i)$.   **26.** $\frac{1}{2}\sqrt{2}(1-i)$.
27. $(a+bi)^{\frac{1}{2}}+(a-bi)^{\frac{1}{2}}$.
35. (i) 7; (ii) $\sqrt{2}$; (iii) 2; (iv) 1025; (v) $\frac{5}{13}$; (vi) 10; (vii) $1\frac{47}{65}$.
37. $1, \dfrac{1+\iota\sqrt{3}}{2}, \dfrac{-1+\iota\sqrt{3}}{2}, -1, \dfrac{-1-\iota\sqrt{3}}{2}, \dfrac{+1-\iota\sqrt{3}}{2}$.
38. $-1, \dfrac{1+\iota\sqrt{3}}{2}, \dfrac{1-\iota\sqrt{3}}{2}$; $\dfrac{1}{\sqrt{2}}(1\pm i), \dfrac{1}{\sqrt{2}}(-1\pm i)$.

40. $x = \dfrac{\pm\sqrt{5}-1}{4} \pm \dfrac{i}{4}\sqrt{10 \pm 2\sqrt{5}}$.

41. $1, \dfrac{1+i}{\sqrt{2}}, i, \dfrac{-1+i}{\sqrt{2}}, -1, \dfrac{-1-i}{\sqrt{2}}, -i, \dfrac{1-i}{\sqrt{2}}$.

EXERCISES XXV.
Rational Functions.

3. $x^{n-1} + x^{n-2} + x^{n-3} \ldots\ldots + x^2 + x + 1$. **4.** 3.
20. $p = 15;\ q = -6;\ r = 1;\ n = -20$.
21. $67 + 61(x-3) + 19(x-3)^2 + 2(x-3)^3$.
22. $x^4 + 6x^3 + 23x^2 + 46x + 56 = 0$.   **23.** $x^3 + 4x^2 + 9x + 18 = 0$.
24. $x^4 + 2x^3 - 25x^2 - 26x = 0$.   **25.** $x^4 - 14x^2 + 9 = 0$.
26. $1 + \sqrt{5},\ 1 - \sqrt{5},\ 1$ and 2.   **27.** $2 + \sqrt{3},\ 2 - \sqrt{3},\ 2$.
28. $1,\ 2,\ \tfrac{1}{2}\{-1 \pm \sqrt{-7}\}$.   **29.** $-2,\ -4,\ -3 \pm 5i$.
30. $2, 4, -5$. **31.** $3, 5, 7$. **32.** $2, -2, 3, 5$. **33.** $-7\tfrac{3}{4}$.
34. $0;\ q/r;\ -2q/r;\ -3/r;\ -q^3 - 3r;\ r;\ q^2/3r$.
35. $p^2 - 2q;\ -2q/r;\ p^4 - 4p^2q + 2q^2 - 4r$.
36. $y^4 - 2y^3 - 32y^2 - 77y - 34 = 0$.   **37.** $y^3 - 3y - 14 = 0$.
38. $y^3 - 3y^2 - 306y - 1512 = 0$.   **39.** $256y^4 + 416y^2 + 544y - 1955 = 0$.
40. $x^4 - 23x^2 + 61x - 45 = 0$.

EXERCISES XXVI.
Graphic Representation of Functions.

2. $x = -2 \pm \sqrt{2},\ y \not< -2$.   **3.** $x = \dfrac{-2 \pm \sqrt{19}}{3},\ y \not< -6\tfrac{1}{3}$.
4. $x = \pm a,\ y \not< -a^2$.   **5.** $x = 3,\ y \not< 0$.   **6.** $y \not< 6$.
7. $y \not< 5\tfrac{2}{3}$.   **8.** $x = 0,\ y \not< 0$.   **9.** $x = 1$ or $3,\ y \not> 1$.
10. $y \not> -2\tfrac{7}{8}$.   **11.** $x = -3 \pm \sqrt{14},\ y \not> 14$.   **12.** $x \not< 1$.
13. (vi) y cannot lie between roots of $y^2 + 14y + 1 = 0$.
14. (vi) Same limits as in 13.
15. (v) y can take all values. Limits are roots of $y^2 - 3y + 1 = 0$.

ANSWERS. **641**

16. (v) Same as 15. 17. (vii) y cannot lie between 1 and 9.
18. (iv) y cannot lie between 1 and $\frac{1}{5}$.
19. (vi) y cannot lie between roots of $y^2-16y+4=0$.
20. (v) y can take all values. Limits are roots of $y^2-y+1=0$.
21. Case (i) $\frac{1}{2}$, $\frac{1}{3}$.   22. Case (ii) $-\frac{1}{3}$, $\frac{1}{3}$.
23. Case (iv) $\frac{1}{2}$, 1. 24. Case (iii) 1·43, $-$·18.
25. Case (i) 5·3, 1·7. 26. Case (iii) $-\frac{1}{3}$, $\frac{1}{10}$.
27. Case (i) 3, $\frac{1}{8}$.   30. No.   31. No.   32. No.
33. No.   34. Yes; between $-\frac{1}{2}$ and 1.
35. Yes; not between $-\frac{1}{4}$ and 1.
36. Yes; between $-\frac{1}{2}$ and $+\frac{1}{2}$.
37. Yes; not between -3 and $+3$.
38. Yes; not between 1 and 5.   39. Yes; $y \not< -\frac{5}{4}$.
40. Yes; not between $-\frac{17}{3}$ and 0.

EXERCISES XXVIII.

The Binomial Theorem for Fractional and Negative Indices.

7. $1+\frac{1}{4}x-\frac{3}{32}x^2+\frac{7}{128}x^3$.   8. $1-\frac{1}{2}x+\frac{3}{8}x^2-\frac{5}{16}x^3$.
9. $1-\frac{1}{4}x-\frac{3}{32}x^2-\frac{7}{128}x^3$.   10. $1+\frac{2}{3}x-\frac{1}{9}x^2+\frac{4}{81}x^3$.
11. $a^{-2}+x^2a^{-4}+x^4a^{-6}+x^6a^{-8}$. 12. $a-\frac{1}{2}x^2a^{-1}-\frac{1}{8}x^4a^{-3}-\frac{1}{16}x^6a^{-5}$.
13. $a^{-\frac{3}{5}}+\frac{3}{10}a^{-\frac{8}{5}}x+\frac{39}{200}a^{-\frac{13}{5}}x^2+\frac{499}{4000}a^{-\frac{18}{5}}x^3\ldots$.
14. $a^{-3}+3a^{-5}x^{-1}+6a^{-7}x^{-2}+10a^{-9}x^{-3}$. 15. $\frac{m}{b}-\frac{mc^4}{2b^3}+\frac{3mc^8}{8b^5}-\frac{5mc^{12}}{16b^7}$.
16. $c^{\frac{3}{2}}-\frac{3}{4}c^{-\frac{1}{2}}x^2-\frac{3}{32}c^{-\frac{5}{2}}x^4-\frac{5}{128}c^{-\frac{9}{2}}x^6$.
17. $\dfrac{1}{a^2}+\dfrac{6x^{\frac{1}{3}}}{a^{\frac{7}{3}}}+\dfrac{21x^{\frac{2}{3}}}{a^{\frac{8}{3}}}+\dfrac{56x}{a^3}$.
18. $a^{-1}+\frac{1}{5}x^5a^{-6}+\frac{3}{25}x^{10}a^{-11}+\frac{11}{125}x^{15}a^{-16}$.
19. $-108x^3$; $(-1)^r(r+1)3^rx^r$.
20. $-120x^7$; $(-1)^r\dfrac{1}{6}(r+1)(r+2)(r+3)x^r$.

ANSWERS.

21. $462x^6$; $\dfrac{1}{120}(r+1)(r+2)(r+3)(r+4)(r+5)x^r$.

22. $(-1)^{r-1} \cdot \dfrac{1.1.3\ldots\ldots(2r-3)}{1.2.3\ldots\ldots r} x^r$. **23.** $2\frac{81}{4}x^{12}$.

24. $\dfrac{1.4.7\ldots\ldots(3r-2)}{3^{r-1}.r!} \cdot \left(\dfrac{x^2}{a^3}\right)^r$.

25. $\dfrac{17.14.11\ldots\ldots(3r-20)}{3^r.r!} \cdot x^{\frac{17-3r}{3}} y^{\frac{34-3r}{6}} z^{\frac{r}{2}}$.

27. $(r+1)(2r+1)$. **30.** The 2nd term. **31.** The 12th term.

32. The 20th term. **33.** The 9th term. **34.** The 2nd term.

35. The 15th term; the 27th term; the 1st term.

37. The 2nd term $= 1\frac{1}{4}$. **38.** The 3rd term $= 7\frac{37}{225}$.

39. The 1st term = the 2nd term = 27.

40. The 4th term = the 5th term = $1\frac{2360}{6591}$. **41.** The 3rd term.

45. The 1st term = the 2nd term = $(\sqrt{2})^{\sqrt{2}}$. **46.** The 2nd term.

53. $(1+x)^2$. **54.** $1+x+x^2+x^3$.

55. $\dfrac{1}{120}(r+1)(r+2)(r+3)(r+4)(r+5)$. **56.** $\dfrac{231}{1024}$.

57. $(-1)^{n-1}\dfrac{1}{24}n(n+1)(n+2)(n+3)$. **58.** $\dfrac{1}{4}n(n+1)(n+2)(n+3)$.

60. $n=3$; $n=r$. **61.** 3. **62.** $4n+1$. **63.** 121.

64. $2^{r-3}(r^2+7r+8)$. **65.** $3n^2+n+1$. **67.** $4368x^5$.

68. $\dfrac{n(n+1)\ldots\ldots(n+r-1)}{1.2.3\ldots\ldots r}a^r$.

72. $\dfrac{1.3.5\ldots\ldots(2p-1)}{2^p.p!}$ where $p = \frac{1}{2}r$ or $\frac{1}{2}(r-1)$, according as r is even or odd.

74. $1+\dfrac{2}{3}x+\dfrac{2}{9}x^2$. **84.** $\dfrac{1}{3}2^{n-4}n(n+2)(n+7)$. **85.** -20.

86. $-\dfrac{1}{2}b+\dfrac{3}{8}a^2$.

87. $a^{-\frac{2}{3}}+\dfrac{2}{3}a^{-\frac{5}{3}}bx+\left(\dfrac{5}{9}a^{-\frac{8}{3}}b^2+\dfrac{2}{3}a^{-\frac{5}{3}}c\right)x^2+\left(\dfrac{40}{81}a^{-\frac{11}{3}}b^3+\dfrac{10}{9}a^{-\frac{8}{3}}bc\right)x^3$.

88. $1+\frac{3}{4}x$. **92.** (a) 15; 18. **95.** 1·9873. **96.** 2·9931.

97. 1·4422. **98.** 1·2599. **99.** 3·0723. **100.** 1·9520.

113. (a) ·974; (b) ·986; (c) ·997.

ANSWERS. 643

EXERCISES XXIX.

Exponential and Logarithmic Series.

5. $x^r = p^q \cdot e^{qx-s}$. 7. 1. 8. $3e$. 9. $\dfrac{1}{2}(e - e^{-1})$.

10. $\frac{3}{2}e$. 11. $e^2 - e$. 13. $e - 1$. 14. $15e - 1$.

17. $(-1)^n \cdot \dfrac{1+n-3n^2}{n!}$. 18. $1 - \dfrac{x^2}{2} - \dfrac{x^3}{3 \cdot 1!} - \dfrac{x^4}{4 \cdot 2!} - \dfrac{x^5}{5 \cdot 3!} - \ldots$

20. $2\frac{17}{14}$. 39. $2\log_e x + \left\{\dfrac{3+2}{x} - \dfrac{1}{2} \cdot \dfrac{3^2+2^2}{x^2} + \dfrac{1}{3} \cdot \dfrac{3^3+2^3}{x^3} - \ldots\right\}$.

40. $2\left\{x - \dfrac{2x^3}{3} + \dfrac{x^5}{5} + \dfrac{x^7}{7} - \dfrac{2x^9}{9} + \dfrac{x^{11}}{11} + \dfrac{x^{13}}{13} - \dfrac{2x^{15}}{15} + \ldots\right\}$.

46. $1\cdot462398$. 47. $\cdot 9542425$. 49. $1 - \log_e 2$.

50. $\frac{3}{4} - \log_e 2$. 51. $\frac{2}{3}\log_e 2 - \frac{5}{12}$.

EXERCISES XXX.

Summation of Series.

1. $\dfrac{n(4n^2-1)}{3}$. 2. $\dfrac{2n(n+1)(2n+1)}{3}$.

3. $\dfrac{n}{6}(2n^2+3n-5)$. 4. $\frac{1}{3}n(n+1)(n+2)$.

5. $\frac{1}{3}n(4n^2+6n-1)$. 6. $\frac{2}{3}n(n+1)(3n^2+n-1)$.

7. $\frac{1}{12}n(n+1)(n+2)(3n+1)$. 8. $\frac{1}{12}n(n+1)(n+2)(3n+13)$.

9. $\frac{1}{4}n(n+1)(n+4)(n+5)$. 10. $\frac{1}{12}n(n+1)(3n^2+11n+4)$.

11. $3\frac{3}{4}$; $\frac{1}{24}n(n+1)(2n-5)$. 14. $\frac{5}{36} - \dfrac{3n+5}{6(n+1)(n+2)(n+3)}$; $\frac{5}{36}$.

15. $\frac{11}{18} - \frac{1}{3}\left(\dfrac{1}{n+1} + \dfrac{1}{n+2} + \dfrac{1}{n+3}\right)$; $\frac{11}{18}$. 16. $\frac{13}{240}$.

17. (i) $\frac{1}{5}n^5 + \frac{1}{2}n^4 - \frac{1}{2}n^2 + \frac{3}{5}n$; (ii) $\frac{1}{6}n(n+1)(4n+11)$.

18. $\dfrac{3^{n+2} - 8n - 9 - 2n^2}{4 \times 3^n}$. 19. $\dfrac{n}{n+1}$. 20. $\dfrac{n}{2n+1}$.

21. $\dfrac{n(n+2)}{3(2n+1)(2n+3)}$. 22. $\frac{1}{4}$. 23. $\frac{1}{18}$.

24. $\dfrac{1}{a+1} - \dfrac{1}{a+n+1}$.   25. $\dfrac{2n}{n+1}$.

26. $\tfrac{1}{2}n(n+1)$; $\tfrac{1}{6}n(n+1)(n+2)$.   28. $n(n+4) + \tfrac{3}{2}(3^n - 1)$.

29. $2^{n+2} + 7n - 4$.   30. $\tfrac{1}{2}n(n+1)$; $\tfrac{1}{6}n(n+1)(n+2)$.

31. $\tfrac{1}{6}n(2n^2 + 15n + 19)$.   32. 13.   33. 785.

EXERCISES XXXI.

DETERMINANTS.

1. 6, 9, 9.   2. $-$; $+$.   3. $+$; $-$.   4. $-$; $-$.
5. 0.   6. -8.   7. $a(a-b)(b-c)$.
8. $(p-q)(q-r)(r-p)$.   9. $2abc$.   10. -24.
11. $-(a-b)(b-c)(c-a)(a+b+c)$.   12. 0.
13. $(a-b)(b-c)(c-a)$.   14. $4abc$.   15. 2288.
16. 0.   17. 0.   24. $a^3 + b^3 + c^3 - 3abc = 0$.
25. $a^2 + b^2 + c^2 - bc - ca - ab = 0$.   26. $x=1$; $y=2$; $z=4$.
27. $x=1$; $y=2$; $z=3$.   28. $x=3$; $y=4$; $z=6$.
29. $3:2:3$.   30. -5.

EXERCISES XXXII.

CONTINUED FRACTIONS.

1. $\dfrac{1}{2+} \dfrac{1}{4+} \dfrac{1}{8+} \dfrac{1}{16}$.

2. $31 + \dfrac{1}{1+} \dfrac{1}{1+} \dfrac{1}{1+} \dfrac{1}{1+} \dfrac{1}{1+} \dfrac{1}{1+} \dfrac{1}{1+} \dfrac{1}{3}$.

3. $8 + \dfrac{1}{1+} \dfrac{1}{1+} \dfrac{1}{1+} \dfrac{1}{1+} \dfrac{1}{7}$.

4. $5 + \dfrac{1}{1+} \dfrac{1}{5+} \dfrac{1}{2+} \dfrac{1}{2+} \dfrac{1}{3}$.

5. $1 + \dfrac{1}{1+} \dfrac{1}{9+} \dfrac{1}{30+} \dfrac{1}{1+} \dfrac{1}{3}$.

6. $\dfrac{1}{1+} \dfrac{1}{1+} \dfrac{1}{4+} \dfrac{1}{1+} \dfrac{1}{2+} \dfrac{1}{1+} \dfrac{1}{3+} \dfrac{1}{2+} \dfrac{1}{3}$.

ANSWERS. 645

7. $2+\dfrac{1}{1+}\dfrac{1}{2+}\dfrac{1}{1+}\dfrac{1}{1+}\dfrac{1}{4+}\dfrac{1}{1+}\dfrac{1}{1+}\dfrac{1}{6+}\dfrac{1}{10+}\dfrac{1}{1+}\dfrac{1}{1+}\dfrac{1}{2}$

8. $3+\dfrac{1}{7+}\dfrac{1}{16+}\dfrac{1}{11}$.

9. $\dfrac{5}{1}, \dfrac{6}{1}, \dfrac{35}{6}, \dfrac{76}{13}, \dfrac{187}{32}, \dfrac{637}{109}$.

10. $1, \dfrac{1}{2}, \dfrac{2}{3}, \dfrac{3}{5}, \dfrac{5}{8}, \dfrac{8}{13}, \dfrac{13}{21}, \dfrac{21}{34}, \dfrac{76}{123}$.

11. $1, \dfrac{1}{2}, \dfrac{5}{9}, \dfrac{6}{11}, \dfrac{17}{31}, \dfrac{23}{42}, \dfrac{86}{157}, \dfrac{195}{356}, \dfrac{671}{1225}$.

12. $1, \dfrac{3}{2}, \dfrac{10}{7}, \dfrac{43}{30}, \dfrac{225}{157}, \dfrac{1393}{972}$.

13. $\dfrac{1}{2}, \dfrac{2}{5}, \dfrac{5}{12}, \dfrac{12}{29}, \dfrac{29}{70}$.

14. $1, \dfrac{3}{2}, \dfrac{31}{21}, \dfrac{530}{359}, \dfrac{13811}{9355}$.

15. The ninth. The eighth exceeds it by more than ·0002 and is greater than 2·7183.

17. Between $\dfrac{1}{3171}$ and $\dfrac{1}{2730}$.

EXERCISES XXXIII.

Elimination.

1. $2a^2 - b = 1.$ 2. $4a^2 - \dfrac{b^2}{a^2} + 4 = 0.$ 3. $x + a = 0$
4. $x = 0.$ 5. $4a^3 + 3a = b.$ 6. $8a^4 + 8a^2 - b + 1 = 0.$
7. $x^2 + y^2 = a^2 + b^2.$ 8. $x^2 + y^2 = a^2 + b^2.$ 9. $x^2 + y^2 = a^2.$
10. $a^3 + b^3 + c^3 - 3abc = 0.$ 11. $a^3 + b^3 + c^3 + abc = 0.$
12. The equations are insufficient for the elimination of x, y, and z, since any one of them can be deduced from the other two.
13. $l^2 + 3m^2 - 2mn + 4nl + 6lm = 0.$ 14. $b^2 = ac.$
15. Either $a = -c$, or $b = 0$ and $a = c.$ 16. $c = 6\tfrac{4}{43}.$
17. $hm - kl + \dfrac{mn}{l} = 0.$

EXERCISES XXXIV.

Indeterminate Equations.

1. $x = 5, y = 3$.   2. $x = 13, y = 8$.   3. $x = 7, y = 15$.
4. 7.   5. $x = 7, y = 7; x = 7 + 12t, y = 7 + 17t$.
6. $x = 17, y = 20; x = 36, y = 11; x = 55, y = 2$.
7. 7.   8. $x = 13, y = 10$.
9. $x = 1, y = 32; x = 9, y = 17; x = 17, y = 2$.
10. $\frac{1}{13}, \frac{1}{17}$ or $\frac{16}{17}, \frac{12}{13}$.   11. $x = 3, y = 2; x = 3 + 16t, y = 2 + 11t$.
12. $323t + 274$.   13. 11, and 2 involving zero values.   14. 2.
15. 6, and a solution involving a zero value.
16. $x = 17, y = 4; x = 8, y = 9$.   17. 2.
18. 4 miles.   19. 38.

EXERCISES XXXV.

1. $\dfrac{1}{x+2} + \dfrac{1}{x+3}$.   2. $\dfrac{2}{x+3} - \dfrac{1}{x+4}$.

3. $\dfrac{1}{x-5} - \dfrac{1}{x-4}$.   4. $\dfrac{3}{x-6} - \dfrac{1}{x+2}$.

5. $\dfrac{1}{x} + \dfrac{1}{x+1} - \dfrac{4}{2x+1}$.   6. $x + \dfrac{1}{2x+1} - \dfrac{3}{x+2}$.

7. $\dfrac{3}{4(x-2)} - \dfrac{3}{4(x+2)} + \dfrac{1}{(x-2)^2}$.

8. $\dfrac{11}{4(1+x)} - \dfrac{11}{4(3+x)} - \dfrac{7}{2(1+x)^2}$.

9. $\dfrac{1}{2x-1} + \dfrac{1}{x-3} + \dfrac{4}{x+3}$.

10. $x + 1 + \dfrac{4}{3(x-1)} - \dfrac{7}{6(2x+1)} + \dfrac{5}{2(2x+1)^2}$.

11. $\dfrac{1}{3}\dfrac{1}{x+2} - \dfrac{1}{3}\dfrac{1}{x-1} - \dfrac{1}{(x+2)^2} + \dfrac{2}{(x-1)^2}$.

12. $\dfrac{1}{1+x} + \dfrac{1}{1-x} + \dfrac{2}{1+x^2}$.

13. $-\dfrac{1}{x+1} + \dfrac{x}{x^2+1}$.   14. $\dfrac{x}{x^2+1} + \dfrac{1-x}{x^2-x+1}$.

15. $x^2 + 4x + 15 + \dfrac{1024}{17}\dfrac{1}{x-4} - \dfrac{1}{17}\dfrac{4x-1}{(x^2+1)}$.

16. $\dfrac{1}{2x} - \dfrac{x}{2(x^2+2)} - \dfrac{x}{(x^2+2)^2}$.

17. $-\dfrac{1}{1-x}+\dfrac{3}{(1-x)^2}-\dfrac{4}{(1-x)^3}+\dfrac{3}{(1-x)^4}.$

18. $(1+2.3^r)x^r.$ 19. $\tfrac{1}{2}\{1-2^{r+3}+3^{r+2}\}x^r.$

20. $\dfrac{a^r-b^r}{a-b}\dfrac{1}{a^r b^r}x^r.$

21. If r is even, r. If $r=4m+1$, $r-1$. If $r=4m-1$, $r+1$.

22. $\left(2r+1+\dfrac{1}{2^{r+1}}\right)x^r.$ 23. If $r=4m$, x^r.
If $r=4m+1$, $-x^r$.
If $r=4m+2$, or
$4m+3$, 0.

24. $\dfrac{1}{x(1-x)}\left\{\dfrac{1}{1+x^{n+1}}-\dfrac{1}{1+x}\right\}.$

25. (1) $1-\dfrac{1}{(n+1)^2}$, (2) $\dfrac{1}{1-a}\left\{\dfrac{1}{1+x}-\dfrac{a^n}{1+a^n x}\right\}.$

26. $\dfrac{1+x^2}{2x}\log(1+x)-\dfrac{1}{2}+\dfrac{x}{4}.$

27. $a^n+\beta^n$
$$=a^n-n.a^{n-2}.b+\dfrac{n(n-3)}{\underline{2}}a^{n-4}.b^2$$
$$-\dfrac{n(n-4)(n-5)}{\underline{3}}.a^{n-6}.b^3, \text{ etc.}$$

where $a=a+\beta$, $b=a\beta$.

Made in United States
North Haven, CT
21 June 2024